Hydromagnetic Waves in the Magnetosphere and the Ionosphere

Recently Published in the ASSL series

Volume 353: *Hydromagnetic Waves in the Magnetosphere and the Ionosphere*, by Leonid S. Alperovich, Evgeny N. Fedorov. Hardbound 978-1-4020-6636-8

Volume 352: *Short-Period Binary Stars: Observations, Analyses, and Results*, edited by Eugene F. Milone, Denis A. Leahy, David W. Hobill. Hardbound ISBN: 978-1-4020-6543-9, September 2007

Volume 351: *High Time Resolution Astrophysics*, edited by Don Phelan, Oliver Ryan, Andrew Shearer. Hardbound ISBN: 978-1-4020-6517-0, September 2007

Volume 350: *Hipparcos, the New Reduction of the Raw Data*, by Floor van Leeuwen. Hardbound ISBN: 978-1-4020-6341-1, August 2007

Volume 349: *Lasers, Clocks and Drag-Free Control: Exploration of Relativistic Gravity in Space*, edited by Hansjörg Dittus, Claus Lämmerzahl, Salva Turyshev. Hardbound ISBN: 978-3-540-34376-9, September 2007

Volume 348: *The Paraboloidal Reflector Antenna in Radio Astronomy and Communication – Theory and Practice*, by Jacob W.M. Baars. Hardbound 978-0-387-69733-8, July 2007

Volume 347: *The Sun and Space Weather*, by Arnold Hanslmeier. Hardbound 978-1-4020-5603-1, June 2007

Volume 346: *Exploring the Secrets of the Aurora*, by Syun-Ichi Akasofu. Hardbound 978-0-387-45094-0, July 2007

Volume 345: *Canonical Perturbation Theories – Degenerate Systems and Resonance*, by Sylvio Ferraz-Mello. Hardbound 978-0-387-38900-4, January 2007

Volume 344: *Space Weather: Research Toward Applications in Europe*, edited by Jean Lilensten. Hardbound 1-4020-5445-9, January 2007

Volume 343: *Organizations and Strategies in Astronomy: Volume 7*, edited by A. Heck. Hardbound 1-4020-5300-2, December 2006

Volume 342: *The Astrophysics of Emission Line Stars*, by Tomokazu Kogure, Kam-Ching Leung. Hardbound ISBN: 0-387-34500-0, June 2007

Volume 341: *Plasma Astrophysics, Part II: Reconnection and Flares*, by Boris V. Somov. Hardbound ISBN: 0-387-34948-0, November 2006

Volume 340: *Plasma Astrophysics, Part I: Fundamentals and Practice*, by Boris V. Somov. Hardbound ISBN 0-387-34916-9, September 2006

Volume 339: *Cosmic Ray Interactions, Propagation, and Acceleration in Space Plasmas*, by Lev Dorman. Hardbound ISBN 1-4020-5100-X, August 2006

Volume 338: *Solar Journey: The Significance of Our Galactic Environment for the Heliosphere and the Earth*, edited by Priscilla C. Frisch. Hardbound ISBN 1-4020-4397-0, September 2006

Volume 337: *Astrophysical Disks*, edited by A. M. Fridman, M. Y. Marov, I. G. Kovalenko. Hardbound ISBN 1-4020-4347-3, June 2006

Volume 336: *Scientific Detectors for Astronomy 2005*, edited by J. E. Beletic, J. W. Beletic, P. Amico. Hardbound ISBN 1-4020-4329-5, December 2005

For other titles see www.springer.com/astronomy

Hydromagnetic Waves in the Magnetosphere and the Ionosphere

Leonid S. Alperovich

Tel Aviv University
Tel Aviv, Israel

Evgeny N. Fedorov

Russian Academy of Sciences
Moscow, Russia

Springer

Prof. Leonid S. Alperovich
Department of Geophysics & Planetary Sciences
Tel Aviv University
Tel-Aviv 69978
Israel
tel. (direct): 972-3-640-7411
leonid@luna.tau.ac.il

Prof. Evgeny N. Fedorov
Bolshaya Gruzinskaya 10
Moscow 123995
Russia
tel.: 7-495-254-8905
ENFedorov1@yandex.ru

Library of Congress Control Number: 2007935109

ISBN 978-1-4020-6636-8 e-ISBN 978-1-4020-6637-5

Printed on acid-free paper.

9 8 7 6 5 4 3 2 1

springer.com

Contents

List of Tables

List of Figures

Preface

This book is concerned with Ultra-Low-Frequency (ULF)-electromagnetic waves observed on the Earth and in Space. These are so-called geomagnetic variations or pulsations. Alfvén's discovery related to the influence of the strong magnetic field on the conducting fluids (magnetohydrodynamics) led to development of the concept that the ULF-waves are magnetospheric magnetohydrodynamic (MHD)-waves.

MHD-waves at their propagation gather information about the magnetosphere, ionosphere, and the ground. There are two applied aspects based on using the ULF electromagnetic oscillations. The first one is the ground-based diagnostics of the magnetosphere. This is an attempt to monitor in the real time the magnetosphere size, distance to the last closed field-lines, distribution of the cold plasma, etc.

The second one is the deep electromagnetic sounding of the Earth. The basis for these studies is the capability of any electromagnetic wave to penetrate a conductor to a finite depth. The ULF-waves can reach the depth of a few hundred kilometers.

Thus, geophysicists now have a unique tool that can be applied to study the solid Earth as well as its gas-plasma shells – the ionosphere and magnetosphere.

The purpose of the book is threefold: first, to study the propagation of the MHD-waves in the inhomogeneous magnetosphere, to understand the main features of coupling between various modes, efficiency of such interaction, behavior within the so-called resonance regions, to trace the MHD-wave from the remote magnetospheric regions to the ionosphere; second, to study the transformation of the MHD-waves in the ionosphere, which shields a ground observer from the magnetospheric emitters; and third, to show the practical value of the developed theoretical concepts to the magnetosphere and the Earth ULF- soundings.

This book attempts to integrate topics pertaining to all scales of the MHD-waves, emphasizing the linkages between the ULF-waves below the ionosphere on the ground and magnetospheric MHD-waves. We intend this book to be the one from which graduate and post-graduate students may study the science of MHD-waves in the magnetosphere and ionosphere by themselves. Only an introductory electromagnetic course and some elements of plasma physics need to be considered as a prerequisite to the reading of this book. It is assumed that the reader is familiar with advanced calculus. If at all possible, we tried to create a single reference for the interested reader and to do so that the mathematical rigour does not overshadow the physical object.

Tel-Aviv, Israel
Moscow, Russia
May 2007

Leonid Alperovich
Evgeny Fedorov

Acknowledgments

We could not have written this book without the help of many people. First and foremost, we would like to thank those, with whom we worked together for a long time, who, in many respects, defined the spectrum of our scientific interests: Prof. Valeria Troitskaya, Prof. Anatoly Guglielmi, Prof. Alexander Povzner, Prof. Dmitri Chetaev (deceased), and Prof. Yuri Galperin (deceased). Many thanks also to Prof. Izhak Chaikovsky, Dr. Boris Fidel and Dr. Vyacheslav Pilipenko for their helpful comments, advices, discussions and great friendship. Prof. Masashi Hayakawa's, Prof. Oleg Molchanov's and Prof. Colin Price's insight and advice as to the experimental ULF aspects of the book were extremely useful and thought provoking. Heartfelt thanks to Dr. Nadezhda Yagova for her assistance on various production tasks as well as her critical comments. She showed great interest in the eventual outcome. Evgeny Alperovich proofread the manuscript and made many helpful suggestions. We are grateful to him for his patience during the long ordeal. We thank Dr C. K. Aanandan for his assistance in approvement of clarity and language of Chapter 7. Special thanks also go to the staff of the Department of Geophysics and Planetary Sciences (Raymond and Beverly Sackler Faculty of Exact Sciences, Tel Aviv University) and Institute of Physics of the Earth (Russian Academy of Sciences) for being so generous with their time and assistance. We would especially like to thank Dr. Evgeny Gurvich for his interest in the book and the valuable assistance that he provided. Our appreciation goes to Zipi Rosen and Sara Rehavi for providing one of us with computer and technical support. We would also like to thank the International Space Science Institute (ISSI) (Bern, Switzerland) that gave us timely support. Throughout our stay at the ISSI, we received strong encouragement and extraordinary assistance from its staff. The staff of the ISSI were also most helpful and cordial and allowed us access to their specialized library, and the use of their computer facilities. L.S.A. thanks ISF (grant 605/05) and E.N.F. thanks ISTC (Project 2990) for their support. Many thanks also to Springer publisher's editorial board, for the patience and enthusiastic support they gave our work from start to finish.

Introduction

Present-day scientific vision of the structure and dynamics of space in the vicinity of the Earth is based on experimental data obtained with both the help of space technology and as a result of purposeful observation of a number of geophysical parameters from the Earth. The Ultra Low-Frequency (ULF) natural electromagnetic oscillations discussed in this book are the most informative in that sense.

The ULF-oscillations, also called 'geomagnetic pulsations' or 'micropulsations', occupy the 10^0-10^{-3} Hz range. The intensity of these pulsations is measured in thousandths (0.001) of 1% of the basic geomagnetic field value. Systematic studies of the pulsations conducted in many countries in accordance with the International Geophysical Year program in 1958, made it possible to construct the general pattern for the space-time behavior of geomagnetic pulsations . The following classification was adopted [1]

Table 0.1. The classification of ULF-Pulsations

Continuous pulsations		Irregular pulsations	
Name	Period (s)	Name	Period (s)
*Pc*1	0.5–5	*Pi*1	1–40
*Pc*2	5–10	*Pi*2	40–150
*Pc*3	10–45		
*Pc*4	45–150		
*Pc*5	150–600		

The pulsations result from complicated plasma processes taking place in the solar wind, in the Earth's magnetosphere or ionosphere. Expanding to the Earth's surface, these oscillations undergo a number of changes, which enabled us, assuming a certain hypothesis of their origin, to get a definite insight into the medium they have passed through.

The method of ground-based monitoring of the near-Earth space, relying on explorations of geomagnetic pulsations, has been designated as 'ground-based MHD-diagnostics'. It provides a realistic foundation for timely reception of regular information on the solar wind and the magnetosphere. Ground-based diagnostics remain at present the main method for exploring the distribution of cold plasma in the magnetosphere, for determining the dimensions of the magnetosphere and the distance to the last closed field-lines.

At the same time, ground-based methods have a natural restriction stemming from the presence of an ionospheric screen that forms a barrier between the observer on the ground and the MHD magnetospheric emitters. In passing through the ionosphere, the signal not only undergoes amplitude changes, but its composition and polarization structure are also distorted. As it is refracted in the high-latitude ionosphere and subsequently propagates to the middle and low latitudes, the ionosphere and the Earth cause significant changes in the initial magnetospheric wave pattern.

It is obvious that the development of ground-based methods and the analysis of the resolving capacity of ground-based diagnostics, i.e. the degree to which factors of ionospheric origin mask the magnetospheric wave pattern of wave perturbations, both require a consistent consideration of the problem of ionosphere's role in MHD-wave dynamics.

Observations of geomagnetic oscillations are also used for electromagnetic exploring of the deep structure of the Earth through some geophysical prospecting methods. At the present stage in the development of these methods, the study of peculiarities in the distribution of MHD-waves in the ionosphere as well as the factors that form the spatial pattern of the alternating geomagnetic field are of primary significance.

This book will introduce the reader to the various aspects of the problem of propagation and generation of natural hydromagnetic waves. We presume that the complexity and intricacy of the problem of ionosphere's role in the dynamics of low-frequency MHD-perturbations call for the use of a complex approach based on exploring the transformation of various kinds of effects. The notion of effect as used here is broad. It embraces MHD-waves of natural origin as well as perturbations caused by large-scale seismo- and gas-dynamic occurrences within the Earth and near its surface capable of exciting MHD-waves. Lastly, this notion includes high-energy anthropogenic effects, such as, for instance, the heating of the ionosphere by a powerful radio wave, industrial and other explosions, causing artificial electromagnetic signals to be excited.

Chapter 1 is devoted to the basic principles of the electrodynamics of partially ionized plasma. We present derivation of equations for the specific dielectric permeability and electrical conductivity of the ionosphere and magnetosphere relating them with the main plasma parameters of the media such as concentration of electrons and neutral particles, their temperatures, and magnitude of the ambient magnetic field. The ULF-range and, hence, the wavelengths are so large that we can simplify the theoretical consideration by using the thin ionosphere approximation and neglecting the detailed structure of the height profile. It allows us to introduce, instead of specific conductivity, a new electrodynamic parameter of the ionosphere, a tensor of the integral conductivity. In the case of arbitrary inclination of the geomagnetic field, one can introduce the generalization of the integral conductivity on the tensor of layered conductivity.

Before proceeding to the investigation of the problem of MHD-wave propagation in the magnetosphere and reflection of the waves from the ionosphere, we thought it necessary to present in Chapter 2 a description of the main electrodynamic properties of the solar wind, of the magnetosphere, the ionosphere, the atmosphere and the Earth.

Chapter 3 is devoted to a cursory review of works dealing with the morphological peculiarities of geomagnetic variations. We did not aim here at an exhaustive description of all the experimental studies of the geomagnetic variations, being aware that at the present stage, a profound and serious comparative description summarizing a vast number of studies must be an independent in-depth work.

At the same time, in that review, an attempt is made to acquaint the reader with the most striking peculiarities of the space-time behavior of pulsations. We presume that they are most vividly manifested in extreme intensity related to the high-latitude zone and to the area of plasmapause projection on the Earth's surface. In recent years, a number of conclusive publications have appeared in which data on spatial distribution, velocities, comparison of electromagnetic oscillations with attendant phenomena, etc., are systematized. We have, therefore, confined ourselves to a rather cursory review of works on the indicated range of problems.

Chapter 4 studies the characteristics of small hydromagnetic wave propagation that is determined in linear approximation by equilibrium disturbances of plasma and of the magnetic field. The main objective of the section is to expose only the basic concepts on equilibrium plasma configurations. The basic properties of these waves, both in homogeneous and in inhomogeneous unbounded plasmas, are briefly described in this chapter.

In Chapter 5 we abandon the assumption of medium unboundedness and consider a model of a bounded MHD medium. We consider the waves within the MHD-box, which enables us to reveal new important peculiarities of hydrodynamic perturbations in magnetospheric plasma. Particularly, it gives the simplest way to understand the principal features of coupling between different MHD-wave modes and the field line MHD-wave resonances.

A general method for analyzing MHD-waves in curve magnetic fields, in particular, in the dipole field, analysis of the electromagnetic field in the vicinity of resonance magnetic shells are given in Chapter 6. The developed methods enabled us to study the field structure including a number of additional factors, e.g. inhomogeneous plasma distribution, variable curvature of magnetic field-lines, and both the Hall and Pedersen conductivities of plasma boundaries (conjugate ionospheres).

Chapters 7–8 describe MHD-wave transformation in the ionosphere in an approximation of the 'thin' ionosphere. This approach allowed almost all the questions concerning the problem of interaction between low-frequency perturbations and the ionosphere to be answered. The problem was solved in stages. First, expressions were found for the reflection and transmission coefficient matrices of a MHD-wave with arbitrary polarization (Alfvén or magnetosonic). The greatest difficulties arise when attempts are made to reconstruct electric and magnetic fields on the Earth and above the ionosphere from sources with a definite dependence on coordinates and time. This is usually solved by using the Fourier integral of the spectral decomposition of the sources with corresponding transmission or reflection matrices with the use of real wavenumbers. This book develops a different approach based on the use of complex wavenumbers.

This method allowed asymptotic solutions to be constructed at long distances from the source. A peculiar wave excited by magnetospheric waves in an atmospheric waveguide was also explored. Such waves are able to propagate over long distances, as far as a round-the-world echo, with velocities close to that of light. The derivation of excitation coefficients, phase and other characteristics for atmospheric waves is presented in Chapter 8. The chapter also discusses the effect of ionospheric and terrestrial conductivity on spatial decrements, horizontal velocities, etc. Theoretical dependencies are shown of the amplitudes and phases of terrestrial and above-ionospheric fields.

While the preceding chapters treat the model of an homogeneous ionosphere, Chapter 9 considers the model of inhomogeneous ionosphere and its influence on geomagnetic variations. We first show the validity of the quasi-DC (direct current) approximation for the description of the ionospheric currents caused by the FMS- and Alfvén waves. Then we discuss applicability of this approximation to the ionosphere containing irregularities of different scales. This type of formulation of the MHD-wave transformation problem is convenient because one can get solutions for the ground fields, even for arbitrary horizontal distributions of the ionospheric conductivity. The transition region between dayside and nightside ionospheres is an example a strong horizontal gradients of the electron concentration and as a sequence of this a strong horizontal anomaly of the conductivity. We obtain an explicit solution for the ground magnetic field caused by the Alfvén wave incident onto the sunrise-sunset ionosphere. Examples of the global distributions of the magnetic variations produced by the field-aligned currents, either of an incident Alfvén wave or electric field of an FMS-wave, are demonstrated for different seasons.

Chapter 10 is devoted to the contribution of stochastic inhomogeneities of ionospheric conductivity to terrestrial and magnetospheric pulsation fields. We briefly review existing theories of effective conductivity relating as to the regular media with the scalar local conductivity as to magnetized media.

Theoretical outcomes are compared with the results of the laboratory experiments on silicic inhomogeneous films placed into the strong magnetic field.

The idea of natural electromagnetic oscillations incident normally on the Earth's surface, underlies the present-day electromagnetic methods for exploring the electric properties of the Earth's interior. Chapter 11 considers the possible contribution of the discovered peculiarities of a magnetospheric wave reaching the Earth's surface to the results of ground-based ULF electromagnetic (magnetotelluric) sounding. The prospects for electromagnetic sounding of the Earth from satellite altitudes are considered here as well.

The conceptions developed in Chapters 7 and 8 provided the basis for the theory of Doppler and Faraday radiobearing of magnetohydrodynamic waves in the ionosphere. Comparison of pulsation observations on the Earth and in the ionosphere is discussed in Chapter12 which describes new ground-based methods for detection of MHD-waves in the ionosphere. The theoretical foundations for such methods are given together with an estimation of the possible effects of hydromagnetic waves.

The following three Chapters are devoted to the problem of artificial generation of MHD-waves. In Chapter 13 we consider the problem of the ionospheric heating by a powerful radio wave. Some of the effects that a powerful radio wave produces on ionospheric plasma is a rise in electron temperature, a change in the frequency of electrons colliding with neutrals and, consequently, the appearance of a region with anomalous electric conductivity. There is another mechanism by which distortions are introduced into ionospheric current systems - due to the rise in electron temperature the ionization - recombination cycle is shifted. The basic regularities are investigated in the behavior of the non-stationary equations system for the chemical kinetics of the ionosphere E-layer, numerical and analytical estimations are presented of the intensity of geomagnetic variations for various powers of the emitting system and various regimes of pump radio frequency amplitude modulation.

In Chapter 14 we describe a mechanism of MHD-perturbation generation connected with the release plasma producing matter into the ionosphere. When plasma producing compounds are injected into the ionosphere at heights of 150-200 km, the injected cloud is scattered. Because the cloud density is high, conditions are typical for ionospheric dynamo-area, i.e. intensive horizontal currents are generated. The currents are non-stationary, they produce MHD-waves and, in particular, Alfvén waves. As it follows from Chapters 4–6, an Alfvén wave can propagate along geomagnetic field-lines over huge distances without essential distortion of the form or significant losses of energy. We examined the excitation of such waves and their penetration into the magnetosphere.

The last chapter is devoted to the theoretical foundations and experimental results of the acoustic method as an effective impact on the ionosphere. The method consists of intentionally creating a powerful acoustic wave on the ground. As it enters the ionosphere, the wave produces macroscopic movements of the E-layer and causes the generation of a magnetospheric MHD-wave. The results of the first experiment of this kind are discussed.

We use Gaussian units. But sometimes, for example in studies of applied problems such as the electromagnetic sounding of Earth, we use the international system of units (SI).

Reference

1. Jacobs J. A., Y. Kato, S. Matsushita, and V. A. Troitskaya, Classification of geomagnetic micropulsations, *J. Geophys. Res.*, **69**, 180, 1964.

1

Partially Ionized Plasma

1.1 Introduction

The Earth's ionosphere and magnetosphere consist of a number of regions filled with an ionized gas immersed into the Earth's magnetic field. Since one of the important motivations of this book is understanding the low frequency electrodynamics of the ionosphere and magnetosphere, it is useful to begin with the basic equations of the partially ionized plasma with the magnetic field. Fundamental equations for the system of neutral and charged particles moving in the internal and external electromagnetic fields are well known. These are the Maxwell's equations and kinetic equations. It is impossible to give here even the basis for such a wide subject as plasma physics, and we restrict ourselves to a short list of basic equations and definitions in this branch of physics. Detailed considerations can be found in numerous books on plasma physics (e.g. [4], [7], [8]) and on the electromagnetic waves in plasmas (e.g. [5], [10]). More detailed consideration will be given for the tensors of conductivity and dielectric permeability in the multi-fluid hydromagnetic approximation.

In this chapter we use the right-handed Cartesian coordinates $\{x, y, z\}$ and let $\hat{\mathbf{x}}$, $\hat{\mathbf{y}}$, $\hat{\mathbf{z}}$ be unit vectors in the directions of the x-, y-, z-axis.

1.2 Comments on the Plasma Dynamics

The Distribution Function

Let us consider the plasma consisting of electrons with the charge $-e$ ($e = 4.8 \times 10^{-10}$ CGSE), multiple charged ions $Z_\alpha e$ and neutrals. The gas state can be described by a distribution function $f_\alpha(\mathbf{r}, \mathbf{V}, t)$. It is defined so that the mean number of particles dN_α in the elementary volume of the phase space

$$\mathrm{d}\mathbf{r}\,\mathrm{d}\mathbf{V} = \mathrm{d}x\,\mathrm{d}y\,\mathrm{d}z\,\mathrm{d}V_x\,\mathrm{d}V_y\,\mathrm{d}V_z$$

is

$$\mathrm{d}N_\alpha = f_\alpha(\mathbf{r}, \mathbf{V}, t)\,\mathrm{d}\mathbf{r}\,\mathrm{d}\mathbf{V}.$$

From the definition of $f_\alpha(\mathbf{r}, \mathbf{V}, t)$, the concentration of particles of the sort α can be written as

$$N_\alpha(\mathbf{r}, t) = \int f_\alpha(\mathbf{r}, \mathbf{V}, t)\,\mathrm{d}\mathbf{V}. \tag{1.1}$$

The products

$$\rho_{m\alpha} = m_\alpha N_\alpha(\mathbf{r}, t), \quad \rho_\alpha = Z_\alpha e N_\alpha(\mathbf{r}, t) \tag{1.2}$$

are the mass density and charge density of particles of the type α; $Z_e = -1$ for electrons and $Z_\alpha = 0$ for neutrals. Total concentration, charge density and mass density are determined as

$$N = \sum_\alpha N_\alpha, \quad \rho = \sum_\alpha Z_\alpha e N_\alpha(\mathbf{r}, t), \quad \rho_m = \sum_\alpha m_\alpha N_\alpha(\mathbf{r}, t).$$

The summation is taken over all plasma species.

The Boltzmann–Vlasov Equation

As for the non-ionized gases, the kinetic equation method is applicable only for low-density plasma in which the mean energy of the interaction of two particles is low in comparison with their mean kinetic energy. In this case, it is possible to consider the plasma as a gas. Neglecting at the first step the existence of the multi-charged ions, the energy of the interaction between two charged particles is $e^2/\langle r\rangle$, where $\langle r\rangle$ is a mean distance between the particles. If N is the total number of particles in a unit volume, then $\langle r\rangle \sim N^{-1/3}$. Under these simplifying assumptions, we write the condition of applicability of the gas approximation in the form

$$kT \gg e^2/\langle r\rangle \sim e^2 N^{1/3}, \tag{1.3}$$

where k is the Boltzmann constant $(k = 1.38 \times 10^{-16}\,\mathrm{erg} \cdot \mathrm{K})$, T is the plasma temperature (in Kelvins, K). This condition can be expressed in terms of the Debye radius

$$r_D = \sqrt{\frac{kT}{8\pi N e^2}}. \tag{1.4}$$

r_D determines the distance of screening of the Coulomb fields in the plasma. Taking into account (1.4), the condition (1.3) can be rewritten as

$$\langle r\rangle^2/8\pi \ll r_D^2, \tag{1.5}$$

i.e. the average distance between particles $\langle r\rangle$ is small in comparison with the Debye radius. This condition can also be written as $Nr_D^3 \gg 1$, i.e. a Debye sphere should include many particles.

For the distribution function f_α of species α of the plasma particles in a collisionless case, we have the Liouville theorem

$$\frac{\mathrm{d\,f}_\alpha}{\mathrm{d\,t}} = \frac{\partial\, f_\alpha}{\partial\, t} + \boldsymbol{\nabla}_{\mathbf{r}} f_\alpha \cdot \frac{\partial\,\mathbf{r}}{\partial\, t} + \boldsymbol{\nabla}_{\mathbf{V}} f_\alpha \cdot \frac{\partial\,\mathbf{V}}{\partial\, t} = 0. \tag{1.6}$$

Here $\mathrm{d}\,/\mathrm{d\,t}$ is differentiating along the particle trajectory in the phase space determined by the dynamic equations. Space and velocity gradients, respectively, $\boldsymbol{\nabla}_{\mathbf{r}} f_\alpha$ and $\boldsymbol{\nabla}_{\mathbf{V}} f_\alpha$, of the distribution function are

$$\begin{aligned}
\boldsymbol{\nabla}_{\mathbf{r}} f_\alpha &= \frac{\partial\, f_\alpha}{\partial\, x}\hat{\mathbf{x}} + \frac{\partial\, f_\alpha}{\partial\, y}\hat{\mathbf{y}} + \frac{\partial\, f_\alpha}{\partial\, z}\hat{\mathbf{z}},\\
\boldsymbol{\nabla}_{\mathbf{V}} f_\alpha &= \frac{\partial\, f_\alpha}{\partial\, V_x}\hat{\mathbf{x}} + \frac{\partial\, f_\alpha}{\partial\, V_y}\hat{\mathbf{y}} + \frac{\partial\, f_\alpha}{\partial\, V_z}\hat{\mathbf{z}}.
\end{aligned}$$

Let $(\partial\, f_\alpha/\partial\, t)_{coll}$ be the rate of change of the distribution function caused by collisions. $(\partial\, f_\alpha/\partial\, t)_{coll}\,\mathrm{d}\mathbf{r}\,\mathrm{d}\mathbf{V}$ is the time variation of the number of particles in a time unit in the phase volume $\mathrm{d}\mathbf{r}\,\mathrm{d}\mathbf{V}$. For a plasma in the self-consistently electromagnetic field the kinetic equations are

$$\frac{\partial\, f_\alpha}{\partial\, t} + \mathbf{V}\boldsymbol{\nabla}_{\mathbf{r}} f_\alpha + \frac{\mathbf{F}_{(\alpha)}}{m_\alpha}\boldsymbol{\nabla}_{\mathbf{V}} f_\alpha = \left(\frac{\partial\, f_\alpha}{\partial\, t}\right)_{coll}, \tag{1.7}$$

where $\mathbf{F}_{(\alpha)}(\mathbf{r},\mathbf{V},t)$ is a force acting on the particles α. For charged particles, $\mathbf{F}_{(\alpha)}$ is the Lorentz force

$$\mathbf{F}_{(\alpha)} = q_\alpha\left(\mathbf{E} + \frac{1}{c}\left[\mathbf{V}\times\mathbf{B}\right]\right), \tag{1.8}$$

where $q_e = -e$ for electrons and $q_\alpha = Z_\alpha e$ for ions; $Z_\alpha e$ is the charge of the multiple charged ions.

Interaction between colliding neutral molecules is strongly localized because the interaction is efficient only for an impact parameter of the order of atomic radius. Between neutral-neutral collisions, the motion of a neutral is determined only by external fields. Contrary, charged particles interact simultaneously and hence collectively with many other nearby charged particles because of long range Coulomb forces. They are shielded at distances of the order of the Debye radius r_D that is, according to (1.5), larger than the spacing $\bar{r}$ between particles. Thus, the interaction of two charged particles is a collective effect from many particles at such distances. These forces should be excluded from the collision integral in the right-hand side of (1.7).

Charged particles interact collectively in plasmas through their electromagnetic fields. To reveal collective interactions, let us present the microscopic electric ($\mathbf{E}_\mu$) and magnetic ($\mathbf{B}_\mu$) fields acting at a charged particle as sums of the mean and the fluctuated fields:

$$\mathbf{E}_\mu = \mathbf{E} + \mathbf{E_1}, \quad \mathbf{B}_\mu = \mathbf{B} + \mathbf{b_1},$$

where **E** and **B** are the macroscopic electric and magnetic fields averaged over the volumes with many particles. Field fluctuations are described with the terms $\mathbf{E}_1$ and $\mathbf{b}_1$ result in fluctuations in the particle movement presented in (1.7) by the collision integral. The **E** and **B** fields determine the Lorentz force (1.8) that is to be substituted into (1.7) for charged particles.

Let ρ_{ext} and $\mathbf{j}_{ext}$ are the densities of the charge and current produced by external drivers. Then Maxwell's equations for the plasma with the external sources are given by

$$\begin{aligned}
\nabla \times \mathbf{B} &= \frac{1}{c}\frac{\partial \mathbf{E}}{\partial t} + \frac{4\pi}{c}\mathbf{j}_{ext} + \frac{4\pi}{c}\sum_\alpha \mathbf{j}_{(\alpha)}, \\
\nabla \times \mathbf{E} &= -\frac{1}{c}\frac{\partial \mathbf{B}}{\partial t}, \\
\nabla \cdot \mathbf{E} &= 4\pi\rho_{ext} + 4\pi\sum_\alpha \rho_\alpha, \quad \nabla \cdot \mathbf{B} = 0,
\end{aligned} \tag{1.9}$$

where

$$\begin{aligned}
\rho_\alpha(\mathbf{r}, t) &= q_\alpha \int f_\alpha(\mathbf{r}, \mathbf{V}, t)\, \mathrm{d}\mathbf{V}, \\
\mathbf{j}_{(\alpha)}(\mathbf{r}, t) &= q_\alpha \int \mathbf{V} f_\alpha(\mathbf{r}, \mathbf{V}, t)\, \mathrm{d}\mathbf{V}.
\end{aligned} \tag{1.10}$$

Hydrodynamic Equations

A treatment of the distribution function and the kinetic equations can be found in a series of handbooks on plasma physics. The complete system of kinetic and field equations is extremely complicated and may not always be practically used. However, the physical pattern of phenomena, and often proper quantity results as well, may be found using an essentially simpler theory. Following Chapman and Cowling [3], we call it an elementary theory. Numerous phenomena related to velocity distribution, for instance, Landau damping, and all the kinetic instabilities in the partially ionized plasma are neglected in the elementary theory. A plasma state is characterized by the directed velocities of an 'mean' electron $\mathbf{v}_e$, ion $\mathbf{v}_{(\alpha)}$ of species α, neutral particle $\mathbf{v}_{(\beta)}$ of species β and by their mean temperatures T_e, T_α, T_β. The electron T_e and ion temperatures T_α are defined as

$$W_e = \frac{3}{2}kT_e, \quad W_\alpha = \frac{3}{2}kT_\alpha,$$

where W_e and W_α denote the energy of chaotic motions of the 'mean' electron and α ion.

The mean velocity $\mathbf{v}_{(\alpha)}$ of the α-th particles is

$$\mathbf{v}_{(\alpha)} \equiv \langle \mathbf{V}_{(\alpha)} \rangle = \frac{1}{N_\alpha} \int \mathbf{V} f_\alpha \, \mathrm{d}\mathbf{V}. \tag{1.11}$$

The mean velocity $\mathbf{v}$ for the gas as a whole of the total density

$$\rho_m = \sum_\alpha N_\alpha m_\alpha$$

is determined as

$$\mathbf{v} = \frac{1}{\rho_m} \sum_\alpha N_\alpha m_\alpha \mathbf{v}_{(\alpha)}. \tag{1.12}$$

The total density of an electric current $\mathbf{j}$ is expressed in terms of mean velocities as

$$\mathbf{j} = -eN_e \mathbf{v}_e + \sum_\alpha eZ_\alpha N_\alpha \mathbf{v}_{(\alpha)}, \tag{1.13}$$

where N_e and N_α are electron and ion concentrations, respectively. The summation is taken over all ion species.

Moments of the Boltzmann's Equation

A velocity averaging of an arbitrary function $Q_{(\alpha)}(\mathbf{V})$ is given by

$$\langle Q_{(\alpha)} \rangle (\mathbf{r}, t) = \frac{1}{N_\alpha(\mathbf{r}, t)} \int Q_{(\alpha)}(\mathbf{V}) \; f_\alpha(\mathbf{r}, \mathbf{V}, t) \, \mathrm{d}\mathbf{V}, \tag{1.14}$$

where the concentration $N_\alpha(\mathbf{r}, t)$ is determined in (1.1).

The equations of conservation of momentum for mean quantities may be called as multi-fluid hydrodynamics, which are close in form to the hydrodynamic equations. They can be obtained by averaging the kinetic equations. By multiplying the kinetic equation (1.7) by $Q_{(\alpha)}(\mathbf{V})\, \mathrm{d}\mathbf{V}$ and integrating over velocities $\mathbf{V}$, we obtain

$$\begin{aligned} \frac{\partial}{\partial t}(N_\alpha \langle Q_{(\alpha)} \rangle) + \frac{\partial}{\partial x_k} \left(N_\alpha \langle Q_{(\alpha)} V_k \rangle \right) - \frac{N_\alpha}{m_\alpha} \langle F_{(\alpha)k} \frac{\partial}{\partial V_k} Q_{(\alpha)} \rangle \\ = \int \left(\frac{\partial f_\alpha}{\partial t} \right)_{coll} Q_{(\alpha)}(\mathbf{V}) \, \mathrm{d}\mathbf{V}, \end{aligned} \tag{1.15}$$

where we have adopted the convention on tensor summation over index $k = 1, 2, 3$ (hereafter summation is adopted over all the double tensor index). The third term is obtained through integration by parts.

Let us assume that the number concentration of the particles of each sort is conserved in collisions. Then

$$\int \left(\frac{\partial f_\alpha}{\partial t} \right)_{coll} \mathrm{d}\mathbf{V} = 0. \tag{1.16}$$

Substitution $Q_{(\alpha)}(\mathbf{V}) = 1$ into (1.15) yields the equation of conservation of the particle numbers

$$\frac{\partial N_\alpha}{\partial t} + \boldsymbol{\nabla} \cdot (N_\alpha \mathbf{v}_{(\alpha)}) = 0, \tag{1.17}$$

and $Q_{(\alpha)}(\mathbf{V}) = m_\alpha \mathbf{V}$ leads to the momentum equation

$$m_\alpha N_\alpha \left(\frac{\partial v_{(\alpha)j}}{\partial t} + (\mathbf{v}_{(\alpha)} \boldsymbol{\nabla}) v_{(\alpha)j} \right) = N_\alpha F_{(\alpha)j} - \frac{\partial}{\partial x_k} P_{(\alpha)jk} + R_{(\alpha)j}. \tag{1.18}$$

Here

$$\mathbf{F}_{(\alpha)} = Z_\alpha e \left(\mathbf{E} + \frac{1}{c} \left[\mathbf{v} \times \mathbf{B} \right] \right) + m_\alpha \mathbf{g},$$

where $\mathbf{g}$ is the acceleration in non-electromagnetic fields (e.g., acceleration of gravity), for neutrals $Z_\alpha = 0$. $P_{(\alpha)jk}$ is the pressure tensor with elements

$$P_{(\alpha)jk} = P_{(\alpha)kj} = m_\alpha N_\alpha \langle (V_j - v_{(\alpha)j})(V_k - v_{(\alpha)k}) \rangle. \tag{1.19}$$

The last term in the right-hand side of (1.18) is the frictional drag

$$\mathbf{R}_{(\alpha)} = \int m_\alpha \left(\mathbf{V} - \mathbf{v}_{(\alpha)} \right) \left(\frac{\partial f_\alpha}{\partial t} \right)_{coll} \mathrm{d}\mathbf{V}. \tag{1.20}$$

Equation (1.18) is the equation of the momentum conservation and it is an analogue of the corresponding hydrodynamic equation. An equation similar to the hydrodynamic energy conservation law can be obtained by taking $Q_{(\alpha)}(\mathbf{V}) = m_\alpha \mathbf{V}^2/2$ in (1.15). These equations become completed just with additional assumptions about the pressure tensor $P_{(\alpha)jk}$ and the friction force $\mathbf{R}_{(\alpha)}$. Strictly speaking, they should be obtained from the initial kinetic equations [2].

In the simplest approximation, pressure tensors are supposed to be isotropic $P_{(\alpha)jk} = P_\alpha \delta_{jk}$, where P_α is pressure (scalar), δ_{jk} is the unit tensor (δ_{jk} is the Kronecker symbol). A more common approximation is often used in plasma applications with diagonal tensors $P_{(\alpha)jk}$ but non-equal components $P_{(\alpha)xx}$, $P_{(\alpha)yy}$, $P_{(\alpha)zz}$. Finally, it is possible to take into account viscous stresses for the pressure tensor. Then $P_{(\alpha)jk} = P_\alpha \delta_{jk} - \tau_{jk}$, where τ_{jk} is a tensor of viscous stresses. A 'minus' sign reflects the fact that the viscous stresses cause momentum losses caused by an internal friction.

Three-Fluid Magnetohydrodynamics

The simplest and often used approximation is an assumption about isotropy of the pressure tensor $P_{(\alpha)jk}$ and reducing it to

$$P_{(\alpha)jk} = P_\alpha \delta_{jk}. \tag{1.21}$$

Henceforth in this book we put $P_\alpha = kN_\alpha T_\alpha$.

The friction $\mathbf{R}_{(\alpha)}$ acting on particles α is connected with their collisions with all other particles. Let $\mathbf{R}_{\alpha\beta}$ be the change of momentum at collisions of particles α with particles β. Then

$$\mathbf{R}_{(\alpha)} = \sum_\beta \mathbf{R}_{\alpha\beta}.$$

It follows from the law of momentum conservation at collisions that

$$\mathbf{R}_{\alpha\beta} = -\mathbf{R}_{\beta\alpha}.$$

Consider, for instance, the force between electrons and one of the ionic species. Electrons within the time of $\sim 1/\nu_{ei}$ lose their relative velocity $\mathbf{v}_e - \mathbf{v}_i$ by collisions. The ions in the process acquire the momentum $m_e(\mathbf{v}_e - \mathbf{v}_i)$ per each electron. This means that the friction $-m_e N_e(\mathbf{v}_e - \mathbf{v}_i)$ acts on electrons. Opposite and equal force acts on ions. Thus, $\mathbf{R}_{(\alpha)}$ can be approximated by expressions proportional to relative mean velocities of the collide particles

$$\mathbf{R}_{(\alpha)} = -N_\alpha \sum_{\beta \neq \alpha} \mu_{\alpha\beta} \nu_{a\beta} (\mathbf{v}_\alpha - \mathbf{v}_\beta), \tag{1.22}$$

where $\nu_{\alpha\beta}$ is the collision frequency of the particle α with all the particles β; the reduced masses

$$\mu_{\alpha\beta} = \frac{m_\alpha m_\beta}{m_\alpha + m_\beta},$$

where m_α, m_β are the mass of particles α, β, respectively. The reduce mass $\mu_{e\beta} \approx m_e$ for collisions of electrons with heavy particles.

Consider a three-component plasma containing electrons, ions and neutrals. Then continuity equation (1.17) gives

$$\frac{\partial N_e}{\partial t} + \boldsymbol{\nabla} \cdot (N_e \mathbf{v}_e) = 0, \tag{1.23}$$

$$\frac{\partial N_i}{\partial t} + \boldsymbol{\nabla} \cdot (N_i \mathbf{v}_i) = 0, \tag{1.24}$$

$$\frac{\partial N_n}{\partial t} + \boldsymbol{\nabla} \cdot (N_n \mathbf{v}_n) = 0, \tag{1.25}$$

and the momentum equation (1.18) with taken into account (1.21) becomes

$$\frac{d\mathbf{v}_e}{dt} = -\frac{e}{m_e}\mathbf{E}'_e - \frac{\nabla P_e}{m_e N_e} - \nu_{ei}(\mathbf{v}_e - \mathbf{v}_i) - \nu_{en}(\mathbf{v}_e - \mathbf{v}_n), \tag{1.26}$$

$$\frac{d\mathbf{v}_i}{dt} = \frac{e}{m_i}\mathbf{E}'_i - \frac{\nabla P_i}{m_i N_i} - \nu_{ie}(\mathbf{v}_i - \mathbf{v}_e) - \nu_{in}(\mathbf{v}_i - \mathbf{v}_n),$$

$$\frac{d\mathbf{v}_n}{dt} = -\frac{\nabla P_n}{m_n N_n} - \frac{m_e N_e}{m_n N_n}\nu_{en}(\mathbf{v}_n - \mathbf{v}_e) - \frac{m_i N_i}{m_n N_n}\nu_{in}(\mathbf{v}_n - \mathbf{v}_i), \tag{1.27}$$

$$P_\alpha = kT_\alpha N_\alpha, \quad \mathbf{E}'_{e,i} = \mathbf{E} + \frac{1}{c}[\mathbf{v}_{e,i} \times \mathbf{B}], \tag{1.28}$$

where

$$\frac{d\mathbf{v}_\alpha}{dt} = \frac{\partial \mathbf{v}_\alpha}{\partial t} + (\mathbf{v}_\alpha \nabla)\mathbf{v}_\alpha, \qquad \alpha = e,\, i,\, n.$$

m_e and $\mathbf{v}_e$, m_i and $\mathbf{v}_i$, m_n and $\mathbf{v}_n$ are, respectively, the mass and the mean velocity of electrons, ions and neutrals; P_e, P_i, P_n are partial pressures of electrons, ions and neutrals. Equations (1.23)–(1.27) are supplemented with Maxwell's equations

$$\nabla \times \mathbf{B} = \frac{4\pi}{c}\mathbf{j} + \frac{1}{c}\frac{\partial \mathbf{E}}{\partial t}, \qquad \nabla \cdot \mathbf{E} = 4\pi\rho, \tag{1.29}$$

$$\nabla \times \mathbf{E} = -\frac{1}{c}\frac{\partial \mathbf{B}}{\partial t}, \qquad \nabla \cdot \mathbf{B} = 0, \tag{1.30}$$

$$\mathbf{j} = e(N_i\mathbf{v}_i - N_e\mathbf{v}_e), \qquad \rho = e(N_i - N_e). \tag{1.31}$$

We must also supplement the energy conservation equation or heat transfer equations which depend on the electron T_e, ion T_i and neutral T_n temperatures. Below, we shall restrict our consideration to the isothermal case $T_e = T_i = T_n = \text{const}$.

Magnetohydrodynamics

Next simplifications in the plasma description is utilization of the magnetohydrodynamics (MHD) equations. Let the typical spatial scale of the system and frequency of the electromagnetic field be denoted as L, and ω, and the particle free path and thermal velocity by l, and v_T, respectively. Also let, $\omega_{ci} = eB/m_i c$ be the ion cyclotron frequency and ν be the collision frequency of the ion with neutrals. Then the conditions for which the generalized 3-fluid hydrodynamics equation reduce to the one-fluid magnetohydrodynamics can be written

$$\begin{aligned} &1.\quad l < L, \\ &2.\quad \omega \ll \nu \sim v_T/l, \\ &3.\quad \omega << \omega_{ci}. \end{aligned} \tag{1.32}$$

Leaving aside details of reducing of 3-fluid hydrodynamics to the one-fluid magnetohydrodynamics, let us write the MHD equations [5]

$$\rho\left(\frac{\partial\,\mathbf{u}}{\partial\,t}+(\mathbf{u}\boldsymbol{\nabla})\mathbf{u}\right)=-\boldsymbol{\nabla}P+\frac{1}{c}[\mathbf{j}\times\mathbf{B}], \tag{1.33}$$

$$\frac{\partial\,\rho}{\partial\,t}+\boldsymbol{\nabla}\cdot(\rho\,\mathbf{u})=0, \tag{1.34}$$

and Maxwell's equations are

$$\boldsymbol{\nabla}\times\mathbf{E}=-\frac{1}{c}\frac{\partial\,\mathbf{B}}{\partial\,t}, \tag{1.35}$$

$$\boldsymbol{\nabla}\times\mathbf{B}=\frac{4\pi}{c}\mathbf{j}, \tag{1.36}$$

$$\mathbf{j}=\sigma_0\left\{\mathbf{E}+\frac{1}{c}[\mathbf{u}\times\mathbf{B}]\right\}. \tag{1.37}$$

Here σ_0 is electrical conductivity which can be expressed as

$$\sigma_0=\frac{Ne^2}{m_e\nu}$$

The mass density ρ, total pressure P and macroscopic velocity $\mathbf{u}$ are given by

$$\rho=N_em_e+N_im_i+N_nm_n, \tag{1.38}$$

$$P=P_e+P_i+P_n, \tag{1.39}$$

$$\mathbf{u}=\frac{1}{\rho}(N_em_e\mathbf{v}_e+N_im_i\mathbf{v}_i+N_nm_n\mathbf{v}_n). \tag{1.40}$$

Equations (1.33), (1.34) are obtained from the general condition for the hydromagnetic approximation: the collision frequencies should be larger than the ion cyclotron frequencies

$$\omega\ll\omega_{ci}\ll\nu. \tag{1.41}$$

Then all plasma components are involved into the motion and from (1.26)–(1.27) follows

$$v_e\approx v_i\approx v_n.$$

The conditions of applicability of MHD-approximation in the strong magnetic field can be essentially weaker than the inequality (1.32). In particular, MHD-approximation in the strong magnetic field can provide correct results even in a collisionless plasma. In this case, instead of small parameter l/L controlling the applicability of hydrodynamics, a new small parameter r_{ci}/L is introduced, where r_{ci} is the Larmour radius $r_{ci}=v_{Ti}/\omega_{ci}$. For the cold plasma it is necessary for the frequency ω to be small in comparison with the ion cyclotron frequency ω_{ci}.

Equations for Disturbances

Suppose that in the equilibrium state, the velocities $\mathbf{v}_a$ and electric field $\mathbf{E}$ vanish and the plasma is embedded into external magnetic field $\mathbf{B}_0$. Unperturbed values of concentration denote as $N_0 = N_{e0} = N_{i0}$, N_{n0}, and pressure $P_{\alpha 0}$. Denote small perturbations of magnetic and electric fields, particle number densities, velocities and pressure by $\mathbf{b}$, $\mathbf{E}$, n_α, $\mathbf{v}_{(\alpha)}$ and p_α.

Substituting into (1.23)–(1.27)

$$\mathbf{B}_0 + \mathbf{b},\ \mathbf{E},\ N_{\alpha 0} + n_\alpha,\ \mathbf{v}_{(\alpha)},\ P_{\alpha 0} + p_\alpha, \tag{1.42}$$

and neglecting all quadratic terms of the perturbations, we obtain linearized equations of 3-fluid hydrodynamics. From now on, if it is not specified, we drop $'0'$ when there was no risk of ambiguity and write N_α instead of $N_{\alpha 0}$, for easy of notation. Linearized continuity equations (1.23)–(1.25) allow us to find perturbations of concentrations of plasma components from divergencies of mean velocities:

$$\frac{\partial n_e}{\partial t} = -N \boldsymbol{\nabla} \cdot \mathbf{v}_e, \quad \frac{\partial n_i}{\partial t} = -N \boldsymbol{\nabla} \cdot \mathbf{v}_i, \quad \frac{\partial n_n}{\partial t} = -N_n \boldsymbol{\nabla} \cdot \mathbf{v}_n. \tag{1.43}$$

From the momentum equations (1.26)–(1.27) we get

$$\frac{\partial \mathbf{v}_e}{\partial t} = -\frac{e}{m_e}\left(\mathbf{E} + \frac{1}{c}[\mathbf{v}_e \times \mathbf{B}_0]\right) - \frac{\boldsymbol{\nabla} p_e}{m_e N_e} - \nu_{ei}(\mathbf{v}_e - \mathbf{v}_i) - \nu_{en}(\mathbf{v}_e - \mathbf{v}_n), \tag{1.44}$$

$$\frac{\partial \mathbf{v}_i}{\partial t} = \frac{e}{m_i}\left(\mathbf{E} + \frac{1}{c}[\mathbf{v}_i \times \mathbf{B}_0]\right) - \frac{\boldsymbol{\nabla} p_i}{m_i N_i} + \frac{m_e}{m_i}\nu_{ei}(\mathbf{v}_e - \mathbf{v}_i) - \nu_{in}(\mathbf{v}_i - \mathbf{v}_n), \tag{1.45}$$

$$\frac{\partial \mathbf{v}_n}{\partial t} = -\frac{\boldsymbol{\nabla} p_n}{m_n N_n} - \frac{m_e}{m_n}\frac{N_e}{N_n}\nu_{en}(\mathbf{v}_n - \mathbf{v}_e) - \frac{m_i}{m_n}\frac{N_i}{N_n}\nu_{in}(\mathbf{v}_n - \mathbf{v}_i). \tag{1.46}$$

A linear approximation to (1.33)–(1.34) is given in Chapter 4, where wave disturbances in the linear magnetohydrodynamics are considered.

1.3 Electromagnetic Field Equations

Let us introduce the electric induction vector $\mathbf{D}(\mathbf{r}, t)$ as

$$\mathbf{D}(\mathbf{r}, t) = \mathbf{E}(\mathbf{r}, t) + 4\pi \int_{-\infty}^{t} \mathbf{j}(\mathbf{r}, t')\, \mathrm{d}t' \tag{1.47}$$

making it possible to combine density of induced charges and currents with displacement currents in the field equations. Thus, Maxwell's equations (1.9) become

$$\nabla \times \mathbf{E} = -\frac{1}{c}\frac{\partial \mathbf{B}}{\partial t}, \qquad \nabla \cdot \mathbf{B} = 0, \tag{1.48}$$

$$\nabla \times \mathbf{B} = \frac{1}{c}\frac{\partial \mathbf{D}}{\partial t}, \qquad \nabla \cdot \mathbf{D} = 0. \tag{1.49}$$

In order to use these equations, the way in which **D** and **j** depend on **E** and **B** must be given.

Theoretical considerations in this book are carried out mainly within the framework of the linear electrodynamics which is valid for the small wave fields, when the conductivity and dielectric permeability are independent on the wave **E** and **B**. The non-linear phenomena appear only in sufficiently strong fields, and will be discussed in Chapter 13 on man-made generation of the ULF-waves.

In the general case, the state of the system depends not only on the field in a given time instant t and at a given point $\mathbf{r}$, but also on their values at previous moments and in remote points. Therefore, generally, it is necessary to use in the material relations which connect **D** and **J** with **E** and **B** taking into account the time and spatial dispersion of the medium. Therefore, the material relations may be written as

$$D_i(\mathbf{r}, t) = \int_{-\infty}^{t} dt' \int \mathrm{d}\mathbf{r}'\, \varepsilon_{ij}(\mathbf{r}, \mathbf{r}', t - t')\, E_j(\mathbf{r}', t'), \tag{1.50}$$

$$j_i(\mathbf{r}, t) = \int_{-\infty}^{t} dt' \int \mathrm{d}\mathbf{r}'\, \sigma_{ij}(\mathbf{r}, \mathbf{r}', t - t')\, E_j(\mathbf{r}', t'), \tag{1.51}$$

where i, j are the tensor indices $1, 2, 3$ corresponding to the axes $x = x_1$, $y = x_2$, $z = x_3$.

These are the most general forms of the linear relations between the electric induction **D** and the electric current **j** with the electric field **E** taking into account the causality principle (the electric induction at a given time point t is determined by the electric field at $t' \leq t$) and the time homogeneity of plasma properties.

In numerous important cases the kernels ε_{ij} and $\boldsymbol{\sigma}_{ij}$ of the integral operators in (1.50), (1.51) vanish fast with distance d in the neighborhood of the point $\mathbf{r}$. Then integrating over $\mathrm{d}\mathbf{r}'$ can be done only at $|\mathbf{r} - \mathbf{r}'| \lesssim d$. If the electric field **E** varies slowly at the scale d, it can be factored out from the integral over $\mathrm{d}\mathbf{r}'$. Then we obtain from (1.50)–(1.51)

$$D_i(\mathbf{r}, t) = \int_{-\infty}^{t} dt'\, \varepsilon_{ij}(\mathbf{r}, t - t')\, E_j(\mathbf{r}, t'), \tag{1.52}$$

$$j_i(\mathbf{r}, t) = \int_{-\infty}^{t} dt'\, \boldsymbol{\sigma}_{ij}(\mathbf{r}, t - t')\, E_j(\mathbf{r}, t'). \tag{1.53}$$

The effects of spatial dispersion are neglected here. This approximation is applicable for plasmas with rather low temperatures (cold plasma approximation). The colder is the plasma, the weaker is spatial dispersion. For example, at low temperatures the terms with pressure can be omitted in (1.44)–(1.46). Then, the electron and ion velocities, the current density $\mathbf{j}$ (1.31) and the induction $\mathbf{D}$ (1.47), depend only on the electric field $\mathbf{E}$ at the point $\mathbf{r}$.

Let us apply the Fourier transform to (1.52)–(1.53). Suppose that

$$\mathbf{E}(\mathbf{r},\omega) = \int \mathbf{E}(\mathbf{r},t)\exp(i\omega t)dt$$

and, similarly, for $\mathbf{D}(\mathbf{r},t)$ and $\mathbf{j}(\mathbf{r},t)$. The same notation $\mathbf{E}$ is used for Fourier transforms $\mathbf{E}(\mathbf{r},\omega)$ and images $\mathbf{E}(\mathbf{r},t)$. This does not lead to misunderstanding because the arguments are given explicitly, the same for other variables. Then, (1.52), (1.53) and (1.47) become

$$\mathbf{D}(\mathbf{r},\omega) = \boldsymbol{\varepsilon}(\mathbf{r},\omega)\,\mathbf{E}(\mathbf{r},\omega), \tag{1.54}$$

$$\mathbf{j}(\mathbf{r},\omega) = \boldsymbol{\sigma}(\mathbf{r},\omega)\,\mathbf{E}(\mathbf{r},\omega), \tag{1.55}$$

$$\boldsymbol{\varepsilon}(\mathbf{r},\omega) = 1 + i\frac{4\pi}{\omega}\boldsymbol{\sigma}(\mathbf{r},\omega), \tag{1.56}$$

where

$$\boldsymbol{\varepsilon}(\mathbf{r},\omega) = \int_{-\infty}^{t} dt'\, e^{i\omega t'}\boldsymbol{\varepsilon}(\mathbf{r},t'), \quad \boldsymbol{\sigma}(\mathbf{r},\omega) = \int_{-\infty}^{t} dt'\, e^{i\omega t'}\boldsymbol{\sigma}(\mathbf{r},t'). \tag{1.57}$$

$\boldsymbol{\varepsilon}(\mathbf{r},\omega)$ is called the tensor of complex dielectric permeability. It can be present as a sum of Hermitian and anti-Hermitian parts

$$\boldsymbol{\varepsilon}(\mathbf{r},\omega) = \boldsymbol{\varepsilon}'(\mathbf{r},\omega) + i\frac{4\pi}{\omega}\boldsymbol{\sigma}'(\mathbf{r},\omega), \tag{1.58}$$

where $\boldsymbol{\varepsilon}'(\mathbf{r},\omega)$ and $\boldsymbol{\sigma}'(\mathbf{r},\omega)$ are Hermitian ones. $\sigma(\mathbf{r},\omega)$ is the tensor of complex conductivity. It can be also presented as a sum of Hermitian and anti-Hermitian parts

$$\boldsymbol{\sigma}(\mathbf{r},\omega) = \boldsymbol{\sigma}'(\mathbf{r},\omega) - i\frac{\omega}{4\pi}\boldsymbol{\varepsilon}'(\mathbf{r},\omega). \tag{1.59}$$

Specific expressions for $\boldsymbol{\varepsilon}(\mathbf{r},\omega)$ and $\boldsymbol{\sigma}(\mathbf{r},\omega)$ for the cold three-component plasma containing electrons, ions and neutrals are obtained in the next section.

1.4 Dielectric Permeability and Conductivity

Drag of the Neutral Gas with Ions

$\boldsymbol{\varepsilon}(r,\omega)$ and $\boldsymbol{\sigma}(r,\omega)$ can be found from (1.44)–(1.46). From (1.46) for the molecule velocity we have

$$\mathbf{v}_n = a\mathbf{v}_e + b\mathbf{v}_i \tag{1.60}$$

with

$$a = \frac{\nu_{ne}}{-i\omega + \nu_n}, \qquad b = \frac{\nu_{ni}}{-i\omega + \nu_n}, \quad \nu_n = \nu_{ne} + \nu_{ni},$$

$$\nu_{ne} = \frac{m_e}{m_n}\frac{N}{N_n}\nu_{en}, \quad \nu_{ni} = \frac{m_i}{m_n}\frac{N}{N_n}\nu_{in}.$$

Substituting (1.60) into (1.44) and (1.45), we obtain

$$-\frac{e}{m_e}\mathbf{E} = (-i\omega + \nu_e - a\nu_{en})\mathbf{v}_e + \frac{e}{m_e c}[\mathbf{v}_e \times \mathbf{B}_0] - (\nu_{ei} + b\nu_{en})\mathbf{v}_i, \quad (1.61)$$

$$\frac{e}{m_i}\mathbf{E} = -(\nu_{ie} + a\nu_{in})\mathbf{v}_e - \frac{e}{m_i c}[\mathbf{v}_i \times \mathbf{B}_0] + (-i\omega + \nu_i - b\nu_{in})\mathbf{v}_i, \quad (1.62)$$

where $\nu_i = \nu_{ie} + \nu_{in}$, $\nu_{ie} = (m_e/m_i)\nu_{ei}$. Equations (1.61) and (1.62) in Cartesian coordinates $\{x, y, z\}$ with the z-axis pointing along $\mathbf{B}_0$ ($\mathbf{B}_0 = B_0\hat{\mathbf{z}}$, for definiteness, $B_0 > 0$), become

$$-\frac{e}{m_e}\mathbf{E} = (-i\omega + \nu_e - a\nu_{en})\mathbf{v}_e + \omega_{ce}\,\mathbf{S}\mathbf{v_e} - (\nu_{ei} + b\nu_{en})\mathbf{v}_i, \quad (1.63)$$

$$\frac{e}{m_i}\mathbf{E} = -(\nu_{ie} + a\nu_{in})\mathbf{v}_e - \omega_{ci}\,\mathbf{S}\mathbf{v_i} + (-i\omega + \nu_i - b\nu_{in})\mathbf{v}_i, \quad (1.64)$$

where the electron and ion cyclotron frequencies are

$$\omega_{ce} = \frac{eB_0}{m_e c}, \quad \omega_{ci} = \frac{eB_0}{m_i c},$$

and

$$\mathbf{S} = \begin{pmatrix} 0 & 1 & 0 \\ -1 & 0 & 0 \\ 0 & 0 & 1 \end{pmatrix}, \quad \mathbf{v}_{e,i} = \begin{pmatrix} \mathbf{v}_{e,i\perp} \\ v_{e,i\parallel} \end{pmatrix} = \begin{pmatrix} v_{e,i\,x} \\ v_{e,i\,y} \\ v_{e,i\,z} \end{pmatrix}.$$

The easiest way to solve (1.63) and (1.64) is turn to the basis of circle polarizations

$$v_e^{(\pm)} = v_{ex} \pm i v_{ey}, \quad v_{e\parallel} = v_{ez},$$

$$v_i^{(\pm)} = v_{ix} \pm i v_{iy}, \quad v_{i\parallel} = v_{iz},$$

$$E^{(\pm)} = E_x \pm i E_y, \quad E_\parallel = E_z.$$

Then (1.63) and (1.64) for the transversal components yield

$$[-i(\omega \pm \omega_{ce}) + \nu_e - a\nu_{en}]\, v_e^{(\pm)} - (\nu_{ei} + b\nu_{en}) v_i^{(\pm)} = -\frac{e}{m_e}E^{(\pm)}, \quad (1.65)$$

$$-(\nu_{ie} + a\nu_{in})\, v_e^{(\pm)} + [-i(\omega \mp \omega_{ci}) + \nu_i - b\nu_{in}]\, v_i^{(\pm)} = \frac{e}{m_i}E^{(\pm)}. \quad (1.66)$$

Hence

$$v_e^{(\pm)} = -i\frac{eE^{(\pm)}}{m_e\Delta_\pm}\left(\omega \mp \omega_{ci} + \frac{\omega}{\omega + i\nu_n} i\nu_{in}\right), \tag{1.67}$$

$$v_i^{(\pm)} = i\frac{eE^{(\pm)}}{m_i\Delta_\pm}\left(\omega \pm \omega_{ce} + \frac{\omega}{\omega + i\nu_n} i\nu_{en}\right), \tag{1.68}$$

where

$$\Delta_\pm = [\omega \pm \omega_{ce} + i(\nu_e - a\nu_{en}]\,[\omega \mp \omega_{ci} + i(\nu_i - b\nu_{in})] + (\nu_{ei} + b\nu_{en})(\nu_{ie} + a\nu_{in}).$$

The magnetic field has no effect on the velocity and current components along $\mathbf{B}_0$. Therefore, in order to find longitudinal velocities, one can put $\omega_{ce} = \omega_{ci} = 0$, then

$$v_{e\|} = -i\frac{eE_\|}{m_e\Delta_\|}\left(\omega + \frac{\omega}{\omega + i\nu_n} i\nu_{in}\right), \tag{1.69}$$

$$v_{i\|} = i\frac{eE_\|}{m_i\Delta_\|}\left(\omega + \frac{\omega}{\omega + i\nu_n} i\nu_{en}\right), \tag{1.70}$$

where

$$\Delta_\| = [\omega + i(\nu_e - a\nu_{en})][\omega + i(\nu_i - b\nu_{in})] + (\nu_{ei} + b\nu_{en})(\nu_{ie} + a\nu_{in}).$$

The electron and ion current densities and corresponding conductivities in the basis of circle polarizations are

$$j_e^{(\pm)} = j_{ex} \pm ij_{ey}, \quad j_{e\|} = j_{ez} \text{ and } j_i^{(\pm)} = j_{ix} \pm ij_{iy}, \quad j_{i\|} = j_{iz},$$
$$\sigma_e^{(\pm)} = \sigma_{exx} \mp i\sigma_{exy}, \quad \sigma_{e\|} = \sigma_{ez} \quad \text{and} \quad \sigma_i^{(\pm)} = \sigma_{ixx} \mp i\sigma_{ixy}, \quad \sigma_{i\|} = \sigma_{iz},$$

and

$$\sigma_{exx} = \frac{1}{2}\left(\sigma_e^{(+)} + \sigma_e^{(-)}\right), \quad \sigma_{ixx} = \frac{1}{2}\left(\sigma_i^{(+)} + \sigma_i^{(-)}\right), \tag{1.71}$$

$$\sigma_{exy} = \frac{i}{2}\left(\sigma_e^{(+)} - \sigma_e^{(-)}\right), \quad \sigma_{ixy} = \frac{i}{2}\left(\sigma_i^{(+)} - \sigma_i^{(-)}\right). \tag{1.72}$$

Here we take into account that $\sigma_{xx} = \sigma_{yy}$, $\sigma_{xy} = -\sigma_{yx}$. The transversal and longitudinal components of the total conductivity is a sum of the electron and ion parts:

$$\sigma^{(\pm)} = \sigma_e^{(\pm)} + \sigma_i^{(\pm)}, \quad \sigma_\| = \sigma_{e\|} + \sigma_{i\|}. \tag{1.73}$$

Substituting (1.67)–(1.70) into

$$j_e^{(\pm)} = -Nev_e^{(\pm)} = \sigma_e^{(\pm)}E^{(\pm)}, \quad j_{e\|} = -Nev_{e\|} = \sigma_{e\|}E_\|, \tag{1.74}$$

$$j_i^{(\pm)} = Nev_i^{(\pm)} = \sigma_i^{(\pm)}E^{(\pm)}, \quad j_{i\|} = Nev_{i\|} = \sigma_{i\|}E_\|. \tag{1.75}$$

gives equations for the electron and the ion conductivities

$$\sigma_e^{(\pm)} = i\frac{\omega_{pe}^2}{4\pi\Delta_\pm}\left(\omega \mp \omega_{ci} + \frac{\omega}{\omega + i\nu_n} i\nu_{in}\right), \tag{1.76}$$

$$\sigma_{e\|} = i\frac{\omega_{pe}^2}{4\pi\Delta_\|}\left(\omega + \frac{\omega}{\omega + i\nu_n} i\nu_{in}\right),$$

$$\sigma_i^{(\pm)} = i\frac{\omega_{pi}^2}{4\pi\Delta_\pm}\left(\omega \pm \omega_{ce} + \frac{\omega}{\omega + i\nu_n} i\nu_{en}\right),$$

$$\sigma_{i\|} = i\frac{\omega_{pi}^2}{4\pi\Delta_\|}\left(\omega + \frac{\omega}{\omega + i\nu_{in}} i\nu_{en}\right). \tag{1.77}$$

where $\omega_{pe}^2 = 4\pi Ne^2/m_e$ and $\omega_{pi}^2 = 4\pi Ne^2/m_i$ are squares of the electron and ion plasma frequencies, respectively. The total conductivities are

$$\sigma^{(\pm)} = i\omega\frac{\omega_{pe}^2}{4\pi\Delta_\pm}G, \quad \sigma_\| = i\omega\frac{\omega_{pe}^2}{4\pi\Delta_\|}G, \tag{1.78}$$

$$G = 1 + \frac{m_e}{m_i} + \frac{\nu_{in} + \frac{m_e}{m_i}\nu_{en}}{-i\omega + \nu_n}. \tag{1.79}$$

Substituting (1.78), (1.79) into (1.56), we obtain the transversal and longitudinal components of the complex dielectric permeability

$$\varepsilon_{xx} \mp i\varepsilon_{xy} = 1 - \frac{\omega_{pe}^2}{\Delta_\pm}G, \quad \varepsilon_{zz} = 1 - \frac{\omega_{pe}^2}{\Delta_\|}G. \tag{1.80}$$

Here, the drag of the neutral component by charged plasma particles is taken into consideration. The drag effect is important only for relatively low frequencies

$$\omega \lesssim \nu_n. \tag{1.81}$$

The characteristic times of the ion-neutral drag can be estimated for the E- and F-layers as

$$2\pi/\nu_n \sim \begin{cases} 5\times 10^4\,\mathrm{s} & E\text{-layer}, \\ 2\times 10^4\,\mathrm{s} & F\text{-layer}. \end{cases}$$

Thus, the drag of the neutrals by charged particles should only be considered for studying hourly variations.

'Motionless' Neutral Gas

If the frequency is much more than the collision frequency of a neutral with all charged particles, i.e. $\omega \gg \nu_n$ (ν_n see in (1.60)), then the contribution of the drag of neutrals into the conductivity becomes negligible and (1.76) and

(1.77) reduce to

$$\sigma_e^{(\pm)} = i\frac{\omega_{pe}^2}{4\pi\Delta_\pm}\left(\omega \mp \omega_{ci} + i\nu_{in}\right), \quad \sigma_{e\|} = i\frac{\omega_{pe}^2}{4\pi\Delta_\|}\left(\omega + i\nu_{in}\right), \tag{1.82}$$

$$\sigma_i^{(\pm)} = i\frac{\omega_{pi}^2}{4\pi\Delta_\pm}\left(\omega \pm \omega_{ce} + i\nu_{en}\right), \quad \sigma_{i\|} = i\frac{\omega_{pi}^2}{4\pi\Delta_\|}\left(\omega + i\nu_{en}\right). \tag{1.83}$$

These equations allow considerable further simplification. $\nu_{ei}\nu_{ie} = \nu_{ei}^2\, m_e/m_i$ can be omitted in $\Delta_\pm$ if $\nu_{in} >> \nu_{ei}\, m_e/m_i$. This inequality holds throughout the ionosphere, except the upper part of the F-layer. Then

$$\begin{aligned}\Delta_\pm &= (\omega \pm \omega_{ce} + i\nu_e)(\omega \mp \omega_{ci} + i\nu_i),\\ \Delta_\| &= (\omega + i\nu_e)(\omega + i\nu_i),\end{aligned}$$

and (1.82)–(1.83) become

$$\sigma_e^{(\pm)} = i\frac{\omega_{pe}^2}{4\pi(\omega \pm \omega_{ce} + i\nu_e)}, \quad \sigma_{e\|} = i\frac{\omega_{pe}^2}{4\pi(\omega + i\nu_e)}, \tag{1.84}$$

$$\sigma_i^{(\pm)} = i\frac{\omega_{pi}^2}{4\pi(\omega \mp \omega_{ci} + i\nu_i)}, \quad \sigma_{i\|} = i\frac{\omega_{pi}^2}{4\pi(\omega + i\nu_i)}. \tag{1.85}$$

By substituting (1.84), (1.85) into (1.71), (1.72), we obtain

$$\sigma_{xx} = c\frac{Ne}{B_0}\left(\frac{X_e}{1+X_e^2} + \frac{X_i}{1+X_i^2}\right), \tag{1.86}$$

$$\sigma_{xy} = c\frac{Ne}{B_0}\left(\frac{1}{1+X_i^2} - \frac{1}{1+X_e^2}\right), \tag{1.87}$$

$$\sigma_\| = c\frac{Ne}{B_0}\left(\frac{1}{X_e} + \frac{1}{X_i}\right). \tag{1.88}$$

where

$$X_e = \frac{\nu_e - i\omega}{\omega_{ce}}, \quad X_i = \frac{\nu_i - i\omega}{\omega_{ci}}.$$

The tensor of dielectric permeability $\boldsymbol{\varepsilon}(\mathbf{r}, \omega)$ could be obtained easily from (1.56).

Ultra Low Frequency Range

Consider Ultra Low Frequency (ULF) range holding the inequality

$$\nu_n \ll \omega \ll \omega_{ci} \ll \omega_{ce}. \tag{1.89}$$

Expand the conductivities (1.86)–(1.88) into the power series over ω/ω_{ce} and ω/ω_{ci}. First we present the following approximate equations

$$\frac{X_\alpha}{1+X_\alpha^2} \approx \frac{\beta_\alpha}{1+\beta_\alpha^2} - \beta_\alpha^2 \frac{\beta_\alpha^2 - 1}{(1+\beta_\alpha^2)^2} \frac{i\omega}{\omega_{c\alpha}}, \tag{1.90}$$

$$\frac{1}{1+X_\alpha^2} \approx \frac{\beta_\alpha^2}{1+\beta_\alpha^2} + \frac{2\beta_\alpha^2}{(1+\beta_\alpha^2)^2} \frac{i\omega}{\omega_{c\alpha}}, \qquad \alpha = e \text{ or } i, \tag{1.91}$$

where dimensionless parameters $\beta_e = \omega_{ce}/\nu_e$ and $\beta_i = \omega_{ci}/\nu_i$ are called the electron and ion magnetization , respectively. The components σ_{xx} and σ_{xy} can be expressed in terms of the Pedersen $\sigma_{xx}(\omega \to 0) = \sigma_P$ and Hall $\sigma_{xy}(\omega \to 0) = -\sigma_H$ conductivities as

$$\sigma_{xx} = \sigma_P - \frac{i\omega}{4\pi} \frac{c^2}{\widetilde{c}_A^2}, \qquad \sigma_{xy} = -\sigma_H. \tag{1.92}$$

It is accounted here that in the ULF-range the imaginary part of σ_{xy} is small compared to its real part. Substituting (1.90), (1.91) into (1.86) and (1.87), we find the Pedersen conductivity

$$\sigma_P = c\frac{Ne}{B_0}\left(\frac{\beta_i}{1+\beta_i^2} + \frac{\beta_e}{1+\beta_e^2}\right) \tag{1.93}$$

and the Hall conductivity

$$\sigma_H = c\frac{Ne}{B_0}\left(\frac{\beta_e^2}{1+\beta_e^2} - \frac{\beta_i^2}{1+\beta_i^2}\right). \tag{1.94}$$

Omitting the terms $\sim m_e/m_i$ and $\sim (m_e/m_i)^{1/2}$, we obtain for the Alfvén velocity modified by the ion collisions $\widetilde{c}_A$ and the collisionless Alfvén velocity c_A:

$$\widetilde{c}_A \approx \frac{c_A}{\beta_i} \frac{(1+\beta_i^2)}{(\beta_i^2 - 1)^{1/2}}, \qquad c_A = \frac{B_0}{(4\pi N m_i)^{1/2}}. \tag{1.95}$$

Usually, $c \gg c_A$ and, therefore, the term with the displacement current in (1.80) is omitted.

It can be shown (e.g. [5]), that $\nu_{in}/\nu_{en} \sim (m_e/m_i)^{1/2}$. Then

$$\left|\frac{X_e}{X_i}\right| = \frac{m_e}{m_i}\left(\frac{\nu_e^2 + \omega^2}{\nu_i^2 + \omega^2}\right)^{1/2} \lesssim \left(\frac{m_e}{m_i}\right)^{1/2} \ll 1. \tag{1.96}$$

The longitudinal conductivity is given by (1.88) which, in view of (1.96), may be written as

$$\sigma_\parallel = \frac{\omega_{pe}^2}{4\pi(\nu_e - i\omega)}. \tag{1.97}$$

Further, the longitudinal resistivity from (1.97) is

$$\sigma_\parallel^{-1} = \sigma_0^{-1} - i\omega\frac{4\pi}{\omega_{pe}^2}, \qquad \sigma_0 = \frac{\omega_{pe}^2}{4\pi\nu_e}. \tag{1.98}$$

It is possible to omit some terms in (1.92)–(1.94) and (1.97) for specific magnetospheric regions and obtain the equations for an estimate of the corresponding electrodynamic plasma parameters.

In the atmosphere and lower D-layer, the collision frequencies are so high that the electron and the ion are non-magnetized, i.e. $\beta_e \ll 1$, $\beta_i \ll 1$. Then the conductivity becomes isotropic. In the lower D-layer, neglecting the ion conductivity, we get

$$\sigma_\parallel \approx \sigma_\perp \approx \sigma_0 = \frac{\omega_{pe}^2}{4\pi\nu_e} \text{ and } \sigma_H \approx 0. \tag{1.99}$$

In the D- and lower E-layers, at the altitudes between ≈80 and ≈100 km, electrons are strongly magnetized $\beta_e \gg 1$ and $\beta_i \ll 1$ but $\beta_e\beta_i < 1$. Here both the Pedersen σ_P and Hall σ_H conductivities are determined by electrons and reduce to

$$\sigma_P = \sigma_{Pe} \approx \sigma_0/{\beta_e}^2, \quad \sigma_H = \sigma_{He} \approx \sigma_0/\beta_e$$

This is an important special case of the different sensitivity of σ_P and σ_H to the magnetic field. It can cause an anomalous effective Pedersen conductivity of the ionosphere containing small random inhomogeneities. This is discussed in Chapter 10.

Above, in the E-layer, there is an intermediate case of low but finite ion magnetization and high electron magnetization, so that $\beta_i < 1$, $\beta_e \gg 1$, and $\beta_i > \beta_e^{-1}$. Then (1.93) and (1.94) become

$$\sigma_P \approx \beta_i c \frac{Ne}{B_0} = \frac{\omega_{pi}^2}{4\pi\nu_i}, \quad \sigma_H \approx \sigma_0/\beta_e. \tag{1.100}$$

This means that the Pedersen conductivity is predominantly ionic and the Hall conductivity is determined by the electrons' drift in crossed electric and magnetic fields.

The polarization conductivity σ_{pol}, which determines the polarization current, is equal to the imaginary part of σ_{xx}. If $\beta_i < 1$, (1.92) becomes

$$\sigma_{pol} = i\mathrm{Im}(\sigma_{xx}) \approx \frac{i\omega}{4\pi}\left(\frac{\omega_{pi}}{\nu_i}\right)^2. \tag{1.101}$$

From this equation and (1.100), the ratio of the polarization current to the Pedersen current is $|\sigma_{pol}/\sigma_P| = \omega/\nu_i$. Joining of the inequalities (1.89) and $\beta_i \ll 1$ gives $\omega \ll \omega_{ci} \ll \nu_i$. Hence $\omega/\nu_i \ll \omega/\omega_{ci} \ll 1$. So that the polarization conductivity can be neglected in comparison with Pedersen conductivity.

Now, let both magnetization parameters be large, $\beta_i \gg 1$ and $\beta_e \gg 1$. These inequalities are met in the F-layer and in the magnetosphere. Omitting small terms $\sim\beta_i^{-2}$, one can see that the modified Alfvén velocity $\tilde{c}_A^2 \approx c_A^2$. Then (1.92) may be rewritten

$$\sigma_{xx} \approx \frac{c^2}{4\pi c_A^2}(\nu_i - i\omega), \quad \sigma_{xy} \approx 0. \tag{1.102}$$

The concentration of neutrals decreases rapidly with altitude and there is no need to take into account the ion-neutral collisions in the upper ionosphere and in the magnetosphere. Thus, it is possible to omit ν_i in (1.102) and write

$$\sigma_{xx} \approx -i\omega \frac{c^2}{c_A^2} .$$

Using (1.58) we find

$$\varepsilon_{xx} = c^2/c_A^2 . \tag{1.103}$$

At the end of this section we consider the longitudinal conductivity. It is convenient to define the electron inertial length as

$$\lambda_e = c/\omega_{pe}. \tag{1.104}$$

Then, for the longitudinal resistivity, (1.98) becomes

$$\sigma_{\parallel}^{-1} = \frac{4\pi}{c^2}\left(\nu_e - i\omega\right)\lambda_e^2. \tag{1.105}$$

If the collisions can be neglected, the inverse longitudinal dielectric permeability is

$$\varepsilon_{\parallel}^{-1} = 1 + k_0^2\lambda_e^2, \tag{1.106}$$

where $k_0 = \omega/c$ is the wavenumber in the free space. Note that (1.105)–(1.106) give a correct estimate only in a very limited altitude range in the upper ionosphere. The collisions can be neglected in comparison with the inertial term only in the upper part of the ionospheric F-layer. On the other hand, at high altitudes the effects of ion and electron temperatures play a key role. The ion Larmour radius ρ_i becomes a typical spatial scale related to the longitudinal conductivity instead of the electron inertial length.

1.5 Dispersion Equation

We have found the equations for the tensor of the dielectric permeability. Note once again that the waves in the frequency range considered are characterized by tremendous spatial scales comparable with the Earth's radius and sometimes essentially exceeding it. What are the properties of the ULF-waves and their propagation velocities in the ionosphere, in the magnetosphere and along the ground surface? How are their properties related to the dielectric permeabilities and conductivities found? These questions can be answered by studying the dispersion equations. Certainly, we need to keep in mind that for the majority of cases the typical scales of plasma inhomogeneities are comparable with those of ULF-wave disturbances and, thus, the analysis of full wave equations in the inhomogeneous plasma is required. Nevertheless, the dispersion equations obtained for the homogeneous or the weakly inhomogeneous plasma allow us to estimate spatial and temporal scales of plasma disturbances.

Let a plane wave be propagated in a homogeneous plasma. It is convenient, as before, to choose the z-axis in the direction of the undisturbed magnetic field $\mathbf{B}_0$. Let all wave components be dependent on ω and x as $\propto \exp(-i\omega t + ik_x x)$ and independent on y. Hence we may symbolically write

$$\frac{\partial}{\partial t} = -i\omega, \qquad \frac{\partial}{\partial x} = ik_x, \qquad \frac{\partial}{\partial y} = 0. \tag{1.107}$$

Equation (1.107) shows that all field quantities contain the factor $\exp(-i\omega t + ik_x x)$. This factor is assumed to be omitted in the same way as the time factor $\exp(-i\omega t)$ is omitted. Then the remaining terms are functions of z and Maxwell's equations (1.48), (1.49) and (1.54)–(1.56) reduce to

$$-\frac{\mathrm{d}E_y}{\mathrm{d}z} = ik_0 b_x, \tag{1.108}$$
$$\frac{\mathrm{d}E_x}{\mathrm{d}z} = ik_0 b_y + ik_x E_z, \tag{1.109}$$
$$ik_x E_y = ik_0 b_z, \tag{1.110}$$
$$-\frac{\mathrm{d}b_y}{\mathrm{d}z} = -ik_0(\varepsilon_\perp E_x + \varepsilon_\top E_y), \tag{1.111}$$
$$\frac{\mathrm{d}b_x}{\mathrm{d}z} = -ik_0(-\varepsilon_\top E_x + \varepsilon_\perp E_y) + ik_x b_z, \tag{1.112}$$
$$ik_x b_y = -ik_0 \varepsilon_\parallel E_z, \tag{1.113}$$

where $\varepsilon_\perp = \varepsilon_{xx}$, $\varepsilon_\top = \varepsilon_{xy}$. Elimination of E_z and b_z from (1.108)–(1.113) gives

$$\frac{\mathrm{d}\mathbf{s}}{\mathrm{d}z} = ik_0 \mathbf{T}\mathbf{s}, \tag{1.114}$$

where

$$\mathbf{T} = \begin{pmatrix} 0 & 0 & 0 & 1 - \dfrac{k_\perp^2}{k_0^2 \varepsilon_\parallel} \\ 0 & 0 & 1 & 0 \\ \varepsilon_\top & \varepsilon_\perp - \dfrac{k_\perp^2}{k_0^2} & 0 & 0 \\ \varepsilon_\perp & -\varepsilon_\top & 0 & 0 \end{pmatrix}, \qquad \mathbf{s} = \begin{pmatrix} E_x \\ -E_y \\ b_x \\ b_y \end{pmatrix}. \tag{1.115}$$

We put here $k_x^2 = k_\perp^2$ because of the axial symmetry with respect to the magnetic field.

In a homogeneous medium the coefficients of (1.115) are independent of the coordinate z and (1.114) can be solved by finding four eigenvalues

$q_j (j = 1, 2, 3, 4)$ and the eigenvectors of the matrix $\mathbf{T}$ which meet the characteristic equation

$$\det(\mathbf{T} - q\mathbf{1}) = 0, \tag{1.116}$$

where $\mathbf{1}$ is the unit 4×4 matrix. Each root q_j has a corresponding eigenvector $\mathbf{s}^j$ of the matrix $\mathbf{T}$ which satisfies

$$\mathbf{T}\mathbf{s}^j = q_j \mathbf{s}^j. \tag{1.117}$$

The longitudinal component of the wavenumber is now defined as $k_\parallel^{(j)} = k_z^{(j)} = k_0 q_j$. Then (1.114) has four solutions $\mathbf{s}^{(j)} \exp(ik_\parallel^{(j)} z)$ corresponding to two wave modes each propagating into the positive and the negative z-direction.

From (1.116) we get

$$\begin{aligned}&\left(k_0^2\varepsilon_\perp(\omega) - k^2\right)\left(k_0^2\varepsilon_\perp(\omega) - \frac{\varepsilon_\perp(\omega)}{\varepsilon_\parallel(\omega)}k_\perp^2 - k_\parallel^2\right)\\&+k_0^2\varepsilon_\top(\omega)\left(k_0^2\varepsilon_\top(\omega) - \frac{\varepsilon_\perp(\omega)}{\varepsilon_\parallel(\omega)}k_\perp^2\right) = 0,\end{aligned} \tag{1.118}$$

where $\varepsilon_\alpha(\omega) = i4\pi\sigma_\alpha(\omega)/\omega$, $\alpha = \perp, \top, \parallel$. Solve the biquadratic equation (1.118) with respect to $k_\parallel^2$:

$$\begin{aligned}k_\parallel^2 &= k_0^2\varepsilon_\perp(\omega) - \left(1 + \frac{\sigma_\perp(\omega)}{\sigma_\parallel(\omega)}\right)\frac{k_\perp^2}{2}\\&\pm\left[\left(1 - \frac{\sigma_\perp(\omega)}{\sigma_\parallel(\omega)}\right)^2\frac{k_\perp^4}{4} - k_0^4\varepsilon_\top^2(\omega) + \frac{\sigma_\perp(\omega)}{\sigma_\parallel(\omega)}k_0^2\varepsilon_\top(\omega)k_\perp^2\right]^{1/2}\end{aligned} \tag{1.119}$$

Equations (1.118) and (1.119) are meaningful only if an explicit dependence of the dielectric permeability on frequency is known. For a cold plasma, $\sigma_\alpha(\omega)$ is determined by (1.86)–(1.88). In a general case of a partially ionized plasma, these equations are rather cumbersome. However, the simple equations for the tensor of dielectric permeability obtained in the previous section allow us to study dispersion equations for almost all the cases in the ULF frequency range.

Equation (1.118) may be written more explicitly for the specific regions of the magnetosphere and the ionosphere. But note once again that we limit our consideration to the cold plasma and neglect kinetic effects, which can be important in the remote regions for the small-scale disturbances.

In the ionospheric D-layer the medium is the isotropic conductor with the conductivity σ_0 (1.99). Substituting $\varepsilon_\perp \approx \varepsilon_\parallel \approx 4\pi\sigma_0/\omega$, where σ_0 is defined in (1.99), and $\varepsilon_\top \approx 0$, into (1.118), we get

$$i\omega = \frac{c^2}{4\pi\sigma_0}k^2, \quad k = \left(k_\perp^2 + k_\parallel^2\right)^{1/2}. \tag{1.120}$$

This is the dispersion equation for the isotropic conductor. The electromagnetic field is described by the diffusion equation with the diffusion coefficient $c^2/4\pi\sigma_0$.

Equation (1.120) gives

$$k = (1 + i)d_s^{-1}, \tag{1.121}$$

where $d_s = c(2\pi\omega\sigma_0)^{-1/2}$ is called a skin depth. The plain wave $\propto \exp(i\mathbf{kr})$ decays at e times at the distance d_s propagating along $\mathbf{r}$.

In the E-layer the polarization conductivity σ_{pol} is small in comparison with Pedersen conductivity σ_P, and the longitudinal conductivity $\sigma_\parallel \approx \sigma_0$ is large in comparison with the Pedersen conductivity:

$$\sigma_\perp \approx \sigma_P, \quad \sigma_\top \approx -\sigma_H \ \text{and} \ \frac{\sigma_P}{\sigma_0} \approx \frac{\nu_e}{\nu_i}\frac{\omega_{pi}^2}{\omega_{pe}^2} \sim \left(\frac{m_e}{m_i}\right)^{1/2} \ll 1.$$

With these conditions for not very large $k_\perp$ we get,

$$\frac{\sigma_P}{\sigma_0}\frac{k_\perp^2}{2} \ll \frac{4\pi\omega\sigma_H}{c^2},$$

The terms with the longitudinal conductivity can be omitted in (1.119) and

$$k_\parallel^2 = \frac{2i}{d_P^2} - \frac{k_\perp^2}{2} \pm \left(\frac{k_\perp^4}{4} + \frac{4}{d_H^4}\right)^{1/2}, \tag{1.122}$$

where

$$d_P = \frac{c}{\sqrt{2\pi\sigma_P\omega}}, \quad d_H = \frac{c}{\sqrt{2\pi\sigma_H\omega}}.$$

The sign '+', in (1.122) corresponds to two normal modes; one propagating to the positive and the other to the negative direction of the z-axes. Attenuation of these modes is determined by Pedersen conductivity σ_P. The sign '−', in (1.122), corresponds to two non-propagating normal modes.

Consider in detail quasi-longitudinal propagation. Let the transversal wavenumbers meet the condition

$$k_\perp d_H \ll 1, \tag{1.123}$$

then expanding the radical expression in (1.122) into the power series over $k_\perp d_H$, we obtain for the propagating normal mode

$$\frac{(k_\parallel d_H)^2}{2} = 1 - \left(\frac{k_\perp d_H}{2}\right)^2 + i\frac{\sigma_P}{\sigma_H}. \tag{1.124}$$

Thus, a weakly decaying wave mode exists in a partly ionized plasma with Pedersen and Hall conductivities at $\sigma_P/\sigma_H \ll 1$ [9].

Finally, we consider the dispersion equation at strong electron and ion magnetizations. Leaving only transversal (1.102) and longitudinal (1.105) conductivities and substituting into (1.119) $\varepsilon_\perp = c^2/c_A^2$, $\varepsilon_\top = 0$ $\varepsilon_\| = -[(k_0\lambda_e)^2(1+i\nu_e/\omega)]^{-1}$, we find that (1.119) reduces to the product of two terms. Equating one of them to zero, we obtain the equation for the so-called isotropic or the fast magnetosonic (FMS) wave:

$$\omega^2 = c_A^2 k^2. \tag{1.125}$$

The second wave mode is called the Alfvén wave and the dispersion equation for it is obtained by equating the second term to zero. Thus, the dispersion equation for Alfvén waves taking into account the attenuation at the longitudinal resistivity is

$$[1+(k_\perp^2\lambda_e^2)^2]\omega^2 + i\omega\nu_e(k_\perp^2\lambda_e^2)^2 - k_\|^2 c_A^2 = 0. \tag{1.126}$$

These two wave modes are considered in details in Chapter 4.

For the majority of ULF-wave phenomena in the Earth's magnetosphere and the upper ionosphere, the role of the ohmic part of the transversal and longitudinal conductivities is insignificant. The media can be considered as an anisotropic dielectric with the transversal dielectric permeability $\varepsilon_\perp$ and zero-loss energy. However, a role of the longitudinal conductivity $\sigma_\|$ and/or dielectric permeability $\varepsilon_\|$ can dominate for small-scale disturbances. For such disturbances the dissipation and dispersion become significant at scales

$$L_\perp = \frac{c}{\omega\varepsilon^{1/2}} = \lambda_e\left(1+i\frac{\nu_e}{\omega}\right)^{1/2}. \tag{1.127}$$

In the ionosphere the electron inertial length $\lambda_e \sim 10-100\,\mathrm{m}$. In this book our concern is only the large-scale disturbances with $L \gg |L_\perp|$ for which the effects of longitudinal resistivity can be ignored.

References

1. Akasofu, S. I. and S. Chapman, *Solar-Terrestrial Physics*, Clarendon Press, Oxford, England, 1972.
2. Braginskii, S. I., Transport in a plasma, in *Reviews of Plasma Physics*, **1**, (Ed. by M. A. Leontovich), Consultants Bureau, New York 1965.
3. Chapman, S. and T. G. Cowling, *Mathematical Theory of Nonuniform Gases*, 3rd Edn., Cambridge: Cambridge University Press, 1970.
4. Clemmow, P. C. and J. P. Dougherty, *Electrodynamics of Particles and Plasmas*, Addison-Wesley, New York, 1990.
5. Ginzburg, V. L., *The Propagation of Electromagnetic Waves in Plasmas*, Pergamon Press, 2d ed., Oxford, 1970.
6. Ginzburg, V. L. and A. A. Rukhadze, *Waves in Magnetoactive Plasma*, Hand Physics, **49**, Springer-New York, 1972.

7. Gurnett, D. A. and A. Bhattacharjee, *Introduction to Plasma Physics: With Space and Laboratory Applications,* Cambridge Univ. Press, Cambridge, 2005.
8. Kulsrud, R. M., *Plasma Physics for Astrophysics*, Princeton Series in Astrophysics, Princeton University Press, Princeton, NJ, 2004.
9. Piddington, J. H., Hydromagnetic waves in ionized gas, *Monthly Notices Roy. Astron. Soc.*, **115**, 671, 1955.
10. Silin, V. P. and A. A. Rukhadze, *Electromagnetic Properties of Plasma and Plasma-like Media*, Moscow, Pub. House, Atomizdat, 1961.

2

Electrodynamic Properties of Space

2.1 The Solar Wind and the Earth's Magnetosphere

The theoretical studies of Chapman [16], Parker ([27], [28] and later papers), Gold [21], and Chamberlain [14] predicted that Sun's very high temperature should generate a permanent plasma flux. Satellite measurements have confirmed the existence of a corpuscular flux called 'solar wind' with a characteristic velocity of 300–800 km/s. Proton concentration in a quiet solar wind is usually between 1 and 10 cm^{-3} and grows by approximately an order of magnitude in disturbed periods. The solar wind drags out a magnetic field from the Sun whose magnitude is 3–7 nT. Under disturbed conditions the field may noticeably increase. Measurements have also shown that the proton temperature is anisotropic and maximal along the interplanetary magnetic field-lines. The temperature anisotropy coefficient is 3. The proton temperature is usually of the order of several eV. Measurements have demonstrated the presence in the solar wind of a percentage of He^{++} ions with the same velocity as protons. The thermal velocity of He^{++} is also about the same as the proton thermal velocity. That means the He^{++} temperature is about 4 times more than the proton temperature. The electron temperature is 10–50 eV. Longitudinal and transversal specific electric conductivities are $10^{14}\,\mathrm{s}^{-1}$ and $10^5\,\mathrm{s}^{-1}$, respectively.

On approaching the Earth, the solar wind interacts with the geomagnetic field, confining it ([2], [3], [4]). The field is encapsulated on the day side at a distance of 5 to 10 Earth's radii R_e and extends for hundreds of Earth's radii on the night side. This region is designated as the Earth's magnetosphere. The shape and size of the magnetosphere depend not only on the velocity and density of the solar wind, but also on the magnitude and direction of the interplanetary magnetic field ([16], [20]).

The earliest data on the magnetosphere boundary were obtained in 1961 from measurements of the magnetic field and of low-energy solar plasma protons by the Explorer-10 satellite [10]. Analysis of the data established that supersonic flow around the geomagnetic region by the solar wind must result in the formation of a departed shock wave in front of the region. Plasma measurements from the same satellite provided data on the connection between the field and the plasma in the transition region. The movement of the magnetosphere boundary was repeatedly observed for 48 h. Field intensity within the magnetosphere was larger than that of a dipole field at the same distances. Multiple subsequent experimental and theoretical investigations not only confirmed the reality of the magnetosphere's existence, but also revealed in it new important structural peculiarities – its tail, neutral layer, plasmapause, cleft, etc.

The magnetosphere is filled with a mixture of hot and cold plasmas. Hot protons form the so-called inner and outer Van Allen radiation belts [31]. The inner belt is filled with high-energy protons with energy of 1–100 MeV and it extends from 1.1 to 3.3 Earth's radii. The spatial distribution of protons with energy higher than 50 MeV has two peaks: one at the geocentric distance of $1.5R_E$ and with $4 \times 10^3\,\mathrm{cm}^{-2}\,\mathrm{s}^{-1}$ flux, the other at $R = 2.2R_E$ and with $10^3\,\mathrm{cm}^{-2}\,\mathrm{s}^{-1}$ flux. The distribution of protons with 4 MeV energy has one maximum at the distance of $1.8R_E$ and $10\,\mathrm{cm}^{-2}\,\mathrm{s}^{-1}$ flux. Protons in the energy range between 200 eV and 1 MeV fill the outer radiation belt. The proton flux with peak energy of 10 keV is about $10^4\,\mathrm{cm}^{-2}\,\mathrm{s}^{-1}$.

The inner electron belt consists of both a natural belt and an artificial belt. The electrons in this region are largely the result of high-altitude nuclear explosions carried out in the late fifties and early sixties. An outer electron belt is located at $4-5R_E$ with a clear maximum in the distribution at 200 keV. At distances less than $9R_E$ there is an increase in the flux of 100 keV electrons ([1], [24]).

The MHD-wave propagation in the magnetosphere is primarily determined by the cold 0.1–1 eV dense background plasma. The ratio of magnetic pressure $(B^2/8\pi)$ to gas pressure (nkT) is significantly greater than one. Therefore the Earth's magnetic field fully determines both the dynamics of the cold plasma and its basic electromagnetic properties. The collision frequencies of charged particles with one another and with neutral particles are considerably lower than electron and ion cyclotron frequencies.

Within the magnetosphere, there is a special region of cold plasma, the plasmasphere, bounded by the plasmapause. Concentration of cold plasma inside the plasmasphere decreases monotonically with radial distance in the equatorial plane from $10^4\,\mathrm{cm}^{-3}$ at $R = 1.2R_E$ to $10^2-10^3\,\mathrm{cm}^{-3}$ at the inner boundary of the plasmapause. The characteristic half-width of the plasmapause is in a few hundred of kilometers. On the plasmapause, the concentration drops abruptly by 1–2 orders, to $N_e = 10^1-10^2\,\mathrm{cm}^{-3}$. The geocentric distance of the plasmapause is controlled by magnetic and solar activity and fluctuates between about $6R_E$ for a quiet magnetosphere and $3R_E$ for disturbed conditions ([12], [13]).

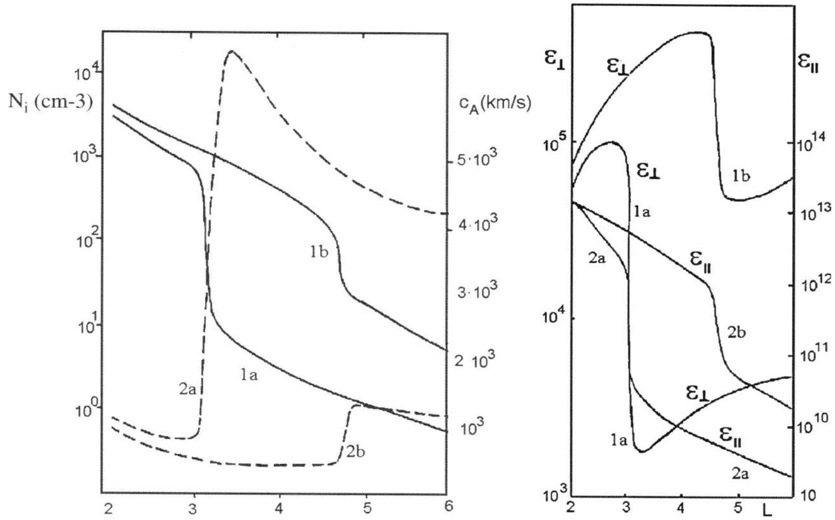

Fig. 2.1. Left frame: the equatorial distributions of the ion concentration N_i (solid line) and Alfvén velocity c_A (dashed line) as functions of the dimensionless radial distance $L = r/R_E$. Curves marked by 'a' respect to the disturbed conditions, and by 'b' is quiet conditions. Right frame: as the left frame for the transversal and longitudinal components of the dielectric permeability

The Alfvén Velocity

It follows from (1.95) that the Alfvén velocity c_A is determined by the cold plasma density and the magnitude of the magnetic field. The left frame of Fig. 2.1 shows the distributions of cold plasma and c_A in the equatorial plane of the magnetosphere versus the radial distance $L = r/R_E$. Line $2a$ refers to $c_A(L)$ for a disturbed and $2b$ for a quiet magnetosphere. Corresponding distributions of cold plasma densities are lines $1a$ and $1b$, respectively.

The transversal dielectric permeability of the magnetosphere is $\varepsilon_m = c^2/c_A^2$ (see (1.103)). Since the longitudinal dielectric permeability $\varepsilon_{\|}$ is greater than ε_m because of the rarity of electron and ion collisions, in the first approximation $\varepsilon_{\|}$ can be set to infinity. Hence the magnetosphere in the ULF range is a strongly anisotropic dielectric with finite permeability ε_m across the magnetic field and infinite $\varepsilon_{\|}$ along it.

2.2 Ionosphere

High frequency radio soundings discovered the stratified structure of the ionosphere and detected layers with different electron concentrations [11].

These are D-, E-, $F1$-, and F-layers (arranged according to altitudes). The D-region occupies heights between 60 and 90 km. It is remarkable that the plasma in this region is a mixture of electrons and negative and positive ions. Concentration of negative ions is comparable to that of positive ions at around 80 km height. At night there is no noticeable ionization at the D-layer. In the E-region, located between 90 and 140 km, the electron concentration is higher by about 2 orders of magnitude than in the D-region, reaching $10^5\,\mathrm{cm}^{-3}$. Above the E-layer, $F1$- and $F2$-layers are located with maximal electron concentration in $F1$ of $N_e \approx 2.5 \times 10^5\,\mathrm{cm}^{-3}$ during years of minimal solar activity and $\approx 4.5 \times 10^5\,\mathrm{cm}^{-3}$ during years of maximum solar activity. The $F2$-region is very sensitive to variations of solar activity. The altitude for the maximal electron concentration in $F2$, which coincides with the maximal electron concentration in the ionosphere, fluctuates between 200 and 400 km and is on average $5 \times 10^5 - 10^6\,\mathrm{cm}^{-3}$.

Figure 2.2 demonstrates the height distribution of electrons and ions, according to the IRI–2000 model [9] for the 50° geomagnetic latitude. Height profiles corresponding to winter noon and winter midnight conditions for sunspot numbers $R = 15$ are shown in the left top and bottom frames of Fig. 2.2, respectively. The same dependencies for the summer noon and summer midnight for the disturbed solar activity ($R = 150$) are shown in the right top and bottom frames of Fig. 2.2.

Collision Frequencies

To calculate the transmission and reflection characteristics of hydromagnetic waves, it is necessary to know the electrical conductivities. Consequently it is necessary to know the collision processes between ionospheric particles. Charge particles gain directed velocity under the action of the wave electric field. At the same time, electrons and ions collide and exchange energy with each other as well as with neutral particles. The energy of a directed particle movement is partially converted into chaotic particle motions at collisions. The energy of hydromagnetic waves is converted into Joule heating of the ionospheric plasma.

Electron-Neutrals

To define the collision frequency (ν_{en}) of electrons (e) with neutral (n) particles ν_{en} it is necessary to know the effective cross-section of an electron scattered on the colliding particles. Using these cross-sections and integrating by velocities, it is possibly to define an effective collision frequency of an electron with molecules ([6], [22]). Approximated expressions for the collision frequencies of an electron with nitrogen N_2 and oxygen O_2 molecules as well as atomic oxygen O, hydrogen H and helium He are given in [22]

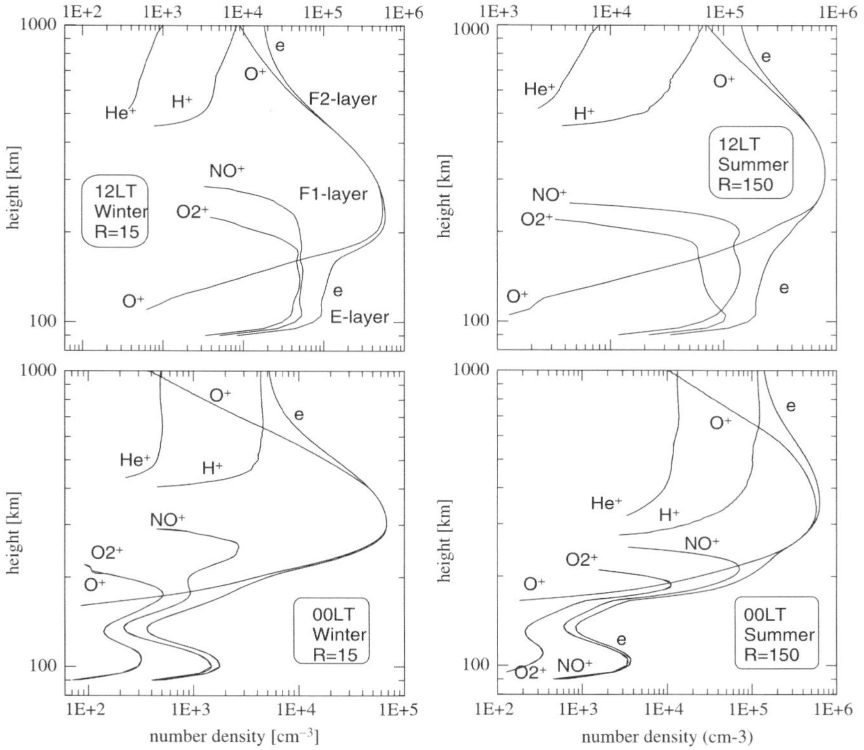

Fig. 2.2. The height distribution of electron and ion number densities, at the 50° geomagnetic latitude for winter (left frame) and summer (right frame) at noon (upper frame) and midnight (lower frame). After IRI–2000 [9]

$$
\begin{aligned}
\nu_{e\,\mathrm{N}_2} &= 2.5 \times 10^{-11} N_{\mathrm{N}_2} T_e \left(1 + 0.93 \times 10^{-2} T_e^{1/2}\right)^{-1}, \\
\nu_{e\,\mathrm{O}_2} &= 1.5 \times 10^{-10} N_{\mathrm{O}_2} T_e^{1/2} (1 + 4.2 \times 10^{-2} T_e^{1/2}), \\
\nu_{e\,\mathrm{O}} &= 2.8 \times 10^{-10} N_{\mathrm{O}} T_e^{1/2}, \\
\nu_{e\,\mathrm{H}} &= 4.5 \times 10^{-10} N_{\mathrm{H}} T_e^{1/2}, \\
\nu_{e\,\mathrm{He}} &= 4.6 \times 10^{-10} N_{\mathrm{He}} T_e^{1/2}.
\end{aligned} \tag{2.1}
$$

Electron-Ions

For the effective electron–ion collision frequency we use the expression [22]

$$
\nu_{ei} = 5.5 \frac{N}{T_e^{3/2}} \ln \left[215 \frac{T_e}{N^{1/3}} \left(\frac{T_i}{T_e + T_i} \right)^{1/3} \right]. \tag{2.2}
$$

Here, T_e and T_i are electron and ion temperatures in Kelvins (K).

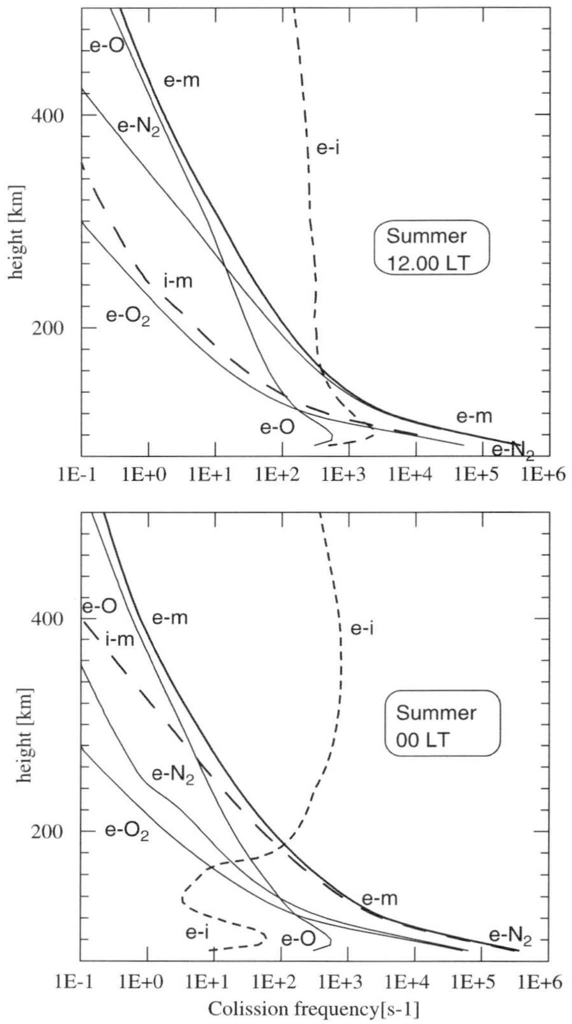

Fig. 2.3. Collision frequencies of electron-ion (dots), electron-neutral (solid lines), ion-neutral (dashed) versus height. Top frame: summer, noon; bottom frame: summer, midnight

Electron collision frequencies with different ionospheric species and total electron collision frequency height dependencies are shown in Fig. 2.3. Two ionospheric models corresponding to the summer nightside and summer dayside models were used in Fig. 2.3. Electron-neutral collisions ν_{en} has a dominant role up to 170 km during the day and 200 km at night. At higher altitudes, the ionospheric plasma is strongly ionized. The main contribution to the total ν_e comes from the electron-ion collisions ν_{ei} .

Table 2.1. Reaction rate β_{in} for the ion-induced molecular polarization[22]

$Q = \beta_{in} \times 10^{10}$ cm^3 / s ion-neutral pairs	Q		Q		Q
O^+, N_2	5.1	NO^+, N_2	9.0	H^+, N_2	2.4
O^+, O_2	4.5	NO^+, O	7.6	H^+, He	4.3
O^+, He	2.2	NO^+, H	21	He^+, O	4.5
O^+, H	2.6	NO^+, He	6.4	He^+, H	7.7
NO^+, O_2	8.3	H^+, O_2	2.7	He^+, N_2	4.1

Ion-Neutrals

The ion-induced dipole interaction dominates during collisions of ions and neutral particles at moderate temperatures $T \propto 300{-}500^o$K [19]. In this case, the collision frequencies of different ions with molecules are independent on temperature and given by [7]

$$\nu_{in} = \beta_{in} N_n,$$

where N_n is concentration of neutral particles per cm^3. Coefficients β_{in} are given in Table 2.1 [22].

Intercharge processes dominate under collisions of ions and molecules of the same kind, $X^+ + X \Rightarrow X + X^+$. An effective collision frequency for this process is found from the approximate expression

$$\nu_{in} = \beta_{in}^0 (T_i + T_n)^{1/2} N_n.$$

β_{in}^0 for different kinds of recharges are shown in Table 2.2 ([7], [22]).

Assuming that ions and neutral gas have the Maxwell's distribution, the effective ion-neutral collision frequency ν_{in} is given by [15]

$$\nu_{in} = 2.6 \times 10^{-9} \, (N_n + N_i) \, M^{-1/2}. \tag{2.3}$$

N_i is ion concentration. Ion mass m_i and molecular mass m_n are presumed to be the same, $M = m_i = m_n$. The line '$i - m$' of Fig. 2.3 shows height

Table 2.2. Reactionrate Coefficients β_{in}^0 for the charge interchange ([7], [22])

ion-molecule pairs	$\beta_{in}^0 \times 10^9$cm^3/s $\times$ eV$^{1/2}$
O^+, O	1.8
O_2^+, O_2	1.2
H^+, H	11
He^+, He	3.3
N_2^+, N_2	2.3
N^+, N	1.8

dependencies of the effective $\nu_{in}(h)$. In the ionospheric E-layer, ν_{in} is defined by the elastic collisions of NO^+ and O^+ ions with molecules, and in the F-layer by the intercharge of O^+.

E-Layer

The most important region determining the reflection and dissipation of magnetospheric MHD-waves as well as the effectiveness of their penetration into the Earth is the highly-conductive E-layer. While higher ionosphere layers and the magnetosphere are anisotropic dielectrics in the ULF-range, the E-layer is similar in its properties to a conductive anisotropic medium. The geomagnetic field is so strong for the ionospheric electrons that they become strongly magnetized, i.e. the dimensionless ratio of their cyclotron frequency ω_{ce} to collision frequency is much larger than one. An electric field applied across the geomagnetic field causes an electron drift in a direction perpendicular to both the ambient magnetic and the applied electric fields. Ions are non-magnetized at these altitudes because they collide with neutrals so often that their collision frequency ν_{in} begins to noticeably exceed the cyclotron frequency ω_{ci}.

The transversal electric conductivities in the E-layer, determined by both electrons and ions, become anisotropic. Therefore, the electric field produces a Pedersen current directed along the field as well as a current directed perpendicularly to both the electric and geomagnetic fields, called the Hall current. The ionosphere and the magnetosphere can be considered to be magnetoactive media. In the coordinate system with the z-axis parallel to the magnetic field $\mathbf{B}_0$, the components of the electric current are $j_z = \sigma_{zz}E_z$, $j_x \pm ij_y = (\sigma_{xx} \mp i\sigma_{xy})(E_x \pm iE_y)$. Relations between $\sigma_{xx}(\omega)$, $\sigma_{xy}(\omega)$, and respectively σ_P, σ_H are given in (1.92). The conductivities in terms of collision and cyclotron frequencies are (1.86)–(1.88). Figure 2.4 presents altitude dependencies $\sigma_{zz} = \sigma_{\|}$, σ_P and σ_H for dayside and nightside models.

The electron and ion magnetization, $\beta_e = \omega_{ce}/\nu_e$ and $\beta_i = \omega_{ci}/\nu_i$, are given in (1.93) and (1.94), respectively. Both σ_P and σ_H consist of two parts – electron and ion. Figure 2.5 shows the height distribution of the ion β_i and electron β_e magnetization parameters. At heights of less than 60 km, both β_e and β_i vanish and we have

$$\sigma_P = \frac{Ne^2}{m\nu_e}, \qquad \sigma_H = 0.$$

Above 70 km electrons rotate around the magnetic field rarely colliding with neutral particles. Parameter β_e increases from 1 in the D-layer and reaches 10^4 at 200 km. It is evident that the electron part of σ_H is larger than the ion part at heights between 70 and 140 km where $\beta_i \lesssim 1$. One can say that the Hall conductivity in the D- and E-layers is determined by electrons. As to the higher levels (F-layer), σ_H vanishes because both electrons

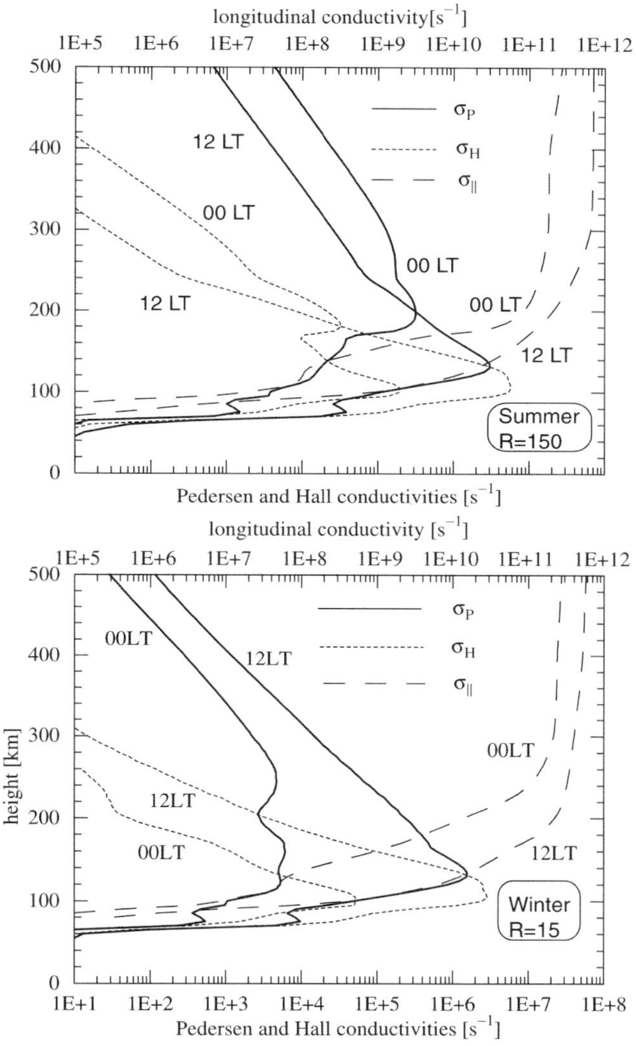

Fig. 2.4. The height distributions of the specific Pedersen σ_P, Hall σ_H and longitudinal $\sigma_{\parallel}$ conductivities

and ions are magnetized ($\beta_e, \beta_i \gg 1$), they drift together in the crossed electric and magnetic field. There is no mutual slipping and therefore there is no Hall current in the high regions.

Comparing electron and ion terms in brackets in (1.93), one notices that the relative contribution of electrons and ions to σ_P is defined by the product $\beta_e\beta_i$. Where $\beta_e\beta_i < 1$, the main part of the conductivity σ_P is defined by electrons. In the low E-layer, between 70 and 100 km, electrons are magnetized $\beta_e \gg 1$ and ions collide so frequently that $\beta_i \ll 1$. Here $\beta_e\beta_i < 1$ and $\sigma_P \approx$

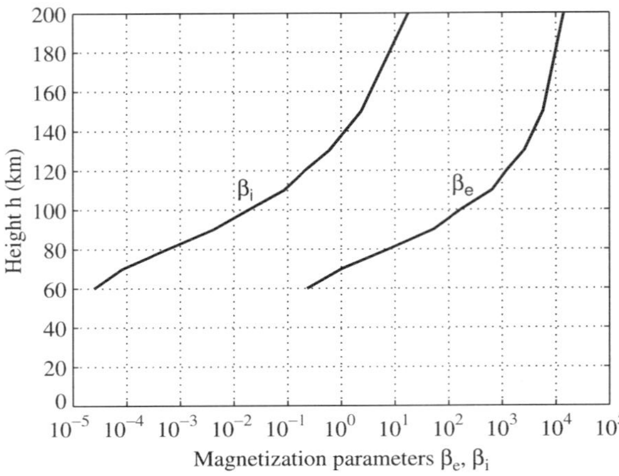

Fig. 2.5. Height dependencies of the electron $\beta_e = \omega_{ce}/\nu_e$ and ion $\beta_i = \omega_{ci}/\nu_i$ parameters

σ_{Pe}. Above the 100 km level, in the E-layer the atmospheric concentration decreases so that the ion-neutral collisions become less frequent, β_i is small but finite, $\beta_e\beta_i > 1$, and $\sigma_P \approx \sigma_{Pi}$. Here, ions move along the applied electric field, electrons are magnetized so that they drift just perpendicular to both the applied electric field and the geomagnetic field. In the F-layer, $\beta_e, \beta_i \to \infty$ and σ_P vanishes.

Layered Conductivity

It can be seen from Fig. 2.4 that $\sigma_\parallel$ exceeds σ_P and σ_H by about 4 orders of magnitude. Due to this fact the ionosphere and the magnetosphere are effectively short-circuited, as the electric field $\mathbf{E}$ excited by large-scale perturbations in both media is transferred with practically no damping from the magnetosphere to the conjugate ionospheres or from the ionosphere of one hemisphere to the other.

Equipotentiality of the magnetic field-lines enables us in many cases to approximate the ionosphere by a thin conductive shell surrounding the Earth. Let us place the origin of the Cartesian coordinates at some point on the shell, the axis x and y point southward and eastward respectively, the z axis points upward. If the magnetic inclination (angle between geomagnetic field $\mathbf{B}_0$ and the horizontal plane) is I, then the current density produced by $\mathbf{E}$ can be written as

$$j_x = (\sigma_\parallel \cos^2 I + \sigma_P \sin^2 I)E_x + \sigma_H E_y \sin I + (\sigma_\parallel - \sigma_P)E_z \cos I \sin I, \quad (2.4)$$

$$j_y = -\sigma_H E_x \sin I + \sigma_P E_y + \sigma_H E_z \cos I, \tag{2.5}$$

$$j_z = (\sigma_\parallel - \sigma_P) E_x \cos I \sin I - \sigma_H E_y \cos I + (\sigma_\parallel \sin^2 I + \sigma_P \cos^2 I) E_z. \tag{2.6}$$

Suppose the ionosphere is an anisotropic sheet with non-conductive upper and lower boundaries. Then vanishing of the vertical component of the current j_z enables us to find E_z from (2.6) in the form

$$E_z = \frac{(\sigma_P - \sigma_\parallel) E_x \cos I \sin I + \sigma_H E_y \cos I}{\sigma_\parallel \sin^2 I + \sigma_P \cos^2 I}. \tag{2.7}$$

Substitution of (2.7) into (2.4) and (2.6) yields [5]:

$$j_x = \sigma_{xx} E_x + \sigma_{xy} E_y,$$
$$j_y = -\sigma_{xy} E_x + \sigma_{yy} E_y,$$

with components of the so-called 'layered conductivity'

$$\sigma_{xx} = \frac{\sigma_\parallel \sigma_P}{\sigma_\parallel \sin^2 I + \sigma_P \cos^2 I},$$

$$\sigma_{xy} = \frac{\sigma_\parallel \sigma_H \sin I}{\sigma_\parallel \sin^2 I + \sigma_P \cos^2 I} = -\sigma_{yx}\ ,$$

$$\sigma_{yy} = \frac{\sigma_P \sigma_\parallel \sin^2 I + (\sigma_P^2 + \sigma_H^2) \cos^2 I}{\sigma_\parallel \sin^2 I + \sigma_P \cos^2 I}. \tag{2.8}$$

These formulae are valid just for large-scale low frequency fields. Since $\sigma_\parallel \gg \sigma_P, \sigma_H$ the field-lines can be considered equipotential for the ULF large-scale perturbations, i.e. the E_x and E_y electric components are height-independent. Integration the equations for the currents j_x, j_y over height gives

$$J_x = \Sigma_{xx} E_x + \Sigma_{xy} E_y,$$
$$J_y = -\Sigma_{xy} E_x + \Sigma_{yy} E_y,$$

where

$$\Sigma_{xx} = \int \sigma_{xx} dz, \qquad \Sigma_{xy} = \int \sigma_{xy} dz, \qquad \Sigma_{yy} = \int \sigma_{yy} dz \tag{2.9}$$

are the height integrated conductivities and

$$J_x = \int j_x dz, \qquad J_y = \int j_y dz$$

are components of the surface ionospheric current.

This is a great step forward. In most of the problems we will present, we simply replace the tensor of the specific ionospheric conductivity $\sigma_{ij}(\mathbf{r}, z)$ by a tensor of the integral conductivity $\Sigma_{ij}(\mathbf{r})$ that depends only on the horizontal coordinates $\mathbf{r}\{x, y\}$. The ionosphere is presented as a thin conductive layer or an interface between the magnetosphere and atmosphere with conductivity Σ_{ij} (model of the 'thin' ionosphere). Magnetic fields above and below the ionosphere are according to the Biot–Savart law a result of integration throughout the volume occupied by electrical currents. In the approach based on the integral conductivities the volume currents are replaced by the surface currents $\mathbf{J}\{J_x, J_y\}$.

Note that great care must be exercised in moving from σ_{ij} to Σ_{ij}. First, doing this, real polarized electric fields arising due to local inhomogeneities can be much greater than the fields calculated on the basis of the integral conductivities. We discuss this problem in Chapter 10 devoted to the effective conductivity of a cloudy ionosphere. The second is the location of the surface currents. Height dependencies of the Pedersen and Hall components of σ_{ij} are distinct from each other. As a consequence of that, the corresponding surface currents should be placed at different heights. For example (see Fig. 2.4), a small additional maximum of the Pedersen conductivity appears at the night F-layer heights. Despite its smallness, its contribution to the Σ_P of the entire thickness of the ionosphere turns out to be roughly the same as that of the E-region. And the last point that should be mentioned is the substitution of $\boldsymbol{\sigma}$ by $\boldsymbol{\Sigma}$ to the loss of the inter ionosphere waveguide that can trap a certain type of MHD-oscillations.

The ionospheric conductivity increases sharply in the auroral zones and near the equator. The high-latitude maximum is caused by precipitations of auroral particles. The integral Pedersen and Hall conductivities are larger here by an order of magnitude than those of the middle-latitude ionosphere.

Geomagnetic Equator

Contrary to the high latitude ionosphere where the conductivity is defined mainly by precipitations of high energy particles, the equatorial conductivity is determined by the geometry of the geomagnetic field. A peculiar phenomenon appears in the vicinity of the geomagnetic equator in the form of a sharp conductivity increase, first described by Cowling ([17], [18]). Its mechanism is associated with the polarized charges arising on the upper and low ionospheric boundaries.

Suppose that an electric field, excited in the equatorial ionosphere, is directed along the equator across the geomagnetic field-lines. The nature of the field is of no significance. It may presumably originate from high-latitude sources producing global fields and currents, or from local near-equatorial perturbations, or from MHD-waves that penetrated into the region of the middle- and low-latitude ionosphere. Let us choose, as before, the y-axis as pointing eastward along the equator. The electric field E_y creates a Pedersen current $\sigma_P E_y$ along the equator.

The Hall current $\sigma_H E_y$ must in this case flow in the vertical direction. The Hall current is trapped between the atmosphere, the lower boundary of the Hall layer, and the $F1$-layer as its upper boundary. Electric charges appear on the boundaries together with a vertical polarization field E_p, which in its turn produces a vertical Pedersen current. The latter compensates for the vertical Hall current. Hence

$$E_P = -E_y \frac{\sigma_H}{\sigma_P}.$$

The horizontal Hall current from the vertical polarization field has the same direction as the initial Pedersen current. The total horizontal current j_y is

$$j_y = \left(\sigma_P + \frac{\sigma_H^2}{\sigma_P} \right) E_y.$$

The quantity σ_C

$$\sigma_C = \sigma_P + \frac{\sigma_H^2}{\sigma_P}$$

that determines the horizontal electric conductivity of the near-equator ionosphere, is called the Cowling conductivity. The altitude dependencies of σ_P and σ_H are different, their extremes are removed from each other, therefore the ratio σ_H^2/σ_P can be quite large.

The global distributions of the layered integral ionospheric conductivity with components Σ_{xx}, Σ_{xy}, Σ_{yy} (see (2.9)) calculated for the summer in the northern hemisphere are shown in Figs. 2.6–2.8. The calculations are performed for the simplest case when the electron concentration is governed only by the solar UV radiation. Recombination and other chemical processes taking place in the upper atmosphere are ignored. One can see that Σ_{yy} increases in the equatorial region. Here Σ_{yy} exceeds its value in the middle-latitude ionosphere by $1-2$ orders of magnitude. In the longitudinal dependence of the conductivities, two anomalies appear in the terminator zone on the boundaries of the dayside and nightside ionospheres. The conductivities here change by an order of magnitude.

2.3 Atmosphere

The atmosphere is a mixture of diverse gases, primarily oxygen O_2, nitrogen N_2, carbon dioxide CO_2 and water vapor H_2O. Other gases are contained in the atmosphere in negligible amounts. Atmospheric air always contains tiny particles of various substances in the solid and liquid states. Their radii do

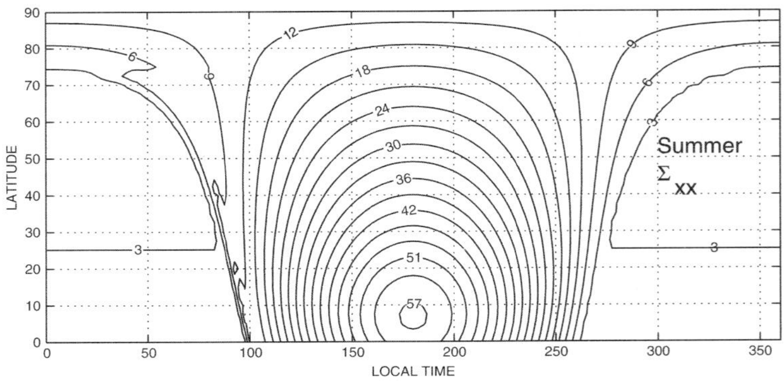

Fig. 2.6. Global latitude–local time distribution of the ionospheric integral layered conductivity Σ_{xx} component for the summer in the northern hemisphere. Isolines of Σ_{xx} are marked by S (Siemens) $= (\mathrm{Ohm})^{-1}$. It is assumed that the electron concentration is defined only by the solar UV radiation intensity of which in turn depends on the zenith angle. x are the southward and y the eastward coordinates

not usually exceed 10–20 µm. The particles are very varied with regard to their chemical composition, amount and physical properties. Some of them may be electrically charged.

Ions are initially created in the atmosphere as a result of the process of gas ionization under the action of hard rays. The main ionizers of the lower atmospheric layers are emissions from radioactive substances contained in the

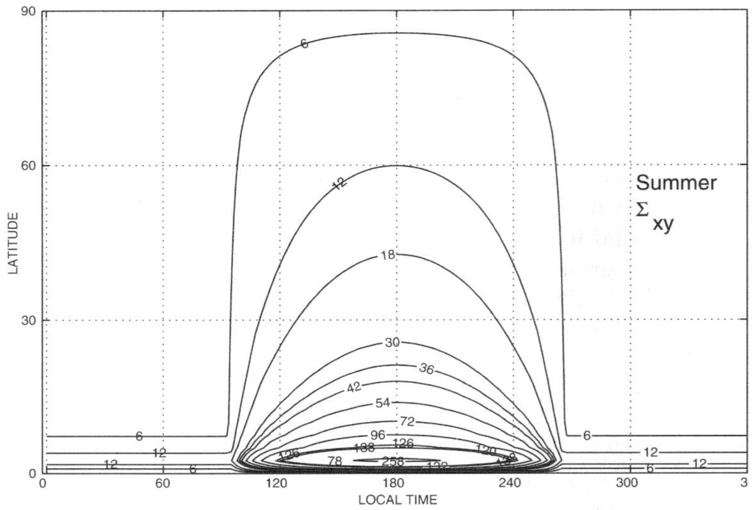

Fig. 2.7. Same as Fig. 2.6 but for Σ_{xy} component

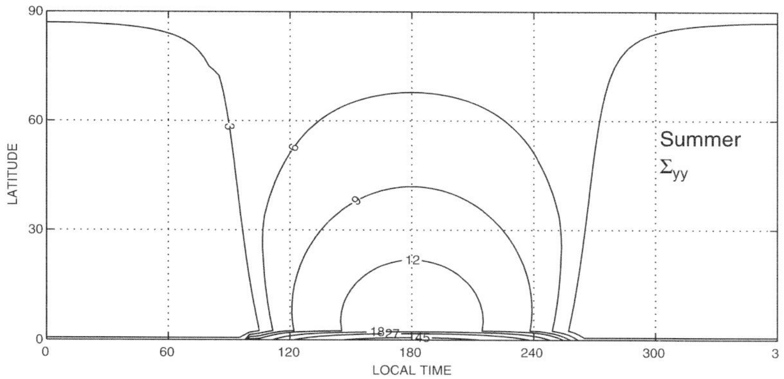

Fig. 2.8. Same as Fig. 2.6 but for Σ_{yy} component

Earth's crust and in the atmosphere as well as cosmic rays. Solar UV rays reaching the lower atmospheric layers (with wavelength $\lambda > 285\,\text{nm}$) have no part in their ionization.

Under conditions of normal atmospheric pressure, electrons are created almost instantly through ionization. Less than $10^{-6}\,\text{s}$ later the electrons attach to neutral atoms and negative ions are created. Such molecular ions exist for fractions of a second, since under the action of polarization forces they are joined by 10–15 molecules from the ambient air. The emerging clusters (light particles) are stable enough, but they can join bigger dust particles and give up their charge to the latter (larger particles). The mobility of the light components exceeds that of the larger particles by at least 3 orders of magnitude. Therefore the atmospheric conductivity is predominantly (90%) determined by the light component.

Henceforth for estimations, we will put the value of the near-ground atmospheric conductivity to be $\sigma_a = 10^{-4}\,\text{s}^{-1}$. Turbulent mixing of the air causes inessential altitude dependency of the conductivity up to ≈3 km height. Above this, the conductivity increases rapidly.

In most known ULF-problems, the near-ground atmospheric conductivity is not essential for obtaining ground distributions of the waves. One can present the near-ground atmospheric layer as a perfect insulator. There is one exception, though. The vertical electric pulsation field associated with vertical electric currents, the value of which directly depends on the atmospheric conductivity. But the electromagnetic wave mode with vertical electric field in the ULF-range usually has an extremely small intensity. Any attempts to observe it experimentally were unsuccessful. $\sigma_a(z)$ may be roughly approximated by an exponent with a scale of the order of the atmospheric scale-height, i.e. $\sigma_a \propto \exp(a/H_\sigma)$, where the scale-height of the atmospheric conductivity is $H_\sigma \approx 8\,\text{km}$. The altitude profile $\sigma_{atm}(z)$ is presented in Fig. 2.9.

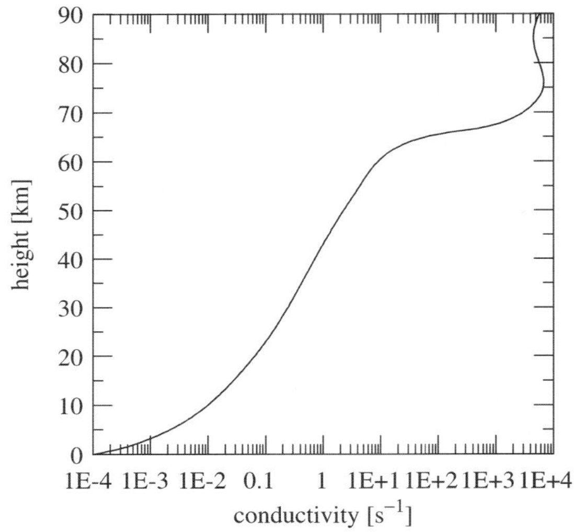

Fig. 2.9. The distribution of the atmospheric conductivity as a function of altitude

2.4 Summary

In general, in problems of interaction of MHD-waves with the Earth-atmosphere-ionosphere-magnetosphere system, this system may be replaced by a layered system, each layer with its own specific properties (see Fig. 2.10). The ground conductivity is so high compared with the ionospheric conductivity, that it can be considered to be a metal conductor. Its conductivity depends both on the depth and on the horizontal coordinates. It is known, for example that there exist two kinds of crust conductivity. The continental crust has a resistivity $l0^3 \Omega \cdot$ m at the depth ≈100 km and underlined by the high conductive layer of a resistivity $10^2 \Omega \cdot$ m from ≈100 to ≈300 km. There is, on the other hand, substantial evidence for a sharp decrease in electrical resistivity up to 1−10 $\Omega\cdot$ m on a global scale at a depth between 400 and 800 km (e.g. [23], [29], [30]).

The atmospheric conductivity differs by 10 orders of magnitude from the ground conductivity. As a rule, the atmosphere can be considered to be an insulator. The one exception to this rule is the problem of excitation and propagation of the atmospheric wave of type TM_0 (see Chapter 8).

Above 60 km, in the ionospheric D-layer, the air density is so rarefied that electrons can gyrate freely in the geomagnetic field and ions collide so frequently that their motion is defined entirely by neutrals. Thus, the electrical conductivity of the D-layer is determined by electrons. The D-layer is a specific region with strong Hall and evanescent Pedersen conductivities.

The E-layer extends approximately from 90 to 140 km. The neutral concentration decreases here so that ions can move by the action of the electric

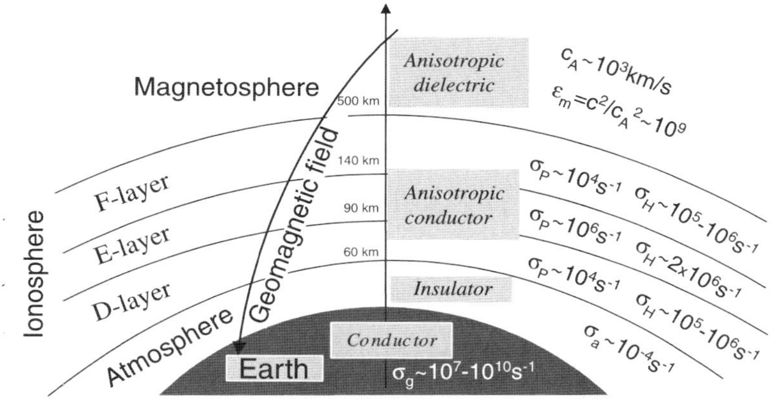

Fig. 2.10. Sketch showing electrodynamic properties of the system Earth-atmosphere-ionosphere-magnetosphere

field. The conductivity exhibits strong σ_P and σ_H. At that, σ_P is determined by ions and σ_H by electrons. This layer can be represented by an anisotropically conductive layer.

In the F-layer motion of electrons and ions is governed by the geomagnetic field. σ_P steeply decreases, while σ_H smoothly approaches zero.

In the outermost layer of the upper atmosphere, in the magnetosphere, both electrons and ions are so strongly magnetized that there is no slipping of electrons with respect to ions. As a result, the real part of the transversal conductivity vanishes. The longitudinal conductivity tends to infinity.

In the ionosphere, except the upper part of the F-layer, (1.102) for the ratio of the polarized current to the Pedersen current, gives $|\sigma_{pol}/\sigma_P| \approx \omega/\nu_i \ll 1$. Thus, in the ULF-band we can restrict our consideration to the altitudes of $\sim$500$-$1000 km only with the Pedersen and Hall conductivities. On the other hand, in the magnetosphere, we can keep only the transversal polarized conductivity. In this case, it is purely imaginary, and the dielectric permeability ε_m is real. Therefore, from here on we shall use ε_m in the characterization of the wave magnetospheric properties. Thus, the tensor of the dielectric permeability $\boldsymbol{\varepsilon_m}$ is diagonal. The transversal component of $\boldsymbol{\varepsilon_m}$ is $\varepsilon_m = c^2/c_A^2$, and the longitudinal is $|\varepsilon_{\|}| \to \infty$.

Such electrodynamic description of the magnetosphere–ionosphere plasma applies only to large-scale perturbations. For the rather small transversal scales $L_\perp$ we need to take into account the longitudinal resistivity. The $L_\perp$ can be estimated from the dispersion equation (1.119). For simplicity, let us omit in (1.119) the Hall conductivity terms leaving only the terms with longitudinal conductivity. Then the left-hand side of (1.119) is a product of two factors. Equating to zero each of them we obtain two equations. The first one does not contain the longitudinal conductivity. And from the second equation we have

$$k_{\parallel}^2 = \frac{\omega^2}{c_A^2} - \frac{\sigma_\perp}{\sigma_\parallel} k_\perp^2 .$$

For low frequencies, when the first term on the right-hand side can be ignored, the above reduces to

$$k_\parallel = \pm i \sqrt{\frac{\sigma_P}{\sigma_\parallel}} k_\perp .$$

It follows from Fig. 2.4 and the last equation that the ratio of a longitudinal scale $L_\parallel$ to the transversal scale $L_\perp$ in the F-layer is of the order of $L_\parallel / L_\perp \sim 300$. This means that the wave magnetospheric perturbations with $L_\perp \gtrsim$ a few kilometers, reach the E-layer almost without decay. But perturbations with $L_\perp < 1\,\mathrm{km}$ decay so much that they cannot reach the E-layer.

References

1. Akasofu, S. I. and S. Chapman, *Solar-Terrestrial Physics*, Clarendon Press, Oxford, England, 1972.
2. Alfvén, H., Theory of magnetic storms, I. *K. svenska Vetensk-Akad. Handl.*, **18**, 1939.
3. Alfvén, H., Theory of magnetic storms. II, III. *K. svenska Vetensk-Akad. Handl.*, **18**, 1940.
4. Alfvén, H., On the electric field theory of magnetic storms and aurorae, *Tellus*, **7**, 50–64, 1955.
5. Baker, W. G. and D. F. Martyn, Electric currents in the ionosphere, I. The conductivity. *Phil. Trans. R. Soc.*, **A246**, 281–294, 1953.
6. Banks, P., Collision frequencies and energy transfer. Electrons, *Plan. Space Sci.*, **14**, 1085–1103, 1966.
7. Banks, P., Collision frequencies and energy transfer. Ions, *Plan. Space Sci.*, **14**, 1105–1122, 1966.
8. Berdichevsky, M. N. and M. S. Zhdanov, Advanced Theory of Deep Geomagnetic Sounding, Elsevier, Amsterdam, 1984.
9. Bilitza, D., IRI 2000, *Radio Science*, **36**, 261–276, 2001.
10. Bonetti, A., H. S. Bridge, A. J. Lazarus, B. Rossi, and F. Scherb, Explorer 10 Plasma Measurements, *J. Geoph. Res.*, **68**, 4017–4063, 1963
11. Breit, G. and Tuve, M. A., A test for the existence of the conducting layer, *Phys. Rev.*, **28**, 554–575, 1926.
12. Carpenter, D. L., Whistler evidence of a 'knee' in the magnetospheric ionization density profile, *J. Geophys. Res.*, **68**, 1675–1682, 1963.
13. Carpenter, D. L. and R. R. Anderson, An ISEE/whistler model of equatorial electron density in the magnetosphere, *J. Geophys. Res.*, **97**, 1097–1108, 1992.
14. Chamberlain, J. W., 1960, Interplanetary gas. II Expansion of a model solar corona, *Astrophys. J.*, **131**, 47–56.
15. Chapman, S., The electrical conductivity of the ionosphere (a review), *Nuovo Cimento*, **4**, suppl., 1385–1412, 1956.

16. Chapman, S., Notes on the solar corona and the terrestrial atmosphere, *Smithsonian Contribution to Astrophysics*, **2**, 1–11, 1957.
17. Cowling, T. G., Magnetism, Solar : The electrical conductivity of an ionized gas in the presence of a magnetic field, *Monthly Notices of the Royal Astronomical Society*, **93** 90, 1932.
18. Cowling, T. G., The Electrical Conductivity of an Ionized Gas in a Magnetic Field, with Applications to the Solar Atmosphere and the Ionosphere, *Proceedings of the Royal Society of London.* Series **A**, Math. and Phys. Sciences, **183**, 453–479, 1945.
19. Dalgarno, A., M. B. McElroy, and R. J. Moffett, Electron temperatures in the ionosphere, *Planet., Space Sci.*, **11**, 463–464, 1963.
20. Dungey, J. W., Interplanetary magnetic field and the auroral zones, *Phys. Rev. Lett.*, **6**, 47–48, 1961.
21. Gold, T., Plasma and magnetic fields in the solar system, *J. Geophys. Res.*, **64**, 1665-1674, 1959.
22. Gurevich, A. V., *Nonlinear Phenomena in the Ionosphere*, Springer, New York, 1978.
23. Hermance, J. F., Electrical Conductivity Models of the Crust and Mantle, *Global Earth Physics, A Handbook of Physical Constants*, AGU Reference Shelf, **1**, 190–205, 1995.
24. Hess, W., *The Radiation Belt and Magnetosphere,* Blaisdell, Waltham, Mass., 1968.
25. Kaufman, A. A. and G. V. Keller, *The Magnetotelluric Sounding*, Elsevier, Amsterdam, 1981.
26. Kelley, M.C., Physics and Chemistry of the Upper Atmosphere, Cambridge University Press, 1989.
27. Parker, E. N., Dynamical instability in an anisotropic ionized gas of low density, *Phys. Rev.*, **109**, 1874–1876, 1958.
28. Parker, E. N., Extension of the solar corona into interplanetary space, *J. Geophys. Res.*, **64**, 1675–1681, 1959.
29. Vanyan, L. L., Electrical Conductivity of the Asthenosphere project, *Int. Assoc. Geomagn. Aeron. News*, 73–84, 1980.
30. Vanyan, L. L., Deep geoelectrical models: geological and electromagnetic principles, *Phys. Earth Planet. Int.*, **25**, 273–279, 1981.
31. Van Allen, J. A. and L. A. Frank, Radiation around the Earth to a radial distance of 107,400 km, *Nature*, **183**, 430–434, 1959.

3

ULF-Waves on the Ground and in Space

3.1 Introduction

Geomagnetic ULF-oscillations are excited by solar wind on the magnetospheric boundary where the wind's velocity vanishes ([15], [16], [17]). The waves arise due to the discontinuity of the tangential velocity (Kelvin–Helmholtz instability). The mechanism for the build-up of oscillation resembles the onset of ripples on a water surface, in a field of corn or of flags fluttering in gusts of wind. Oscillations evolve mainly on the flanks of the magnetosphere. Like any surface waves, they decay as they penetrate deeper into the magnetosphere. Perturbations in the Alfvén mode are guided and propagate with practically no distortions from the magnetosphere boundary to the high-latitude ionosphere.

The peculiarities of wave generation and its relation to the morning and evening sectors of the magnetosphere must manifest in the different pulsation characteristics. In the daily variation of long-period oscillation frequency dependency, two anomalies ought be observed, one in the morning and one in the evening. In addition the horizontal vector is expected to change is rotation direction at these times as well.

An intensity maximum must exist at high latitudes in the region of magnetopause projection on the Earth's surface.

Starting with the 1st International Geophysical Year, the MHD-wave nature was confirmed experimentally both in observatory explorations over many years of regular observations and during specially arranged experiments in synchronous observations of oscillations by meridional and latitudinal chains of magnetometer stations ([3], [4], [12], [18], [36], [46], [14], [56], [57]).

In addition to large-scale hydrodynamic instabilities within the magnetosphere and on its boundary, pulsations can also be caused by the kinetic micro-instabilities of magnetospheric plasma and/or of the solar wind ([9], [12], [25], [36], [108]). Pulsations can also be excited by solar wind inhomogeneities reaching the magnetosphere. Lastly, the ionosphere is another possible region for the emergence of geomagnetic pulsations.

Pulsations are usually global in character. Certain simultaneous oscillations are traced with confidence over half of the globe. Oscillations observed in one of the hemispheres are often visible in the opposite hemisphere ([3], [5], [50], [54], [64], [84], [88], [114]). They have been discovered everywhere in the magnetosphere, from its inner regions up to the magnetopause on the dayside and up to distances of many tens of Earth radii in the magnetotail ([9],[13], [18], [43], [44], [51], [63], [70], [102]). They were also observed in solar wind, as well as in a shock wave that departed from the magnetosphere. We refer the reader to the papers and monographs devoted to morphological properties and physical aspects of generation of the geomagnetic ULF-variations (see, e.g., [28], [30], [39], [90]). The monograph [89] covered the wide-ranging volume important aspects of ULF-waves, their main properties, comparisons of ground and satellite observations within the magnetosphere, in the solar wind, and in the magnetosphere tail, simultaneous occurrence at conjugate points, tracing the oscillations from the ground through the ionosphere to the magnetosphere and the solar wind. Topics presented in this monograph include numerical and experimental studies aimed at quantifying more precisely the locations and mechanics of the ULF-waves.

The improving of theoretical and practical knowledge on nonlinear aspects of ULF-oscillations was presented and exploited in [31].

3.2 The Physical Pattern

Magnetosphere

Let a certain source of MHD-waves with arbitrary polarization emerge in the peripheral regions of the magnetosphere or in the solar wind. Radiation can depart from the generation region as an Alfvén wave or as a fast magnetosonic wave (FMS). The Alfvén wave will be guided by field-lines and propagate directly from the generation region to the ionosphere.

The perturbations are carried inside the magnetosphere by FMS-waves as well. Four regions are distinguished in the magnetosphere with regard to FMS-wave propagation in it:

(a) the propagation region;
(b) the reflection region;
(c) the damping region behind the reflection point and, lastly,
(d) the region of resonance interaction.

The basic parameter determining the character of propagation is the Alfvén velocity $c_A = B_0/(4\pi\rho_0)^{1/2}$ (B_0 is a magnitude of the geomagnetic field, ρ_0 is the density of the background cold plasma). The geomagnetic field and the magnetospheric plasma are inhomogeneous and therefore the Alfvén velocity distribution is significantly inhomogeneous as well. This leads to the coupling of FMS- and Alfvén waves.

Usually, the transfer of wave energy from an FMS- to an Alfvén wave is not effective. But the coupling between these modes comes into particular prominence within the resonance region ([10], [11], [90], [113]).

The FMS-wave encountering inhomogeneities on its way causes a guided Alfvén wave, which, propagating along a field-line, can reach one of the conjugate ionospheres. Here the wave penetrates through the ionosphere to the ground, is partially reflected and returns to the conjugate ionosphere. The Alfvén waves are then confined between Northern and Southern hemispheres. If the double traveltime between the ends of the field line coincides with the period of the exciting external source or one of its harmonics, then the wave is resonantly amplified. This resonance is called a 'field line resonance' (FLR) [24].

Thus, there are two channels for an FMS-perturbation to reach the Earth. The decayed FMS-wave is reradiated into an Alfvén wave guided along a field-line. Then the perturbation can be amplified by the resonance. If the FMS-wave is excited by a source located on the magnetospheric boundary, it can reach the ground directly but only in a very narrow target angle of about 1° because of strong refraction [93].

A Source on the Magnetospheric Boundary

A sketch of propagation of an FMS-wave caused by a local source from the magnetospheric boundary to the ground is shown in Fig. 3.1. There are three characteristic regions:

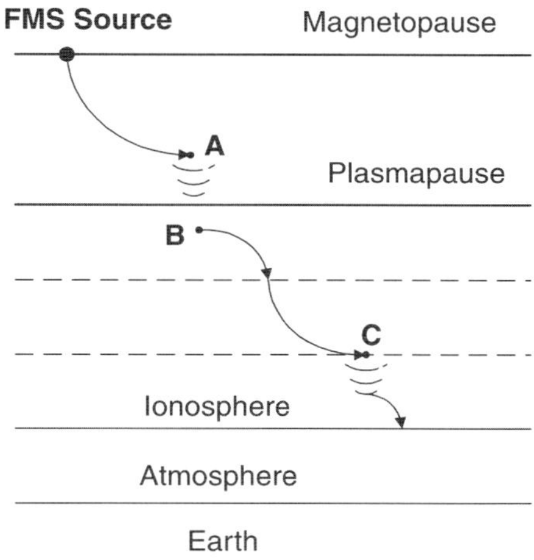

Fig. 3.1. A sketch of an FMS-propagation from a local source on the magnetospheric boundary to the Earth

- One is located between the reflection point A and the magnetopause.
- The wave is decayed behind it and gets into the BC region between the plasmapause and the internal reflection point C.
- Penetrating to the other side of the turning point C, the wave again decays exponentially as it gets the equatorial ionosphere.

FMS-wave transformation into the Alfvén wave is intensified on the FLR-frequencies. If the spectrum of the initial source contains a wide spectrum of FMS-oscillations, each of them finds its own resonant field-line. In this case, it is possible that global oscillations with lattitude dependent frequencies will appear. If initially, the FMS-source radiates in a narrow frequency band, then the wave is amplified only within a single resonance shell. In this case, oscillations with the same frequency should be observed. Intensity of such oscillations decreases with latitude and increases sharply within a narrow region.

Alfvén Resonances

We distinguish two types of resonance interaction. One of them, the resonance of FMS-waves with standing oscillations, was already mentioned. The other is the resonance with traveling Alfvén waves, when the FMS-wave phase velocity is equal to the local Alfvén velocity on a certain field-line. If the phase displacement between the waves remains unchanged, the Alfvén wave amplitude is increased by resonance. This type of resonance is noticeably manifested if the waves are propagating together for a sufficiently long time. This is possible only if the Alfvén wavelength is small in comparison with the length of the field-line.

Some peculiarities of this type of resonance are:

1. Reflecting walls are not necessary for the wave coupling. Resonance intensification occurs on traveling waves. This means in particular that the Alfvén wave moving in only one direction, say towards the Northern Hemisphere, is intensified. The excited traveling Alfvén wave reaches the Northern Ionosphere, is reflected, and goes back to the conjugate ionosphere without resonance amplification. Oscillations with noticeably differing amplitudes must then be observed at the two ends of the field-line.

The latitude dependent amplitudes on the ground are determined for both resonances by 4 factors:

1. general attenuation of the FMS-wave as it penetrates deep into the magnetosphere;
2. its spatial spectrum;
3. the effectiveness of FMS-wave transformation and
4. ionospheric attenuation.

With the decrease in geocentric distance, the Alfvén velocity grows on average. As FMS-waves propagate towards Earth, they are sorted by their wavenumbers. Small-scale perturbations are reflected in the outer parts of the magnetosphere and return to its boundary. Closer to the Earth, the FMS-wave spectrum is depleted of small-scale perturbations. After the reflection point the FMS-wave intensity exponentially decays at distance of the order of azimuthal scale. In the inner parts of the plasmasphere, the FMS-wave frequency becomes comparable to the frequency of the standing Alfvén oscillations and its spatial scale become comparable to the length of the corresponding field-line. A resonance surface emerges in this region of non-dissipative attenuation.

Cavity Resonance

Dungey predicted the emergence of global axially symmetric resonance oscillations on the FMS with discrete eigenfrequencies ([15], [17]). Cavity oscillations appear between the reflection (turning) point and the magnetopause. Their properties are identical with the axially symmetric oscillations of an acoustic resonator.

For asymmetric perturbations, when FMS and Alfvén modes couple, the notions of cavity resonance oscillations remain correct on the whole. However, additional energy losses appear because the wave energy is transferred from a global large-scale FMS-mode to local small-scale Alfvén oscillations. The most effective transformation takes place in the vicinity of the field line where the frequencies of its FLR and of the cavity mode harmonics coincide. Experiments with laboratory plasma confirm that there is a significant attenuation of cavity modes due to excited Alfvén resonances. On large tokamaks noticeable warm-up was observed during excitation and subsequent dissipation of FMS-cavity modes ([62], [91]).

Unlike laboratory plasma in Tokamak-type systems, the magnetosphere is an open system for FMS-waves, which results in radiation losses of cavity-mode energy. And this peculiarity might seem the strongest argument against the hypothesis of cavity resonances, especially for the low-frequency part of the spectrum. Indeed, high-frequency waves with wavelength smaller than the geocentric distance to the plasmapause can be captured in the plasmasphere. A low-frequency wave must go to the tail of the magnetosphere and then leave it. However, if such conditions emerge on the magnetopause, in the magnetosphere tail that velocity along the magnetopause will noticeably change on smaller scales than the transverse dimensions of the magnetosphere, i.e. about $10{-}20\mathrm{R_E}$, that region will effectively reflect FMS-waves. The magnetosphere can then be likened to an open resonator.

Oscillations of such systems are known to be accompanied by the radiation into free space. An essential peculiarity of open resonators is the large difference in damping rates between the main (corresponding to small mode numbers) and the higher modes. Among the resonance modes, a mode appears with a high Q-factor. As the transverse wavenumber rises, radiation losses increase rapidly. The low Q-factor of the high frequency modes results in effective rarefaction of the open resonator's spectrum. Additional selection,

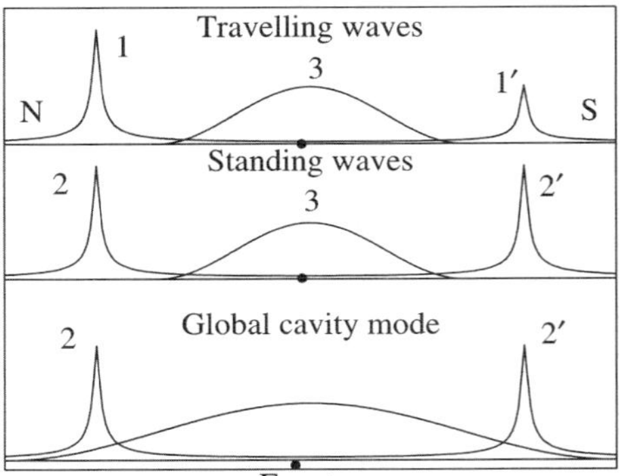

Fig. 3.2. A sketch of spatial amplitude distributions on the ground caused by various resonances. Index 1 refers to the high-frequency resonance on traveling waves. Indices 2 and 2′ mark resonances on standing Alfvén waves. 3 refers to an FMS-wave that has reached the ground directly

i.e. intensified suppression of all types of oscillation except the basic one, can occur due to variable properties of the magnetosphere tail boundary.

Continuous oscillations should occur in the open magnetosphere when there is a strong reflection of the waveguide modes moving from the dayside magnetosphere to its tail and reflected from its 'open' ends. Such an effect is observed in acoustics during resonance oscillations of open tubes, whistles, etc. If, for instance, the magnetosphere tail is regarded as a cylindrical waveguide, then, according to McKenzie (1970), the plasma sheet can oscillate with a period of 6 min.

Figure 3.2 demonstrates a stylized picture of the latitudinal dependencies for various resonances:

1. resonance on travelling waves;
2. resonance on standing waves;
3. global cavity mode.

Index 1 in Fig. 3.2 marks the high-frequency resonance on traveling waves. This resonance intensification can be observed at high latitudes. The difference in the magnitudes of intensity maxima at conjugate points 1 and 1′ is connected with the choice of the preferential northern propagation for an FMS-wave. Indices 2 and 2′ mark resonances on standing Alfvén waves. The highest cavity mode harmonics excite corresponding FLR-harmonics. 3 refers to an FMS-wave that has reached the ground directly. Oscillations of the same frequency could in this case be recorded as wide apart in the latitude and longitude observation points. The amplitude decreases to middle and low latitudes is concerned with FMS-wave attenuation as it penetrates into the magnetosphere and with non-resonance excitation of Alfvén waves.

Ionosphere

The described path for the wave propagation from the magnetopause to inner L-shells is not the only one possible. The waves that have reached the high-latitude ionosphere can cause perturbations propagating along the ionosphere to the middle and low latitudes. The ULF-waves in the ionosphere possess a tremendous skin depth scale many times larger than the characteristic vertical scale of the ionosphere.

There is an essential difference between the mechanisms for ionospheric transformation of an Alfvén wave and an FMS-wave. The Alfvén wave brings a longitudinal current to the ionosphere, which does not penetrate into the atmosphere because of its insignificant conductivity and spreads over the ionosphere. The total current system of the reflected and incident waves is arranged so that the magnetic effect from it under the ionosphere vanishes. Only the magnetic effect of the Hall currents can be observed, while above the ionosphere, the magnetic field is determined by a system of longitudinal currents closed in the ionosphere by the Pedersen current [20]. For this reason the magnetic components under the ionosphere must be turned by $\pi/2$ if the source is an Alfvén wave. The ground signature of the Alfvén wave incident onto the ionosphere manifests itself in the FMS-polarization. It is important to note that these concepts can also be extended to the wide range of oscillation periods from several seconds to several days.

On the contrary, the ionosphere is almost transparent for the FMS-mode causing two electrical currents - ionospheric and terrestrial. The relative contribution of both depends on a number of factors, but primarily on the ratio of the ionospheric to the Earth's conductivities.

3.3 ULF-Waves on the Ground and in Space

Pc5 Pulsations

$Pc5$ pulsations are basically a high-latitude phenomenon. They are characterized not only by long periods (150–600 s), but by tremendous amplitudes as well. Their amplitudes are ≈40–100 nT at high latitudes, and under conditions of high activity, it can rise to 400–600 nT.

$Pc5$ are usually observed at latitudes of 60–75°. Figure 3.3 [42] shows the latitudinal dependency of the Fourier amplitudes and relative phases. The curves illustrate the characteristics typical for field line resonances (FLRs) ([10], [90]). The latitude of maximum amplitude decreases with increasing frequency. For all frequencies, the phases change with $\approx\pi$ near the corresponding amplitude maxima, which also is a typical FLR feature.

The oscillations occur more often in the morning and afternoon hours. For strong geomagnetic activity ($Kp \geq 5$) excitation of $Pc5$ oscillations is about equally probable in the interval from 6 to 18LT [3].

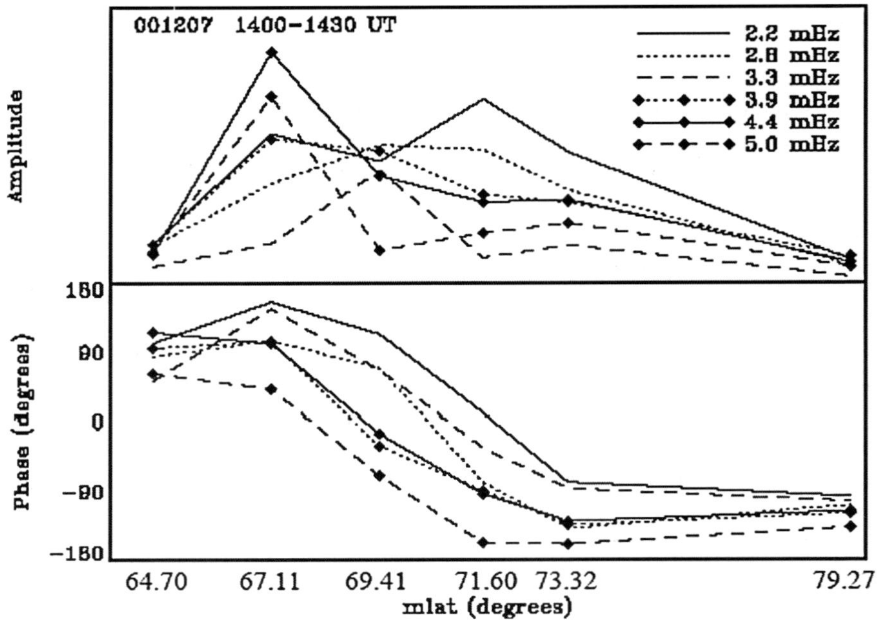

Fig. 3.3. Latitudinal variations of the Fourier amplitudes and phases of the NS ULF-magnetic component along the CANOPUS Churchill line. After [42]

Amplitude distributions of individual cases of $Pc5$ are strongly elongated along the parallel ($\delta\lambda = 60°$) and bounded in latitude [81]. At distance of $\delta\varphi \approx 1°$, a sharp change can occur in the pulsation period. The latitude of the $Pc5$ intensity maximum decreases with as the Kp-index increases, and the main period falls from 500 to 300 s when the maximum is shifted from the latitude of $\varphi° \approx 70°$ to $\varphi° \approx 60°$. Such a connection of the period with geomagnetic latitude indicates that $Pc5$ pulsations emerge on field-lines ever closer to Earth as magnetospheric perturbation increases.

Figure 3.4 shows the direction of horizontal magnetic vector rotation constructed on data from a Canadian station network [77]. A change of the rotation sign is clearly visible during transition from morning to evening oscillations.

A comparison of records of such pulsations made by a satellite and two stations at approximately footprint points showed that they are clearly visible both on the ground and in space ([26], [47], [87]). Usually, variations of the H-component (North-South) at the conjugate points occur in phase, while those of the D-component (East-West) occur in antiphase ([5], [54]). The horizontal vector rotates in opposite directions in two hemispheres. Figure 3.5 presents simultaneous magnetograms from two observatories: Byrd (Antarctic) and Pole (Arctic)[34] . It can be seen that separate $Pc5$ splashes have a kind of corresponding trigger pulses at the start of a pulsation train.

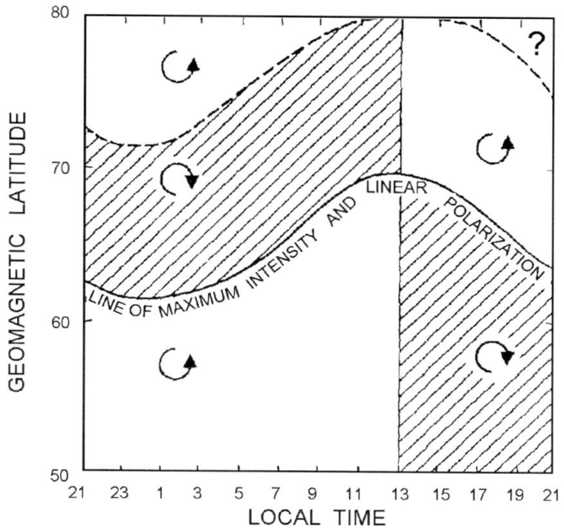

Fig. 3.4. The variation of amplitude and the sense of polarization of $Pc5$ seen at high latitudes with latitude and longitude. After [77]

$Pc5$ pulsations were discovered in the ionospheric plasma with the help of polar radars, which made it possible to see the patterns for velocity distribution in ionospheric plasma and in electric fields. Figure 3.6 presents an instance of meridional distribution of pulsations at the ionosphere level obtained with the help of the STARE radar system [106]. The system monitors the drift of electron density perturbations in the E-layer under the action of the wave electric field. The upper panel of Fig. 3.6 presents the coordinate dependence of the value of the northern component of the electric field (circles), while the lower part shows the same for the phase (circles) [106]. The solid line presents the fitted curves of the amplitude and phase. The region of resonance enhancement of oscillation is clearly visible. The half-width of the resonance region for $Pc5$ oscillations at high latitudes is about 0.5° or the first tens of kilometers.

High-latitude profile observations of $Pc5$ geomagnetic pulsations [56] as well as radar measurements of ionospheric plasma drifts point to the existence of a line spectrum. The 1.8, 2.4, 3.1 mHz frequencies dominated ([60], [107]). Similar results were obtained by comparing F-layer drifts and ground pulsations [77]. $Pc5$ appeared mostly in the local morning. The typical duration of the oscillations was 3, 4 h with surprisingly stable frequencies, changes in which did nor exceed 5–10% and did not change from day to day and from month to month ([77], [106]). The power spectra of the Doppler velocities had four peaks: on 1.3 mHz frequencies near 71.5°, on 1.8–1.9 mHz at 70.5°, on 2.7 mHz at 69.75° and 3.3–3.4 mHz at 69.0S°.

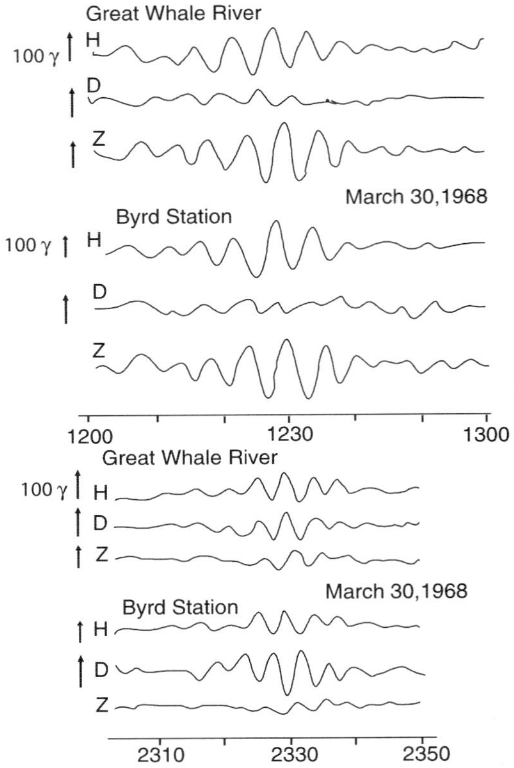

Fig. 3.5. Example of two simultaneous *Pc*5 wavetrains recorded at two conjugate points (north obs. Great Whale River and south obs. Pole). After [34]

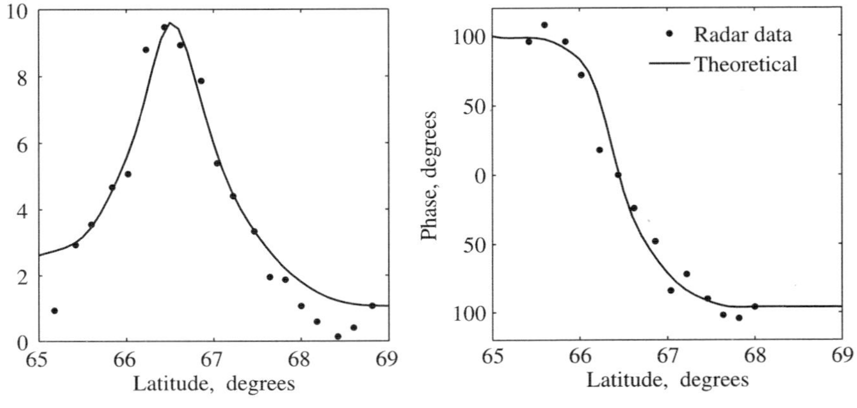

Fig. 3.6. Electric field of the FLR-shell oscillations in the ionosphere. After [106]

The allocated latitude band where the spectral peaks are localized was interpreted as excitation of FLR by cavity or waveguide mode. From these frequencies it is possible to make a rough estimate of the Alfvén velocity as $c_A \approx 40\,\mathrm{km/s}$. Such velocity requires too large hydrogen concentrations as compared to those found by satellite measurements. These results may point to a noticeable presence of O^+ ions near the magnetopause [69].

The results of ground, radar and satellite measurements can also be coordinated by assuming that FMS-wave reflection occurs not on the magnetopause, but on the bow shock, i.e. in a region of very low Alfvén velocity [33].

There is another possibility to account for the observed low frequencies of *Pc*5 pulsations in the framework of the cavity resonance mechanism. The frequency of the lowest cavity mode can be determined not by oscillations between the turning point and the magnetopause (or bow shock), but by conditions in the magnetosphere tail or on its boundary. Satellite observations have shown that compressional *Pc*5 appear mainly on the nightside on $L >$ 7–8. Such standing waves with periods of 3–9 min are constantly present in the magnetosphere tail as well at distances of 10–22R_E ([1], [82], [83]).

Traveling Magnetospheric Twin-Vortices

Twin-vortices ionospheric current systems in the *Pc*5 range moving tailward at the velocity of 2–5 km/s were found in ([19], [22]). These systems were attributed to local inhomogeneities of the solar wind leading to changes in the dynamic pressure on the magnetopause and in longitudinal currents. The characteristic spatial scale of the vortices at ionospheric altitudes is ≈2000 km. They are transmitted towards the ionosphere by an Alfvén wave.

The large changes in the solar wind dynamic pressure in the shape of an unipolar pressure jump (see Fig. 3.7(a)) produce a Field-Aligned Current (FAC) proportional to the derivative of the pressure perturbations with respect to the azimuthal coordinate [22]. The signs of the derivative is opposite on the two sides of the pressure pulse, i.e. the excited FACs are of bipolar character. The stepwise pressure change is also sufficient for the excitation of a bipolar system (Fig. 3.7b) [45]. In this case, the FAC is proportional to the second derivative of the pressure perturbation.

A couple upward and downward FAC spaced along the latitude at about 2000 km are closed in the ionosphere by Pedersen currents. The ground magnetic field of the ionospheric currents under the polar homogeneous ionosphere is solely due to the Hall current system ([20], [21]) that is the bipolar Hall current vortices produce vortices in the ground magnetic field.

Figure 3.8 (b) present *NS*- and *EW*-components from the Scandinavian magnetometer array of stations (Fig. 3.8(a)). The main peculiarity of the records is the isolated *NS*-component bipolar pulse as well as a unipolar pulse in the *EW*-component. There are no noticeable variations in the vertical component. Figure 3.8 clearly shows earlier onsets of variations at

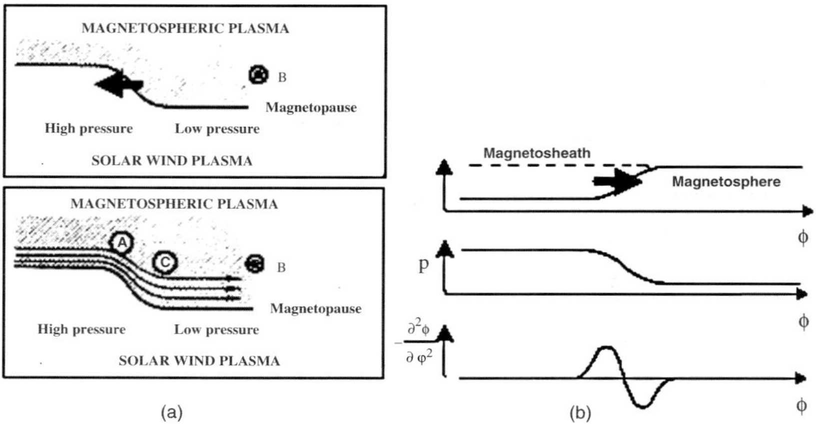

Fig. 3.7. Suggested scenario in which the large changes in the solar wind dynamic pressure produces a unipolar FAC proportional to the azimuthal derivative of the pressure perturbations (a) After Glassmeier *et al.* (1989). (b) A stepwise pressure change and a bipolar system of FAC is proportional to the second derivative of the pressure perturbation. After Kivelson and Southwood (1991)

eastern stations which corresponds to the tailward motion of the convective perturbation.

The equivalent current system constructed from these data and supplemented by records from the Greenland station chain is shown in the upper

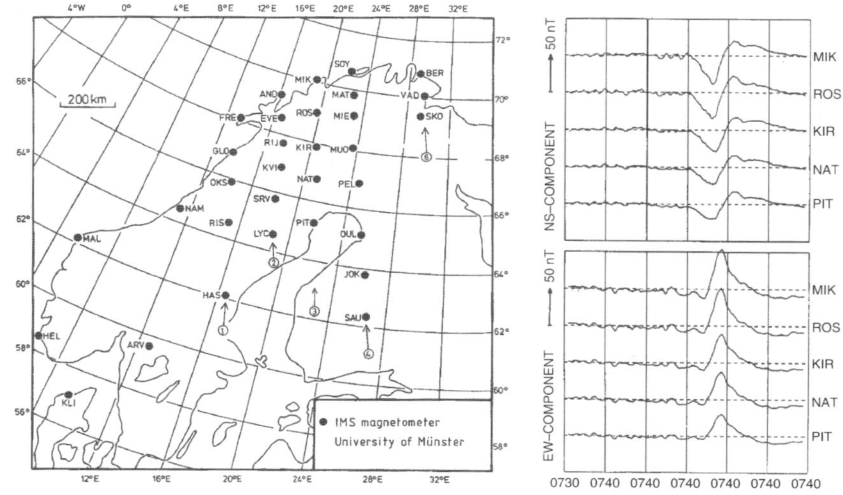

(a) (b)

Fig. 3.8. Simultaneous *NS*- (upper part of panel (b)), and *EW*- magnetograms (lower part) from the Scandinavian magnetometer array of stations (a) [23]

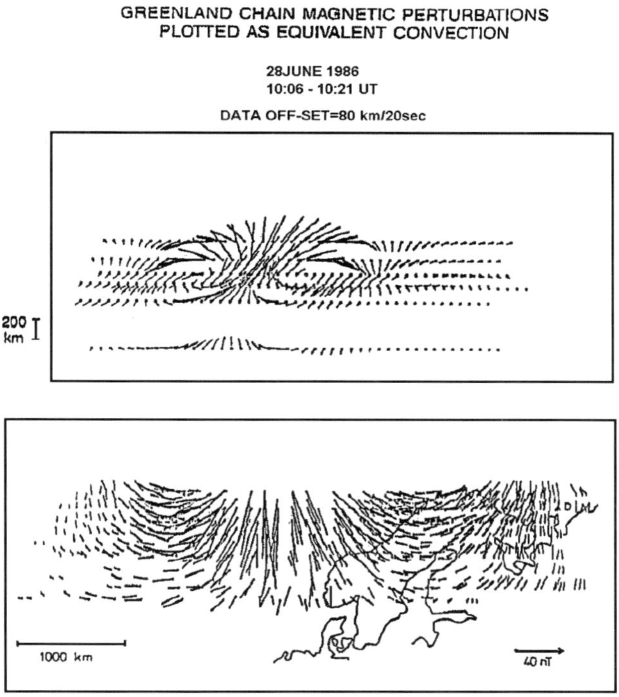

Fig. 3.9. The equivalent current system constructed on the Greenland station chain data shown in Fig. 3.8 (upper panel). The lower panel presents the result of the spatial interpolation reproduced from Fig. 3.8 data [23]

part of Fig. 3.9. The lower panel presents the result of the spatial interpolation of these data [23]. Mapping of the ground velocity of the vortices to the magnetopause gives 220 km/s. This testifies to a possible connection of the source with variations of solar wind parameters. Measurements of drift velocities carried out in the ionosphere and results of satellite observations showed the reliability of the scenario about traveling vortices being generated by solar wind inhomogeneities ([53], [105], [41]).

*Pc*3, 4 Pulsations

Dayside *Pc*3, 4 pulsation amplitudes are strongly localized in latitude. Simultaneous observations of pulsations demonstrated that the latitude of maximum amplitude depends on the wave period, and the phase of the *NS*-magnetic component varies abruptly at the scale of the order of 100 km across the latitudinal maximum [27]. The pulsations were observed at a meridional chain of 5 stations in Great Britain located approximately along -140° longitude at distances of 120−170 km from one another for all the five observatories

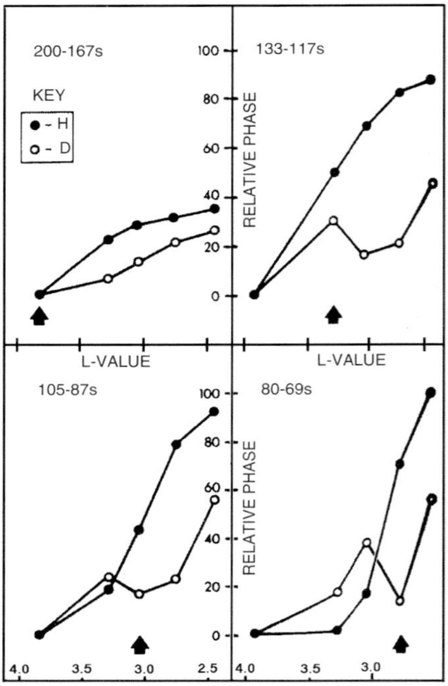

Fig. 3.10. Average wave phases relative to the phase at the northern station versus L-value. The amplitude maximum of the latitudinal profile is marked by the solid arrow. After [27]

($L = 2.44$–3.83). The plasmapause during all the three $Pc3, 4$ series was to the north of the station array because of the low magnetic activity ($Kp \leq 3+$). Figure 3.10 presents the latitudinal dependence of phase angles for various frequencies in the pulsation spectrum calculated for three 2-hour series of $Pc3, 4$ oscillations at each station measured relative to the phase at the most northern observatory. It can be seen that the position of the maximum in the meridional distribution of pulsations is shifted southwards with the increase in frequency in accordance with the FLR-theory. The FLR-theory predicts larger changes in H-component than in D-component across the amplitude maximum. Moreover, phase was predicted to increase with decreasing latitude. Examination of Fig. 3.10 indicates that the phase changes are in agreement with these predictions.

The question of the lowest latitudes for which a noticeable role of FLR in ground pulsations is worth mentioning ([59], [86], [95]). It is clear that at low latitudes, the Q-factor of FLR decreases as a field line plunges into the high conductive ionosphere. The resonance frequencies $\omega_R(\varphi)$ have a minimum at middle and low latitudes and grow near the equatorial region. Experimentally,

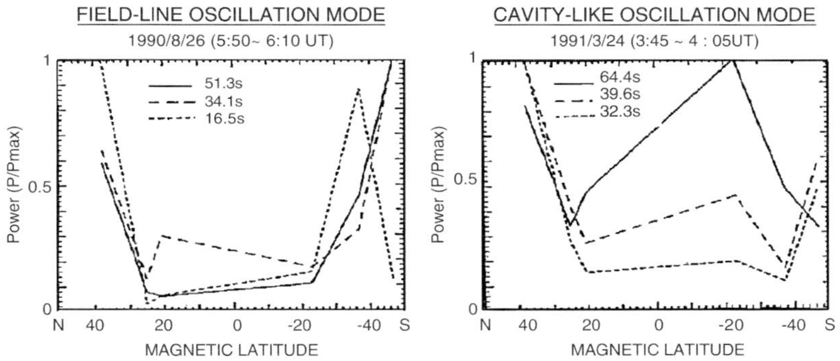

Fig. 3.11. Typical latitudinal profiles of normalized power spectral densities; (left) FLR-oscillations and (right) cavity mode-like oscillations obtained at the 210°MM chain stations of $L = 1.4{-}2.13$. After [112]

FLR in the $Pc3, 4$ range was successfully discovered from the latitudes ≈25° ($L = 1.2$) up to the plasmapause. Outside, at high latitudes, this type of pulsations apparently results from FMS-wave resonance with a traveling Alfvén wave.

Energy from outer sources must be absorbed most intensively at frequency of the cavity mode. When this frequency coincides with the FLR-frequency on some magnetic shell, an effective energy transfer occurs from cavity to local Alfvén oscillations. Despite the considerable number of theoretical and computation works devoted to cavity resonance, its existence cannot be regarded as definitively established. Additional the questions of the Q-factor of an open resonator and of the paths of FMS-wave propagation towards Earth, remain to be proved.

The experimental evidence for $Pc3, 4$ cavity resonance was obtained by Yumoto [110]. The idea of searching for cavity resonance reduces to finding out global oscillations at discrete frequencies caused by rapid changes in solar wind pressure. Power spectral densities of pulsation intervals at 210° magnetic meridional chain stations [111] were compared for 20 min just before, just after, and plus 20 min after Sc/Si (geomagnetic Sudden commencement/Sudden pulse). Four cases of cavity mode were found, localized in the dayside sector from 9 to 15 LT with high magnetic activity $Kp \geq 7+$.

After Sc the levels of pulsation power spectra grew by 10 times as compared to the level before Sc. In pulsation spectra at different stations identical components appeared with periods of $T = 64.4s$, 39.6 s and 32.2 s. The found high magnitude oscillations were of global character, occupying the latitude belt from $\phi = -46°.7$ to $\phi = 37°.8$. The intensity maximum was located at low latitudes (Fig. 3.11).

Connection of the $Pc3, 4$ Frequency with the IMF

The dependence of the $Pc3, 4$ pulsation period at middle-latitude observatories on the magnitude and direction of Interplanetary Magnetic Field (IMF) was discovered in 1971 [101]. Inverse dependence of $Pc2-4$ pulsation frequencies on the magnitude of the IMF was shown: Direct measurements of wave frequencies and IMF values in the upstream region ([35], [73]) confirmed the results of the ground-based studies:

$$f\,(\text{mHz}) \approx 5.7 \times B_{\text{IMF}}\,(\text{nT})\,.$$

It was suggested that the origin of these pulsations is extra-magnetospheric. In the proposed generation mechanism, the ion downstream of the solar wind flowing around the magnetosphere is reflected from the bowshock. An arising upstream ion flow propagates along the IMF and excites ion cyclotron instability. The instability causes FMS-waves, whose frequency is determined by the ion cyclotron frequency in the solar wind ([29], [48]). Correlation analysis of the low-latitude $Pc3$ pulsations ($L = 1.3-1.8$) and the oscillations of the same frequency range detected by the synchronous satellite GOES 2 ($L = 6.67$) also confirmed the linear dependency of the frequency on the strength of the IMF (B_{IMF}) in the form

$$f\,(\text{mHz}) \approx 6.0 \times B_{\text{IMF}}\,(\text{nT})\,.$$

Satellite Measurements of the $Pc3$ Waves

Simultaneous events of $Pc3, 4$ pulsations on the ground and in the magnetosphere were found in ([65], [92], [103], [109]). The waves were observed at the same time in Space and on the Earth at the same frequency and with similar wave forms. Both compressional and transverse oscillations were identified in the satellite magnetic measurements. A close correspondence between the compressional component and the ground signals was observed. These results confirmed that $Pc3, 4$ waves were transported through the magnetosphere by the FMS-waves.

Takahashi *et al.* [96] traced the $Pc3$ wave package from the outer magnetosphere to the ground and observed a case of simultaneous $Pc3$ oscillations inside the magnetosphere and at a meridional chain of high-latitude stations. There a time interval was selected during which the AMPTE CCE satellite was located near the subsolar magnetopause. The data came from five stations: three in Iceland ($L \approx 6$), a station conjugate to Iceland in Antarctica, and another station at the average cusp latitude. The stations were located almost at the basis of the filed line on which the satellite was placed.

Observation results are on the whole in agreement with the above-described scenario for the generation and propagation of hydromagnetic waves in the magnetosphere. It was pointed out the similarity of the amplitudes and structures of wave packages in the H-component at stations in Iceland to the longitudinal component detected by the CCE [96]. Spectral analysis showed that $Pc3$ pulsations have a peak at frequency $f \sim 60\,\mathrm{mHz}$ at CCE. This frequency is close to the estimate by [101]. A peak on the same frequency was also discovered at the magnetometer points.

At the same time at the higher latitude station ($\Phi = 78.9°\mathrm{N}$) and at the conjugate station located in the Southern Hemisphere, the pulsation amplitude in the $Pc3$ range is considerably smaller. The existence of a high-latitude maximum in the $Pc3$ range at geomagnetic latitudes $\Phi \approx 65°-70°$ and a sharp fall in pulsation amplitude with northward movement are well known facts [38].

Takahashi et al [96] relate the intensity maximum at $\Phi \leq 70°$ by an Alfvén wave passing along the magnetopause to the cusp boundary towards Earth. The high-latitude station of $\Phi = 78.9°$ is then to the North of the cusp. This is why the lack of oscillations in this region can serve as an indirect indication to the most probable path of the Alfvén wave into the high-latitude region—from the generation area within the upstream region to the magnetosphere, through the magnetopause, by transformation of FMS into an Alfvén wave and finally propagation along the external force tubes on the inner boundary of the cusp towards the ionosphere and the Earth.

We believe that the transformation of FMS- into Alfvén perturbations does not occur on the high harmonics of standing Alfvén oscillation, but on Alfvén waves traveling towards the Northern Ionosphere. With this excitation mechanism a natural explanation is found for the fact of asymmetry in pulsations recorded at conjugated stations. At the station in the Southern Hemisphere, the pulsation amplitude is considerably smaller than at the stations conjugated with it. This difference was explained in [96] by the asymmetry of ionospheres in the Northern and Southern Hemispheres.

$Pc3$ pulsations were observed in the very high latitude ionosphere, in the so-called ionospheric cusp, in the magnetosheath and on the ground [58]. The data used in the ionospheric analysis were obtained by the HF radars of SuperDarn and near the magnetopause by the Geotail satellite.

If the initial wave packet of FMS-waves is supposed to propagate northwards and the condition of wave synchronism with traveling Alfvén waves is fulfilled, then the excited Alfvén wave package will also move towards the Northern Hemisphere. It will be demonstrated in Chapter 7 that during transition from day to night, the magnetosphere wave resistance is close to the ionosphere resistance and the Alfvén wave reflection coefficient proves to be small. It follows from estimates that it is of the order of the ratio of H-component amplitudes at conjugate stations. That may mean that the Alfvén package, approaching first the Northern Ionosphere, was weakly reflected from it and left for the Southern Hemisphere with a small amplitude.

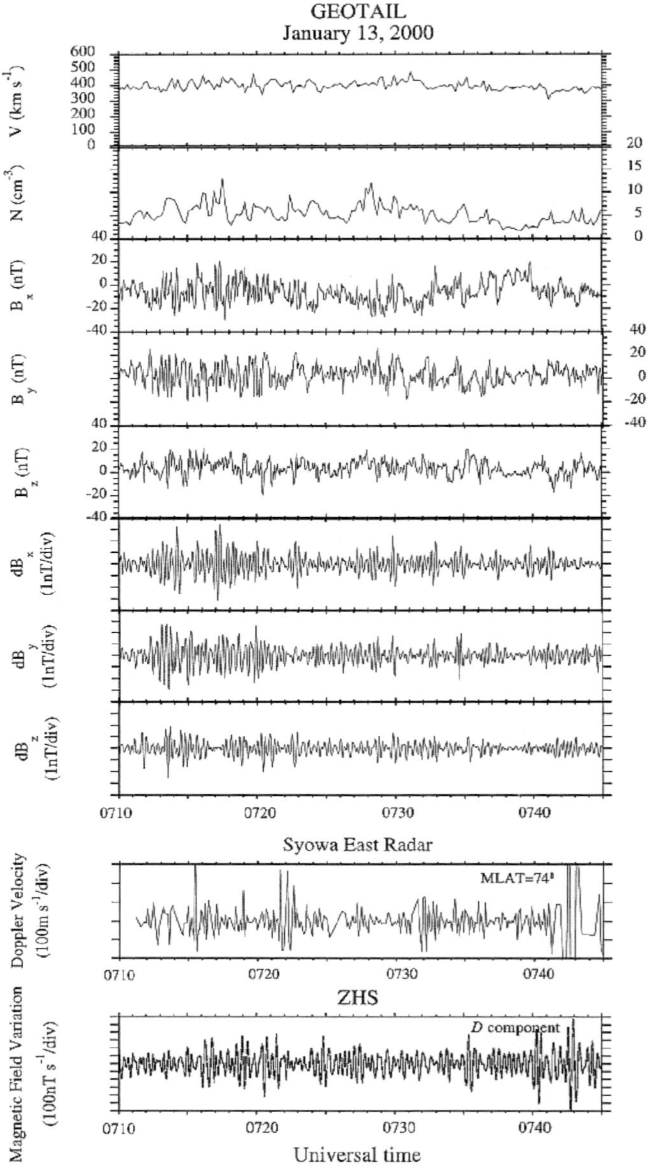

Fig. 3.12. Magnetic field data observed by Geotail in the magnetosheath and magnetic field on the ground at RANK (∼74°MLAT) during 1505–1545 UT, January 24, 1999. After [58]

*Pi*2 Pulsations

Irregular pulsations of the *Pi* type are closely linked to a magnetospheric substorm [66]. Pulsations of this type have the characteristic shape of a damping oscillation train with periods of 60–100 s and a duration of 5–10 min. The most important feature of *Pi*2 is the onset of these oscillations in the explosive phase of the substorm. An abrupt *Pi*2 splash is a kind of signal for the onset of the explosive phase. A sequence of *Pi*2 can be observed in the course of substorm evolution. The appearance of beam-shaped auroras is characteristic of the generation period of these oscillations, which testifies to the injection of electron flows. With increasing brightness of the auroras, their amplitudes grow as well. However, at a certain value of their brightness, their amplitudes begin to decrease and a kind of oscillation quenching takes place.

These oscillations are a typical nighttime phenomenon with a clear maximum near the auroral zone (63–70°) and a comparatively small intensification near the plasmapause ([32], [38], [44]). On the basis of these facts, it was assumed in earlier works that the primary excitation process takes place on the auroral magnetic field-lines, while the plasma density gradient on the plasmapause only facilitates the secondary resonance oscillation enhancement in this region.

The oscillations embrace at least 1/3 of the globe. Their intensity maximum occurs approximately in the midnight hours. Comparison of data shows that oscillations appear virtually simultaneously at points several thousand kilometers apart. Phase delays between identically designated components at a middle-latitude ($\Phi \approx 50°$) station chain stretching along the geomagnetic parallel for about 60° are presented for two individual cases in Fig. 3.13. The azimuthal wave number for *Pi*2 oscillations changed from 0 to 5. The characteristic phase velocity along the parallel is within the first hundreds of km/s. The major axis of the ellipse is directed towards the intensity maximum region at high latitudes [6].

Contours in Fig. 3.13 demonstrate the intensity distribution of the total horizontal vector of *Pi*2 oscillations. The positions of the recording stations are marked by dots. Intensity increases monotonically towards the high latitudes and the midnight meridian (dashed line). The isolines stretch approximately along parallels, embracing virtually all the Earth's nightside. The maximum at high latitudes is local, with oscillation intensity in it about 15 nT.

The influence of major ionosphere conductivity anomalies on *Pi*-oscillations was revealed by Shimizu and Yanagihara [85] who discovered the intensification of nighttime high-latitude *Pi*-oscillations on the dayside in the equatorial electrojet region. The result was confirmed in ([63], [94]) and it was shown that these oscillations are constantly present at low latitudes and near the equator ($L < 2$).

One of the hypotheses about the excitation of *Pi*-oscillation is the release of magnetic field energy in the magnetosphere tail in the course of magnetic reconnection processes with subsequent resonance excitation of the field-lines

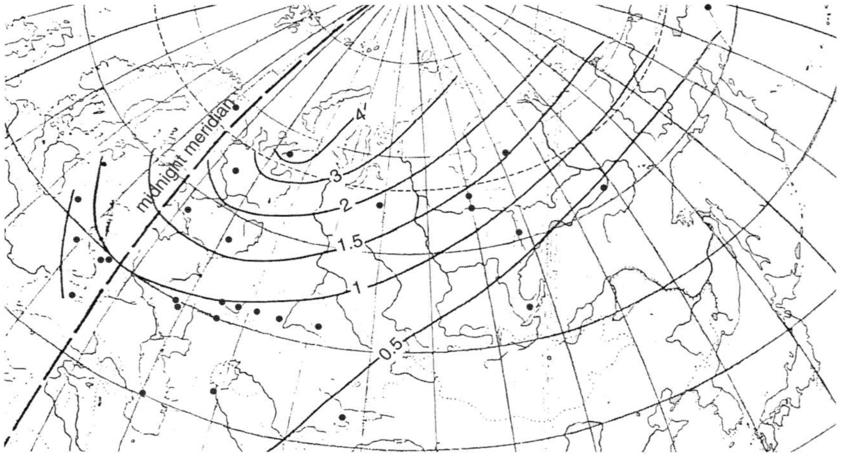

Fig. 3.13. Spatial distribution of the total horizontal vector of *Pi*2 oscillations is shown by intensity contours. The locations of the stations are marked by dots. Dashed line is the midnight meridian. The maximum at high latitudes is local, with oscillation intensity in it about 15 nT

in the nighttime sector of the magnetosphere ([68], [75], [76]). It was discovered the close coupling between the westward traveling surge (WTS) and *Pi*2 pulsations during the substorm onset [71]. The oscillation intensity maximum coincides with the equatorward electrojet boundary, i.e. the oscillation generation region coincides with the closed field-lines.

Virtually all the noted peculiarities can be explained by uniting two mechanisms [72]. The first is the feedback instability [2], the other is the appearance of an oscillatory regime as ionospheric conductivity changes sharply ([55], [8]). Local spills into the ionosphere with a stationary electric field result in a sharp increase of electron concentration and conductivity due to the first mechanism. The background electric field is diminished as a result. The field change will travel with Alfvén velocity along the magnetic field-lines and, being reflected from the conjugated ionosphere, can dampen *Pi*2-type geomagnetic pulsations. The effectiveness of the mechanism is high enough, since the background electric field in the surge can diminish by an order of magnitude within a short time and reach 5–10 mV/m. The magnetic field in the Alfvén magnetospheric wave will then equal 10–200 nT.

The discovery of daytime *Pi*2 demonstrates the possibility of oscillations penetrating into the low latitudes and into the vicinity of the equator through the ionosphere not only in the shape of spread-out currents, and also testifies to the possibility of global cavity modes with very low azimuthal numbers [94]. One of the regions where resonance oscillations can appear is the inner boundary of the plasma layer in the magnetosphere tail [74].

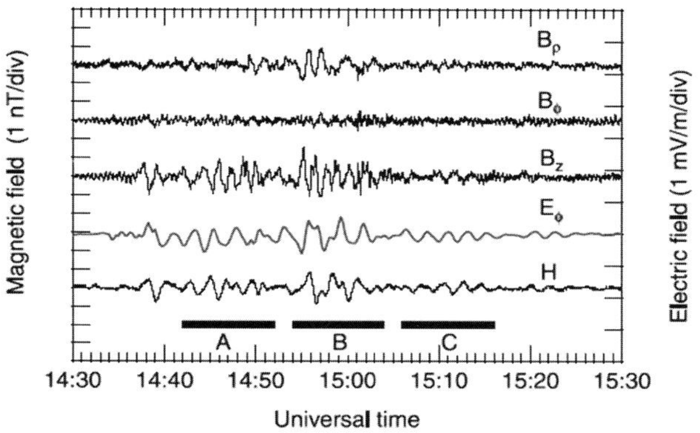

Fig. 3.14. CRRES and Kakioka ($L = 1.26$) data for the selected *Pi*2 events. The CRRES field components are respectively B_ρ: radial; B_φ: azimuthal, positive eastward; B_z: northward. The CRRES $E_\varphi$electric field component is dusk-to-dawn. Bottom: The Kakioka magnetic field. After [99]

Multisatellite observations of substorm onsets on the nightside at $4 < L < 7$ revealed transversal waves with frequency depending on L that can be attributed to the FLR. About 30% of the *Pi*2 events on the Earth are accompanied by nightside toroidal pulsations [98].

Analysis of low latitude magnetic data from the Combined Release and Radiation Effects Satellite (CRRES) and at low latitude station Kakioka ($L = 1.26$) shows that the *Pi*2 pulsations appear when both CRRES and Kakioka were on midnight. At CRRES the pulsations were detected in the FMS-mode. An example of the *Pi*2 pulsations simultaneously observed in space and on the ground is shown in Fig. 3.14 [99].

Using one of the reference works by Anderson [1], we cite only the tabulated results of observations of MHD-waves in the magnetosphere. Basing the classification of magnetospheric perturbations on their periods and polarizations, Anderson distinguishes 6 pulsation types: compressional *Pc*5, poloidal *Pc*4, compressional *Pc*3, toroidal field line harmonics, toroidal *Pc*5 and incoherent noise (see Table 3.1). Here $b_\nu, b_{||}$, and b_φ mark, respectively, the radial, longitudinal and azimuthal magnetic components.

Notes:

1. Compressional *Pc*3, 4 relate to wave-particle interaction in the foreshock and shock.
2. Toroidal *Pc*3, 4 or multi-harmonics are field line resonance harmonics in *Pc*3, 4 range. Compressional *Pc*3, 4 is a driver (coupling with the fundamental toroidal mode).

Table 3.1. Typical characteristics of various *Pc*-pulsations observed in the magnetosphere. See: http://www.oulu.fi/ spaceweb/textbook/ulf.html

	Compressional $Pc3,4$	Poloidal $Pc5$	Compressional $Pc5$	Toroidal $Pc3,4$	Toroidal $Pc5$
τ (s)	20–30	≥100	300–600	20–40	300–600
Polarization		b_ν	b_ν, $b_\parallel$	b_φ	b_φ
Amplitude	≤5 nT	∼10 nT	$\delta B/B_0 \sim$ 0.2–0.5		5–10 nT
LT	6–16	11–24		Day	6;18
L-shell	1.8–10	6–8	16–21; 3–6 7–8; >10 tail, 10–22 [83]	6.6–9	equatorial node
m	2–5	50–100 ([52], [96])	20 [115]		
Ground/ Space	Simultan.				
Resonance		2nd FLR-harmonics	Cavity Open resonator		
Source	Upstream waves	Drift-mirror instab.	Mirror instab. [52]	Upstream waves	Kelv.–Helm. instab. [70]

3. Poloidal *Pc*4 are related to injections of energetic plasma and subsequent low activity or convection electric field; occurring at the second harmonic field line resonance frequency.
4. Compressional *Pc*5 are related to of high beta plasma (ion injections).
5. Toroidal *Pc*5 are the fundamental mode field line resonances; source in the flanks.
6. Incoherent noise increases with magnetic activity.

Excluded from the table are noise oscillations with a broadband spectrum. According to [1], they are visible on both the morning and evening sides of the magnetosphere in the $L \approx 5-9$ interval and make up a considerable part – about 25% of all the observed pulsations. A general tendency has been revealed: a rise in the level of noise waves correlated with an increase in magnetic activity.

References

1. Anderson, B. J., An overview of spacecraft observations of 10 s to 600 s period magnetic pulsations in the Earth's magnetosphere. In *Solar Wind Sources of Magnetospheric Ultra-Low-Frequency Waves*, eds. M. J. Engebretson, K. Takahashi, and M. Scholer, AGU Geophysical Monogr. Ser., **81**, 25–43, American Geophysical Union, Washington, DC, 1994.

2. Atkinson, G., Auroral arcs: Result of the interaction of a dynamic magnetosphere with the ionosphere, *J. Geophys. Res.*, 4746–4754, 1970.
3. Baker, G. J., Donovan E. F., and Jackel B. J., A comprehensive survey of auroral latitude *Pc*5 pulsation characteristics *J. Geophys. Res.*, **108 (A10)**, 384, 2003
4. Ballatore, P., Lanzerotti L. J., and Maclennan C. G., Multistation measurements of *Pc*5 geomagnetic power amplitudes at high latitudes, *J. Geophys. Res.*, **103 (A12)**, 29455–29465, 1998
5. Ballatore, P., *Pc*5 micropulsation power at conjugate high-latitude locations *J. Geophys. Res.*, **108 (A4)**, 1146, 2003
6. Baransky, L. N., Y. E. Belokris, M. B. Gohberg, E. N. Fedorov, and C. A. Green, Restoration of the meridional structure of geomagnetic pulsation fields from gradient measurements, *Planet. Space Sci.*,**37**, 859–864, 1989.
7. Baransky, L. N., V. A. Troitskaya, I. V. Sterlikova, M. B. Gohberg, N. A. Ivanov, I. P. Kravtchenko, J. W. Munch, and K. Wilhelm, The analysis of simultaneous observations of nighttime *Pi* pulsations on an east-west profile, *J. Geophys.*, **48**, 1–6, 1980.
8. Baumjohann, W., and K.-H. Glassmeier, The transient response mechanism and *Pi*2 pulsations at substorm onset – Review and outlook, *Planet. Space Sci.*, 1361–1370, 1984.
9. Chaston, C. C., L. M. Peticolas, C. W. Carlson, J. P. McFadden, F. Mozer, M. Wilber, G. K. Parks, A. Hull, R. E. Ergun, R. J. Strangeway, M. Andre, Y. Khotyaintsev, M. L. Goldstein, M. Acuna, E. J. Lund, H. Reme, I. Dandouras, A. N. Fazakerley, and A. Balogh, Energy deposition by Alfvén waves into the dayside auroral oval: Cluster and FAST observations, *J. Geophys. Res.*, **110 (A2)**, A02211, 2005
10. Chen, L. and A. Hasegawa, A theory of long-period magnetic pulsations, 1. Steady state excitation of field line resonance, *J. Geophys. Res.*, **79**, 1024–1032, 1974.
11. Cheng, C. C., Chao J. K., Hsu T, S., Evidence of the coupling of a fast magnetospheric cavity mode to field line resonances, *Earth and Planet. Sci*, **50 (8)**, 683–697 1998
12. Chisham G., Mann, I. R., A *Pc*5 ULF-wave with large azimuthal wavenumber observed within the morning sector plasmasphere by Sub-Auroral Magnetometer Network *J. Geophys. Res.*, **104 (A7)**, 14717–14727, 1999
13. Cramm, R., Glassmeier K. H., Othmer C., Fornacon K. H., Auster H. U., Baumjohann W., and Georgescu E., A case study of a radially polarized *Pc*4 event observed by the Equator-S satellite, *Ann. Geophys.*, **18 (4)**, 411–415, 2000
14. De Lauretis, M., Francia P., Vellante M., Piancatelli A., Villante U., and Di Memmo D., ULF-geomagnetic pulsations in the southern polar cap, Simultaneous measurements near the cusp and the geomagnetic pole *J. Geophys. Res.*, **110 (A11)**, A11204, 2005
15. Dungey, J. W. Electrodynamics of the outer exosphere, *The Physics of the Ionosphere,* Phys. Soc., London, 229–236, 1955.
16. Dungey, J. W., *Cosmical Electrodynamics*, Cambridge Univ. Press, London 1958.
17. Dungey, J. W., *The structure of the exosphere, or, adventures in velocity in Geophysics*, The Earth's Environment (Proceedings of the 1962 Les Houches Summer School), pp. 503–550, C. DeWitt, J. Hieblot and A. Lebeau, editors, Gordon and Breach, NY, 1963.

18. Engebretson, M. J., Cobian R. K., Posch J. L., and Arnoldy R. L., A conjugate study of *Pc*3,4 pulsations at cusp latitudes, Is there a clock angle effect? *J. Geophys. Res.*, **105 (A7)**, 15965–15980, 2000
19. Friis-Christensen, E., M. A. McHenry, C. R.Clauer, and S. Vennerstrom, Ionospheric travelling convection vortices observed near the polar cleft: A triggered response to sudden changes in the solar wind, *Geophys. Res. Lett.*, **15** 253–256, 1988.
20. Fukushima, N., Equivalence in ground geomagnetic effect of Chapman–Vestine's and Birkeland–Alfvén's electric current-systems for polar magnetic storms, *Rep. Ionos. Space Res.*, **90**, 4069, 1965.
21. Fukushima, N., Electric currents systems for polar magnetic storms and their magnetic effect below and above the ionosphere, *Radio Sci.*, **6**, 269–275, 1971.
22. Glassmeier, K. H., M. Honisch, and J. Untiedt, Ground-based and satellite observations of travelling magnetospheric convection twin-vortices, *J. Geophys. Res.*, **94**, 2520–2528, 1989.
23. Glassmeier, K. H. and C. Heppner, Travelling magnetospheric convection twin-vortices: Another case study, global characteristics, and a model, *J. Geophys. Res.*, **97**, 3977–3992, 1992.
24. Glassmeier, K. H., Othmer C., Cramm R., Stellmacher M., and Engebretson M., Magnetospheric field line resonances: A comparative planetology approach, *Surveys in Geoph*, **20 (1)**, 61–109 1999
25. Glassmeier, K. H., Buchert S., Motschmann U., Korth A., Pedersen A., Concerning the generation of geomagnetic giant pulsations by drift-bounce resonance ring current instabilities, *Ann. Geophys.*, **17**, (3), 338–350, 1999
26. Glassmeier, K. H. and Stellmacher M., Concerning the local time asymmetry of *Pc*5 wave power at the ground and field line resonance widths, *J. Geophys. Res.*, **105 (A8)**, 8847–18855, 2000
27. Green, C., Meridional characteristics of *Pc*4 micropulsations events in the plasmasphere, *Planet. Space Sci.*, **26**, 955, 1978
28. Guglielmi, A. V. and V. A. Troitskaya, *Geomagnetic Pulsations and Diagnostics of the Magnetosphere*, Nauka, Moscow, 1973.
29. Guglielmi, A. V., Diagnostics of the magnetosphere and interplanetary medium by means of pulsations, *Space Sci. Rev.*, **16**, 331, 1974
30. Guglielmi, A. V., *MHD Waves in the Plasma Environment of Earth,* Nauka, Moscow (in Russian), 1979
31. Guglielmi, A. V., and O. A. Pokhotelov, Nonlinear problems of the geomagnetic pulsations, *Space Sci. Rev.*, **65**, 5–57, 1994.
32. Han, D. S., T. Iyemori, Y. F. Gao, Y. H. Sano, F. X. Yang, W. S. Li, M. Nose, Local time dependence of the frequency of *Pi*2 waves simultaneously observed at 5 low-latitude stations, *Earth and Planet. Space Sci.*, **55 (10)**, 601–612, 2003
33. Harold, B. G. and Samson J. C. Standing ULF modes of the magnetosphere: A theory, *Geophys. Res. Lett.*, **19**, 1811–1814, 1992.
34. Hirasawa, T., Long-period geomagnetic pulsations (*Pc*5) with typical sinusoidal waveforms, *Rep. Ionos. Space Res. Japan*, **24**, 66–79, 1970.
35. Hoppe, M. and C. T. Russel, Particle acceleration at planetary bow shock waves, *Nature*, **235**, N5844, 41–42, 1982.
36. Howard, T. A. and F. W. Menk, Ground observations of high-latitude *Pc*3,4 ULF-waves, *J. Geophys. Res.*, **110 (A4)**, A04205, 2005.

37. Hughes, W. J., Magnetospheric ULF Waves: A Tutorial with a Historical Perspective, In *Solar Wind Sources of Magnetospheric Ultra-Low-Frequency Waves*, eds. M. J. Engebretson, K. Takahashi, and M. Scholer, AGU Geophysical Monogr. Ser., 81, 1–11, American Geophysical Union, Washington, DC, 1993.
38. Jacobs, J. A., and K. Sinno, World-wide characteristics of geomagnetic micropulsations, *Geophys. J.*, **3**, 333–353, 1960.
39. Jacobs, J. A. *Geomagnetic micropulsations.* In *Physics and Chemistry in Space*, **1**, J. G. Roederer and J. Zahringer, eds. Springer, Berlin, Heidelberg, New York, 1970.
40. Kangas, J., Guglielmi A., Pochotelov O. Morphology, and physics of short-period magnetic pulsations, *Space Sci. Rev.*, **83**, p. 432, 1998.
41. Kataoka, R., H. Fukunishi, K. Hosokawa, H. Fujiwara, A. S. Yukimatu, N. Sato, and Y.-K. Tung, Transient production of F-region irregularities associated with TCV passage, *Ann. Geophys.*, **21**, 1531–1541, 2003.
42. Kauristie, K., T. I. Pulkkinen, O. Amm, A. Viljanen, M. Syrjasuo, P. Janhunen, S. Massetti, S. Orsini, M. Candidi, J. Watermann, E. Donovan, P. Prikryl, I. R. Mann, P. Eglitis, C. Smith, W. F. Denig, H. J. Opgenoorth, M. Lockwood, M. Dunlop, A. Vaivads, and M. Andre, Ground-based and satellite observations of high-latitude auroral activity in the dusk sector of the auroral oval, *Ann. Geophys.*, **19**, 1683–1696, 2001.
43. Keiling, A., Wygant J. R., Cattell C., Kim K. H., Russell C. T., Milling DK, Temerin M, Mozer FS, and Kletzing CA., *Pi*2 pulsations observed with the Polar satellite and ground stations: Coupling of trapped and propagating fast mode waves to a midlatitude field line resonance *J. Geophys. Res.*, **106 (A11)**, 25891–25904, 2001.
44. Kim, K. H., K. Takahashi, D. H. Lee, P. R. Sutcliffe, and K. Yumoto, *Pi*2 pulsations associated with poleward boundary intensifications during the absence of substorms *J. Geophys. Res.*, **110 (A1)**, A01217, 2005.
45. Kivelson, M. and D. J. Southwood, Ionospheric traveling vortex generation by solar wind buffeting the magnetosphere, *J. Geophys. Res.*, **96**, 1661–1667, 1991.
46. Kleimenova, N. G., P. Francia, U. Villante, O. V. Kozyreva, J. Bitterly, J. J. Schott, The temporal and spatial variations of low frequency geomagnetic pulsations at polar cusp and cap latitudes *Ann. Geophys.*, **42 (4)**, 675–682, 1999.
47. Kokubun, S., McPherron R. L., Russell C. T., *J. Geophys. Res.*, **81**, 5141, 1976.
48. Kovner, M. S., Lebedev V. V., Plyasova-Bakunina T. A., and V. A Troitskaya, On the generation of low frequency waves in the solar wind in the front of the bow shock, *Planet. Space Sci.*, **24**, 261, 1976.
49. Kuwashima, M., Wave characteristics of magnetic *Pi*2 pulsations in the auroral region - Spectral and polarization studies, *Mem. Nat. Inst. Polar Res.*, Series **A**, n.15, 1978.
50. Lanzerotti, L. J., A. Shono, H. Fukunishi, C. G. Maclennan, Long-period hydromagnetic waves at very high geomagnetic latitudes *J. Geophys. Res.*, **104 (A12)**, 28423–28435, 1999.
51. Lessard, M. R., Hudson M. K., Samson J. C., and Wygant J. R., Simultaneous satellite and ground-based observations of a discretely driven field line resonance, *J. Geophys. Res.*, **104 (A6)**, 2361–12377, 1999
52. Lin, C. S., and J. N. Barfield, Azimuthal propagation of storm time *Pc*5 waves observed simultaneously by geostationary satellites GOES 2 and GOES 3, *J. Geophys. Res.*, **90**, 11075–11077, 1985.

53. Luhr, H., W. Blawert, Ground signatures of travelling convection vortices, In *Solar Wind Sources of Magnetospheric Ultra-Low-Frequency Waves*, eds. M. J. Engebretson, K. Takahashi, and M. Scholer, AGU Geophysical Monogr. Ser., **81**, 231–251, American Geophysical Union, Washington, DC, 1993.
54. Liu, Y. H., Fraser B. J., Liu R. Y., and Ponomarenko P. V., Conjugate phase studies of ULF waves in the *Pc*5 band near the cusp, *J. Geophys. Res.*, **108 (A7)**, 1274, 2003.
55. Maltzev, Yu. P., S. Leontiev, and W. B. Lyatsky, *Pi*2 pulsations as a result of evolution of an Alfvén pulse originating in the ionosphere during a brightening of aurora, *Planet. Space Sci.*, **22**, 1519–1533, 1974.
56. Mathie R. A., F. W. Menk, I. R. Mann, D. Orr, Discrete field line resonances and the Alfvén continuum in the outer magnetosphere, *Geophys. Res. Lett.*, **26**, (6), 659–662, 1999.
57. Matsuoka, H., K. Takahashi, S. Kokubun, K. Yumoto, T. Yamamoto, S. I. Solovyev, and E. F. Vershinin, Phase and amplitude structure of *Pc*3 magnetic pulsations as determined from multipoint observations *J. Geophys. Res.*, **102 (A2)**, 2391–2403, 1997.
58. Matsuoka, M., A. S. Yukimatu, H. Yamagishi, N. Sato, G. J. Sofko, B. J. Fraser, P. Ponomarenko, R. Liu, and T. Goka, Coordinated observations of *Pc*3 pulsations near cusp latitudes, *J. Geophys. Res.*, **107**, NO. A11, 1400, doi:10.1029/2001JA000065, 2002.
59. Menk F. W., C. L. Waters, B. J. Fraser, field line resonances and waveguide modes at low latitudes 1. Observations, *J. Geophys. Res.*, 105 (A4), 7747–7761, 2000.
60. McDiamid, D. R. and W. Allan, Simulation and analysis of auroral radar signatures generated by a magnetospheric cavity mode, *J. Geophys. Res.*, **95**, 20911–20922, 1990.
61. McKenzie, J. F., Hydromagnetic oscillations of geomagnetic tail and plasma sheet, *J. Geophys. Res.*, **75**, 5331–5339, 1970.
62. Messiaen, A. M. and P. E. M. Vandenplas, in *Plasma Physics and Controlled Nuclear Fussion Research*, (Int. Atomic Energy Agency, Vienna), **III**, p. 319, 1975.
63. Nose, M., Iyemori T., Nakabe S., Nagai T., Matsumoto H., Goka T., ULF pulsations observed by the ETS-VI satellite, Substorm associated azimuthal *Pc*4 pulsations on the nightside *Earth and Planet. Sci.*, **50 (1)**, 63–80, 1998.
64. Obana, Y., A. Yoshikawa, J. V. Olson, R. J. Morris, B. J. Fraser, and K. Yumoto, North-south asymmetry of the amplitude of high-latitude *Pc*3–5 pulsations: Observations at conjugate stations, *J. Geophys. Res.*, **110 (A10)**, 2005.
65. Odera, T. J., D. Van Swol, C. T. Russel, and C. A. Green, *Pc*3, 4 magnetic pulsations observed simultaneously in the magnetosphere and at multiple ground stations, *Geophys. Res. Lett.*, **18**, 1671–1674, 1991.
66. Olson, J. V., *Pi*2 pulsations and substorm onsets, A review *J. Geophys. Res.*, **104 (A8)**, 7499–17520, 1999
67. Osaki, H., K. Yumoto, K. Fukao, K. Shiokawa, F. W. Menk, B. J. Fraser, Characteristics of low-latitude *Pi*2 pulsations along the 210 degrees magnetic meridian, *Journ. Geom. Geoel.*, **48 (11)**, 1421–1430, 1996
68. Osaki, H., K. Takahashi, H. Fukunishi, T. Nagatsuma, H. Oya, A. Matsuoka, and D. K. Milling, *Pi*2 pulsations observed from the Akebono satellite in the plasmasphere, *J. Geophys. Res.*, 103 (A8), 17605–17615, 1998.

69. Poulter, E. M., W. Allan, J. G. Keys, and E. Nielsen, Plasmathrough ion mass densities determined from ULF-pulsation eigenperiods, *Planet. Space. Sci.,* **32**, 1069–1078, 1984.
70. Rae, I. J., E. F Donovan, I. R. Mann, F. R. Fenrich, C. E. J. Watt, D. K. Milling, M. Lester, B. Lavraud, J. A. Wild, H. J. Singer, H. Reme, and A. Balogh, Evolution and characteristics of global *Pc*5 ULF waves during a high solar wind speed interval, *J. Geophys. Res.*, **110 (A12)**, A12211, 2005
71. Rostoker, G. and J. C. Samson, Polarization characteristics of *Pi*2 pulsations and implications for their source mechanisms: Location of source regions with respect to the auroral electrojets, *Planet. Space Sci.*, **29**, 225–247, 1981.
72. Rothwell, P. L., M. B. Silevitch, and L. P. Block, *Pi*2 pulsations and the westward travelling surge, *J. Geophys.Res.*, **91**, 6921–6928, 1986.
73. Russel, C. T., and M. M. Hope, The dependence of upstream wave periods on the interplanetary magnetic field strength, *Geophys. Res. Lett.*, **8**, 615–617, 1981.
74. Saito, T., Sakurai T., Koyama Y., Mechanism of association between *Pi*2 pulsation and magnetospheric substorm, *J. Atmos. Terr. Phys.*, **38**, 1265, 1976.
75. Saka, O., Akaki H., Watanabe O., Baker D. N., Ground-satellite correlation of low-latitude *Pi*2 pulsations: A quasi-periodic field line oscillation in the magnetosphere *J. Geophys. Res.*, **101 (A7)**, 15433–15440, 1996
76. Saka, O., K. Okada, O. Watanabe, D. N. Baker, G. D. Reeves, R. D. Belian, *Pi*2 associated particle flux and magnetic field modulations in geosynchronous altitudes *J. Geophys. Res.*, **102 (A6)**, 11363–11373, 1997.
77. Samson, J. C., J. A. Jacobs, and G. Rostoker, Latitude dependent characteristics of long-period geomagnetic micropulsations, *J. Geophys. Res.*, **76**, 3675–3683, 1971.
78. Samson, J. C., R. A. Greenwald, J. M. Ruohoniemi, T. J. Hughes, and D. D. Wallis, Magnetometer and radar observations of MHD cavity modes in the Earth's magnetosphere, *Can. J. Phys.*, **69**, 929–937, 1991.
79. Samson, J. C., B. G. Harold, J. M. Ruohoniemi, R. A. Greenwald, and A. D. M. Walker, field line resonances associated with MHD-waveguides in the Earth's magnetosphere, *Geophys. Res. Lett.*, **19**, 441–444, 1992.
80. Samson, J. C., D. D. Wallis, T. J. Hughes, F. Greutzberg, J. M. Ruohoniemi, and R. A. Greenwald, Substorm intensifications and field line resonances in the nightside magnetosphere, *J. Geophys. Res.*, **97**, 8495–8518, 1992.
81. Samson, J. C., Cogger L. L., Pao Q., Observations of field line resonances, auroral arcs, and auroral vortex structures, *J. Geophys. Res.*, **101 (A8)**, 7373–17383, 1996.
82. Sarafopolous, D. V, and E. T. Savris Long-period standing waves at the plasma sheet boundary layer region observed by ISEE-1, *Ann. Geophys.,* **9**, 333–347, 1991.
83. Sarafopolous, D. V., E. T. Savris, Quite-time *Pc*5 pulsations in the Earth's magnetotail: IMP-8, ISEE-1 and ISEE-3 simultaneous observations, *Ann. Geoph.*, **12**, 121–138, 1994.
84. Shields, D. W., E. A. Bering, A. Alaniz, S. E. M. Mason, W. Guo, R. L. Arnoldy, M. J. Engebretson, W. J. Hughes, D. L. Murr, L. J. Lanzerotti, and C. G. Maclennan, Multistation studies of the simultaneous occurrence rate of *Pc*3 micropulsations and magnetic impulsive events, *J. Geophys. Res.*, **108 (A6)**, 1225, 2003.

85. Shimizu, N. and K. Yanagihara, Equatorial enhancement of micropulsations *Pi*2, *Mem. Kakioka Mag. Obs.*, **12**, 57, 1966.
86. Shinohara, M., K. Yumoto, N. Hosen, A Yoshikawa., H. Tachihara, O. Saka, T. I. Kitamura, N. B. Trivedi, J. M. Da Costa, N. J. Schuch, Wave characteristics of geomagnetic pulsations across the dip equator, *J. Geophys. Res.*, **103 (A6)**, 11745–11754, 1998.
87. Sinha, A. K., T. K. Yeoman, J. A. Wild, D. M. Wright, S. W. H. Cowley, A. Balogh, Evidence of transverse magnetospheric field line oscillations as observed from Cluster and ground magnetometers, *Ann. Geophys.*, **23 (3)**, 919–929, 2005.
88. Sinitsin, V. G., Y. M. Yampolski, A. V. Zalizovski, K. M. Groves, M. B Moldwin, Spatial field structure and polarization of geomagnetic pulsations in conjugate areas, *J. of Atm. Terr. Phys.*, **65 (10)**, 1161–1167, 2003.
89. *Solar Wind Sources of Magnetospheric Ultra-Low-Frequency Waves*, eds. M. J. Engebretson, K. Takahashi, and M. Scholer, AGU Geophysical Monogr. Ser., **81**, American Geophysical Union, Washington, DC, 1993.
90. Southwood, D. J., Some features of field line resonance in the magnetosphere, *Planet. Space. Sci.*, **22**, 483–491, 1974.
91. Stix, T.H., Bull. Am.Phys.Soc., **21**, 1134, 1976.
92. Su, S. Y., K. Y. Chen, Wu J. M., Yeh H. C., Chao C. K., ROCSAT observation of the field line resonance effect in a plasma pulsation at topside ionosphere, *J. Geophys. Res.*, **110 (A1)**, A01303, 2005.
93. Sugiura, M., Propagation of hydromagnetic waves in the magnetosphere, *Radio Sci.*, **69D**, 1133, 1965.
94. Sutcliffe, P. R. and K. Yumoto, Dayside *Pi*2 pulsations at low latitudes, *Geophys. Res. Lett.*, **16**, 887–890, 1989.
95. Tanaka, Y. M., K. Yumoto , A. Yoshikawa, M. Shinohara, H. Kawano, and T. I. Kitamura, Longitudinal structure of *Pc*3 pulsations on the ground near the magnetic equator, *J. Geophys. Res.*, **109 (A3)**, A03201, 2004.
96. Takahashi, K., P. R. Higbie, and D. N. Baker, Azimuthal propagation and frequency characteristic of compressional *Pc*5 waves observed at geostationary orbit, *J. Geophys. Res.*, **90**, 1473–1485, 1985.
97. Takahashi, K., B. J. Andersen, P. T. Newell, T. Yamamoto, and N. Sato, Propagation of compressional *Pc*3 pulsations from space to the ground: A registered case study using multipoint measurements. In *Solar Wind Sources of Magnetospheric Ultra-Low-Frequency Waves*, Geophysical Monograph Series, **81**, 355–373, 1994.
98. Takahashi, K., B. J. Anderson, and S.-I. Ohtani, Multisatellite study of nightside transient toroidal waves, *J. Geophys. Res.*, **101**, 24,815–24,825, 1996.
99. Takahash, K., R. R. Anderson, and W. J. Hughes, *Pi*2 pulsations with second harmonic: CRRES observations in the plasmasphere, *J. Geophys. Res.*, **108**, NO. A6, 1242, doi:10.1029/ 2003JA009847, 2003.
100. Toivanen, P. K., D. N. Baker, W. K. Peterson, H. J. Singer, J. Watermann, J. R. Wygant, C. T. Russell, and C. A. Kletzing, Polar observations of transverse magnetic pulsations initiated at substorm onset in the high-latitude plasma sheet, *J. Geophys. Res.*, **108**, NO. A7, 1267, 2003.
101. Troitskaya, V. A., T. A. Plyasova-Bakunina, and A. V. Gul'elmi, Relationship between *Pc*2 – 4 pulsations and the interplanetary field, *Dokl. Akad. Nauk USSR*, **197**, 1312, 1971.

102. Verkhoglyadova, O., A. Agapitov, A. Andrushchenko, V. Ivchenko, S. Romanov, and Y. Yermolaev, Compressional wave events in the dawn plasma sheet observed by Interball-1, *Ann. Geophys.*, **17 (9)**, 1145–1154, 1999.
103. Vellante, M., H. Luhr, T. L. Zhang, V. Wesztergom, U. Villante, M. De Lauretis, A. Piancatelli, M. Rother, K. Schwingenschuh, W. Koren, and W. Magnes, Ground//satellite signatures of field line resonance: A test of theoretical predictions, *J. Geophys. Res.*, **109 (A6)**, A06210, 2004.
104. Villante, U., P. Francia, S. Lepidi, *Pc*5 geomagnetic field fluctuations at discrete frequencies at a low latitude station, *Ann. Geophys.*, **19 (3)**, 321–325, 2001
105. Vogelsang, H., H. Lühr, H. Voelker, J. Woch, T. Bösinger, T. A. Potemra, and P. A. Lindqvist, An ionospheric travelling convection vortex event observed by ground-based magnetometers and by Viking, *Geophys. Res. Lett.*, 20, 2343–2346, 1993.
106. Walker, A. D. M., R. A. Greenwald, Pulsation structure in the ionosphere derived from auroral radar data, *J. Geom. Geoelectr.*, **32**, Suppl.II, 111, 1980.
107. Warnecke, J., H. Luhr, and K. Takahashi, Observational features of field line resonances excited by solar wind pressure variations on 4 September, 1984, *Planet. Space Sci.*, **38**, 1517–1531, 1990.
108. Wright, D. M., T. K. Yeoman, I. J. Rae, J. Storey, A. B. Stockton-Chalk, J. L. Roeder, and K. J. Trattner, Ground-based and Polar spacecraft observations of a giant (*Pg*) pulsation and its associated source mechanism , *J. Geophys. Res.*, **106 (A6)**, 10837–10852, 2001.
109. Yumoto, K. and T. Saito, Relation of compressional HM waves at GOES 2 to low-latitude *Pc*3 magnetic pulsations, *J. Geophys. Res.*, **88**, 10041–1052,1983.
110. Yumoto, K., T. Saito, B. T. Tsurutani, E. J. Smith, and S. I. Akasofu, Relationship between the IMF magnitude and *Pc*3 magnetic pulsations in the magnetosphere, *J. Geophys. Res.*, **89**, 9731–9740, 1984.
111. Yumoto, K., Y. Tanaka, K. K. Oguti, K. Y. Shiokawa, A. Isono, B. J. Fraser, F. W. Menk, K. J. W. Lynn, M. Seto, and 210° MM Magnetic Observation Group, Globally coordinated magnetic observations along the 210° magnetic meridian SDTEP period: 1. Preliminary results for low-latitude *Pc*3's , *J. Geomag. Geoelectr.*, **44**, 261–276, 1992.
112. Yumoto, K., A. Isono, K. Shiokawa, H. Matsuoka, Y. Tanaka, F. W. Menk, and J. Fraser, Global-cavity Mode-like and localized field line *Pc*3,4 oscillations stimulated by interplanetary pulses (Si/Sc): Initial results from the 210 MM magnetic observations, In *Solar Wind Sources of Magnetospheric Ultra-Low-Frequency Waves*, Geophysical Monograph Series, **81**, 335–344, 1994.
113. Yumoto, K., Generation and propagation of low-latitude magnetic pulsations: A review, *J. Geophys.*, **60**, 79, 1986.
114. Zanandrea, A., J. M Da Costa, S. L. G Dutra, N. B. Trivedi, T. Kitamura, K. Yumoto, H. Tachihara, M. Shinohara, O. Saotome, *Pc*3,4 geomagnetic pulsations at very low latitude in Brazil Planet. Space Sci., **52 (13)**, 1209–1215, 2004.
115. Zhu, X. and M. Kivelson, Compressional ULF waves in the outer magnetosphere: 2. A case study of *Pc*5 type wave activity, *J. Geophys. Res.* **99**, 241–252, 1994.

4

Magnetohydrodynamic Waves

4.1 MHD Equations

Basic Equations

In Chapter 1 the macroscopic electrodynamic parameters of plasma were obtained in the framework of multifluid hydrodynamics. Taking into account the multicomponent composition of partly ionized ionospheric plasma is necessary because many important peculiarities of wave perturbations cannot be studied in the approximation of one-fluid magnetohydrodynamics. In the magnetosphere, on the other hand, a one-fluid approximation is usually sufficient and it is more important to consider the finiteness of plasma pressure than its complex composition.

The magnetospheric plasma will be regarded here as a single hydrodynamic medium characterized by a single hydrodynamic velocity $\mathbf{u}(\mathbf{r}, t)$ and plasma density $\rho\,(\mathbf{r}, t)$. Then their relations with partial velocities $\mathbf{v}_k(\mathbf{r}, t)$ and concentrations $N_k(\mathbf{r}, t)$ are given by

$$\mathbf{u}\,(\mathbf{r}, t) = \sum \frac{m_k N_k(\mathbf{r}, t)}{\rho(\mathbf{r}, t)} \mathbf{v}_k(\mathbf{r}, t),$$

$$\rho(r, t) = \sum m_k N_k(\mathbf{r}, t).$$

The variables ρ and $\mathbf{u}$ obey the hydromagnetic equations:

- Equation of mass conservation

$$\frac{\partial \rho}{\partial t} + \boldsymbol{\nabla} \cdot (\rho \mathbf{u}) = 0\,; \tag{4.1}$$

- Equation of motion for conductive gas

$$\rho \left(\frac{\partial \mathbf{u}}{\partial t} + (\mathbf{u} \boldsymbol{\nabla})\, \mathbf{u} \right) = -\boldsymbol{\nabla} P + \frac{1}{c} \left[\mathbf{j} \times \mathbf{B} \right]. \tag{4.2}$$

Here P is the gas kinetic pressure, $\mathbf{j}$ is the current density.

The hydrodynamic equations should be supplemented by Faraday's induction law

$$\nabla \times \mathbf{E} = -\frac{1}{c}\frac{\partial \mathbf{B}}{\partial t}, \qquad (4.3)$$

and by Ampere's law

$$\nabla \times \mathbf{B} = \frac{4\pi}{c}\mathbf{j}. \qquad (4.4)$$

Applying operator $\nabla \cdot$ to (4.3), we obtain

$$\frac{\partial}{\partial t}\nabla \cdot \mathbf{B} = 0, \qquad (4.5)$$

so that it is sufficient to accept the condition of magnetic field solenoidality

$$\nabla \cdot \mathbf{B} = 0 \qquad (4.6)$$

as an initial condition. The electric and magnetic fields are fully determined by (4.3), (4.4) and by Ohm's law

$$\mathbf{j} = \sigma\left(\mathbf{E} + \frac{1}{c}\left[\mathbf{u} \times \mathbf{B}\right]\right), \qquad (4.7)$$

where σ is plasma conductivity. For an ideal plasma, when $\sigma \to \infty$, it follows from (4.7) that

$$\mathbf{E} = -\frac{1}{c}\left[\mathbf{u} \times \mathbf{B}\right]. \qquad (4.8)$$

Equations (4.1)–(4.8) should be supplemented by the state equation

$$P = P(\rho, T), \qquad (4.9)$$

as well as by the heat transfer equation, which for an ideal medium, when energy dissipation can be neglected, is given by

$$\frac{ds}{dt} = \frac{\partial s}{\partial t} + (\mathbf{u}\nabla)\, s = 0, \qquad (4.10)$$

where s is the entropy density. Since $P = P(\rho, s)$, then s can be found as a function of P and ρ and (4.10) for the isentropic conditions becomes

$$\frac{\partial P}{\partial t} = c_s^2\frac{\partial \rho}{\partial t}, \qquad c_s^2 = \left(\frac{\partial P}{\partial \rho}\right)_{s=\text{const}}, \qquad (4.11)$$

where c_s is the adiabatic sound velocity. The set of equations (4.1)–(4.11) form a complete system allowing the description and comprehension of most of the effects arising in the propagation of ULF-waves in the magnetosphere and in the upper ionosphere.

Equilibrium Configurations

The characteristics of hydromagnetic wave propagation are determined in linear approximation by equilibrium distributions of plasma and of the magnetic field. The main objective of the present section is to expound only the basic concepts of equilibrium plasma configurations. We shall not consider many important questions arising in the study of equilibrium plasma configurations, referring the reader to numerous works where problems of plasma equilibrium in a magnetic field are discussed (e.g. [14], where the mathematical treatment of the magnetic equilibrium and its applications to the sun and planets with a magnetic field are given).

According to (4.2), plasma can be in equilibrium only if its pressure gradient is balanced by Ampere's force

$$\boldsymbol{\nabla} P = \frac{1}{c}\left[\mathbf{j} \times \mathbf{B}\right]. \tag{4.12}$$

Expressing $\mathbf{j}$ in terms of $\mathbf{B}$ with the help of (4.4) and using the vector identity

$$\frac{1}{2}\boldsymbol{\nabla}\left(\mathbf{B}\cdot\mathbf{B}\right) = \left(\mathbf{B}\boldsymbol{\nabla}\right)\mathbf{B} + \left[\mathbf{B}\times\left[\boldsymbol{\nabla}\times\mathbf{B}\right]\right],$$

the equilibrium equation (4.12) can be written as

$$\boldsymbol{\nabla}\left(P + P_m\right) - \frac{1}{4\pi}\left(\mathbf{B}\boldsymbol{\nabla}\right)\mathbf{B} = 0. \tag{4.13}$$

Here

$$P_m = \frac{B^2}{8\pi}$$

can be interpreted as a magnetic pressure and

$$\frac{1}{4\pi}(\mathbf{B}\boldsymbol{\nabla})\mathbf{B}$$

as a tension along field-lines.

A geometric presentation of equilibrium equations can be obtained if a unit vector $\mathbf{h} = \mathbf{B}/B$, directed along the magnetic field, is introduced. Then (4.13) becomes

$$\boldsymbol{\nabla} P + \boldsymbol{\nabla}_{\perp}\frac{B^2}{8\pi} = \frac{B^2}{4\pi}\left(\mathbf{h}\boldsymbol{\nabla}\right)\mathbf{h} = \frac{B^2}{4\pi R}\mathbf{n}, \tag{4.14}$$

where $\mathbf{n}$ is normal to the field-line, R is its curvature and $\boldsymbol{\nabla}_{\perp} = \boldsymbol{\nabla} - \mathbf{h}(\mathbf{h}\boldsymbol{\nabla})$ is the transverse gradient. It follows from (4.14) that the magnetic field exerts pressure $B^2/8\pi$ in the transverse direction and creates an additional force in the direction of the field line concavity as a result of their tension.

Consider a simple equilibrium configuration often used in modeling wave processes in the magnetosphere as well as in other plasma objects. Let the magnetic field-lines be straight. Axis z in the Cartesian coordinate system $\{x,\ y,\ z\}$ is then directed parallel to the field-lines.

Denote by $\hat{\mathbf{x}}, \hat{\mathbf{y}}, \hat{\mathbf{z}}$ the unit vectors of the coordinate system. Since the magnetic field has only one component parallel to axis z, $\mathbf{B} = B\hat{\mathbf{z}}$, the solenoidality $\boldsymbol{\nabla} \cdot \mathbf{B} = 0$ yields

$$\frac{\partial \mathbf{B}}{\partial z} = 0, \qquad \mathbf{B} = \mathbf{B}\,(x, y)\,.$$

Then the equilibrium equation (4.14) reduces to

$$\boldsymbol{\nabla}_{\perp}\left(P + \frac{B^2}{8\pi}\right) = 0, \qquad \frac{\partial P}{\partial z} = 0. \tag{4.15}$$

It follows from the second equation in (4.15) that pressure P is constant along the field-line, $P = P(x, y)$, and the first equation in (4.15) yields

$$P(x, y) + \frac{B^2(x, y)}{8\pi} = \text{const}\,. \tag{4.16}$$

Condition (4.16) is valid only if Ampere's force is balanced by the plasma pressure gradient. If forces of a different nature, for instance, gravitational, are taken into account, plasma density and pressure can change along the field-lines.

Linear Approximation

Most effects linked to ULF-propagation in the magnetosphere can be understood within the limits of linear approximation. Consider small perturbations of plasma near the equilibrium. Denote by $\mathbf{B}_0$ the magnetic field equilibrium value, by P_0 – plasma pressure and by ρ_0 – plasma density. The undisturbed electric field $\mathbf{E}_0$ and plasma velocity $\mathbf{u}_0$ will be taken as zero. Put

$$\mathbf{B} = \mathbf{B}_0 + \mathbf{b}, \quad P = P_0 + p, \quad \rho = \rho_0 + \rho_1 .$$

$\mathbf{b}$, $\mathbf{E}$, p, ρ_1, $\mathbf{u}$ are small perturbations near equilibrium. Substituting these expressions into (4.1)–(4.4), (4.6), (4.8), taking into account equilibrium conditions and neglecting all the degrees of perturbations except the first, we obtain

$$\frac{\partial \rho_1}{\partial t} + \boldsymbol{\nabla} \cdot (\rho_0 \mathbf{u}) = 0, \tag{4.17a}$$

$$\rho_0 \frac{\partial \mathbf{u}}{\partial t} = -\boldsymbol{\nabla}\left(c_s^2 \rho_1\right) + \frac{1}{4\pi}\left[\left[\boldsymbol{\nabla} \times \mathbf{B}_0\right] \times \mathbf{b}\right] + \frac{1}{4\pi}\left[\left[\boldsymbol{\nabla} \times \mathbf{b}\right] \times \mathbf{B}_0\right], \tag{4.17b}$$

$$\frac{\partial \mathbf{b}}{\partial t} = \boldsymbol{\nabla} \times \left[\mathbf{u} \times \mathbf{B}_0\right], \qquad \boldsymbol{\nabla} \cdot \mathbf{b} = 0, \tag{4.17c}$$

$$\mathbf{E} = -\frac{1}{c}\left[\mathbf{u} \times \mathbf{B}_0\right], \tag{4.17d}$$

where $c_s = c_s(\rho_0)$ is sound velocity in an equilibrium configuration.

System (4.17a)–(4.17d) can be rewritten in a convenient form in terms of displacements $\boldsymbol{\xi}$, given by

$$\mathbf{u} = \frac{\partial \boldsymbol{\xi}}{\partial t}. \tag{4.18}$$

Substituting this expression for $\boldsymbol{\xi}$ into (4.17a) and (4.17c) and integrating the obtained ratios with respect to time, give density and magnetic field perturbations in the form

$$\rho_1 = -\boldsymbol{\nabla} \cdot (\rho_0 \boldsymbol{\xi}) \,, \qquad \mathbf{b} = \boldsymbol{\nabla} \times [\boldsymbol{\xi} \times \mathbf{B}_0] \,. \tag{4.19}$$

The first relation means that the plasma density at the certain point diminishes proportionally to the quantity of liquid flowing out of the volume.

Transform the second equation in (4.19). Present the displacement vector as the sum

$$\boldsymbol{\xi} = \boldsymbol{\xi}_{||} + \boldsymbol{\xi}_{\perp},$$

where $\boldsymbol{\xi}_{||}$ is the displacement component parallel to $\mathbf{B_0}$ and $\boldsymbol{\xi}_{\perp}$ is perpendicular to $\mathbf{B}_0$. As the vector product of the collinear vectors $\boldsymbol{\xi}_{||}$ and $\mathbf{B}_0$ is zero, only $\boldsymbol{\xi}_{\perp}$ remains in the expression for $\mathbf{b}$. If to apply $\boldsymbol{\nabla}\times$ to the equation for $\mathbf{b}$ in (4.19), then we find

$$\mathbf{b} = -\mathbf{B}_0 \boldsymbol{\nabla} \cdot \boldsymbol{\xi}_{\perp} - (\boldsymbol{\xi}_{\perp} \boldsymbol{\nabla})\, \mathbf{B}_0 + (\mathbf{B}_0 \boldsymbol{\nabla})\, \boldsymbol{\xi}_{\perp}, \tag{4.20}$$

where it is accounted that $\boldsymbol{\nabla} \cdot \mathbf{B}_0 = 0$. The first term here is proportional to the density perturbations ρ_1. Equation (4.20) shows that as plasma is compressed across the initial magnetic field $\mathbf{B}_0$, field-lines are frozen into the plasma and are displaced so that the magnetic field in the direction of $\mathbf{B}_0$ is intensified proportionally to the plasma compression. As displacement $\boldsymbol{\xi}_{\perp}$ is changed in the direction of $\mathbf{B}_0$, the field-lines frozen into the plasma are bent. The magnetic field acquires a transverse component $(\mathbf{B}_0 \boldsymbol{\nabla})\boldsymbol{\xi}_{\perp}$ as a result.

Substitution of (4.19) and (4.20) into (4.17b) yields one vector equation for displacement

$$\begin{aligned} \rho_0 \frac{\partial^2 \boldsymbol{\xi}}{\partial t^2} &= \boldsymbol{\nabla} \{ c_s^2 \boldsymbol{\nabla} \cdot (\rho_0 \boldsymbol{\xi}) - \frac{1}{4\pi} \mathbf{B}_0 \cdot [\boldsymbol{\nabla} \times (\boldsymbol{\xi} \times \mathbf{B}_0)] \} \\ &+ \frac{1}{4\pi} \{ [\boldsymbol{\nabla} \times (\boldsymbol{\xi} \times \mathbf{B}_0) \cdot \; \boldsymbol{\nabla}]\, \mathbf{B}_0 + (\mathbf{B}_0 \cdot \boldsymbol{\nabla})\, \boldsymbol{\nabla} \times (\boldsymbol{\xi} \times \mathbf{B}_0) \} \,. \end{aligned} \tag{4.21}$$

This equation is rather cumbersome and requires special mathematical methods for investigating hydromagnetic wave propagation in sufficiently general equilibrium configurations. However, many important effects can result by studying hydromagnetic oscillations and waves in the simplest model. In such model the field-lines of an equilibrium magnetic field are assumed to be straight and embedded into the plasma with density dependent on one transverse coordinate only. The rest of this chapter and Chapter 5 are devoted to such 1D problems. The investigation of the general case when the curvature of the field-lines as well as the 2D or 3D inhomogeneity of the plasma are taken into account is performed in Chapter 6.

Suppose that the magnetic field $\mathbf{B}_0 = B_0\hat{\mathbf{z}}$. Then, if no external forces are acting on the plasma, the pressure is $P_0 = P_0(x, y)$ and the magnetic field is $\mathbf{B}_0 = \mathbf{B}_0(x, y)$. Using the vector equality

$$\nabla \times [\boldsymbol{\xi}_\perp \times \mathbf{B}_0] = B_0 \frac{\partial \boldsymbol{\xi}_\perp}{\partial z} - \hat{\mathbf{z}} \left[B_0 \nabla \cdot \boldsymbol{\xi}_\perp + (\boldsymbol{\xi}_\perp \cdot \nabla_\perp) B_0 \right]$$

we reduce (4.21) to

$$\begin{aligned} \mathbf{L}\boldsymbol{\xi}_\perp &= -\frac{1}{\rho_0 c_A^2} \nabla_\perp \Big[\rho_0 \left(c_A^2 + c_s^2\right) \nabla \cdot \boldsymbol{\xi}_\perp \\ &\quad + \boldsymbol{\xi}_\perp \cdot \left(c_s^2 \nabla_\perp \rho_0 + \nabla_\perp \frac{\rho_0 c_A^2}{2} \right) + \rho_0 c_s^2 \frac{\partial \boldsymbol{\xi}_{||}}{\partial z} \Big], \end{aligned} \tag{4.22}$$

$$\mathbf{L}_s \boldsymbol{\xi}_{||} = -\frac{1}{\rho_0 c_s^2} \frac{\partial}{\partial z} \left[c_s^2 \, \nabla \cdot (\rho_0 \boldsymbol{\xi}_\perp) + \boldsymbol{\xi}_\perp \cdot \nabla_\perp \frac{\rho_0 c_A^2}{2} \right], \tag{4.23}$$

where $c_A^2 = B_0^2/4\pi\rho_0$ is the Alfvén velocity,

$$\mathbf{L} = \frac{\partial^2}{\partial z^2} - \frac{1}{c_A^2} \frac{\partial^2}{\partial t^2},$$

$$\mathbf{L}_s = \frac{\partial^2}{\partial z^2} - \frac{1}{c_s^2} \frac{\partial^2}{\partial t^2}.$$

4.2 Homogeneous Plasma

Basic Equations

Propagation of small-amplitude hydromagnetic waves in homogeneous plasma is discussed in most textbooks on plasma physics and magnetohydrodynamics ([5], [6], [15], [18]), and even more so in works specially devoted to electromagnetic wave propagation in plasma ([9], [19]). The basic properties of these waves are briefly described in this section. Hydromagnetic waves in the homogeneous case will be analyzed in a way convenient for generalization on inhomogeneous media.

In a homogeneous medium $\nabla\rho_0, \nabla c_A$, and ∇c_s vanish and (4.22) – (4.23) reduce to

$$\mathbf{L}\boldsymbol{\xi}_\perp = -\nabla_\perp \left[(1+\beta) \; \nabla \cdot \boldsymbol{\xi}_\perp + \beta \frac{\partial \boldsymbol{\xi}_{||}}{\partial z} \right], \tag{4.24}$$

$$\mathbf{L}_s \boldsymbol{\xi}_{||} = -\frac{\partial}{\partial z} \; \nabla \cdot \boldsymbol{\xi}_\perp, \tag{4.25}$$

where $\beta = c_s^2/c_A^2$.

Perturbations of hydrodynamic variables ρ_1, $\mathbf{u}$ and electromagnetic field components $\mathbf{E}\left\{\mathbf{E}_\perp, E_\parallel\right\}$ and $\mathbf{b}\left\{\mathbf{b}_\perp, b_\parallel\right\}$ are expressed in terms of displacement $\boldsymbol{\xi}$ as

$$\begin{aligned} \frac{\rho_1}{\rho_0} &= -\boldsymbol{\nabla}\cdot\boldsymbol{\xi}, & \frac{\mathbf{b}_\perp}{B_0} &= \frac{\partial \boldsymbol{\xi}_\perp}{\partial z}, & \frac{b_\parallel}{B_0} &= -\boldsymbol{\nabla}\cdot\boldsymbol{\xi}_\perp, \\ \mathbf{u} &= \frac{\partial \boldsymbol{\xi}}{\partial t}, & \mathbf{E}_\perp &= -\frac{1}{c}\mathbf{u}\times\mathbf{B}_0, & E_\parallel &= E_z = 0. \end{aligned} \tag{4.26}$$

Here $\mathbf{b}_\perp$, $\mathbf{E}_\perp$ and $b_\parallel$, $E_\parallel$ are magnetic and electric wave components perpendicular and parallel to the ambient magnetic field $\mathbf{B}_0$.

Alfvén Waves

A wave with zero perturbations of density ρ_1 and field-aligned displacement $\xi_\parallel$ can be excited in plasmas. As a consequence, the wave field-aligned magnetic component $\mathbf{b}_\parallel = -\mathbf{B}_0\boldsymbol{\nabla}\cdot\xi_\perp$ also vanishes. In this case the wave should carry with it a non-zero vortex

$$\boldsymbol{\Omega}_\parallel = [\boldsymbol{\nabla}\times\mathbf{u}]_\parallel$$

and a non-zero longitudinal component of current density

$$\mathbf{j}_\parallel = \frac{c}{4\pi}[\boldsymbol{\nabla}\times\mathbf{b}]_\parallel ,$$

where

$$[\boldsymbol{\nabla}\times\mathbf{b}]_\parallel = \left([\boldsymbol{\nabla}\times\mathbf{b}]\cdot\frac{\mathbf{B}_0}{B_0}\right).$$

This wave mode is called the Alfvén wave [1]. Equations for the Alfvén waves can be found from (4.24)–(4.25). Transverse displacement $\boldsymbol{\xi}_\perp$ in an Alfvén wave is found from

$$\mathbf{L}\boldsymbol{\xi}_\perp = \frac{\partial^2\boldsymbol{\xi}_\perp}{\partial z^2} - \frac{1}{c_A^2}\frac{\partial^2\boldsymbol{\xi}_\perp}{\partial t^2} = 0. \tag{4.27}$$

In addition, since $\rho_1 = 0$, then it follows from (4.26) that $\boldsymbol{\xi}_\perp$ must be solenoidal

$$\boldsymbol{\nabla}\cdot\boldsymbol{\xi}_\perp = 0. \tag{4.28}$$

For an individual field-line, (4.27) coincides with the elastic string equation. According to (4.27), transverse displacement in an Alfvén wave propagating in the positive z direction is

$$\boldsymbol{\xi}_\perp(x,y,z,t) = \boldsymbol{\xi}_+(x,y,\zeta), \quad \zeta = z - c_A t, \tag{4.29}$$

where $\boldsymbol{\xi}_+(x,y,\zeta)$ at each fixed ζ is an arbitrary solenoidal 2D vector field orthogonal to $\mathbf{B}_0$. From $\boldsymbol{\nabla}\cdot\boldsymbol{\xi}_\perp = 0$ it follows that $\boldsymbol{\xi}_\perp$ can be expressed in terms of a scalar potential $\Psi(x,y,\zeta)$:

$$\boldsymbol{\xi}_{\perp} = \boldsymbol{\nabla} \times \Psi \hat{\mathbf{z}}.$$

Due to the plasma being frozen into the magnetic field $\mathbf{B}_0$, transverse displacements result in the appearance of the transverse magnetic component $\mathbf{b}$, given by

$$\frac{\mathbf{b}_{\perp}}{B_0} = \frac{d}{d\zeta}\boldsymbol{\xi}_{+}(x,y,\zeta)\,. \tag{4.30}$$

The electric field $\mathbf{E}$ in the Alfvén wave is

$$\mathbf{E} = -\frac{c_A}{c}\left[\hat{\mathbf{z}} \times \mathbf{b}_{\perp}\right]. \tag{4.31}$$

As an example, consider two kinds of Alfvén waves. The first is a linearly polarized Alfvén wave of frequency ω with displacements, say, in plane (x,z), the potential Ψ is independent on x and

$$\xi_x = \frac{\partial \Psi(y,\varsigma)}{\partial y}, \qquad \xi_y = 0.$$

Non-zero components of displacement, velocity, magnetic and electric fields and field-aligned current are, correspondingly,

$$\begin{aligned}
\xi_x &= \xi_0(y)\cos\left(\frac{\omega}{c_A}z - \omega t\right), \quad v_x = \omega\xi_0(y)\sin\left(\frac{\omega}{c_A}z - \omega t\right),\\
\frac{b_x}{B_0} &= -\frac{\omega}{c_A}\xi_0(y)\sin\left(\frac{\omega}{c_A}z - \omega t\right), \quad c\frac{E_y}{B_0} = -\omega\xi_0(y)\sin\left(\frac{\omega}{c_A}z - \omega t\right),\\
\frac{4\pi}{c}\frac{j_{\parallel}}{B_0} &= -\frac{\omega}{c_A}\xi_0(y)\sin\left(\frac{\omega}{c_A}z - \omega t\right),
\end{aligned} \tag{4.32}$$

where ξ_0 is a given distribution of displacements ξ along y. The field-lines disturbed by a linearly polarized Alfvén wave are shown in Fig. 4.1a.

Let $\{\varrho,\varphi,z\}$ be the cylindrical coordinate system with z axis directed along an external magnetic field $\mathbf{B}_0$. For the torsional wave, the potential Ψ is axially symmetric and it depends only on the radial distance ϱ. Then the displacement is pure azimuthal $\xi_\varphi(\varrho) = -\partial\Psi/\partial\varrho$ and the harmonics are

$$\xi_\varphi(\varrho) = \xi_0(\varrho)\cos\left(\frac{\omega}{c_A}z - \omega t\right).$$

The disturbance of the magnetic field is also purely azimuthal

$$\frac{b_\varphi}{B_0} = -\frac{\omega}{c_A}\xi_0(\varrho)\sin\left(\frac{\omega}{c_A}z - \omega t\right).$$

The field-lines lie on the coaxial cylindrical surfaces (see Fig. 4.1b).

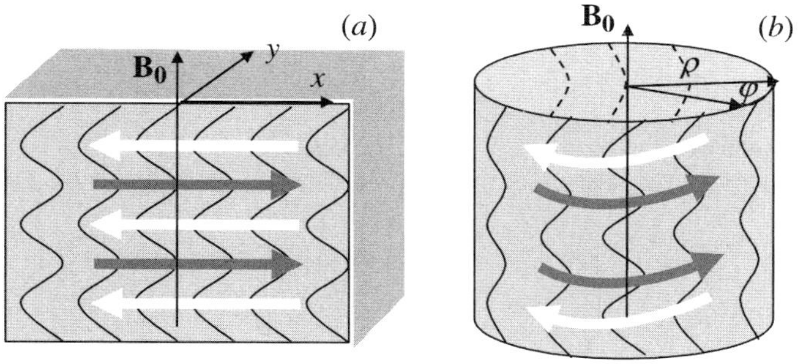

Fig. 4.1. Sketches of the magnetic field-lines disturbed by (*a*) linearly polarized shear Alfvén wave and (*b*) by a torsional shear Alfvén wave

Arbitrary initial perturbation is carried along the field-lines at Alfvén velocity with constant shape and without spreading in the transverse direction. Guiding by field-lines is probably the most important property of Alfvén waves which is unchanged in inhomogeneous plasmas and even under finite curvature of field-lines.

An arbitrary Alfvén perturbation in homogeneous plasma can be presented as a superposition of disturbances of non-interacting field-lines. For a field line with coordinates (x_0, y_0) displacement can be written as

$$\boldsymbol{\xi}_+ (x, y) = \boldsymbol{\nabla} \times \delta (x - x_0,\, y - y_0)\, \hat{\mathbf{z}},$$

where $\delta(x,\, y)$ is the Dirac delta-function. However, in an inhomogeneous plasma, interaction arises between perturbations propagating along adjacent field-lines and a bundle of interacting field-lines is excited as a result. Guiding along field-lines is conserved in this case, as well.

Magnetosonic Waves

Consider two other types of waves with vortex $\boldsymbol{\Omega}_{\|}$ and longitudinal current $\mathbf{j}_{\|}$ vanish. The longitudinal magnetic field $\mathbf{b}_{\|}$, transversal compression $\boldsymbol{\nabla} \cdot \boldsymbol{\xi}_{\perp}$ and longitudinal displacement $\boldsymbol{\xi}_{\|}$ are non-zero. It is convenient to use the perturbation of the longitudinal magnetic field and longitudinal displacement in the description of these waves. Applying transverse operator $\boldsymbol{\nabla}_{\perp} \cdot$ to (4.24) and (4.25), we obtain for $b_{\|}$ and $\xi_{\|}$:

$$\left(\boldsymbol{\nabla}^2 - \frac{1}{c_A^2} \frac{\partial^2}{\partial t^2} \right) \frac{b_{\|}}{B_0} = -\beta\, \boldsymbol{\nabla}_{\perp}^2 \left(\frac{b_{\|}}{B_0} - \frac{\partial \xi_{\|}}{\partial z} \right), \tag{4.33}$$

$$\left(\frac{\partial^2}{\partial z^2} - \frac{1}{c_s^2} \frac{\partial^2}{\partial t^2} \right) \xi_{\|} = \frac{\partial}{\partial z} \frac{b_{\|}}{B_0}. \tag{4.34}$$

In low-pressure plasma,

$$\beta = \frac{c_s^2}{c_A^2} << 1.$$

Therefore, neglecting all terms proportional to β, (4.33) and (4.34) reduce to

$$\left(\boldsymbol{\nabla}^2 - \frac{1}{c_A^2}\frac{\partial^2}{\partial t^2}\right)\frac{b_{\|}}{B_0} = 0, \tag{4.35}$$

$$\left(\frac{\partial^2}{\partial z^2} - \frac{1}{c_s^2}\frac{\partial^2}{\partial t^2}\right)\xi_{\|} = \frac{\partial}{\partial z}\frac{b_{\|}}{B_0}. \tag{4.36}$$

Equation (4.35) describes the FMS-waves. Medium elasticity in a FMS-wave is created by magnetic pressure $P_m = B_0^2/8\pi$. Basing on analogy with adiabatic gas-dynamic sound velocity

$$c_s^2 = \gamma\frac{P_0}{\rho_0}$$

(γ is the adiabatic index) write the Alfvén velocity as

$$c_A^2 = \frac{B_0^2}{4\pi\rho_0} = \gamma_m\frac{P_m}{\rho_0},$$

where $\gamma_m = 2$. The condition $\gamma_m = 2$ is linked to the field being frozen into the plasma, owing to which the magnetic field at $\xi_{\|} = 0$ is changed in proportion to plasma density, while its pressure is proportional to plasma density squared. The longitudinal displacement $\xi_{\|}$ is small in the FMS-wave in low β plasmas. It follows from (4.36) that $\mathbf{b}_{\|}$ propagating at Alfvén velocity produces $\xi_{\|}$ proportional to β, since perturbations $\mathbf{b}_{\|}$ move with Alfvén velocity $c_A \gg c_s$.

Perturbations with

$$\mathbf{b}_{\|} = 0 \quad \text{and} \quad \boldsymbol{\nabla}_{\perp}\cdot\boldsymbol{\xi}_{\perp} = 0$$

are described by the uniform equation (4.36). These are ion sound waves propagating at velocity c_s along the magnetic field. Plasma perturbations occur only along the magnetic field.

Dispersion Equation

Three hydromagnetic wave modes can propagate in homogeneous plasma:

1. Alfvén waves: plasma density is undisturbed and field-lines are twisted due to plasma transverse displacements.
2. Fast magnetosonic (FMS) waves: compressions and rarefactions of field-lines accompanied by plasma density variations. For low-temperature plasma, when $\beta << 1$, longitudinal displacements of plasma in FMS-waves are small, $\xi_{\|} \propto \beta$.

3. Slow magnetosonic waves: ion sound which at $\beta << 1$ almost does not disturb the magnetic field. The plasma is seemingly 1D-density perturbations caused by longitudinal plasma displacements are transmitted along the magnetic field at sound velocity c_s.

If the parameter β is finite, magnetosonic waves cannot be precisely separated into the FMS and ion sound. Coupling equations (4.33) and (4.34) must be investigated. Let us apply Fourier transform decomposing the signal into its harmonic components

$$\xi_{\perp} \exp(-i\omega t + i\mathbf{kr}) \qquad \text{and} \qquad b_{\|} \exp(-i\omega t + i\mathbf{kr}).$$

Equations (4.27), (4.28) for Alfvén waves and (4.33), (4.34) for FMS-waves reduce to:

Alfvén waves

$$\left(k_{\|}^2 - \frac{\omega^2}{c_A^2}\right) \xi_{\perp} = 0, \quad (\mathbf{k} \cdot \boldsymbol{\xi}_{\perp}) = 0. \tag{4.37}$$

Magnetosonic waves

$$ik_{\|} k_{\perp}^2 \beta \xi_{\|} + \left(\frac{\omega^2}{c_A^2} - k^2 - \beta k_{\perp}^2\right) \frac{b_{\|}}{B_0} = 0, \tag{4.38a}$$

$$\left(\frac{\omega^2}{c_s^2} - k_{\|}^2\right) \xi_{\|} - ik_{\|} \frac{b_{\|}}{B_0} = 0, \tag{4.38b}$$

where $k_{\perp}$ and $k_{\|}$ are the values of the perpendicular and parallel to the magnetic field $\mathbf{B}_0$ components of the wave vector $\mathbf{k}$.

A uniform linear system of algebraic equations has a non-trivial solution only in the case of the zero determinant Δ of this system. Since the determinant $\Delta \equiv \Delta(\omega, k)$ is a function of ω and $\mathbf{k}$, this condition results in a functional relation between frequency ω and wave vector $\mathbf{k}$, designated as the dispersion equations which are given by

Alfvén waves

$$\omega^2 - k_{\|}^2 c_A^2 = 0. \tag{4.39}$$

Magnetosonic waves

$$\omega^4 - \left(c_A^2 + c_s^2\right) k_{\|}^2 \omega^2 + c_A^2 c_s^2 k_{\|}^2 k^2 = 0. \tag{4.40}$$

A presentation of magnetohydrodynamic waves in homogeneous plasma can be given in the form of an angle dependency of phase velocity $V_{ph} = \omega/k$ on angle θ between $\mathbf{k}$ and $\mathbf{B}_0$. Then substitution $k_{\|} = k \cos\theta$ into (4.39) and (4.40) leads to

$$V_{ph}^2 = V_A^2 = c_A^2 \cos^2\theta$$

for the Alfvén waves and

$$V_{\pm}^4 - \left(c_A^2 + c_s^2\right) V_{\pm}^2 + c_A^2 c_s^2 \cos^2\theta = 0$$

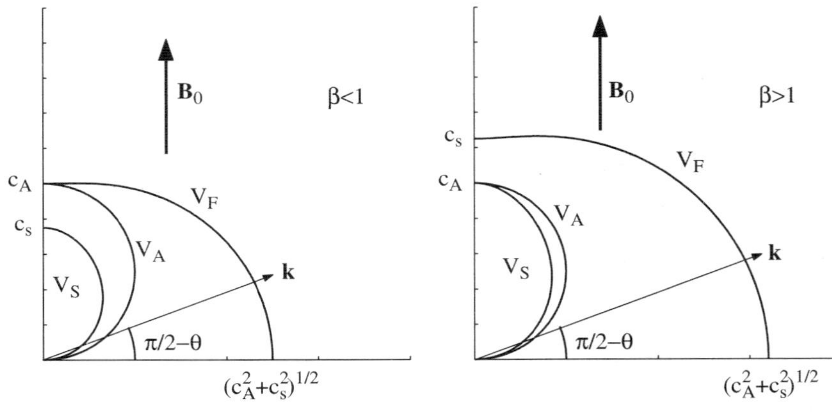

Fig. 4.2. Polar diagrams of phase velocities for the Alfvén and magnetosonic waves

for the magnetosonic waves. Solutions for these equations are

$$V_{ph} = V_A = c_A \left|\cos\theta\right| , \tag{4.41a}$$

$$V_{\pm} = c_A \left[\frac{1+\beta^2}{2} \pm \sqrt{\left(\frac{1-\beta^2}{2}\right)^2 + \beta^2 \sin^2\theta} \right]^{1/2} . \tag{4.41b}$$

The root $V_F = V_+$ given by the upper sign corresponds to the FMS-wave, and by the lower sign $V_S = V_-$ to the Slow MagnetoSonic (SMS) wave.

Dependencies of phase velocities of an Alfvén wave V_A, an FMS-wave V_F and an SMS wave V_S on the angle between the wave vector and the external magnetic field are given by the polar diagrams in Fig. 4.2 (left panel at $c_A > c_s$, right panel at $c_A < c_s$). The polar for an FMS-wave is an oval compressed along the magnetic field direction, and for an SMS-wave it has the shape of two osculating ovals compressed transversely. At $\beta \to 0$ the FMS-polar is transformed into a circle of radius c_A, while the SMS-polar turns into two osculating circles of radius c_s. In the limiting case of incompressible liquid, when $\beta \to \infty$, the polar diagram for the SMS is transformed into two osculating circles of diameter c_s. A similar diagram corresponds to an Alfvén wave at any β.

4.3 Inhomogeneous Plasma

Basic Equations Cold Plasma

If a wavelength is significantly less than the characteristic scale of spatial inhomogeneity in plasma, the wave propagation can be described within the ray approximation, which is sufficiently universal and applicable to many interesting wave phenomena. However, a number of important effects observed in space and laboratory plasmas cannot be studied in the ray approximation. Full-wave equations must be used in order to comprehend these effects. We shall concentrate on the analysis of the full-wave equation mainly on one-dimensional problems, where sufficiently simple and obvious results can be obtained by using elementary mathematical methods. In Chapter 5 a 2D problem will be considered, in the elementary case of straight field-lines.

The z-axis of the Cartesian coordinate system $\{x, y, z\}$ is chosen along the external magnetic field $\mathbf{B}_0 = B_0\hat{\mathbf{z}}$. At $\beta \ll 1$ and $B_0(x, y) = \text{const}$, the equations (4.22), (4.23) for an inhomogeneous low-pressure plasma become

$$\mathbf{L}\boldsymbol{\xi}_\perp = -\boldsymbol{\nabla}_\perp(\boldsymbol{\nabla} \cdot \boldsymbol{\xi}_\perp), \tag{4.42a}$$

$$\mathbf{L}_s\boldsymbol{\xi}_\parallel = -\frac{1}{\rho_0 c_s^2}\frac{\partial}{\partial z}\left[c_s^2\boldsymbol{\nabla} \cdot (\rho_0\boldsymbol{\xi}_\perp)\right]. \tag{4.42b}$$

Solutions of (4.42a) for an inhomogeneous plasma are coupling Alfvén and FMS-waves, and (4.42b) are ion sound waves. Just as in the homogeneous case, transversal compression of plasma in FMS-wave results in a small longitudinal displacement $\xi_\parallel$ which is found as a forced solution to inhomogeneous equation (4.42b). Since $\beta \ll 1$, the ion sound wave propagation velocity is much less than Alfvén velocity, and from (4.42b) it follows that $\xi_\parallel \propto \beta$.

The longitudinal displacement $\xi_\parallel$ in Alfvén and FMS-waves vanishes.

1D Case

Let Alfvén velocity depends only on x-coordinate:

$$c_A = c_A(x).$$

Consider unbounded plasma. Turn to the Fourier presentation both in time t and in the $\{y, z\}$ coordinates. Fourier harmonics of transverse displacement is

$$\xi_\perp(x)\exp(-i\omega t + ik_y y + ik_\parallel z)$$

and longitudinal perturbation of the magnetic field is

$$b_\parallel(x)\exp(-i\omega t + ik_y y + ik_\parallel z).$$

Equation (4.42a) becomes

$$\frac{db_\parallel}{dx} = B_0\left(k_A^2(x) - k_\parallel^2\right)\xi_x, \tag{4.43}$$

$$\frac{d\xi_x(x)}{dx} = -ik_y\xi_y(x) - \frac{b_\parallel(x)}{B_0}, \tag{4.44}$$

$$\left(k_A^2(x) - k_\parallel^2\right)\xi_y(x) = ik_y\frac{b_\parallel(x)}{B_0}, \tag{4.45}$$

where $k_A^2(x) = \omega^2/c_A^2$.

Uncoupled Waves Phase Mixing

We first consider the case $k_y = 0$. Then the coupled equations (4.43)–(4.45) split into two non-interacting subsystems

$$\frac{db_\parallel}{dx} = B_0 \left(k_A^2 - k_\parallel^2\right) \xi_x, \tag{4.46}$$

$$\frac{d\xi_x}{dx} = -\frac{b_\parallel}{B_0} \tag{4.47}$$

for the FMS-wave and

$$\left(k_A^2(x) - k_\parallel^2\right) \xi_y = 0. \tag{4.48}$$

for the Alfvén wave.

Uncoupled equation (4.48), despite its extreme simplicity, describes an important phenomenon arising in Alfvén wave propagation – the so-called phase mixing, essential for the comprehension of wave processes in space and laboratory inhomogeneous plasma. The essence of the effect linked to Alfvén wave guiding in inhomogeneous plasma is as follows. Alfvén waves propagate independently along individual field-lines, at its phase velocity $c_A(x)$ on each line. The phase difference between the Alfvén waves propagating on close field-lines increases proportionally to distance z.

This results in the growth of the transverse wavenumber k_x in the direction of the inhomogeneity and in the decrease of the transverse scale. The scale diminishes until either the dissipative effects (e.g. viscosity and electric conductivity) lead to wave damping, or the transversal dispersion results in the transformation of large-scale Alfvén oscillations into small-scale modes. Both dissipation and dispersion increase with k_x. Heating by an Alfvén wave dissipating due to phase mixing is often adduced for explaining the heating of space and laboratory plasma. For instance, solar coronal heating can be connected with phase mixing in an Alfvén wave arising due to subphotospheric convection and propagating along the magnetic field of the coronal hole.

In the simple case of non-interacting Alfvén and FMS-waves phase mixing can be obtained from (4.48) in the space-time domain

$$\frac{\partial^2 \xi_y}{\partial z^2} - \frac{1}{c_A^2}\frac{\partial^2 \xi_y}{\partial t^2} = 0. \tag{4.49}$$

Suppose that a source of Alfvén waves is turned on in plane $z = 0$ at time $t = 0$. If the plasma density is independent of y, then the displacement ξ_y in the wave traveling towards positive z can be written as

$$\xi_y(x, y, t) = \xi_{y0}(x) f_+(z - c_A t),$$

where the function f_+ is determined by the time dependence of a signal in the source. For harmonic perturbations

$$\xi_y(x, y, t) \propto \exp[i(-\omega t + \Phi_0(x) + k_A(x) z)],$$

where $\Phi_0(x)$ is the phase in the source at $t = 0$, $z = 0$,

$$k_A(x) = \frac{\omega}{c_A(x)}$$

is the Alfvén wavenumber. At distance x from the source, the phase of the Alfvén wave is

$$\Phi(x, y, t) = -\omega t + \Phi_0(x) + k_A(x)z.$$

The transverse wavenumber

$$k_x(x) = \frac{\partial \Phi}{\partial x} = k_x^{(0)} - k_A\,(x)\,z\frac{d}{dx}\ln c_A\,(x) \tag{4.50}$$

varies in proportion to z. Here $k_x^{(0)} = \partial\Phi_0(x)/\partial x$.

In the magnetosphere, $c_A \sim 10^8$ cm/s and the scale-size of Alfvén velocity change is about Earth's radius $R_E \approx 6 \times 10^8$ cm. As is known, the finiteness of the Larmour radius and the electron inertia cause the transversal dispersion of Alfvén waves. For the ion temperature $T_i \approx 0.1$ eV, the ion thermal velocity is $v_{Ti} \approx 10^5$ cm/s. The ion cyclotron frequency ω_{ci} in the magnetosphere is $\omega_{ci} \sim 10$ Hz. Then the Alfvén wavenumber for which the transversal dispersion is essential, can be estimated as $k_A = \omega_{ci}/c_A \approx 10^{-9}\,\mathrm{cm}^{-1}$. Then the transverse wavenumber varies as

$$\Delta k_x \approx \frac{10^{-9}}{R_E} z \approx 2 \times 10^{-18} z.$$

Let $z = 5 \times 10^9$ cm be the length of a mid-latitudinal field-line. Then $\Delta k_x \sim 2 \times 10^{-18} \times 5 \times 10^9\,\mathrm{cm}^{-1} = 10^{-8}\,\mathrm{cm}^{-1} = 10^{-3}\,\mathrm{km}^{-1}$ and the scale size of the phase mixing is $L_\perp \approx 10^3$ km.

The transversal dispersion near the equatorial plane of the magnetosphere becomes significant at transversal scales of about the proton cyclotron radius, i.e. $\sim 10^6$ cm. The dispersion scale is determined by the electron inertial length λ_e which is estimated as a ratio of the light velocity c to the electron plasma frequency ω_{pe}. Near the ionosphere $\lambda_e = c/\omega_0 \sim 10^4$ cm. The dissipation caused by the longitudinal conductivity becomes significant at scales less than 10^5 cm. Thus, the phase mixing is unimportant at middle latitudes. However, at high latitudes, especially for field-lines going to the geomagnetic tail, the phase mixing can become important.

If the properties of the medium vary in the direction of plasma displacement, then eliminating $b_\parallel$ from (4.46) and (4.47), we have a 2D wave equation for displacements ξ_x

$$\frac{\partial^2 \xi_x}{\partial x^2} + \frac{\partial^2 \xi_x}{\partial z^2} - \frac{1}{c_A^2}\frac{\partial^2 \xi_x}{\partial t^2} = 0. \tag{4.51}$$

It is evident that a variation of c_A in the plane of polarization does not produce phase mixing. Recall that dependence on y is determined by the choice of the source and is not connected to the peculiarities of propagation.

Coupled Waves Phase Mixing

At $k_y \neq 0$ interaction takes place between Alfvén and FMS-waves. From (4.43) and (4.45) we obtain

$$\xi_x(x) = \frac{1}{k_A^2(x) - k_\|^2} \frac{\partial}{\partial x} \frac{b_\|(x)}{B_0},$$

$$\xi_y(x) = \frac{ik_y}{k_A^2(x) - k_\|^2} \frac{b_\|(x)}{B_0}.$$

Then (4.44) becomes

$$\frac{d}{dx} \frac{1}{k_A^2(x) - k_\|^2} \frac{db_\|(x)}{dx} + \frac{k_A^2(x) - k_y^2 - k_\|^2}{k_A^2(x) - k_\|^2} b_\|(x) = 0 \tag{4.52}$$

or

$$\frac{d^2 b_\|(x)}{dx^2} - \frac{dk_A^2/dx}{k_A^2(x) - k_\|^2} \frac{db_\|(x)}{dx} + \left(k_A^2(x) - k_y^2 - k_\|^2\right) b_\|(x) = 0. \tag{4.53}$$

In order to study the coupling between these two modes, let us consider first the problem of wave incidence on a stratified plasma medium. Suppose Alfvén velocity $c_A(x)$ increases monotonically from c_1 at $x \to -\infty$ to c_2 at $x \to +\infty$, with Alfvén wave number $k_A(x)$ decreasing from k_1 to k_2 (Fig. 4.3, top left panel). Let the incident wave amplitude be b_0. Introduce reflection and transmission coefficients R and T. Then at $x \to -\infty$, the wave magnetic field $\mathbf{b}$ is the sum of the incident (i) and reflected (r) waves

$$b_\| = b_\|^{(i)} + b_\|^{(r)},$$

where

$$b_\|^{(i)} = \exp\left(-i\omega t + ik_x^{(1)}x + ik_y y + ik_\| z\right) b_0,$$

$$b_\|^{(r)} = R\exp\left(-i\omega t - ik_x^{(1)}x + ik_y y + ik_\| z\right) b_0$$

and at $x \to +\infty$

$$b_\|^{(t)} = T\exp\left(-i\omega t + ik_x^{(2)}x + ik_y y + ik_\| z\right) b_0$$

with $k_x^{(1,2)} = \sqrt{k_{1,2}^2 - k_y^2 - k_\|^2}$.

The problems of oblique incidence of FMS-waves on a layer and oblique incidence of an electromagnetic wave of H-polarization on a plane-layered medium are similar ([11], [12]). The electric field $\mathbf{E}$ in the H-polarized wave is within the plane of incidence, while the magnetic field is perpendicular to

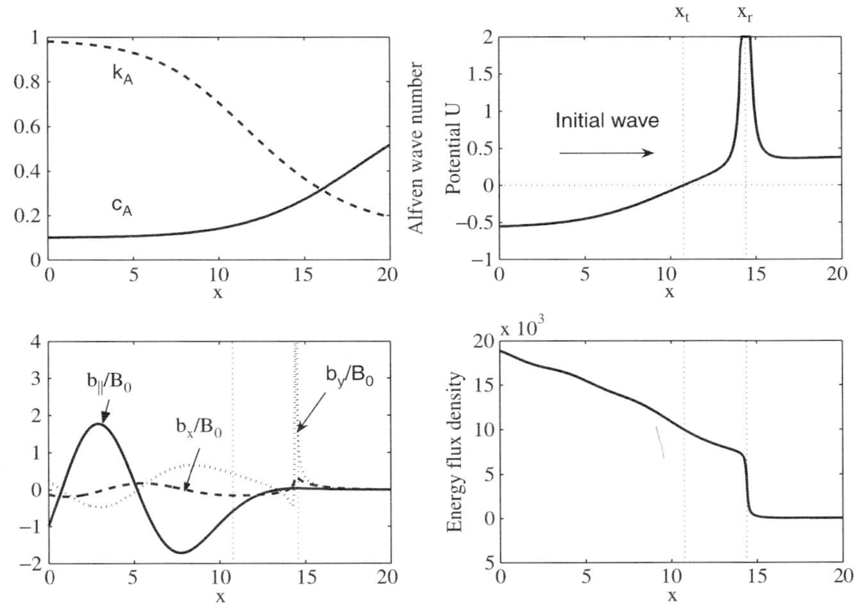

Fig. 4.3. Left top frame: the Alfvén velocity (solid line) and Alfvén wave number (dashed line) as a function of x. Right top frame: Effective potential $U(x)$. The initial wave propagates from left to right, from the region with low Alfvén velocity to high velocity. Two vertical dotted lines show the location of the turning point (x_t) and resonance point (x_r). Left bottom: wave magnetic field near a resonance surface. Right bottom: Energy flux to resonance surfaces

it. The incidence of an H-polarized wave onto a layered medium has been considered and solved completely in ([7], [8]).

The qualitative pattern of the fields can be obtained if we write (4.53) in the form of the Schrödinger wave equation. Substitution $b_{\|} = u\sqrt{k_A^2 - k_{\|}^2}$ into (4.53) gives

$$\frac{d^2u(x)}{dx^2} - U(x)u(x) = 0, \tag{4.54}$$

where

$$U = k_y^2 + k_{\|}^2 - k_A^2 - \frac{1}{2}\frac{d^2k_A^2}{dx^2}\frac{1}{k_A^2 - k_{\|}^2} + \frac{3}{4}\left(\frac{dk_A^2}{dx}\right)^2 \frac{1}{\left(k_A^2 - k_{\|}^2\right)^2}.$$

The left top panel of Fig. 4.3 presents the Alfvén velocity $c_A(x)$ and wavenumber $k_A(x)$. Dependence of the effective potential $U(x)$ on the transversal coordinate x is shown in the right top panel. The coordinate x and $k_A(x)$ are normalized on a scale $l_{\perp}/2$. The function $U(x)$ is calculated at frequency

$\omega = 0.1\,\mathrm{s}^{-1}$, $k_y = 0.5$ and $k_{\|} = 0.4$. Chosen parameters roughly correspond to magnetospheric conditions for the McIllwain parameter $L \sim 2$–5.

The wave coming from $-\infty$ reaches the turning (reflection) point at $x = x_t$ determined from condition $U(x_t) = 0$ and is reflected in it. In magnetospheric problems $x_t \approx \widetilde{x}_t$, which can be found from the approximate equation

$$k_A^2\left(\widetilde{x}_t\right) = k_y^2 + k_{\|}^2. \tag{4.55}$$

In the example in Fig. 4.3 $x_t = 10.76$ and $\widetilde{x}_t = 10.53$. Below x_t and $\widetilde{x}_t$ will not be distinguished.

The wave to the left of the reflection point is a superposition of the propagating incident and reflected waves. In the region to the right of the reflection point there is the exponential wave damping to the resonance point $x = x_r$ determined by the equation

$$k_A^2(x_r) = k_{\|}^2. \tag{4.56}$$

At the field line $x = x_r = 14.41$ the condition of resonance coupling between FMS and Alfvén waves is fulfilled (FLR). Resonance field-lines form the resonance magnetic surface being a plane $x = x_r$ for $1D$-geometry with homogeneous magnetic field $\mathbf{B}_0$. The local Alfvén velocity $c_A(x_r)$ coincides with phase velocity projection on the magnetic field

$$V_{ph} = \frac{\omega}{k_{\|}}, \qquad c_A\left(x_r\right) = V_{ph}.$$

When this point is approached from the left (from the region of small c_A values), magnetic field b_x and b_y increases and under zero dissipation approaches infinity at point x_r. Actually, the magnetic wave component remains finite due to dissipation and dispersion.

For the purpose of clarity, we begin with the qualitative exposition of the results. At oblique wave incidence ($k_y \neq 0$) on the layer in the vicinity of the reflection point x_t the field behaves in the same way as with $k_y = 0$ (see Fig. 4.3). There is a standing wave to the left of the reflection point, while to the right the wave dampens exponentially. Then the amplitude is increasing (Fig. 4.3) with the wave propagating into the layer and approaching the resonance point.

The growth is limited by the dissipation or dispersion. However, the longitudinal magnetic field $b_{\|}$ proves to be limited in the case of straight field lines, even under ideal conditions. On the other hand, we show in Chapter 6 that taking into account the curvature of magnetic field lines may result in unlimited growth of the $\mathbf{b}_{\|}$ component. The magnetic field distributions in Fig. 4.3 are calculated with the account of dissipation caused by the transversal conductivity. The Alfvén wavenumber is, in this case, complex, $\omega/c_A + i\kappa$.

The singularity disappears at $k_y = 0$. But under small dissipation the resonant field increases with k_y becoming progressively less expressed. This is because of the distance between the reflection point x_t and the resonance point x_r increases with the growth of k_y (see 4.55). Since the field outside the resonance region decreases exponentially, with decrement proportional to k_y, the FLR-amplitude decreases at large k_y. Inasmuch as the effect vanishes at $k_y = 0$ and $k_y = \infty$, it is most vividly expressed at intermediate values of k_y.

Phase mixing for Alfvén waves noninteracting with FMS-waves was studied above in (4.2). Suppose here that ξ_x, ξ_y and their derivatives are given in plane $z = 0$ and ξ_x, $\xi_y \propto \exp(-i\omega t + ik_y y)$. Waves are sought that depart in the positive direction of axis z. Note that an analogy exists between the stationary problem of perturbation propagation along axis z and the temporal problem (4.43) – (4.45). The latter is considered closely in ([4], [16], [17])

It is satisfactory here to take advantage of the results obtained in the study of the time evolution of the initial perturbation. We shall describe the asymptotic behavior of waves at large z. Omitting the details (see [16], [17]), we present the conclusions ensuing from this analogy

1. At $z \to \infty$ the component ξ_x can be presented as the sum of two parts. The first are, usually, a slowly exponentially damping collective modes. The second are the modes diminishing by the power law
$$\xi_x = 0(1/z) \quad .$$
Each of them propagates along its respective field line with phase velocity $c_A(x)$. The collective mode arises as a surface wave if there is a surface near which the wave is concentrated and along which it propagates. It can appear also as a cavity mode which fills the whole volume.
2. Asymptotically the component ξ_y decays weakly
$$\xi_y\,(z \to \infty) = \xi_y = 0(1).$$
It can be presented as a sum of the collective mode and the modes propagating along each field line with Alfvén velocity $c_A\,(x)$.
3. Alfvén waves are guided by the magnetic field and can propagate over large distances.
4. A scale shortening is the common property of MHD-waves propagating in inhomogeneous plasma.

These results obtained for a 1D-system remain valid for more complex plasma configurations as well, when the many-dimensional inhomogeneity of the Alfvén velocity, the curvature and convergence of the magnetic field-lines are essential.

Fields in the Vicinity of a Resonance Shell

Consider the behavior of the wave in the vicinity of a resonance field line proceeding from the system of equations (4.43) - (4.45). The solution will be sought by the Frobenius method (e.g.,[6], [20]). From (4.45) we readily see that

$$\xi_y(x) = \frac{ik_y}{k_A^2 - k_{\|}^2} \frac{b_{\|}(x)}{B_0}.$$

Substituting this expression into (4.44) results in

$$\frac{d}{dx}\frac{b_{\|}(x)}{B_0} = \left(k_A^2 - k_{\|}^2\right)\xi_x(x),$$
$$\frac{d}{dx}\xi_x(x) = \left(-1 + \frac{k_y^2}{k_A^2 - k_{\|}^2}\right)\frac{b_{\|}(x)}{B_0}. \tag{4.57}$$

Let the condition of FLR (4.56) be fulfilled at $x = x_1$, i.e. function $k_A^2(x) - k_{\|}^2$ has a simple zero at $x = x_1$. Then the point x_1 is a regular singular point of (4.57). Rewrite (4.57) in the canonical form. For this multiply each equation of the system by $x - x_1$. The obtained equations in matrix form are

$$(x - x_1)\frac{d\mathbf{U}}{dx} = \mathbf{T}(x)\mathbf{U}(x), \tag{4.58}$$

where

$$\mathbf{T}(x) = \begin{pmatrix} 0 & (x - x_1)\left(k_A^2(x) - k_{\|}^2\right) \\ -(x - x_1) + \dfrac{k_y^2(x - x_1)}{k_A^2(x) - k_{\|}^2} & 0 \end{pmatrix},$$
$$\mathbf{U} = \begin{pmatrix} b_{\|}/B_0 \\ \xi_x \end{pmatrix}.$$

If Alfvén velocity changes smoothly near the resonance point, then $\mathbf{T}(x)$ can be expanded into a power series

$$\mathbf{T}(x) = \mathbf{T}_0 + \mathbf{T}_1(x - x_1) + \mathbf{T}_2(x - x_1)^2 + \cdots,$$
$$\mathbf{T}_1 = \begin{pmatrix} 0 & 0 \\ -1 + q_1 & 0 \end{pmatrix}, \quad \mathbf{T}_n = \begin{pmatrix} 0 & s_n \\ q_n & 0 \end{pmatrix}, \quad n = 0, 2, 3, \ldots,$$

where

$$q_0 = \frac{k_y^2}{k_A^2(x_1)'}, \quad q_1 = -k_y^2 \frac{k_A^2(x_1)''}{2\left(k_A^2(x_1)'\right)^2},$$
$$q_2 = k_y^2 \left\{ \frac{\left(k_A^2(x_1)''\right)^2}{4\left(k_A^2(x_1)'\right)^3} - \frac{k_A^2(x_1)'''}{6\left(k_A^2(x_1)'\right)^2} \right\},$$
$$s_0 = 0, \quad s_1 = 0, \quad s_2 = k_A^2(x_1)', \; k_A^2(x_1)' = \left.\frac{dk_A^2(x_1)}{dx}\right|_{x=x_1}.$$

The fundamental matrix solution of (4.58) can be presented as (e.g. [20])

$$\mathbf{U}(x) = \mathbf{P}(x)\exp(\mathbf{T_0}\ln(x - x_1)),$$

where

$$\mathbf{P}(x) = \mathbf{1} + (x - x_1)\mathbf{P}_1 + (x - x_1)^2\mathbf{P}_2 + \cdots$$

is a regular matrix function of x and $\mathbf{1}$ is a unit matrix. By virtue of equality,

$$\mathbf{T}_0^2 = \mathbf{0}$$

($\mathbf{0}$ is a zero matrix) in the expansion of the exponent into a power series all the powers above the first turn into zero and we obtain

$$\begin{aligned}\mathbf{U}(x) = &\left(\mathbf{1} + (x - x_1)\mathbf{P}_1 + (x - x_1)^2\mathbf{P}_2 + \cdots\right)\\ &\times \left(\mathbf{1} + \mathbf{T}_0 \ln(x - x_1)\right). \end{aligned} \tag{4.59}$$

Calculations shown in Chapter 6 yield

$$\frac{b_\parallel(x)}{B_0} = D_1 k_A^2(x_1)' R_1(x) + D_2 \left\{ S_1(x) + \widetilde{R}_1^2 \right\}, \tag{4.60a}$$

$$\xi_x(x) = D_1 R_2(x) + D_2 \left\{ S_2(x) + \frac{\widetilde{R}_2^2}{k_A^2(x_1)'} \right\}, \tag{4.60b}$$

$$\xi_y(x) = D_1 i k_y R_3(x) + D_2 \frac{i k_y}{k_A^2(x_1)'} \left\{ \frac{1}{x - x_1} + S_3(x) + \widetilde{R}_3^2 \right\}. \tag{4.60c}$$

Here we denote $\widetilde{R}_i^2 = k_y^2 R_i(x) \ln\left[k_y(x - x_1)\right]$, $i = 1, 2, 3$. Functions $R_n(x)$ and $S_n(x)$ have the form

$$R_1(x) = \frac{(x - x_1)^2}{2} + \cdots, \qquad R_2(x) = 1 + k_y^2 \frac{(x - x_1)^2}{4} + \cdots,$$

$$R_3 = \frac{x - x_1}{2} + \cdots, \qquad S_1(x) = 1 - k_y^2 \frac{(x - x_1)^2}{4} + \cdots,$$

$$S_2(x) = -\left(1 + \frac{k_y^2 k_A^2(x_1)''}{2\left(k_A^2(x_1)'\right)^2}\right)(x - x_1) + \cdots, \quad S_3(x) = -\frac{k_A^2(x_1)''}{2k_A^2(x_1)'} + \cdots.$$

The coefficients D_1 and D_2 are arbitrary constants which can be found from the boundary conditions. Relations connecting electric $\mathbf{E}$ and magnetic $\mathbf{b}$ fields with displacement $\boldsymbol{\xi}$ reduce for a plane wave $\exp(i\omega t - i\mathbf{kr})$ to (see 4.32)

$$\frac{E_x}{B_0} = i\frac{\omega}{c}\xi_y, \quad \frac{E_y}{B_0} = -i\frac{\omega}{c}\xi_x,$$

$$\frac{b_x}{B_0} = i k_\parallel \xi_x, \quad \frac{b_y}{B_0} = -i k_\parallel \xi_y, \quad \frac{b_\parallel}{B_0} = -\boldsymbol{\nabla} \cdot \boldsymbol{\xi}_\perp.$$

The main singularity of the field near the resonance point x_1 can be found without determining the coefficients D_1 and D_2. Note that at $x \to x_1$ perturbations of the longitudinal magnetic field $b_\parallel$ tend to a constant b_0 and have a logarithmic singularity

$$\frac{b_{\parallel}}{B_0} \approx \frac{b_0}{B_0} \left[1 + k_y^2 \frac{(x - x_1)^2}{2} \ln \left[k_y (x - x_1) \right] \right] . \tag{4.61}$$

Then plasma displacements are given by

$$\xi_x \approx \frac{b_0}{B_0} \frac{k_y^2}{k_A^2(x_1)'} \ln \left[k_y (x - x_1) \right] , \quad \xi_y \approx \frac{b_0}{B_0} \frac{i k_y}{k_A^2(x_1)'} \frac{1}{x - x_1} . \tag{4.62}$$

Contrary to the longitudinal magnetic component which is finite when approaching the resonance field-line, transversal displacements, electric and magnetic components tend to infinity.

Let us include a small dissipation in order to determine the rules for bypassing the singular point and for estimating energy losses at it. Consider, for instance, plasma with small but finite transverse conductivity $\sigma_{\perp}$. The Alfvén wavenumber in a lossless plasma is

$$k_A^2(x) = \frac{\omega^2}{c_A^2(x)} .$$

Take into account transverse conductivity replacing $k_A^2(x)$ by

$$k_A^2 + i\kappa^2 ,$$

with

$$\kappa^2 = \omega \frac{4\pi \sigma_{\perp}}{c^2} .$$

FLR-condition (4.56) can be rewritten

$$k_A^2(x) + i\kappa^2 = k_{\parallel}^2 . \tag{4.63}$$

Condition (4.63) cannot be satisfied with real ω, $k_{\parallel}$ and x. If ω and $k_{\parallel}$ are real variables, then (4.63) has no solution under real x and it is necessary to extend the function $k_A^2(x)$ from the real axis x onto the complex plane $w = x + iv$. Thus, there should exist a complex function $k_A(w)$ such that

$$k_A(w \to x) \to k_A(x) .$$

Substitute $w_1 = x_1 + i\delta$ into (4.63), assuming κ^2 and δ small. Expanding $k_A^2(x_1 + i\delta)$ in a power series of δ and leaving only terms of the first order, we obtain

$$k_A^2(x_1 + i\delta) \approx k_A^2(x_1) + i \left. \frac{dk_A^2(x)}{dx} \right|_{x=x_1} \delta , \tag{4.64}$$

and from (4.63) find

$$\delta \approx - \frac{\kappa^2}{\left. \frac{dk_A^2(x)}{dx} \right|_{x=x_1}} = \frac{\kappa^2 c_A^4(x_1)}{\omega^2 \left. \frac{dc_A^2(x)}{dx} \right|_{x=x_1}} . \tag{4.65}$$

Replacing x_1 by $x_1 + i\delta$ in (4.60a) – (4.60c) and (4.61) – (4.62), we obtain expressions for plasma displacements and magnetic field near the resonance point x_1. For example, from (4.62) we get

$$\xi_y \propto 1/[x-(x_1+i\delta)] \quad \text{and} \quad b_y \propto 1/[x-(x_1+i\delta)].$$

The direction for bypassing the singular point with small losses is determined by the sign of the Alfvén velocity derivative. If $c_A(x)$ increases with x, then $\delta > 0$ and the singular point is bypassed from below. But, if $c_A(x)$ decreases with x, then $\delta < 0$ and the singular point is bypassed from above. Phases of ξ_y, E_x and b_y change rapidly through $+\pi$ with increasing $c_A(x)$ while passing the resonance point in the positive direction of x and in the opposite case of the decreasing of $c_A(x)$, the phase change is $-\pi$.

Energy Losses

Let us find the relation between energy losses on a resonance magnetic surface, and the amplitude of the longitudinal magnetic field on this surface. Consider the most interesting case of low losses in the medium. It is important that as small as we wish conductivity results in finite energy dissipation at a resonance field-line. Energy losses can be found by calculating the difference between the Pointing vector ($\mathbf{S}$) components normal to the resonance surface to its left and right.

We shall find here the energy dissipation on the resonance surface by direct integrating of Joule losses. Consider a cylinder crossed by the resonance surface and the cross-section of a unit area. Integrate the averaged Joule losses over the cylinder. Losses caused by the longitudinal conductivity vanish at $\sigma_{\parallel} \to \infty$. The averaged density of Joule dissipation caused by the transversal conductivity is given by

$$\frac{1}{2}\left(\mathbf{j}_{\perp} \cdot \mathbf{E}_{\perp}^{*}\right) = \frac{\sigma_{\perp}}{2}\left(\mathbf{E}_{\perp} \cdot \mathbf{E}_{\perp}^{*}\right).$$

The E_y component having a logarithmic singularity at point x_1 does not contribute to losses at $\sigma_{\perp} \to 0$. Therefore, the Joule dissipation at small δ is

$$\Delta S = \frac{1}{2}\int_{-\infty}^{+\infty} \sigma_{\perp} |E_x|^2 \, dx \quad = \frac{\sigma_{\perp}}{2}\left\{\frac{\omega}{c}|b_0|\frac{k_y}{k_A^2(x_1)'}\right\}^2 \int_{-\infty}^{+\infty} \frac{dx}{(x-x_1)^2+\delta^2}.$$

The scale of plasma density variation at $x = x_1$ can be estimated as

$$a(x_1) = \frac{\rho(x_1)}{\rho'(x_1)},$$

where the prime denotes the derivation with respect to x. Substituting the value of the integral and

$$k_A^2(x_1)' \approx \frac{k_A^2(x_1)}{a} = \frac{k_{\parallel}^2}{a}$$

in terms of α into the expression for $\varDelta S$, we obtain

$$\varDelta S \approx \frac{\omega\,|a(x_1)|}{8}\frac{k_y^2}{k_{||}^2}\,|b_0|^2\,. \tag{4.66}$$

The amplitude b_0 and location of the FLR-point x_1 depend, in (4.66), on frequency ω and wavenumber k_y. For a complete solution of the problem either a numerical or an analytical solution of (4.53) or (4.58) are necessary. The right bottom panel of Fig. 4.3 shows the results of numerical calculations of the Pointing vector. The wave energy loss is equal to the divergence of the Pointing vector.

It is important, that the energy flux is finite also behind the turning point. The tunneling effect supports the energy input to the resonance surface. A finite energy per unit area equal to the Pointing vector discontinuous jump at the resonance surface dissipates in a thin resonance layer. Behind the resonance surface the energy flux becomes small and it vanishes at $\sigma_\perp \to 0$. Under non-monotonous $c_A(x)$ distribution more than one resonance surfaces can exist and the energy flux behind the first resonance surface is finite.

References

1. Alfvén, H., Ark. Mat., *Astr. Fysik*, **B 29**(2), 1942.
2. Alfvén, H., C.-G. Falthammar, *Cosmical electrodynamics*, 2nd ed., Oxford, Clarendon Press, 1963.
3. Boyce, W. E. and R. C. DiPrima, *Elementary Differential Equations and Boundary Value Problems*, 7th edn., John Wiley & Sons, Inc., New York, 2001.
4. Chen, L. and A. Hasegawa, A theory of long-period magnetic pulsations, 1. Steady state excitation of field line resonance, *J. Geophys. Res.*, **79**, 1024–1032, 1974.
5. Chandrasekhar, S. *Hydrodynamic and Hydromagnetic Stability*, Oxford University Press,Oxford, England, 1961.
6. Cowling, T. G., *Magnetohydrodynamics*, 2nd edn., Adam Hilger, London, 1976.
7. Denisov, N. G., On the one feature of the electromagnetic wave propagating in the inhomogeneous plasma., *JETP*, **31**, 609, 1956.
8. Försterling, K, and Wüster, N. O., Über die Entstehung von Oberwellen in der Ionospheäre, *J. Atmos. Terr. Phys.*, **2**, 22, 1951.
9. Ginzburg, V. L., *The Propagation of Electromagnetic Waves in Plasmas*, Pergamon, Oxford, 1970.
10. Ginzburg, V. L., Rukhadze A. A., *Waves in Magnetoactive Plasma*, Hand Physics, **49**, Springer-Verlag, New York, 1972.
11. Guglielmi, A.V. and V. A. Troitskaya, *Geomagnetic Pulsations and Diagnostics of the Magnetosphere*, Nauka, Moscow, 1973.
12. Guglielmi, A.V., Diagnostics of the magnetosphere and interplanetary medium by means of pulsations, *Space Sci. Rev.*, **16**, 331, 1974.
13. Nishida, A., *Geomagnetic Diagnosis of the Magnetosphere*, Springer-Verlag, New York, 1978.

14. Parker, E. N., *Cosmical Magnetic Fields.* Their origin and their activity, Claredon Press, Oxford, 1979.
15. *Handbook of Plasma Physics*, eds. M. N. Rosenbluth and R. Z. Sagdeev, **2**: *Basic Plasma Physics*, II, ed. by A. A. Galeev and R. N. Sudan, North-Holland Physics Publishers, Amsterdam, 1984.
16. Sedlachek, Z., Electrostatic oscillations in cold inhomogeneous plasma, 1. Differential equation approach, *J. Plasma Phys.*, **5**, 239, 1971.
17. Sedlachek, Z., Electrostatic oscillations in cold inhomogeneous plasma, 2. Integral equation approach, *J. Plasma Phys.*, **6**, 187, 1971.
18. Spitzer, L., *Physics of Fully Ionized Gases*, 2nd edn., Interscience Publ., New York-London, 1962.
19. Stix, T. H. *Theory of plasma waves*, McGraw-Hill Book Company, Inc. New York, 1963.
20. Wasow, W., *Asymptotic Expansions for Ordinary Differential Equations,* Robert E. Krieger Publishing Company, Huntington, New York, 1976.

5

Hydromagnetic Resonators

5.1 Model and Basic Equations

'Box' Model

In the previous chapter we have been mainly concerned with the behavior of MHD-waves in inhomogeneous plasma for the special case of an unbounded medium. This led to important simplification in the theory and enabled us to study some peculiarities of the electromagnetic field and plasma displacements in the vicinity of a resonance shell. A substantial feature of the real magnetosphere, disregarded in Chapter 4, is the finiteness of the magnetospheric system in the direction of the magnetic field-lines. In this chapter we abandon the assumption of medium unboundedness and consider a model of a bounded MHD-medium. We follow ([17], [19]) who proposed a simple model of an MHD-box to study the interaction between MHD-waves of different kinds. We shall consider the waves within the MHD-box in Cartesian coordinates, with significant reduction in algebraic complexity. The model enables us to reveal new important peculiarities of hydrodynamic perturbations in magnetospheric plasma. Particularly, it gives the simplest way to understand the principal features of field-line resonance (FLR).

In the box model, the dipole geomagnetic field is replaced by a uniform field. Thus, we treat it as a straightened geomagnetic field. The transfer from the dipole field to the hydromagnetic box is shown on two panels of Fig. 5.1. Consider then the rectangular coordinate system $\{x, y, z\}$ with z directed along the uniform external magnetic field $\mathbf{B}_0$. Axis x corresponds to the radial direction and y corresponds to the azimuthal direction. Cold magnetized plasma ($\beta \ll 1$) is inside a parallelepiped with ribs l_x, l_y, and l_z. Plasma density and Alfvén velocity depend only on the x coordinate, i.e. $\rho_0 = \rho_0(x)$ and $c_A = c_A(x)$. Coordinate x corresponds to the radial coordinate in the equatorial plane of the magnetosphere, the face $x = 0$ is the equatorial region of the ionosphere, the face $x = l_x$ is the outer boundary of

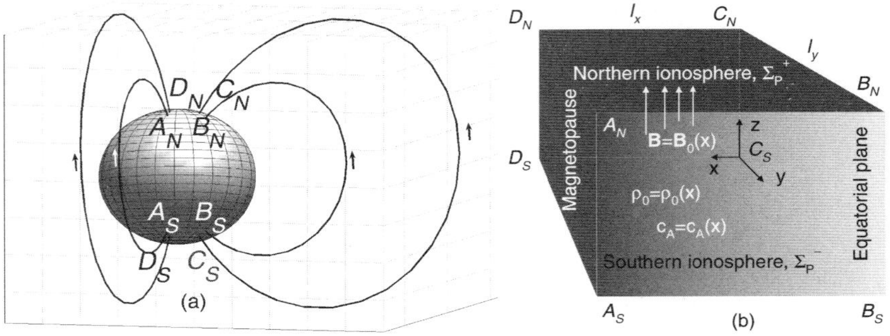

Fig. 5.1. Sketch of the transformation of the dipole geomagnetic field (a) to the hydromagnetic 'box' model (b) in which the external magnetic field $\mathbf{B}_0$ depends only on the x-coordinate corresponding to the radial direction. Plasma density ρ and hence the Alfvén velocity c_A also depend only the x-coordinate. The face $x = 0$ corresponds to the equatorial region of the ionosphere, the face $x = l_x$ is the outer boundary of the magnetosphere – magnetopause, the faces $z = 0$ and $z = l_z$ are the 'southern and northern ionospheres'. Conjugate ionospheres are characterized by the integral Pedersen conductivities $\Sigma_P^{\pm} = \int \sigma_P^{\pm} \mathrm{d}z$. The sign '$-$' refers to the southern and '+' to the northern ionospheres. Panel (a) is plotted using a Matlab code [1]

the magnetosphere-magnetopause, the faces $z = 0$ and $z = l_z$ are the 'southern and northern ionospheres'. Ionospheres are characterized by the integral Pedersen conductivity $\Sigma_P^{\pm} = \int \sigma_P^{\pm} \mathrm{d}z$. $\Sigma_P^{\pm}$ are the integral conductivity of the southern '$-$' and northern '+' ionospheres, $X = 4\pi\Sigma_P/c$ is the dimensionless ionospheric Pedersen conductivity,

$$\Sigma_A = \frac{c\,\varepsilon_m^{1/2}}{4\pi} = \frac{c^2}{4\pi c_A}$$

is the magnetospheric wave conductivity, ε_m is the transversal dielectric permeability.

In this chapter two assumptions are made:

- We assume that the ionosphere can be replaced by a thin conductive layer with integral Pedersen conductivity Σ_P.
- The Hall conductivity Σ_H is neglected because we are interested primarily in the FLR-related effects. It will be shown later in Chapter 6 that Σ_H scarcely affect the FLR. Nevertheless, Σ_H plays a significant role in field formation near the ionosphere and on the Earth.

The interaction of the magnetospheric Alfvén waves with the ionosphere is characterized by the dimensionless parameter proportional to the ratio of the ionospheric Pedersen conductivity to the wave conductivity of the magnetosphere:

$$\bar{\Sigma}_P = \frac{\Sigma_P}{\Sigma_A}. \tag{5.1}$$

This parameter is principal in the Alfvén wave-ionosphere interaction. Thus, we give it in SI, as well:

$$\bar{\Sigma}_P = \frac{[\Sigma_P]_{SI}}{[\Sigma_A]_{SI}}, \quad [\Sigma_A]_{SI} = \frac{1}{\mu_0 [c_A]_{SI}},$$

where $[\Sigma_P]_{SI}$ is the integral ionospheric conductivity in SI, $[\Sigma_A]_{SI}$ is the wave conductivity of the magnetosphere in SI, $\mu_0 = 4\pi \times 10^{-7}$ H/A^2 is the magnetic permeability.

Plasma oscillations can be excited by sources located both inside and outside the box. In the first case, the oscillations are described by (4.42a) with uniform boundary conditions which are supplemented by extraneous currents caused by field sources. In the latter case, following Southwood (1974), let us consider excitation of FLR-oscillations by external sources, setting the field on the outer boundary of the magnetosphere.

Boundary Conditions

If there are no sources of magnetic field near the ionospheres, then the horizontal electric and magnetic components are in linear ratio. Let $X_- = 4\pi\Sigma_P^-/c$ and $X_+ = 4\pi\Sigma_P^+/c$. Then the impedance boundary conditions at the ionosphere can be written as

$$E_x = \mp b_y / X_\mp, \quad E_y = \pm b_x / X_\mp.$$

So, on the box faces $z = 0,\, l_z$, the conditions reduce to

$$\{b_x \mp X_\mp E_y\}_{z=0;l_z} = 0, \quad \{b_y \pm X_\mp E_x\}_{z=0;l_z} = 0, \tag{5.2}$$

where the upper sign corresponds to the southern ionosphere ($z = 0$) and the lower sign corresponds to the northern ionosphere ($z = l_z$). $\Sigma_P^\pm$ is assumed to be constant at each ionosphere. Let us assume that the magnetospheric waves are forced by monochromatic sources $\propto \exp(-i\omega t)$. Combining (4.17d), (4.18), (4.20), and (5.2) gives

$$\left(\frac{\partial \xi_\perp}{\partial z} \pm i\frac{\omega}{c} X_\mp \xi_\perp \right)_{z=0,l_z} = 0. \tag{5.3}$$

We consider the equatorial ionosphere ($x = 0$) free from the sources. Since the fields produced by sources near the outer magnetospheric boundary decay exponentially when penetrating into the magnetosphere, a specific form of the boundary conditions is not important. For definiteness, let

$$E_y \propto \xi_x = 0 \quad \text{or} \quad b_\parallel = 0. \tag{5.4}$$

At the 'magnetospheric' boundary ($x = l_x$) let us put inhomogeneous boundary conditions, for example

$$\xi_x = \xi_0(y, z) \quad \text{or} \quad b_\parallel = b_0(y, z). \tag{5.5}$$

It is appropriate to our problem to use the periodicity condition on coordinate y corresponding to geomagnetic longitude

$$\xi_\perp(x, y, z)\big|_{y=0} = \xi_\perp(x, y, z)\big|_{y=l}\,. \tag{5.6}$$

We shall discuss the applicability of the boundary conditions (5.3), (5.4), and (5.5) in Chapter 7.

Box Equations

As usual, using the methods of Fourier analysis, we can decompose the signal into its sinusoidal components in time and space and study each component separately. Let us consider perturbations $\propto \exp(ik_y y)$ with $k_y = 2\pi m/l_y$, $m = 0, \pm1, \pm2\ldots$. Equation (4.42a), with the field sources and displacements (ξ_x, ξ_y) reduces to

$$L_A \xi_y = ik_y \frac{b_\|}{B_0} - \frac{4\pi}{c}\frac{j_x^{(d)}}{B_0}, \tag{5.7a}$$

$$\frac{\partial}{\partial x}\frac{b_\|}{B_0} = L_A \xi_x - \frac{4\pi}{c}\frac{j_y^{(d)}}{B_0}, \tag{5.7b}$$

$$\frac{\partial}{\partial x}\xi_x = -\frac{b_\|}{B_0} - ik_y \xi_y, \tag{5.7c}$$

where $\mathbf{j}^{(d)}$ are driving currents. Recall that the operator L_A is

$$L_A = \frac{\partial^2}{\partial z^2} - \frac{1}{c_A^2(x)}\frac{\partial^2}{\partial t^2} \tag{5.8}$$

and for monochromatic oscillations $\exp(-i\omega t)$

$$L_A = \frac{\partial^2}{\partial z^2} + k_A^2(x), \qquad k_A^2(x) = \frac{\omega^2}{c_A^2(x)}.$$

A perturbation propagating from a harmonic source in the hydromagnetic box can be presented as a superposition of normal oscillations. Consider first of all the problem of Alfvén oscillations of a field-lines with the periodical conditions (5.6).

5.2 Dungey's Problem

In order to study the standing free oscillations, we assume that exciting sources are absent, that is $\mathbf{j}^{(d)} = 0$ in (5.7a) and $k_y = 0$. Denote by $Q(x)$ the amplitude distribution of displacement in the standing Alfvén oscillations. At a fixed frequency ω, this function is determined from the boundary problem on the eigenvalues of the equation

$$\left(\frac{\mathrm{d}^2}{\mathrm{d}z^2} + k^2\right) Q(z) = 0, \tag{5.9}$$

with the boundary condition

$$\left(Q(z) \mp i\frac{c}{\omega X_{\mp}}\frac{\mathrm{d}Q(z)}{\mathrm{d}z}\right)_{z=0;\, l_z} = 0. \tag{5.10}$$

Here the upper sign refers to $z = 0$, the lower to $z = l_z$.

For the perfect conductive walls ($X_{\mp} \to \infty$ in (5.10)) (5.9) with conditions (5.10) is the self-adjoint Sturm-Liouville problem. It is known (see, e.g., [10], [15]) that in this case the spectrum consists of a discrete set of real positive eigenvalues k_n^2. The corresponding normalized eigenfunctions $Q_n(z)$ are mutually orthonormal functions:

$$\langle Q_n Q_m \rangle = \delta_{nm},$$

where δ_{nm} is the Kronecker symbol,

$$\delta_{nm} = \begin{cases} 1, & n = m, \\ 0, & n \neq m \end{cases}$$

and

$$\langle Q_n Q_m \rangle = \int_0^{l_z} Q_n(z) Q_m^*(z) \mathrm{d}z,$$

where the asterisk denotes complex conjugation. An arbitrary function $F(z)$ satisfying boundary conditions (5.10) can be presented as convergent series

$$F(z) = \sum_{n=1}^{\infty} F_n Q_n(z),$$

where F_n are Fourier coefficients.

At finite Σ_P, the boundary problem (5.9)–(5.10) is non-self-conjugate. It has a discrete spectrum of complex eigenvalues k_n^2 and eigenfunctions $Q_n(z)$, $n = 0, 1, 2, \ldots$. The eigenfunctions in this case are not mutually orthogonal. However, there exist a set of bi-orthonormal functions $\{G_m(z)\}$: $\langle Q_n G_m \rangle = \delta_{nm}$. Functions $G_m(z)$ are found from the conjugate boundary problem. It is obtained in this case from (5.9)–(5.10) by replacing i in (5.10) by $-i$:

$$\left(\frac{\mathrm{d}^2}{\mathrm{d}z^2} + \mu\right) G(z) = 0, \quad \left(G(z) \pm i\frac{c}{\omega X_{\mp}}\frac{\mathrm{d}G(z)}{\mathrm{d}z}\right)_{z=0;\, l_z} = 0.$$

Eigenvalues k_n^2, μ_n and eigenfunctions $G_n(z)$, $Q_n(z)$ are complex conjugate:

$$\mu_n = \left(k_n^2\right)^*, \quad G_n(z) = Q_n^*(z),$$

and the conditions of bi-orthonormality

$$\langle Q_n G_m \rangle = \int_0^{l_z} Q_n(z) Q_m(z) \mathrm{d}z = \delta_{mn} \tag{5.11}$$

are fulfilled between them.

5.3 Explicit Eigenmodes

Let us obtain explicit expressions for $Q_n(z)$ and k_n^2. The solution of (5.9) is

$$Q(z) = \frac{\sin kz}{k} + i \frac{c}{\omega X_-} \cos kz \tag{5.12}$$

It satisfies the boundary condition (5.10) at $z = 0$. Substituting this function into the boundary condition at $z = l_z$, we obtain

$$\left(1 + (\pi q)^2 \alpha_- \alpha_+\right) \frac{\sin \pi q}{\pi q} + i (\alpha_+ + \alpha_-) \cos \pi q = 0, \tag{5.13}$$

where

$$q = \frac{k l_z}{\pi}, \qquad \alpha_\mp = \frac{c}{\omega l_z X_\mp}.$$

Consider two limiting cases of high and low conductivities $\Sigma_P^\pm$. For the perfectly conducting ionospheres, i.e. $\Sigma_P \to \infty$, the parameters $\alpha_\pm = 0$. Then (5.13) has only real roots

$$q_n^\infty = n, \quad k_n^\infty = \frac{n\pi}{l_z}, \qquad n = 1, 2, 3, \ldots .$$

The eigenfunctions of (5.12) reduce to

$$Q_n^\infty(z) = \sqrt{\frac{2}{\pi}} \sin(k_n z). \tag{5.14}$$

At small but finite $\alpha_\pm$ the first-order correction on $\alpha_+ + \alpha_-$ can be found by substituting $q_n = n + \delta q_n$ ($k_n = k_n^\infty + \delta k_n$) into (5.13). Expanding it into a series by δq_n and $\alpha_\pm$ and leaving only linear terms, we obtain

$$\delta q_n \approx -in(\alpha_+ + \alpha_-), \; k_n \approx \frac{n\pi}{l_z} - in\frac{\pi}{l_z}(\alpha_+ + \alpha_-), \; n = 1, 2, 3, \ldots . \tag{5.15}$$

In the opposite case of small conductivities $\Sigma_P^\pm$, we find that (5.13) in the first approximation by $1/\alpha_\pm$ for q_n becomes

$$\pi q \tan(\pi q) = -i \left(\frac{1}{\alpha_-} + \frac{1}{\alpha_+} \right).$$

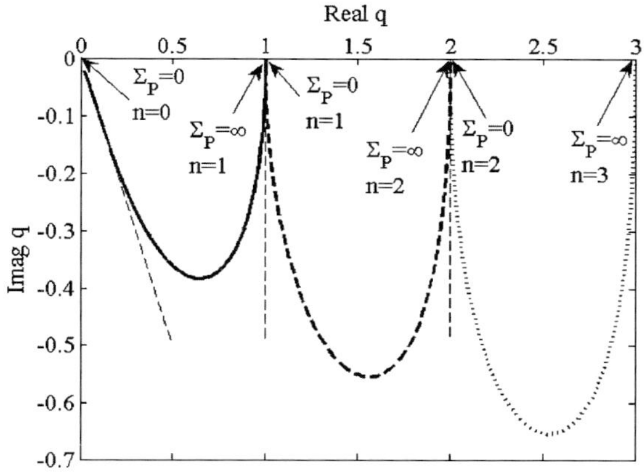

Fig. 5.2. Hodographs of the normalized wavenumber $q_n = k_n l_z/\pi$; n = 1, 2, 3; $\Sigma_P = \Sigma_P^- = \Sigma_P^+$. Each curve (from right to left) corresponds to the change of Σ_P from ∞ to 0. Thin dashed lines are the approximate dependencies (5.16) and (5.17)

Hence, for the fundamental harmonic, i.e. $n = 0$, we have

$$k_{(0)} = \frac{\pi q_0}{l_z} = \frac{1}{l_z} e^{-i\pi/4} \left(\frac{1}{\alpha_-} + \frac{1}{\alpha_+} \right)^{1/2} \tag{5.16}$$

and for the higher harmonics

$$k_n = \frac{\pi n}{l_z} - i \frac{1/\alpha_- + 1/\alpha_+}{\pi n l_z}, \quad n = 1, 2, 3, \ldots . \tag{5.17}$$

Thus, the magnitude of the imaginary part of a wavenumber k_n depends on Σ_P non-monotonically. It is first rises with Σ_P decreasing, reaches a maximum and then declines. The real part of k_n decreases monotonically with the change in Σ_P from ∞ to 0. A harmonic with number n at $\Sigma_P = \infty$ passes into $n-1$ harmonic at $\Sigma_P = 0$. These peculiarities are illustrated in Fig. 5.2. It presents the hodographs of $q_n = k_n l_z/\pi$ for the first three harmonics. The curves are calculated for the symmetrical ionospheres for which $\Sigma_P^+ = \Sigma_P^-$, and the parameter $\alpha = \alpha_+ = \alpha_-$ changes from 0 to ∞. The approximate dependencies (5.16) and (5.17) are shown by thin dashed lines.

5.4 Field-Line Resonance (FLR) Frequencies

The boundary problem (5.9)–(5.10) differs from the classical Dungey's problem [9]. Dungey posed the problem of determining Field Line Resonance (FLR) frequencies of the geomagnetic field-lines in a dipole approximation. Problem (5.9)–(5.10), as distinct from Dungey's problem, consists of determining the longitudinal wavenumber

$$k_n = k_n(\omega)$$

at a fixed oscillation frequency ω.

For perfectly reflecting ionospheres, i.e. Σ_P^-, $\Sigma_P^+ \to \infty$, $X_\pm \to \infty$ and $Q|_{z=0;l_z} = 0$, frequency ω is not included in the boundary conditions. Then both problems in fact coincide. In this case, wavenumber $k_n = k_n^\infty$ is independent of ω and the frequency of the n-th harmonic ω_{An} is given by

$$\omega_{An}(x) = k_n c_A(x).$$

If the integral conductivity at least one of the conjugate ionospheres is finite, i.e. $\bar{\Sigma}_P^- \neq 0$ or $\bar{\Sigma}_P^+ \neq 0$, then the wave energy dissipates in the ionospheres and FLR-frequencies become complex. Denote the frequency of the FLR-n-th harmonic by

$$\omega_n(x) = \omega_{An}(x) - i\gamma_n(x),$$

where $\gamma_n(x)$ is the decrement. $\omega_n(x)$ can be found from the problem obtained by replacing the parameter k^2 by $\omega^2/c_A^2(x)$ in (5.9). The resulting boundary problem is the so-called generalized boundary problem on eigenfrequency ω. It can be shown that the oscillation frequency spectrum $\{\omega_n\}$ will be discrete just as for the ordinary problem.

The FLR-frequencies $\{\omega_n\}$ will be found as follows. From the boundary problem (5.9)–(5.10) we find the dependence of an n-th wavenumber k_n on frequency ω. At the known function $k_n(\omega)$, resonance oscillation frequencies of the x-th field-lines can be found from

$$\omega = \pm k_n(\omega) c_A(x), \tag{5.18}$$

following from the definition of $k_n(\omega)$. Substituting into (5.13)

$$q = q_n(\bar{\omega}) = \frac{k_n l_z}{\pi} = \pm\bar{\omega},$$

where $\bar{\omega} = \omega/\omega_A, \omega_A = \pi c_A/l_z$, after a simple algebra we get

$$\exp(-2i\pi\bar{\omega}) = R_S R_N, \tag{5.19}$$

where

$$R_S = \frac{1-\bar{\Sigma}_P^-}{1+\bar{\Sigma}_P^-} \quad \text{and} \quad R_N = \frac{1-\bar{\Sigma}_P^+}{1+\bar{\Sigma}_P^+},$$

where subscript '−' and '+' refer to the 'Southern' and 'Northern' ionospheres, respectively. Note, that the expressions for R_S and R_N yield reflection coefficients of the Alfvén wave from the N- and S-ionospheres defined as the ratio of the electric component in the incident wave to the same in the reflected one. According to (5.1), $\bar{\Sigma}_P^\pm$ is the ratio of the corresponding integral Pedersen conductivity Σ_P to the wave magnetospheric conductivity Σ_A. $R_{S,N}$ decrease from 1 to −1 with Σ_P changing from 0 to $+\infty$. Two types of resonance modes are possible determined by the sign of $R_S R_N$:

- The half-wave mode, $R_S R_N > 0$. In this case, an integer number of half-waves lies on a field-lines. The resonance frequencies are

$$\frac{\omega_n}{\omega_A} = n - i\frac{\gamma_n}{\omega_A}, \qquad n = 0, \pm 1, \pm 2, \ldots . \tag{5.20}$$

The relative decrement is independent of the harmonic number n and

$$\frac{\gamma_n}{\omega_A} = -\frac{1}{2\pi} \log\left(R_S R_N\right). \tag{5.21}$$

- The quarter-wave mode, $R_S R_N < 0$. The field-lines length is 1/4, 3/4, 5/4, ... of the wavelength. The resonance frequencies are

$$\frac{\omega_n}{\omega_A} = n - \frac{1}{2} - i\frac{\gamma_n}{\omega_A}, \qquad n = 0, \pm 1, \pm 2, \pm 3, \ldots \tag{5.22}$$

and the relative decrement is

$$\frac{\gamma_n}{\omega_A} = -\frac{1}{2\pi} \log\left|R_S R_N\right|. \tag{5.23}$$

In the two limiting cases:

- Small conductivities, $\bar{\Sigma}_P^- \ll 1$ and $\bar{\Sigma}_P^+ \ll 1$, the decrement is

$$\frac{\gamma_n}{\omega_A} \approx \frac{1}{\pi}\left(\bar{\Sigma}_P^- + \bar{\Sigma}_P^+\right). \tag{5.24}$$

- Large conductivities, $\bar{\Sigma}_P^- \gg 1$ and $\bar{\Sigma}_P^+ \gg 1$, the decrement is

$$\frac{\gamma_n}{\omega_A} \approx \frac{1}{\pi}\left(\frac{1}{\bar{\Sigma}_P^-} + \frac{1}{\bar{\Sigma}_P^+}\right). \tag{5.25}$$

In the symmetrical case at $\bar{\Sigma}_P = \bar{\Sigma}_P^- = \bar{\Sigma}_P^+$, and $R = R_S = R_N = \left(1 - \bar{\Sigma}_P\right) / \left(1 + \bar{\Sigma}_P\right)$, we find from (5.21) that

$$\frac{\gamma_n}{\omega_A} = -\frac{1}{\pi} \log\left|\frac{1 - \bar{\Sigma}_P}{1 + \bar{\Sigma}_P}\right| \quad \text{with} \quad \omega_A = \frac{\pi c_A}{l_z}.$$

The dependence of the decrement γ_n and the Q-factor $Q_n = \omega_A / 2\gamma_n$ of the fundamental mode on $\bar{\Sigma}_P$ are presented in Fig. 5.3. For dayside conditions at moderate solar activity the normalized $\bar{\Sigma}_P \approx 10$, and for nighttime conditions $\bar{\Sigma}_P \approx 0.5$. It can be seen from Fig. 5.3 that the Q-factor varies from 9.8 at the daytime to 1.4 at nighttime.

Below, we shall present fields expanded by eigenfunctions (5.12). However, it is cumbersome to find a series the eigenfunctions of the generalized boundary problem. We therefore prefer to regard (5.9)–(5.10) as a problem on wave eigennumbers, keeping for it the designation of 'Dungey's problem'.

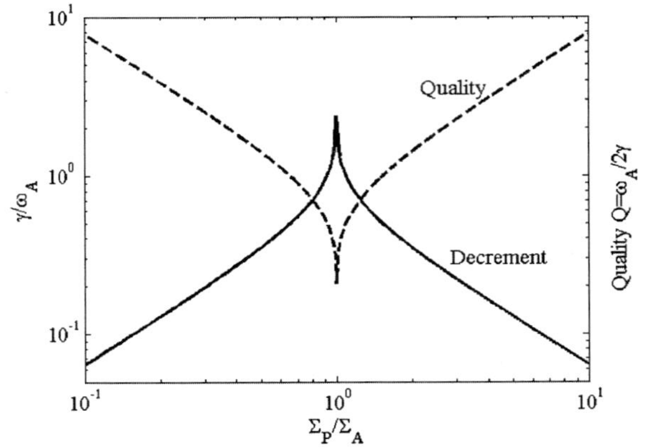

Fig. 5.3. The decrement γ_1/ω_A and the Q-factor $Q_1 = \omega_A/2\gamma_1$ of the fundamental harmonic depending on the normalized integral conductivity

5.5 FLR-Equations

Let us discuss now the propagation of a hydromagnetic wave excited by external sources. The displacement $\xi_\perp$ and longitudinal magnetic field $b_\parallel$ are completely determined by (5.7a), (5.7b), (5.7c) and boundary conditions (5.3), (5.6). We assume that field sources in (5.7a)–(5.7c) are near the magnetopause, while the inner regions of the magnetosphere are free of external currents. The simplest way to deduce the solution is by expanding it into a series of the eigenfunctions $Q_n(x)$. Substituting the series

$$\xi_y = \sum_m a_m(x)Q_m(z), \quad \xi_x = \sum_m c_m(x)Q_m(z),$$
$$\frac{b_\parallel}{B_0} = \sum_m b_m(x)Q_m(z),$$

into (5.7a)–(5.7c) and noting that

$$L_A Q_m(z) = \left(k_A^2(x) - k_m^2\right) Q_m(z),$$

we obtain

$$\sum_m \left(k_A^2(x) - k_m^2\right) a_m(x)Q_m(z) = ik_y \sum_m b_m(x)Q_m(z) - \frac{4\pi}{c}\frac{j_x^{(d)}}{B_0},$$
$$\sum_m b_m'(x)Q_m(z) = \sum_m \left\{\left(k_A^2(x) - k_m^2\right) c_m(x)Q_m(z)\right\} - \frac{4\pi}{c}\frac{j_y^{(d)}}{B_0},$$
$$\sum_m c_m'(x)Q_m(z) = -\sum_m \left\{ik_y a_m(x) + b_m(x)\right\} Q_m(z),$$

where the prime denotes differentiation with respect to x.

Multiplying each equation of this system by $Q_n(x)$ and integrating the obtained equalities over z from 0 to l_z, we obtain equations for mode amplitudes a_n, b_n, and c_n. In the 1D-box model ($c_A = c_A(x)$) oscillations of field-lines do not interact with one another, and the bi-orthogonality condition (5.11) is fulfilled. Then the infinite system of coupled equations is split into uncoupled finite-dimensional equations of the form

$$\left(k_A^2(x) - k_n^2\right) a_n(x) = ik_y b_n(x) - \frac{4\pi}{c}\frac{j_{xn}^{(d)}}{B_0}, \tag{5.26}$$

$$b_n'(x) = \left(k_A^2(x) - k_n^2\right) c_n(x) - \frac{4\pi}{c}\frac{j_{yn}^{(d)}}{B_0}, \tag{5.27}$$

$$c_n'(x) = -ik_y a_n(x) - b_n(x), \tag{5.28}$$

where

$$\mathbf{j}_n^{(d)} = \langle \mathbf{j}_n^{(d)}, Q_n(z) \rangle.$$

Consider the one-mode propagation. We have from (5.26)

$$a_n(x) = \frac{1}{k_A^2(x) - k_n^2}\left(ik_y b_n(x) - \frac{4\pi}{c}\frac{j_{xn}^{(d)}}{B_0}\right).$$

Substituting it into (5.27) and (5.28) yields

$$b_n'(x) = \left(k_A^2(x) - k_n^2\right) c_n(x) - \frac{4\pi}{c}\frac{j_{yn}^{(d)}}{B_0}, \tag{5.29}$$

$$c_n'(x) = \left(-1 + \frac{k_y^2}{k_A^2(x) - k_n^2}\right) b_n(x) + ik_y \frac{4\pi}{c}\frac{1}{k_A^2(x) - k_n^2}\frac{j_{xn}^{(d)}}{B_0}. \tag{5.30}$$

These equations differ from (4.43)–(4.45) only in $k_\parallel$ being replaced by k_n and in the explicit account taken of the field sources. Therefore, many results obtained for unbounded plasma in Chapter 4 can be transferred to the box model by simply replacing the continuously changing wavenumber $k_\parallel$ by the quantized wavenumber k_n. The coordinate x_n of a resonance field-lines of the n-th harmonics is found from

$$k_A(x) = \pm k_n. \tag{5.31}$$

In the non-dissipative case, the wavenumber $k_n = n\pi/l_z$ is real and (5.31) can have a real root x_n corresponding to the resonant point. The point is a regular singularity of (5.29)–(5.30).

If dissipation is not negligible, then

$$k_n = \frac{n\pi}{l_z} + i\kappa_n$$

is a complex value. In our simple model k_n is independent of the x-coordinate. With dissipation taken into account, (5.31) cannot be satisfied at real x and

ω. Let ω be real. Suppose that the function $k_A(x)$ is continued on the complex plane $w = x + iu$ and search for $w_n = x_n + i\delta_n$ from (5.31). It will be shown in the next section that the field distribution in the vicinity of the resonant point has a form of Lorentz's curve and the parameter δ_n is the half-width of this resonance curve.

At $\delta_n \neq 0$ the singularity of (5.29)–(5.30) shifts from the real axis into the complex plane and while integrating of (5.29)–(5.30) over the real axis x the solution remains finite. Estimate δ_n at small losses. Use Taylor-series expansion

$$\begin{aligned} k_A(x_n + i\delta_n) &= k_A(x_n) + i\delta_n \frac{\mathrm{d}}{\mathrm{d}x} k_A(x_n) + \cdots \\ &= k_A(x_n)\left(1 + i\frac{\delta_n}{\Lambda_n} + \cdots\right), \end{aligned} \tag{5.32}$$

where

$$\begin{aligned} \Lambda_n &= k_A(x_n)/k'_A(x_n) = -\omega_{An}(x_n)/\omega'_{An}(x_n) \\ &= -c_A(x_n)/c'_A(x_n) = \frac{1}{2}\rho(x_n)/\rho'(x_n) \end{aligned} \tag{5.33}$$

and $|\Lambda_n|$ is a scale of the FLR-frequency at point x_n. It is taken into account here that the external magnetic field $\mathbf{B}_0$ is independent of x. Substitution of the Taylor expansion into (5.31), yields in the first approximation

$$k_A(x_n) = \pm\frac{n\pi}{l_z}. \tag{5.34}$$

The sign here is determined by the sign of frequency ω. In the next approximation we have an equation for the determination of δ_n:

$$k_A(x)\frac{\delta_n}{\Lambda_n} \approx \kappa_n.$$

For definiteness, put a plus sign in (5.34). Suppose the function $k_A(x) - n\pi/l_z$ has a simple zero at the point x_n, then

$$\delta_n \approx \Lambda_n \frac{\kappa_n(x_n)}{k_A(x_n)}. \tag{5.35}$$

By means of (5.15) and (5.17), with $\omega = \omega_A(x_n)$ we obtain

$$\begin{aligned} \kappa_n &\approx -\frac{1}{l_z}\left(\frac{1}{\bar{\Sigma}_P^-} + \frac{1}{\bar{\Sigma}_P^+}\right) \quad \text{at} \quad \bar{\Sigma}_P^\pm \to \infty, \\ \kappa_n &\approx -\frac{1}{l_z}\left(\bar{\Sigma}_P^- + \bar{\Sigma}_P^+\right) \quad \text{at} \quad \bar{\Sigma}_P^\pm \to 0. \end{aligned}$$

Substituting these expressions into (5.35) yields

$$\delta_n \approx -\frac{\Lambda_n}{n\pi}\left(\frac{1}{\bar{\Sigma}_P^-} + \frac{1}{\bar{\Sigma}_P^+}\right) \quad \text{at} \quad \bar{\Sigma}_P^{\pm} \to \infty \tag{5.36}$$

and

$$\delta_n \approx -\frac{\Lambda_n}{n\pi}\left(\bar{\Sigma}_P^- + \bar{\Sigma}_P^+\right) \quad \text{at} \quad \bar{\Sigma}_P^{\pm} \to 0. \tag{5.37}$$

Comparing (5.36) and (5.37) with (5.24) and (5.25) we get a relationship of the half-width δ_n with decrement γ_n:

$$\delta_n \approx -\frac{\gamma_n}{\omega_{An}}\Lambda_n = \gamma_n \left(\left.\frac{\mathrm{d}\omega_{An}}{\mathrm{d}x}\right|_{x=x_n} \right)^{-1}. \tag{5.38}$$

We obtained (5.38) using the explicit expressions for γ_n and δ_n. Moreover, these formulae can be applied in a rather wider field than could been assumed when deriving them. Let us present other approach independent on the form of the explicit solution.

Curved Field-Line

Let us consider field-lines that are not obligatory straight. Let their position be determined by one coordinate of a curvilinear coordinate system. Generally, a complex resonant frequency $\omega_n(x)$ corresponds to each field-line. These frequencies are found from the dispersion equation $\Delta(\omega, x) = 0$ (see Chapter 6). The latter, as in the case considered above, can be continued to the complex plane $w = x + iu$. Then, assuming low dissipation, we first find the resonance point $x_n(\omega)$ at a given frequency ω, and then determine the real function $\omega_A(x)$. For a weak dissipation, we have

$$w_n(\omega) = x_n(\omega) + i\delta_n(\omega),$$

where ω is real and

$$\omega_n(x) = \omega_{An}(x) - i\gamma_n(x).$$

Functions $\delta_n(\omega)$ and $\gamma_n(x)$ are small. Expanding Δ by δ_n at a fixed x, we have

$$\delta_n = i\frac{\Delta(\omega_A, x_n)}{\partial\Delta(\omega_{An}, x_n)/\partial x},$$

and expanding by γ_n

$$\gamma_n = -i\frac{\Delta(\omega_{An}, x_n)}{\partial\Delta(\omega_{An}, x_n)/\partial\omega}.$$

Hence

$$\frac{\delta_n}{\gamma_n} = -\frac{\partial \Delta(\omega_{An}, x_n)/\partial \omega}{\partial \Delta(\omega_{An}, x_n)/\partial x}.$$

Combining this expression and making use the formula for differentiation of an implicit function

$$\frac{\mathrm{d}\omega_n}{\mathrm{d}x} = -\frac{\partial \Delta(\omega_{An}, x_n)/\partial x}{\partial \Delta(\omega_{An}, x_n)/\partial \omega},$$

we obtain (5.38). Then from (5.38) one can directly find the relationship between the variation of the wave phase (it is determined by the sign of δ) and the sign of the derivative of the FLR-frequency. In a dissipative system the decrement is positive and, thus, the signs of δ and $\mathrm{d}\omega_{An}/\mathrm{d}x$ coincide.

5.6 FLR-Field Structure

Basic Equations

The method of solving (5.29)–(5.30) near a singular point $x_n + i\delta_n$ by means of Frobenius series [6] is the same as in Chapter 4 (see Section 4.2). It follows from (4.59) that the principal terms in the presentation of electromagnetic fields can be written as

$$\begin{aligned}
&\frac{b_\parallel}{b_0} \approx \left(1 + \frac{\bar{x}^2}{2}\ln\bar{x}\right) Q_n(z), \\
&\frac{b_x}{b_0} \approx p\ln\bar{x}\,\frac{\mathrm{d}Q_n(z)}{\mathrm{d}z}, \qquad \frac{b_y}{b_0} \approx i\frac{p}{\bar{x}}\frac{\mathrm{d}Q_n(z)}{\mathrm{d}z}, \\
&\frac{E_x}{b_0} \approx -\frac{\omega}{c}\frac{p}{\bar{x}} Q_n(z), \qquad \frac{E_y}{b_0} \approx -i\frac{\omega}{c} p\ln\bar{x}\, Q_n(z), \\
&\frac{\xi_x}{b_0/B_0} \approx p\ln(\bar{x})\, Q_n(z), \qquad \frac{\xi_y}{b_0/B_0} \approx i\frac{p}{\bar{x}} Q_n(z),
\end{aligned} \tag{5.39}$$

where

$$\bar{x} = k_y\left[x - (x_n + i\delta_n)\right], \quad p = 2k_y^2 \Lambda_n / k_A^2(x_n).$$

and b_0 is the longitudinal wave magnetic field on the resonance line $x = x_n$.

Numerical Example

In order to estimate the possible effects in the magnetosphere, we employ the parameters in (5.39) close to the magnetospheric parameters. 'Box' model is only a very rough approximation and it cannot give a detailed quantitative description of FLR. However, choosing the parameters of the model, the principal effects can be illustrated.

Consider a magnetic shell crossing the magnetic equator at a geocentric distance LR_E (R_E is the Earth's radius). The origin of the rectangular coordinate system is taken at the top of a field-line lying on the selected L-shell. Let x-axis be directed along the equatorial radius, y is azimuthal, z is directed along the field-line. Consider the wave disturbances with an azimuthal wavenumber m. Then near the equatorial plane, the wavenumber is

$$k_y \approx m/LR_E.$$

The main geomagnetic field $B_0 \propto L^{-3}$, and let the plasma density $\rho_0 \propto L^{-\nu}$, then the Alfvén velocity $c_A \propto L^{-3+\nu/2}$. The field-line length is $l_z \approx \pi LR_E$ in a quasi-dipolar field and the wavenumber is $k_A \sim \pi/l_z \approx 1/LR_E$. Then the frequency of the fundamental harmonic $\omega_{A1} \propto L^{-4+\nu/2}$ and its radial derivative $\omega'_{A1} \propto -(4-\nu/2)L^{-5+\nu/2}$. We denote their ratio by

$$\Lambda_1 = -\frac{\omega_{A1}}{\omega'_{A1}} \approx \frac{LR_E}{4-\nu/2}.$$

Substituting Λ_1 and k_y into p (5.39), we have

$$p = m^2\Lambda_1 = m^2\frac{LR_E}{4-\nu/2}, \quad pk_A = \frac{m^2}{4-\nu/2}.$$

The half-width δ_1 is

$$\delta_1 = -\frac{\gamma_1}{\omega_{A1}}\Lambda_1,$$

where γ_1/ω_{A1} is determined in (5.23).

Near the ionosphere (let it be the northern ionosphere for definiteness) relations (5.39) with the boundary condition (5.10) are given by

$$\begin{aligned} b_\parallel &\approx b_0\left(1+\frac{\bar{x}^2}{2}\ln\bar{x}\right), \\ b_x &\approx ib_0pk_A\bar{\Sigma}_P\ln\bar{x}, \quad b_y \approx -b_0\frac{pk_A}{\bar{x}}\bar{\Sigma}_P, \\ E_x &\approx \frac{c}{4\pi}\frac{b_y}{\Sigma_P}, \quad E_y \approx -\frac{c}{4\pi}\frac{b_x}{\Sigma_P}, \end{aligned} \tag{5.40}$$

with $\bar{\Sigma}_P = \Sigma_P/\Sigma_A$.

The two upper frames of Fig. 5.4 show the distributions of amplitudes (left) and phases (right) of b_x and b_y components near the resonance shell calculated according to the approximate relations (5.40). b_y is normalized by the condition $b_y(x_1) = 1$ at the resonance point $x = x_1$. The chosen parameters roughly correspond to the magnetospheric conditions for a middle-latitude magnetic shell. Two bottom frames of Fig. 5.4 are results of numerical integration of (5.29)–(5.30) showing the field distribution produced by a monochromatic source located at the magnetospheric boundary. At $x = 0$, the electric field $E_y(x=0) = 0$. We take E_y at $x = l_x$ such as to satisfy the condition $b_y(x = x_1) = 1$ at the resonance point.

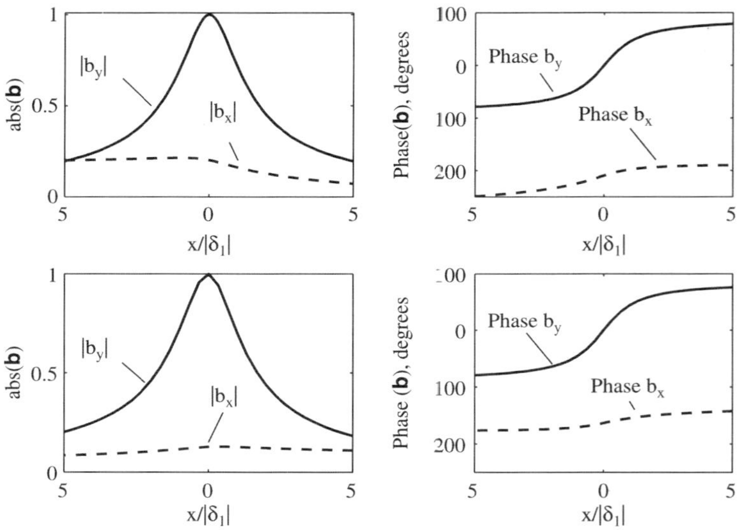

Fig. 5.4. Amplitude and phase of the magnetic field near the FLR-point. Upper panels show the results of calculation according to (5.40) at $f = 10^{-2}$ Hz, normalized integral Pedersen conductivity $\bar{\Sigma}_P = 10$ and Alfvén velocity at the resonance point $c_A = 1500\,\mathrm{km/s}$, $pk_A = -3$. The position of the resonance point is at $x = x_1 = 0$. Results of numerical integration of (5.7a)–(5.7b) under the same parameters at the resonance point are shown in the bottom frames

It can be seen from Figure 5.4 that

- The amplitude of the magnetic field resonance component b_y has a peak of a Lorentz form near the FLR; the phase of b_y changes abruptly at π when passing the resonance point in the direction of the decrease of the resonance frequency.
- Comparison of the upper (according to (5.40)) and bottom (numerical integration of (5.7a)–(5.7c)) panels of Fig. 5.4 demonstrates a good agreement of amplitude and phase distribution of the resonance components b_y. The behavior of non-resonant components is not so well described by (5.40). This is due to the fact that the regular part of the solution can be comparable with its logarithmically singular part.
- The transversal magnetic field is linearly polarized at the x_p point being displaced at a distance slightly less than the resonance half-width from the resonance point. In the numerical example x_p is displaced to the right at $0.22|\delta_1|$. At $x = x_p$, the polarization ellipse changes the sign of rotation.
- At $x > x_p$ the vector rotates counterclockwise when looking along the z-axis and clockwise at $x < x_p$. Thus, the sign of polarization changes near the maximum and the transversal electric and magnetic fields become approximately linear in it.

- The amplitude of the longitudinal magnetic component $b_{\|}$ is approximately

$$\frac{k_y a}{k_A^2 x_1 |\delta_1|}$$

times smaller than the b_y amplitude.

Energy Dissipation

Now we consider energy losses in the FLR-region. The time average energy flux carried by the wave along axis x is given by the projection of the real part of the complex Poynting vector

$$S_x = \frac{c}{8\pi} \left[\mathbf{E} \times \mathbf{b}^* \right]_x = \frac{c}{8\pi} E_y b_{\|}^* . \tag{5.41}$$

When the wave propagates through the resonance region, energy losses can be determined as a difference between the Poynting vectors left and right of the resonance region

$$\Delta S_x = \mathrm{Re} \left(S_x(x_n - 0) - S_x(x_n + 0) \right),$$

where x_n is the coordinate of a resonance point. Substituting E_y and $b_{\|}$ from (5.39) into (5.41), we find

$$\begin{aligned} \Delta S_x \approx & \frac{\omega}{8\pi} \frac{k_y^2 \Lambda_n \left| b_0^2 \right|}{k_A^2(x_n)} \\ & \times \mathrm{Im} \left\{ \ln \left[k_y \left(x_- - x_n - i\delta_n \right) \right] - \ln \left[k_y \left(x_+ - x_n - i\delta_n \right) \right] \right\} \left| Q_n \right|^2 , \end{aligned} \tag{5.42}$$

where $b_0 = b_{\|}(x = x_n)$.

Equation (5.42) proves to be essentially more accurate than can be expected from the comparison of the numerical simulation and singular part of the solution (see (5.40)) shown in Figure 5.4. The regular component of the solution is responsible for the large error in estimates of fields based just on the terms with logarithmic singularity. This component is continuous in the resonance range and its contribution to the discontinuity of the Poynting vector vanishes with the resonance half-width decreasing.

Analytical continuation of the logarithm from the right-hand semi-axis $x - x_n > 0$ onto the left-hand one $x - x_n < 0$ must be carried out in a complex plane $w = x + iu$ with the singular point bypassed from below at $\delta_n > 0$ and from above at $\delta_n < 0$. Therefore at $|\delta_n| \ll |x - x_n|$, the logarithms in (5.42) are

$$\begin{aligned} & \ln \left| k_y \left(x_+ - x_n \right) \right| , \quad x_+ > x_n , \\ & \ln \left| k_y \left(x_- - x_n \right) \right| + i\pi \cdot \mathrm{sign}(\delta_n), \quad x_- < x_n . \end{aligned}$$

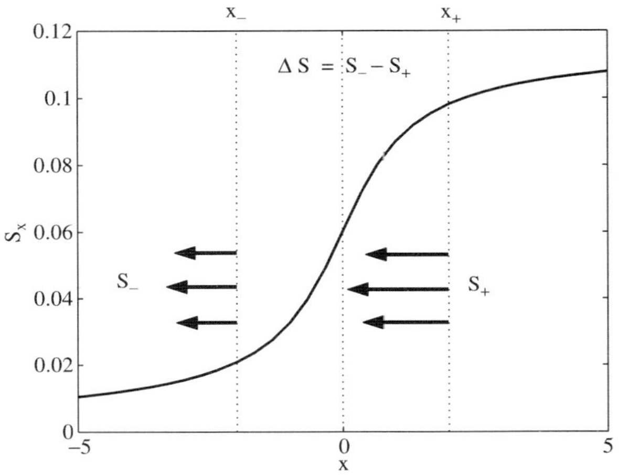

Fig. 5.5. Poynting vector discontinuity near the resonance point. The difference between Poynting vectors at $x = x_-$ and $x = x_+$ corresponds to the wave energy dissipation in the FLR

The imaginary part in (5.42) is $\pi\,\mathrm{sign}(\delta_n) = -\mathrm{sign}(\Lambda)$ (see 5.38). Thus energy losses on a unit area within the resonance region can be estimated as

$$\Delta S_x \approx -\frac{\omega}{8\pi}\frac{k_y^2\,|\Lambda_n|\,|b_0|^2}{k_A^2(x_n)}\,|Q_n(z)|^2 . \tag{5.43}$$

The total energy dissipation W on the entire resonance magnetic shell is found by integrating (5.43) over the field line and summing up on all the lines for which the FLR-condition is satisfied for the set frequencies ω, i.e. on the box cross-section $x = \text{const}$. With the normalization requirement (5.11) we can define the total energy W-expending on the plasma heating within the resonance region as

$$W \approx \frac{\omega}{8\pi}\frac{k_y^2\,|\Lambda_n|\,|b_0|^2}{k_A^2(x_n)}\,l_y . \tag{5.44}$$

At numerical calculations it is convenient to use the expression

$$S_x(x) = i\frac{\omega}{8\pi}\left|b_{\parallel}(x_n)\right|^2 \frac{c(x)b^*(x)}{|b(x_n)|^2}\,|Q_n(z)|^2 , \tag{5.45}$$

where $c(x)$ and $b(x)$ are obtained from (5.29)–(5.30).

Let us compare the energy loss in the resonance point calculated from (5.43) and (5.45). Figure 5.5 shows the x-component of the time averaged Poynting vector as a function of coordinate x near the resonance point. The fields are normalized by the longitudinal magnetic field in the resonance point.

The parameters of the model are the same as for Fig. 5.4. One can see that the Poynting vector decreases step-wise at distances of about a half-width of the resonance shell. The difference ΔS_x between the energy fluxes at the cross-sections $x_- = x_1 - 2\delta_1$ and $x_+ = x_1 + 2\delta_1$ shown in Fig. 5.5 by vertical dashed lines is $\Delta S_x \approx 7.3 \times 10^{-2} |Q_n(z)|^2$. The estimate according (5.45) gives a close value $\approx 6.8 \times 10^{-2} |Q_n(z)|^2$.

5.7 Global and Surface Oscillation Modes

Evolution of MHD-Perturbations

The study of the evolution of initial perturbations and of effects observed when external sources act upon the magnetosphere is important for the comprehension of the space-time behavior of pulsations. This range of questions is of great significance not only in the physics of magnetospheric pulsations, but also in the study of many other astrophysical objects. Similar problems question the context of coronal heating, warming arising in laboratory plasma as well, for instance, when it is warmed up. That problem has a precise analogy in the problem of electrostatic oscillations in cold plasma [24] which was investigated in depth by Barston [5] and by Sedlachek ([21], [22]).

Chen and Hasegawa ([7], [8]) investigated, on the basis of the works by Sedlachek ([21], [22]), the interaction of Alfvén and surface waves on the plasmapause and showed that a weakly damping collective mode must exist on the plasmapause. They applied this theory to the explanation of long-period pulsations excited by pulse sources on the plasmapause. Kivelson and Southwood [11] explored the global oscillation mode arising in the interaction of FMS and Alfvén waves. Zhu and Kivelson [26] found that interaction of FMS and Alfvén waves results in the existence of global weakly damping collective modes. Those modes allow to explain of many observable properties of hydromagnetic waves in the magnetosphere.

Also essential is the analogy between the evolution of the initial perturbation in time t and the propagation of a monochromatic hydromagnetic wave along a magnetic field-line. There is no need to consider these problems separately. The solution of the Cauchy problem over time t can be applied to the problem of propagation along axis z with time t replaced by the spatial coordinate z.

Now let us consider the evolution of an initial perturbation in a simple model of a 1D box with ribs l_x, l_y, l_z. The periodic boundary conditions (5.6) are assumed to be fulfilled on y. Therefore present all the perturbations as a sum of Fourier harmonics $\propto \exp(ik_y y)$ with $k_y = 2\pi m/l_y$. Again, from (4.42a) we obtain equations (5.7a)–(5.8) connecting transversal plasma displacements $\boldsymbol{\xi}_\perp(\xi_x, \xi_y)$ with field-aligned magnetic component $b_\parallel$.

Assume there are no plasma displacements normal to boundaries $x = 0$ ('equatorial ionosphere') and $x = l_x$ ('magnetopause'), i.e.

$$\xi_x|_{x=0;\,l_x} = 0. \tag{5.46}$$

For ideal plasma the latter condition is equivalent to the requirement that the tangential electric components vanish:

$$E_y|_{x=0;\,l_x} = 0.$$

The impedance boundary conditions are satisfied on the lower and upper boundaries of the box:

$$\left(\frac{\partial \xi_\perp}{\partial z} \mp \frac{X_\mp}{c}\frac{\partial \xi_\perp}{\partial t}\right)_{z=0;\,l_z} = 0. \tag{5.47}$$

At the initial moment $t = 0$ plasma displacements and velocities are given by

$$\xi_\perp|_{t=0} = \xi_{\perp 0}(x,z), \qquad \left.\frac{\partial \xi_\perp}{\partial t}\right|_{t=0} = \xi_{\perp 1}(x,z). \tag{5.48}$$

Laplace Transform

The Laplace transform from displacement $\xi_\perp(x,z,t)$ and $b_\parallel(x,z,t)$ is determined in the usual way. Write it in the likewise Fourier form:

$$\xi_{\perp\omega}(x,z) = \int\limits_0^\infty \xi_\perp(x,z,t)\exp(i\omega t)\,\mathrm{d}t, \tag{5.49}$$

where ω is complex in general. Applying this transform to (5.7a)–(5.8), boundary and initial conditions (5.46), (5.47) and (5.48) yields a system of non-uniform equations:

$$L_\omega \xi_{y\omega} = ik_y \frac{b_{\parallel\omega}}{B_0} + f_{y\omega}, \tag{5.50}$$

$$\frac{\mathrm{d}}{\mathrm{d}x}\frac{b_{\parallel\omega}}{B_0} = L_\omega \xi_{x\omega} + f_{x\omega}, \tag{5.51}$$

$$\frac{\mathrm{d}}{\mathrm{d}x}\xi_{x\omega} = -\frac{b_{\parallel\omega}}{B_0} - ik_y\xi_{y\omega}, \tag{5.52}$$

with boundary conditions

$$\xi_x|_{x=0;\,l_x} = 0, \tag{5.53}$$

$$\left(\frac{\mathrm{d}\xi_{\perp\omega}}{\mathrm{d}z} \pm iX_\mp \cdot \frac{\omega}{c}\xi_{\perp\omega}\right)_{z=0;\,l_z} = 0. \tag{5.54}$$

Here

$$f_{x\omega} = -\frac{i\omega\xi_{x0}(x,z) - \xi_{x1}(x,z)}{c_A^2(x)} - \frac{4\pi}{c}\frac{j_{y\omega}^d}{B_0}, \tag{5.55}$$

$$f_{y\omega} = -\frac{i\omega\xi_{y0}(x,z) - \xi_{y1}(x,z)}{c_A^2(x)} - \frac{4\pi}{c}\frac{j_{x\omega}^d}{B_0}, \tag{5.56}$$

$$L_\omega = \frac{\mathrm{d}^2}{\mathrm{d}z^2} + \frac{\omega^2}{c_A^2(x)}. \tag{5.57}$$

At the next step, let us proceed to presenting perturbations as superpositions of oscillations of individual field-lines. For that expand $\xi_{x\omega}$, $\xi_{y\omega}$ and $b_{\|\omega}$ with respect to eigenfunctions $Q_n(z)$ of Dungey's problem (5.9)–(5.10). We have

$$\xi_{x\omega}(x,z) = \sum_{n=0}^{\infty} \xi_{x\omega n}(x) Q_n(z)$$

and analogous expansions for $\xi_{y\omega}$ and $b_{\|\omega}$. Substitute these series into (5.55)–(5.57) and multiply each of the equations by $Q_n(z)$. Integrate the expressions obtained over change interval z and take into account the biorthogonality conditions (5.11), and after performing all these operations, we obtain uncoupled equations for normal oscillations. For n-th mode from (5.55)–(5.57) follows

$$L_{\omega n}\xi_{y\omega n} = ik_y\frac{b_{\|\omega n}}{B_0} + f_{y\omega n}, \tag{5.58}$$

$$\frac{\mathrm{d}}{\mathrm{d}x}\frac{b_{\|\omega n}}{B_0} = L_{\omega n}\xi_{x\omega n} + f_{x\omega n}, \tag{5.59}$$

$$\frac{\mathrm{d}}{\mathrm{d}x}\xi_{x\omega n} = -\frac{b_{\|\omega n}}{B_0} - ik_y\xi_{y\omega n}, \tag{5.60}$$

$$\xi_{xn}\big|_{x=0;l_x} = 0, \tag{5.61}$$

where

$$f_{xn}(x) = \int_0^{l_z} f_x(x,z)Q_n(z)\mathrm{d}z, \qquad f_{yn}(x) = \int_0^{l_z} f_y(x,z)Q_n(z)\mathrm{d}z.$$

Differential operator $L_{\omega n}$ reduces to a multiplication operator

$$L_{\omega n} = -k_n^2 + \frac{\omega^2}{c_A^2(x)}.$$

We find from (5.59) and (5.58) that ξ_x and ξ_y are

$$\xi_{x\omega n} = L_{\omega n}^{-1}\frac{\mathrm{d}}{\mathrm{d}x}\frac{b_{\|\omega n}}{B_0} - L_{\omega n}^{-1}f_{x\omega n}, \tag{5.62a}$$

$$\xi_{y\omega n} = ik_yL_{\omega n}^{-1}\frac{b_{\|\omega n}}{B_0} + L_{\omega n}^{-1}f_{y\omega n}. \tag{5.62b}$$

Substituting these relations into (5.60), we obtain the equation for $b_{\|}$

$$\frac{\mathrm{d}}{\mathrm{d}x}\frac{1}{L_{\omega n}}\frac{\mathrm{d}b_{\|\omega n}}{\mathrm{d}x}+\frac{L_{\omega n}-k_y^2}{L_{\omega n}}b_{\|\omega n}=L_{\omega n}^{-1}\phi_{\omega n} \tag{5.63}$$

with boundary conditions

$$\left.\frac{\mathrm{d}b_{\|\omega n}}{\mathrm{d}x}\right|_{x=0;l_x}=0, \tag{5.64}$$

where

$$\phi_{\omega n}=B_0\frac{\mathrm{d}v\left(L_{\omega n}^{-1}f_{x\omega n}\right)}{\mathrm{d}x}-ik_yB_0L_{\omega n}^{-1}f_{y\omega n}.$$

Find $\xi_{\perp n}(x,t)$ and $b_{\|n}(x,t)$ by applying the inverse Laplace transform to $\xi_{\perp\omega n}(x)$ and $b_{\|\omega n}(x)$. For example, the longitudinal magnetic component is given by

$$b_{\|}(x,t)=\frac{1}{2\pi}\int\limits_{-\infty+i\sigma_0}^{+\infty+i\sigma_0}b_{\|\omega n}(x)\cdot\exp(-i\omega t)\,\mathrm{d}\omega \tag{5.65}$$

and similar expressions for displacements $\xi_{\perp}$. In (5.65) $\sigma_0>0$, the integration path $\Gamma(-\infty+i\sigma_0,+\infty+i\sigma_0)$ passes above all the peculiarities of function $b_{\|\omega n}(x)$ in the complex plane ω.

Find the solution of (5.63) with the right-hand side equal to $\delta(x-\bar{x})$, i.e. find the Green function $G_{\omega n}(x,\bar{x})$ of the boundary problem (5.63), (5.64). For an arbitrary $\phi_{\omega n}(\bar{x})$, perturbations $b_{\|\omega n}(x)$ are determined by the Green integral

$$b_{\|\omega n}(x)=\int\limits_0^{l_x}G_{\omega n}(x,\bar{x})\cdot f_{\omega n}(\bar{x})\mathrm{d}\bar{x}.$$

According to the general theory $G_{\omega n}(x,\bar{x})$ can be expressed in terms of two linearly independent solutions $\varphi_{\omega n}^{(1)}(x)$ and $\varphi_{\omega n}^{(2)}(x)$ of the uniform equation (5.63). The first solution satisfies boundary condition (5.64) at $x=0$ and the other at $x=l_x$. The following formula is valid:

$$G_{\omega n}(x,\bar{x})=J_{\omega n}^{-1}\begin{cases}\varphi_{\omega n}^{(1)}(x)\,\varphi_{\omega n}^{(2)}(\bar{x}) & \text{for } x<\bar{x},\\ \varphi_{\omega n}^{(1)}(\bar{x})\,\varphi_{\omega n}^{(2)}(x) & \text{for } x>\bar{x},\end{cases} \tag{5.66}$$

where

$$J_{\omega n}=\frac{W_{\omega n}(x)}{L_{\omega n}(x)},$$

and it is independent of x.

$$W_{\omega n}(x)=\varphi_{\omega n}^{(1)}(x)\frac{\mathrm{d}\varphi_{\omega n}^{(2)}(x)}{\mathrm{d}x}-\varphi_{\omega n}^{(2)}(x)\frac{\mathrm{d}\varphi_{\omega n}^{(1)}(x)}{\mathrm{d}x}$$

is Wronskian of functions $\varphi_{\omega n}^{(1)}(x)$ and $\varphi_{\omega n}^{(2)}(x)$.

Consider the evolution of the initial perturbation in the simple case of $k_y=0$, when Alfvén and FMS-waves do not interact and we have uncoupled equations.

5.8 Uncoupled Alfvén and FMS-Modes

Alfvén Modes

At $k_y = 0$ from (5.62b) for the Alfvén waves we obtain

$$\xi_{y\omega n} = L_{\omega n}^{-1} \; f_{y\omega n}.$$

Applying to it the inverse Laplace transform, we have

$$\xi_{yn}(x,t) = \frac{1}{2\pi} \int\limits_{-\infty+i\sigma_0}^{+\infty+i\sigma_0} \frac{i\omega\, \xi_{yn0}(x) - \xi_{yn1}(x)}{\omega^2 - k_n^2(\omega)\; c_A^2(x)} \exp\left(-i\omega t\right)\, \mathrm{d}\omega. \tag{5.67}$$

Here the integral external currents are omitted and only initial plasma perturbations are left. For $t > 0$ the integral is along the path $\Gamma(-\infty + i\sigma_0, \infty + i\sigma_0)$ (5.67). If we close the integration path by semicircle Γ' in the low half-plane and consider the integral over the contour $\Gamma + \Gamma'$, then it can be solved using the residuum calculus. Tend the semicircle radius to ∞, then the contribution on the semicircle Γ' vanishes exponentially. The integral is equal to the sum of residues of the integrand in poles which are determined by zeros of the denominator of integrand expression (5.67):

$$\omega^2 - k_n^2(\omega)\; c_A^2(x) = 0. \tag{5.68}$$

A solution of (5.68) at some x has two sets of roots $\omega_n^{(\pm)}(x)$. Then (5.67) reduces to

$$\begin{aligned}
\xi_{yn}(x,t) &= \xi_{yn}^{+}(x,t) + \xi_{yn}^{-}(x,t), \\
\xi_{yn}^{(\pm)}(x,t) &= -i \frac{\left[i\omega_n^{(\pm)} \xi_{yn0}(x) - \xi_{yn1}(x)\right] c_A\,(x) \exp\left(-i\omega_n^{(\pm)}\,(x)\, t\right)}{2\omega_n^{(\pm)}\,(x) \left[1 - c_A\,(x)\, \left.\dfrac{\mathrm{d}k_n\,(\omega)}{\mathrm{d}\omega}\right|_{\omega=\omega_n^{(\pm)}(x)}\right]},
\end{aligned} \tag{5.69}$$

where $\omega_n^{(\pm)}$ are found in (5.20), (5.25) and $k_n\,(\omega)$ is obtained from (5.13). For instance, at $\Sigma_P \to \infty$ from (5.15) we find that

$$1 - c_A\,(x)\, \left.\frac{\mathrm{d}k_n\,(\omega)}{\mathrm{d}\omega}\right|_{\omega=\omega_n^{(\pm)}(x)} \approx 1 - i\frac{2n}{\pi \bar{\Sigma}_P} \left(\frac{\omega_A}{\omega}\right)^2.$$

Then Alfvén oscillations in the box may be considered as a superposition of non-interacting damped oscillations. Each elementary oscillation can be presented as

$$\xi_{yn}(x,z,t) = Q_n(z)\; u_n(x,t),$$

with

$$u_n(x,t) = \xi_{yn}^{+}(x,t) + \xi_{yn}^{-}(x,t).$$

The oscillation caused by an individual field line $Q_n(z)$ can be studied as an eigenvalue of the Dungey's problem (5.9)–(5.10).

For a non-dissipative 1D system $u_n(x,t)$ is found from the equation of the oscillations of an ideal harmonic oscillator

$$\frac{\mathrm{d}^2}{\mathrm{d}t^2}u(x,t) + \omega_{An}^2(x)u(x,t) = -\frac{4\pi}{c}\frac{j_{x\omega}^d(x,t)}{B_0}.$$

where u is the initial displacement and $\mathrm{d}u/\mathrm{d}t$ is plasma velocity at $t = 0$.

Standing Alfvén oscillations of a field line with account taken of the losses at the confined boundaries, are similar to oscillations of a stretched string with energy absorption at the clamped end points. The analogy between field line Alfvén oscillations and oscillations of a string is not only formal in character, but it has a clear physical sense. The restoring force in Alfvén oscillations arises due to the tension of magnetic field-lines (see (4.14)). With dissipation taken into account, the evolution of n-th harmonic of the field line oscillation can be found from the equation of an oscillator with losses

$$\frac{\mathrm{d}^2}{\mathrm{d}t^2}u(x,t) + 2i\gamma_n\frac{\mathrm{d}}{\mathrm{d}t}u(x,t) + \omega_{An}^2(x) = -\frac{4\pi}{c}\frac{j_{x\omega}^d(x,t)}{B_0},$$

where the resonance frequency ω_{An} and the decrement are determined by (5.20)–(5.21).

Cavity Modes

In a cold homogeneous plasma, the absolute value of phase velocity of FMS-waves is independent of the angle between the main magnetic field (see, e.g., (4.41b)). It follows that FMS-waves are not guided by the field-lines and can fill the whole resonance region. Therefore, the normal modes of FMS-wave resonator are often called cavity modes.

Differentiating (5.59) and substituting (5.60) for displacements $\xi_{x\omega n}$, we obtain the equation for FMS-waves in the box model:

$$\frac{\mathrm{d}^2\xi_{x\omega n}}{\mathrm{d}x^2} + \left(\frac{\omega^2}{c_A^2(x)} - k_n^2(\omega)\right)\xi_{x\omega n} = -f_{\xi\omega n}, \tag{5.70}$$

$$\xi_{x\omega n}|_{x=0;\,l_x} = 0. \tag{5.71}$$

The corresponding uniform boundary problem

$$\frac{\mathrm{d}^2\xi_{x\omega n}}{\mathrm{d}x^2} + \left(\frac{\omega^2}{c_A^2(x)} - k_n^2(\omega)\right)\xi_{x\omega n} = 0, \tag{5.72}$$

$$\xi_{x\omega n}|_{x=0;\,l_x} = 0. \tag{5.73}$$

is the generalized Sturm–Liouville problem for determining the resonance frequencies of the cavity modes.

In the non-dissipative systems, the wavenumbers k_n^2 are real and frequency ω independent. Then the boundary problem (5.72)–(5.73) is the usual Sturm–Liouville problem. Its eigenvalues are real, its eigenfunctions form an orthonormalized system, and FMS-plasma perturbations can be presented as a superposition of non-damping normal oscillations. With dissipation taken into account, k_n^2 depends on ω and $\mathrm{Im}k_n^2 \neq 0$. The problem (5.70)–(5.72) has non-zero solutions only at some discrete values of frequency $\omega_{nl} = \omega'_{nl} - i\gamma_{nl}$, $l = 1, 2, 3, \ldots$, with non-zero decrement γ_{nl}.

Let us denote by $V_{nl}(x)$ the eigenfunctions corresponding to resonance frequencies ω_{nl} of (5.71), (5.72) and introduce the designations

$$N_{nl}^{(1)} = \int\limits_0^{l_x} \frac{V_{nl}^2(x)}{c_A^2(x)}\,\mathrm{d}x, \quad N_{nl}^{(2)} = \int\limits_0^{l_x} V_{nl}^2(x)\,\mathrm{d}x.$$

The Green function $G_{\omega n}(x - \bar{x})$ of (5.70)–(5.71) is determined by

$$\frac{\mathrm{d}G_{\omega n}}{\mathrm{d}x^2} + \left(\frac{\omega^2}{c_A^2(x)} - k_n^2(\omega)\right) G_{\omega n} = \delta(x - \bar{x}),$$
$$G_{\omega n}|_{x=0;\, l_x} = 0,$$

and an expression analogous to (5.66), where $J_{\omega n}^{-1}$ is replaced by Wronskian $W_{\omega n}$ of $\varphi_{\omega n}^{(1)}(x)$, $\varphi_{\omega n}^{(2)}(x)$ which are the solutions of (5.72). Let these functions satisfy the boundary conditions

$$\varphi_{\omega n}^{(1)}(x)\Big|_{x=0} = 0, \qquad \frac{\mathrm{d}\varphi_{\omega n}^{(1)}(x)}{\mathrm{d}x}\Bigg|_{x=l_x} = 1, \tag{5.74}$$

$$\varphi_{\omega n}^{(2)}(x)\Big|_{x=l_x} = 0, \qquad \frac{\mathrm{d}\varphi_{\omega n}^{(2)}(x)}{\mathrm{d}x}\Bigg|_{x=l_x} = 1. \tag{5.75}$$

$\varphi_{\omega n}^{(2)}(x)$ is found from the solution of the Cauchy problem (5.72), (5.75) and can therefore be found at all ω. The problem (5.72), (5.74) is not a Cauchy problem. Therefore it may have solutions only at some ω, more precisely at such ω at which a non-trivial solution of (5.72) satisfies the boundary conditions

$$\varphi_{\omega n}^{(1)}(x)\Big|_{x=0} = 0, \qquad \frac{\mathrm{d}\varphi_{\omega n}^{(1)}(x)}{\mathrm{d}x}\Bigg|_{x=l_x} = 0. \tag{5.76}$$

In a sufficiently close neighborhood of eigenfrequencies ω_{nl}, we have $\mathrm{d}\varphi_{\omega n}^{(1)}(l_x)/\mathrm{d}x \neq 0$. Therefore, the conditions (5.71) at $\omega = \omega_{nl}$ can be satisfied by selecting norm $N_{nl}^{(1)}$ of eigenfunctions $V_{nl}(x)$ in such a way as to satisfy the boundary condition (5.74) and by setting

$$\varphi_{\omega n}^{(1)}(x) = V_{nl}(x) \quad \text{at} \quad \omega = \omega_{nl}.$$

Inasmuch as $\varphi^{(1)}_{\omega n}(x)$ is continuous in the vicinity of ω_{nl}, the function $\varphi^{(1)}_{\omega n}(x)$ satisfying (5.74) can be found at some sufficiently small neighborhood of frequency ω_{nl}. The normalization selected here is convenient for further calculations, because functions $\varphi^{(1)}_{\omega n}(x)$ and $\varphi^{(2)}_{\omega n}(x)$ are identical at $\omega = \omega_{nl}$.

Just as for Alfvén waves, calculate the inverse Laplace transform, closing the integration contour in the lower complex half-plane of ω and applying the theorem on residues. The Wronskian of (5.72) is independent of x. Since the solutions $\varphi^{(1)}_{\omega n}(x)$ and $\varphi^{(2)}_{\omega n}(x)$ depend analytically on ω, then $W_{\omega n}$ also depends analytically on ω. Poles of the Green function are determined from

$$W_{\omega n} = 0 \tag{5.77}$$

and coincide with eigenfrequencies ω_{nl}. Indeed, if (5.77) is valid for some ω_*, then

$$\varphi^{(1)}_{\omega n}(x) = \text{const} \cdot \varphi^{(2)}_{\omega n}(x).$$

Here const $= 1$ due to the boundary conditions (5.74), (5.75). Since $\varphi^{(1)}_{\omega n}(0) = \varphi^{(2)}_{\omega n}(x) = 0$ and $\varphi^{(1)}_{\omega n}(l_x) = \varphi^{(2)}_{\omega n}(l_x) = 0$, then ω_* is the eigenfrequency, and $\varphi^{(1)}_{\omega n}(x) = \varphi^{(2)}_{\omega n}(x)$ is the eigenfunction.

Thus displacement $\xi_{x\omega n}$ caused by a point source is given by

$$\xi_{xn}(x,t) = \frac{1}{2\pi} \int\limits_{-\infty + i\sigma_0}^{+\infty + i\sigma_0} G_{\omega n}(x) \cdot \exp(-i\omega t)\, \mathrm{d}\omega .$$

For $t > 0$ close the contour integral in the lower half-plane and obtain an expression for $\xi_{xn}(x,t)$ as the sum of residues

$$\xi_{xn}(x,t) = -i \sum_{l=1}^{\infty} V_{nl}(x)\, V_{nl}(\bar{x}) \left(\frac{\mathrm{d}}{\mathrm{d}\omega}\, W_{\omega n}|_{\omega = \omega_{nl}} \right)^{-1} \exp(-i\omega_{nl} t) . \tag{5.78}$$

Now we should find derivatives of the Wronskian. Substituting the values of functions from boundary conditions (5.74) and (5.75) into the definition of $W_{\omega n}$ at $x = l_x$, we obtain

$$W_{\omega n} = \varphi^{(1)}_{\omega n}(l_x).$$

The equations for V_{nl} and $\varphi^{(1)}_{\omega n}$ are

$$\frac{\mathrm{d}^2 V_{nl}}{\mathrm{d}x^2} + K(x, \omega_{nl}) V_{nl} = 0,$$
$$\frac{\mathrm{d}^2 \varphi^{(1)}_{\omega n}}{\mathrm{d}x^2} + K(x, \omega_n) \varphi^{(1)}_{\omega n} = 0,$$

where $K(x,\omega) = -k_n^2(\omega) + \omega^2/c_A^2$. Multiplying the first equation by $\varphi^{(1)}_{\omega n}$ and the other by V_{nl}, integrating over x from 0 to l_x and subtracting one obtained equality from the other, we have

$$\int_0^{l_x} \left(\frac{\mathrm{d}^2 V_{nl}}{\mathrm{d}x^2} \varphi_{\omega n}^{(1)} - V_{nl} \frac{\mathrm{d}^2 \varphi_{\omega n}^{(1)}}{\mathrm{d}x^2} \right) \mathrm{d}x + \int_0^{l_x} \left(K(x, \omega_{nl}) - K(x, \omega) \right) V_{nl} \, \varphi_{\omega n}^{(1)} \mathrm{d}x = 0.$$

Integration by parts with boundary conditions (5.74), (5.75) yields

$$\varphi_{\omega n}^{(1)}(l_x) - V_{nl}(l_x) = - \int_0^{l_x} \left(K(x, \omega_{nl}) - K(x, \omega) \right) V_{nl} \varphi_{\omega n}^{(1)} \mathrm{d}x.$$

Dividing both parts of the relation over $\omega - \omega_{nl}$ and tending $\omega \to \omega_{nl}$, we have, for the derivative of the Wronskian,

$$\left. \frac{\mathrm{d}W_{\omega n}}{\mathrm{d}\omega} \right|_{\omega_{nl}} = \left. \frac{\varphi_{\omega n}^{(1)}(l_x)}{\partial \omega} \right|_{\omega_{nl}} = 2 \left(\omega_{nl} N_{nl}^{(1)} - k_n(\omega_{nl}) \left. \frac{\mathrm{d}k_n(\omega)}{\mathrm{d}\omega} \right|_{\omega_{nl}} N_{nl}^{(2)} \right)$$

and for displacements

$$\xi_{xn}(x, t) = -\frac{i}{2} \sum_{l=1}^{\infty} \frac{V_{nl}(x) V_{nl}(\bar{x}) \exp\left(-i\omega_{nl} t\right)}{\omega_{nl} N_{nl}^{(1)} - k_n(\omega_{nl}) \left. \frac{\mathrm{d}k_n(\omega)}{\mathrm{d}\omega} \right|_{\omega_{nl}} N_{nl}^{(2)}}.$$

The relationship of the longitudinal magnetic field and plasma displacement is given by

$$\frac{b_{\parallel \omega}}{B_0} = -\frac{\mathrm{d}}{\mathrm{d}x} \xi_{x\omega}.$$

For the Alfvén FLR-periods $T_A(x)$ equal to those shown in Figure 5.9, the cavity resonance periods of the first two harmonics are $T_1 = 64.38\,\mathrm{s}$, $T_2 = 31.16\,\mathrm{s}$. The amplitude distribution of these cavity resonance oscillations is shown in Fig. 5.6. Their decrements $\gamma_1 \approx 0.41 \times 10^{-3}\,\mathrm{s}^{-1}$ and $\gamma_2 \approx 0.18 \times 10^{-3}\,\mathrm{s}^{-1}$ and the Q-factor are 120 and 575, correspondingly.

Such Q-factors are at least 1.5–2 orders higher than the observed Q-factors of the resonance magnetospheric cavity modes. Within the Box model the dissipation is mostly caused by the Joule dissipations at the ends of field-lines (this corresponds to the Joule dissipation in the ionospheres due to Pedersen conductivity). This model is adequate to estimate the Q-factor of FLR, with the disturbances localized in the vicinity of resonance field-lines.

As for the cavity modes, these oscillations cover the whole magnetosphere or a significant part. Azimuthal propagation leads to the dissipation in the magnetotail. The energy also lost itself at the magnetopause, as it only partly reflects the waves and they transmit through the magnetopause into the magnetosheath and then into the interplanetary space.

Thus, FMS-oscillations non-interacting with Alfvén waves are global and fill the whole magnetosphere. In a non-dissipative medium they appear as a

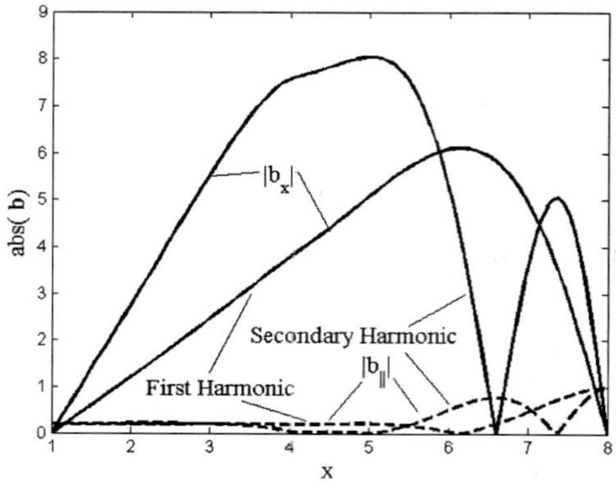

Fig. 5.6. Equatorial distribution of the magnetic field of the first two cavity harmonics

sum of non-damping normal cavity modes, and in a dissipative medium as a sum of damping modes. Since a simple enough pattern of the evolution of plasma perturbations in the hydromagnetic box at $k_y = 0$ was obtained as a result of rather cumbersome calculations demonstrated in the present chapter, we shall conclude it with a brief summary of the main results:

- With perturbations along the coordinate y being constant, plasma oscillations in the box appear as a superposition of non-interacting cavity (FMS) and Alfvén modes.
- In the cavity mode the longitudinal magnetic field and plasma displacement in the direction of the inhomogeneity gradient are finite. Displacement perpendicular to the gradient of the Alfvén velocity vanishes, $\xi_y = 0$.
- In the Alfvén mode component ξ_y is finite, while $b_{\|}$ and ξ_x vanish.
- The properties of cavity modes in the hydromagnetic box are identical to the TE mode in an ordinary electrodynamic resonator filled with an inhomogeneous dielectric. Equation (5.70) is a usual 2D wave equation. Cavity modes have a discrete spectrum of eigenfrequencies ω_{k1}. Arbitrary perturbation with $b_{\|}$ and ξ_x can be presented as a sum of modes, each of which changes harmonically in time. With dissipation taken into account, the harmonics are damped exponentially in time.
- For symmetric harmonics (field-aligned wavenumbers k are odd) (see Figure 5.7) the transverse electric fields and longitudinal magnetic field perturbations have an antinode and the transverse magnetic field has a node in the equatorial plane. For antisymmetric harmonics (k are even) the transverse electric fields and longitudinal magnetic field perturbations have a node, whereas the transverse magnetic field has an antinode in the equatorial plane.

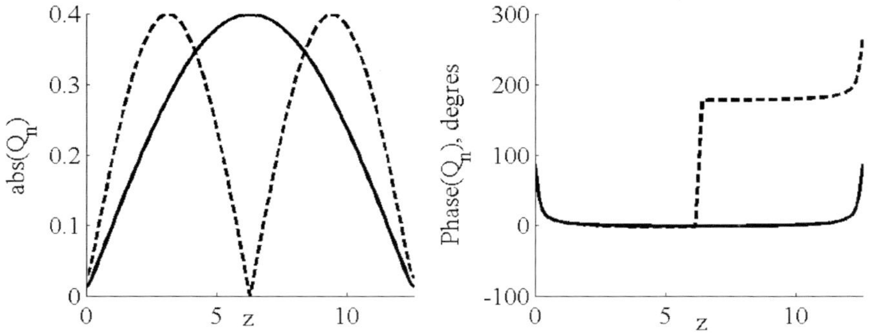

Fig. 5.7. Field-aligned distribution for the first (solid line) and second (dashed line) FLR-harmonics

- Alfvén perturbations are the superposition of non-interacting oscillations of a magnetic shells. All field-lines at the shell x = const and $k_y = 0$ oscillate synchronously. Therefore at each x it is sufficient to describe the oscillations of one of the field-lines at the shell. Each field line can be assigned a corresponding taut elastic string dead at two ends. Perturbation propagates along the string with the Alfvén velocity of the corresponding field-line. Absorption at the ends corresponds to energy losses at string-clamped ends. The oscillation spectrum of each field line (string) is discrete and has a complex value.
- For the case of high Pedersen conductivity peculiar to the dayside ionospheric conditions, displacements ξ_y turn into zero at the ends of the field-line. The same is for rigidly fixed strings. While energy losses at the end points can be disregarded, resonance frequencies are real, and the frequency of the fundamental oscillation harmonics is given by

$$\omega_1(x) = \omega_A(x) = \pi \frac{c_A(x)}{l_z}.$$

The harmonics are equidistant, the resonance frequency of the n-th harmonic is

$$\omega_n^{\pm}(x) = \pm n\omega_A(x), \quad n = 1, 2, 3, \ldots.$$

- At $\Sigma_P \to 0$ (nightside ionosphere) derivative of displacements ξ_y along the field line vanish at the ionospheres. The resonance frequencies $\omega_n^{\pm}(x) = \pm n\omega_A(x)$, $n = 0, 1, 2, 3, \ldots$ are real as well. Displacements are symmetrical with respect to the equatorial plane at even n and antisymmetric at odd n.
- Under finite Σ_P the resonator has complex eigenfrequencies. Each resonance frequency ω_k^+ with $\mathrm{Re}\omega_k^+ > 0$ and $\mathrm{Im}\omega_k^+ < 0$ has a corresponding resonance frequency ω_k^- with $\mathrm{Re}\omega_k^- = -\mathrm{Re}\omega_k^+ < 0$ and $\mathrm{Im}\omega_k^- = \mathrm{Im}\omega_k^+ < 0$.

5.9 Coupling of Alfvén and FMS-Waves

Features

Turn now to the case of $k_y \neq 0$, when Alfvén and FMS-waves interact. This results in the appearance of a number of peculiarities in the behavior of the MHD-waves, the principal of which is the energy transfer from the FMS-mode to the Alfvén mode. When the FMS-oscillation frequency matches the local FLR-frequency, energy transfer between the modes becomes more effective. The spectrum of the MHD-oscillation contains, in the general case, discrete and continuous parts. Energy of normal modes of the discrete part are conserved because of zero dissipation. If $\omega_A(x) = \pi c_A(x)/l_x$ is everywhere a non-constant function, the continuous spectrum includes ω, satisfying at some x to the equation $\omega^2 = \omega_A^2(x)$. The frequency content of the continuous spectrum is independent of k_y, i.e. continuous spectrums for a finite azimuthal wavenumber $k_y \neq 0$ and $k_y = 0$ coincide with each other. An arbitrary disturbance of $b_\|$ can be presented in the coupling case as the integral over the continuous spectrum. An asymptotic time dependence is given by [18]

$$b_\| \propto t^{-1} \exp[\pm i\omega_A(x)t].$$

The behavior in the uncoupling case is quite different. A frequency of the normal mode ω_n of the discrete spectrum can coincide with a frequency of the continuous spectrum. Then, two types of oscillations exist for long times: first, each magnetic shell oscillates at its own frequency $\omega_A(x)$ and goes down as $1/t$. And the second, global oscillations of the whole cavity exist at the resonant frequency ω_n, damping only due to dissipation. However, it can be shown that at $k_y \neq 0$ there is no normal mode transferred into the uncoupled cavity mode. A similar effect was found in electrostatic oscillations in cold inhomogeneous plasma ([21], [22]). A more complicated MHD-model is considered in ([7], [14], [26], [27]).

The problem of the time evolution of an initial disturbance is solved with the Laplace transform over time. In order to perform such calculations, it requires the analysis of the Green function in a manner similar to one applied for the uncoupling case in Section 5.8. In the coupling case the Green function is many-valued and its Riemann surface is many-sheeted. If, on the complex plane ω, there is a cut along the intervals of the continuous spectrum, a one-valued branch of the Green function is found at each sheet. A physical sheet of the Riemann surface is determined by the boundary conditions at this sheet. The poles of the Green function correspond to the resonance frequencies of the global modes. At $k_y \neq 0$ at the physical sheet, there are no poles of the Green function transferring into the poles of the Green function of the uncoupling mode (see Section 5.8).

However, at the non-physical sheet adjacent to the physical sheet, there is a pole of the Green function near the spectral cut at $\omega = \omega_m$, and under $k_y \to 0$ it approaches the cavity mode eigenfrequency. If this pole is near the cut, it is possible to estimate its influence with the help of contour integrating. As a result, exponentially damping disturbances are added to the disturbances damping as $1/t$.

One can consider that these exponentially damping disturbances are small in comparison with ones damping with the power law. However, it is not so. First, these exponential disturbances correspond to global oscillations of the whole cavity. And, on the other hand, their decrement can be small. These modes can exist for a long time before the energy is transferred from a cavity mode to toroidal oscillations of field-lines each oscillating with its own frequency.

Temporal Evolution of Poloidal Alfvén Waves

Following Mann and Wright [14], we shall describe the time evolution of the poloidal Alfvén waves. At $k_y \to 0$, $b_z \to 0$, $\xi_x \to 0$, the azimuthal displacement ξ_y is caused by the toroidal Alfvén waves, and at $k_y \to \infty$, $b_z \to 0$, $\xi_y \to 0$ the displacement ξ_x is caused by the poloidal Alfvén waves. Consider the case of large but finite k_y. Suppose that the displacement $\xi_x(x, t = 0)$ is given at $t = 0$. The longitudinal magnetic field $b_\|$ is not disturbed at the initial time instant, $b_\| \big|_{t=0} = 0$. So we can write

$$\left.\frac{\partial \xi_x}{\partial x}\right|_{t=0} + ik_y \xi_y \left|_{t=0}\right. = 0.$$

Let the plasma at $t = 0$ be stationary. The displacements ξ_x and ξ_y at the next time instants are shown in Figure 5.8 (left frames). The time scale τ is obtained from the condition of the equality of the toroidal and poloidal energies. It is clearly seen from the Fig. 5.8a that the phase mixing exists in both components. At $t = 3\tau$ polarization becomes mostly toroidal. The energy transfer from the poloidal component (solid line) to the toroidal component (dashed line) is shown in Fig. 5.8 (right frame).

5.10 Summary

The main purpose of the box model study is to get the qualitative picture of the penetration of MHD-waves from the magnetospheric boundary into its inner regions and to study the principal features of global resonance and FLR. A penetration pattern of the hydromagnetic waves into the magnetospheric inner regions is contained in (5.7a)–(5.7c).

A scheme of these processes without quantitative details can be summarized as follows. A disturbance propagates as an FMS-wave from the magnetospheric boundary to the turning point x_t, obtained from the condition (4.55) with $k_\| = k_n$

$$k_A^2(x_t) = k_y^2 + k_n^2.$$

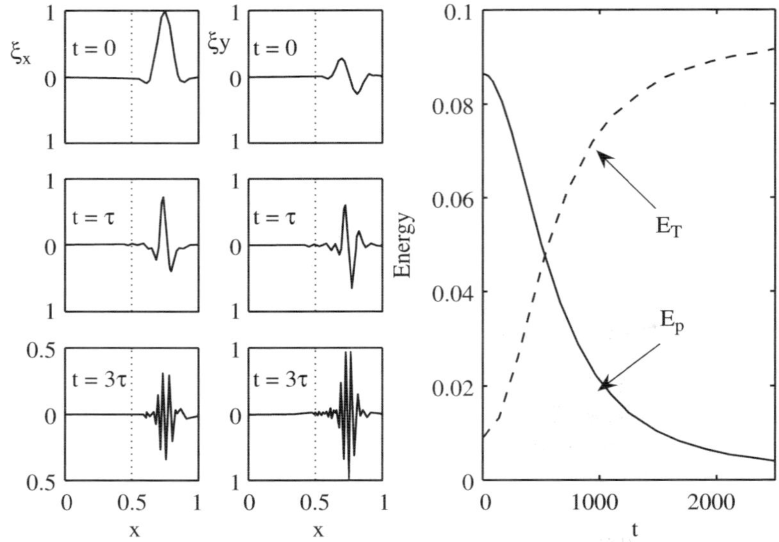

Fig. 5.8. (Left frames) Consecutive couples of figures demonstrate temporal evolution of x-distributions of ξ_x (radial) and ξ_y (azimuthal) displacements. The top panel shows the initial displacement ($t = 0$), the middle is at $t = \tau$ and the bottom is at $t = 3\tau$. The time scale τ corresponds to the time when the energies of toroidal and poloidal components are equal. (Right frame) The time evolution of the poloidal E_p and toroidal E_T energy densities. After Mann and Wright [13]

Behind the turning point FMS-waves damp exponentially up to the magnetic shell with the coordinate x_n where the frequency of the initial FMS-wave becomes equal to the FLR-frequency of the Alfvén oscillations. In the vicinity of $x = x_n$ the amplitude of the wave is growing, and the amplitude distribution tends to the Lorentzian function. This picture can be more complicated because of the existence of several turning points or several FLR points. Due to the tunnel effect, waves propagate through the region of exponential damping up to the next turning point and again gets to the transmission region.

The distribution of FLR-periods shown in Figure 5.9 approximately corresponds to the dependence of the resonance period of the fundamental harmonic on the magnetospheric parameter L. For the disturbances of the period T shown with the horizontal dot-line, there are three FLR-points (vertical solid lines). Thin horizontal line segments shows the region of the wave damping.

The resonant increase of the amplitude of oscillations is also possible at frequencies close to the frequencies of global resonances. Within the box' model, the Q-factor of the cavity spectral line is high, and in the real magnetosphere it may be significantly lower because of losses in the outer magnetosphere and energy transport through the magnetic tail and the magnetopause into the interplanetary space.

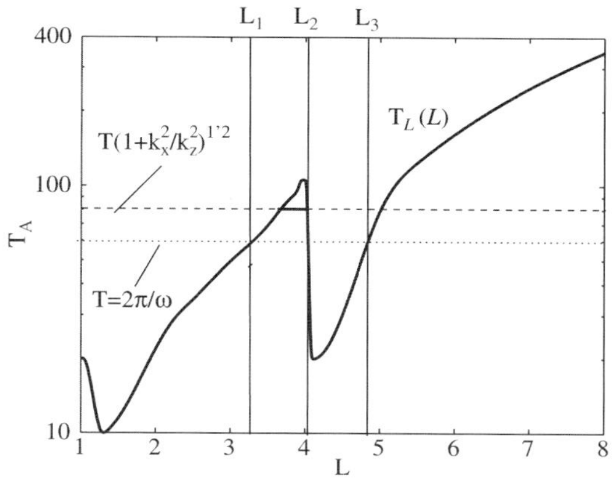

Fig. 5.9. Period T_A of fundamental FLR-harmonics versus the L-shell

References

1. Abokhodair, A. (King Fahd Univ. of Petroleum & Minerals, Saudi Arabia) `http://www.mathworks.com/matlabcentral/fileexchange/loadFile.do?objectId=9397&objectType=FILE` available in 30.01.2007
2. Allan, W., Quarter-wave ULF pulsations, *Planet. Space Sci.*, **31**, 323, 1983.
3. Allan, W. and A. N. Wright, Hydromagnetic wave propagation and coupling in a magnetotai waveguidel, *J. Geophys. Res.*, **103**, 2359, 1998.
4. Allan, W. and A. N. Wright, Magnetotaqil waveguide: Fast and Alfvén waves in the plasma sheet boundary layer and lobe, *J. Geophys. Res.*, **105**, 317, 2000.
5. Barston, E. M., Electrostatic oscillations in inhomogeneous cold plasmas, *Ann. Phys.*, **29**, 282, 1964.
6. Boyce, W. E. and R. C. DiPrima, *Elementary Differential equations and Boundary Value Problems*, 7th edn., John Wiley & Sons, New York, 2001.
7. Chen, L. and A. Hasegawa, A theory of long-periodic magnetic pulsations, 1. Steady state excitation of field line resonance, *J. Geophys. Res.*, **79**, 1024–1032, 1974.
8. Chen, L. and A. Hasegawa, A theory of long-periodic magnetic pulsations, pulse excitation of surface eigenmode, *J. Geoph. Res.*, **79**, 1033–1037, 1974.
9. Dungey, J. W., The structure of the exosphere, or, adventures in velocity in Geophysics, *The Earth's Environment Proceedings of the 1962 Les Houches Summer School*, pp. 503–550, C. DeWitt, J. Hieblot and A. Lebeau, editors, Gordon and Breach, NY, 1963.
10. Kamke, E., Differenrial Gleichungen: Losungsmethoden und Losungen, Teubner, Leipzig, 1959.
11. Kivelson, M. G. and D. J. Southwood, Resonant ULF waves: a new interpretation, *Geophys. Res. Lett.*, **12**, 49, 1985.
12. Kivelson, M. G. and D. J. Southwood, Coupling of global magnetospheric MHD eigenmodes to field line resonances, *J. Geophys. Res.*, **91**, 4345, 1986.
13. Mann, I. R. and A. N. Wright, Finite lifetime of ideal poloidal Alfvén waves, *J. Geophys. Res.*, **100**, 677, 1995.

14. Mann, I. R., A. N. Wright, and P. S. Cally, Coupling of magnetospheric cavity modes to field line resonances: A study of resonance widths, *J. Geophys. Res.*, **100**, 441, 1995.
15. Naimark, M. A., *Linear differential operators*, 'Nauka', Moscow, 1969.
16. Newton, R. S., D. J. Southwood, and W. J. Hughes, Damping of geomagnetic pulsations by the ionosphere, *Planet. Space Sci.*, **26**, 201, 1978.
17. Radoski, H. R., Magnetic toroidal resonances and vibrating field-lines, *J. Geophys. Res.*, **71**, 1891, 1966.
18. Radoski, H. R., Highly asymmetric MHD resonances: The guided poloidal mode, *J. Geophys. Res.*, **72**, 4026, 1967.
19. Radoski, H. R., A note on oscillating field-lines, *J. Geophys. Res.*, **72**, 418, 1967.
20. Radoski, H. R., A theory of latitude dependent geomagnetic micropulsations: the asymptotic fields, *J. Geophys. Res.*, **79**, 596, 1974.
21. Sedlachek, Z., Electrostatic oscillations in cold inhomogeneous plasma, 1. Differential equation approach, *J. Plasma Phys.*, **5**, 239, 1971.
22. Sedlachek, Z., Electrostatic oscillations in cold inhomogeneous plasma, 2. Integral equation approach, *J. Plasma Phys.*, **6**, 187, 1971.
23. Southwood, D. J., Some features of field line resonance in the magnetosphere, *Planet. Space Sci.*, **22**, 483, 1974.
24. Uberoi, C. Alfvén waves in inhomogeneous magnetic fields, *Phys. Fluids*, **15,** 1673, 1972.
25. Wright, A. N., W. Allan, R. D. Elpninstone, and L. L. Cogger, Phase mixing and phase motion of Alfvén waves on tail-like and dipole-like magnetic field-lines, *J. Geophys. Res.*, **104**, 10159, 1999.
26. Zhu, X. M. and M. G. Kivelson, Analytic formulation and quantitative solutions of the coupling ULF problem, *J. Geophys. Res.*, **93**, 8602, 1988.
27. Zhu, X. M. and M. G. Kivelson, Global mode ULF pulsations in magnetosphere with nonmonotonic Alfvén velocity profile, *J. Geophys. Res.*, **94**, 1479, 1989.

6

FLR in Plasma Configurations

6.1 Introduction

In Chapter 5 the FLR-theory was formulated within the framework of a 'plasma box' model: with 1D inhomogeneity across straight field-lines. Despite the seeming simplicity of the model, the spectral properties of the relevant system of MHD equations turned out to be non-trivial. Such a magnetospheric model is a rare example of a spatially confined physical system in which a continuous spectrum appears. The eigenmode equation for standing toroidal shear Alfvén waves has been derived for the dipole case by Dungey [17], Cummings et al. [14] and by many other authors. Basic ideas of FLR-theory that guided much of the subsequent research of the FLR in complex plasma configurations can be found in ([7], [50]), the basic mathematics is given in [30]. Numerous papers are devoted to consideration of 2-D and 3-D cases and semi-kinetic approaches (see [2], [9], [25], [26], [27], [31], [35], [49], [51], [55], [59], [64], [65] and references therein).

A general method for analyzing singularities near resonance magnetic shells is based on expansion of the wave field into series over the distance to a resonance field line $x - x_0$. The series may include power and logarithmic singularities ([40], [59], [65]). This method is in fact a direct generalization of the Frobenius method (see, e.g., [6], [63]) applied to multidimensional systems. The method enables the field structure to be studied allowing for a number of additional factors, i.e. inhomogeneous plasma distribution, variable curvature of magnetic field-lines, both the Hall and Pedersen conductivities of plasma boundaries (conjugate ionospheres). In particular, the axially symmetric case with field-lines orthogonal to the ionospheres was considered in ([40], [65]) and for inclined lines in [21].

Another approach to the construction of the FLR-theory is based on the presentation of wave fields as superposed oscillations of field line harmonics. Amplitudes of the harmonics are determined by the infinite system of coupled ordinary differential equations. This method is similar to the method of modulated normal modes in the theory of radio-wave propagation.

The interaction of FMS-waves with Alfvén and slow magnetosonic modes in a straight magnetic field has been studied in (see, e.g., [33], [45]) and in the general case in ([10], [12], [28], [52]). It was shown that within the framework of linear magnetohydrodynamics with a finite plasma pressure, MHD-disturbances still remain singular at the field-lines where the resonance requirements hold.

6.2 2D Inhomogeneous Plasma in a Uniform Magnetic Field

Basic Equations

Consider linear oscillations of a cold plasma embedded in a box. Let the z-axis of the rectangular coordinate system be oriented along a homogeneous magnetic field $\mathbf{B}_0$. Plasma inhomogeneity in two directions ($\rho_0 = \rho_0(x, z)$) is confined in a box with dimensions l_x, l_y, l_z. Maxwell's equations and the linearized system of the ideal MHD-equations lead to the equation system (4.43)–(4.45). For the Fourier harmonic of displacement

$$(\xi_x, \xi_y)\exp(-i\omega t + ik_y y),$$

the system reduces to

$$\mathbf{L}\xi_y = -ik_y b, \tag{6.1}$$

$$\frac{\partial b}{\partial x} = -\mathbf{L}\xi_x, \tag{6.2}$$

$$\frac{\partial \xi_x}{\partial x} = b - ik_y\xi_y. \tag{6.3}$$

Here

$$\mathbf{L} = \frac{\partial^2}{\partial z^2} + \frac{\omega^2}{c_A^2(x, z)} \tag{6.4}$$

$c_A(x, z) = B_0/\sqrt{4\pi\rho_0(x, z)}$ is an Alfvén velocity, $b = b_z/B_0$ is dimensionless longitudinal wave magnetic field, and k_y is a value of the azimuthal wave vector. Recall that in the box model the faces $z = 0$, and $z = l_z$ simulate the southern and northern ionospheres, respectively. Neglecting dissipation in the ionosphere, we put that tangential components of the electric field be vanish on these faces. Then (6.1)–(6.3) should be supplemented by the boundary conditions

$$\xi_x = 0 \qquad \xi_y = 0, \qquad \text{at} \qquad z = 0,\ l_z. \tag{6.5}$$

Matrix Presentation of the Box Equations

First, we shall in this section, present the 2D boundary problem (6.1)–(6.3) and (6.5) in a matrix form. Then, we obtain an infinite system of matrix equations which we shall reduce to a finite equation system. Dungey's boundary problem, i.e. the problem of Alfvén eigenoscillations between two highly conductive ionospheres, can be written

$$\left(\frac{\mathrm{d}^2}{\mathrm{d}z^2} + \frac{\omega_n^2}{c_A^2(x,z)}\right) Q_n(x,z) = 0,$$
$$Q_n(x,0) = Q_n(x,l_z) = 0.$$

Since the eigenfunctions of this problem form an orthonormal basis, i.e.

$$\left\langle \frac{Q_n(x,z)Q_m(x,z)}{c_A^2} \right\rangle = \int\limits_0^{l_z} \frac{Q_n(x,z)Q_m(x,z)}{c_A^2} \mathrm{d}z = \delta_{nm}, \tag{6.6}$$

the solutions of (6.1)–(6.3) and (6.5) may be sought in the form of a decomposition over this basis

$$\xi_y = a_m\,(x)\,Q_m(x,z), \quad \xi_x = c_m\,(x)\,Q_m(x,z), \quad b = b_m\,(x)\,Q_m(x,z). \tag{6.7}$$

The repeated index convention for summation is systematically used here and below in this chapter. By substituting series (6.7) into (6.1)–(6.3) with the allowance for

$$\mathbf{L}Q_m = \frac{\omega^2 - \omega_m^2}{c_A^2} Q_m,$$

where the operator $\mathbf{L}$ is defined in (6.4), we have

$$\frac{\omega^2 - \omega_m^2}{c_A^2} a_m Q_m = -ik_y b_m Q_m, \tag{6.8}$$

$$b'_m Q_m = -\left\{\frac{\omega^2 - \omega_m^2}{c_A^2} c_m Q_m + b_m Q'_m\right\}, \tag{6.9}$$

$$c'_m Q_m = -ik_y a_m Q_m + b_m Q_m - c_m Q'_m, \tag{6.10}$$

where the prime indicates differentiation with respect to x.

In order to derive expressions $a_n(x)$ in terms of $b_n(x)$ and to resolve the system of equations with respect to $b'_n(x)$ and $c'_n(x)$, let us multiply (6.8) by $Q_n(x,z)$, while (6.9) and (6.10) multiply by $Q_n(x,z)/c_a^2(x,z)$. Then by integrating the obtained expressions from 0 to l_z with conditions (6.6) we find the relation of a_n and b_m

$$a_n = -\frac{ik_y}{\omega^2 - \omega_n^2(x)} \langle Q_n Q_m \rangle b_m, \tag{6.11}$$

and a system for b_n and c_n

$$b_n' = -\left\langle \frac{Q_n Q_m')}{c_A^2(x,z)} \right\rangle b_m - (\omega^2 - \omega_m^2(x)) \left\langle \frac{Q_n Q_m}{c_A^4(x,z)} \right\rangle c_m,$$

$$c_n' = -\left\{ \delta_{mn} - \frac{k_y^2}{\omega^2 - \omega_n^2(x)} \langle Q_n Q_m \rangle \right\} b_m - \left\langle \frac{Q_n Q_m'}{c_A^2(x,z)} \right\rangle c_m. \tag{6.12}$$

In a simple model of a one-dimensional plasma box ($c_A = c_A(x)$), the different harmonics of a field line (i.e., different Fourier components) do not interact with each other. Therefore, the infinite system of coupled equations splits into independent equations (5.26)–(5.28). In a 2D case, i.e. $c_A = c_A(x,z)$, however, harmonics interact and an infinite system of linked equations is obtained.

Let us assume that the condition of an Alfvén resonance is satisfied at $x = x_0$, i.e., for some harmonic with number s, the function $\omega^2 - \omega_s^2(x)$ has a simple zero at x_0. Then the point x_0 is a regular singularity of the system (6.12). Let us multiply (6.12) by $x - x_0$ and rewrite the obtained equations in the matrix form:

$$(x - x_0)\,\mathbf{u}'(x) = \mathbf{T}(x)\,\mathbf{u}(x), \quad \mathbf{u}(x) = \begin{pmatrix} \mathbf{u}_1(x) \\ \mathbf{u}_2(x) \end{pmatrix},$$

$$\mathbf{T} = \begin{pmatrix} (x - x_0)\,\mathbf{R} & (x - x_0)\,\mathbf{S} \\ \mathbf{Q} & (x - x_0)\,\mathbf{R} \end{pmatrix}. \tag{6.13}$$

$\mathbf{u}_1(x)$ and $\mathbf{u}_2(x)$ are column vector functions with elements $b_1(x)$, $b_2(x)$, ... and $c_1(x)$, $c_2(x)$, ... , respectively; $\mathbf{T}(x)$ denotes a matrix function of x which is regular at $x - x_0$; and $\mathbf{Q}$, $\mathbf{R}$, and $\mathbf{S}$ are matrices with elements

$$Q_{nm} = (x - x_0)\left(\delta_{nm} - \frac{k_y^2}{\omega^2 - \omega_n^2(x)} \langle Q_n Q_m \rangle \right),$$

$$R_{nm} = -\left\langle \frac{Q_n Q_m}{c_A^2} \right\rangle,$$

$$S_{nm} = -(\omega^2 - \omega_m^2(x)) \left\langle \frac{Q_n Q_m}{c_A^4} \right\rangle.$$

We expand matrix $\mathbf{T}(x)$ into Taylor series in the vicinity of the point x_0

$$\mathbf{T}(x) = \mathbf{T}_0 + (x - x_0)\mathbf{T}_1 + (x - x_0)^2\mathbf{T}_2 + \cdots.$$

Here

$$\mathbf{T}_0 = \mathbf{T}(x_0) = \begin{pmatrix} \mathbf{0} & \mathbf{0} \\ \mathbf{Q}_0 & \mathbf{0} \end{pmatrix}, \qquad \mathbf{T}_1 = \begin{pmatrix} \mathbf{R}_0 & \mathbf{S}_0 \\ \mathbf{Q}_1 & \mathbf{R}_0 \end{pmatrix}, \ldots;$$

$\mathbf{0}$ is a zero matrix,

$$\mathbf{Q}_0 = \mathbf{Q}(x_0) \qquad (\mathbf{Q}_0)_{nm} = q_m \delta_{ns},$$

where

$$q_m = k_y^2 \frac{\langle Q_s(x_0, z) Q_m(x_0, z) \rangle}{\dfrac{\mathrm{d}}{\mathrm{d}x} \omega_s^2(x_0)}. \tag{6.14}$$

It is important that $\mathbf{T}_0^2 = \mathbf{0}$. The vector equation (6.13) can be replaced by the matrix differential equation

$$(x - x_0)\mathbf{U}'(x) = \mathbf{T}(x)\,\mathbf{U}(x). \tag{6.15}$$

Here $\mathbf{U}(x)$ denotes a matrix in which every vector column is the solution of the system (6.13).

Field Pattern Close to FLR-Shell

First of all let us consider the particular case

$$\mathbf{T}(x) = \mathbf{T}_0, \quad \text{i.e.} \quad \mathbf{T}_1 = \mathbf{T}_2 = \ldots = \mathbf{0}.$$

and make change of variables

$$\tau = \ln[k_y(x - x_0)]$$

that transforms (6.15) into equation with constant coefficients

$$\frac{d\mathbf{U}}{dt} = \mathbf{T}_0 \mathbf{U}.$$

A fundamental matrix of this equation can be written as

$$\mathbf{U} = \exp(\mathbf{T}_0 \tau) = \exp\left\{\mathbf{T}_0 \left[k_y(x - x_0)\right]\right\},$$

where, as usual,

$$\exp(\mathbf{T}_0 \tau) = \mathbf{1} + \mathbf{T}_0 \tau + \frac{\mathbf{T}_0^2 \tau^2}{2} + \cdots.$$

In the general case, a solution of (6.15) can be presented in the form [63]:

$$\mathbf{U}(x) = \mathbf{P}(x) \exp\left\{\mathbf{T}_0 \ln\left[k_y(x - x_0)\right]\right\}, \tag{6.16}$$

where

$$\mathbf{P}(x) = \mathbf{P}_0 + (x - x_0)\,\mathbf{P}_1 + (x - x_0)^2\,\mathbf{P}_2 + \cdots$$

is a regular matrix function. By virtue of the equality $\mathbf{T}_0^2 = \mathbf{0}$, we have

$$\mathbf{U}(x) = \left[\mathbf{P}_0 + (x - x_0)\,\mathbf{P}_1 + (x - x_0)^2\,\mathbf{P}_2 + \cdots\right] \times \left\{\mathbf{1} + \mathbf{T}_0 \ln\left[k_y(x - x_0)\right]\right\}. \tag{6.17}$$

The obtained general solution of the system (6.15) fully proves the main results of the theory of FLR for a 2D inhomogeneous plasma: a wave field has a singularity near a resonance field-line.

The scheme of calculations of coefficient matrices in expansion (6.17) is the following. Inserting a solution in the form of (6.5) into (6.15) and equating coefficients at the same powers of $x - x_0$, we obtain a system of recurrent equations for coefficients $\mathbf{P}_n$:

$$n\mathbf{P}_n = [\mathbf{T}_0\mathbf{P}_n] + \sum_{\nu=0}^{n-1} \mathbf{T}_{n-\nu}\mathbf{P}_\nu, \qquad n = 0, 1, 2, \ldots . \tag{6.18}$$

Here the equality

$$[\mathbf{AB}] = \mathbf{AB} - \mathbf{BA}.$$

is the commutator of matrices $\mathbf{A}$ and $\mathbf{B}$. When $n = 0$ the relationship (6.18) becomes

$$[\mathbf{T}_0\mathbf{P}_0] = \mathbf{0}.$$

It can be put here that $\mathbf{P}_0 = \mathbf{1}$; this choice also ensures the invertibility of the matrix $\mathbf{U}$. Next, the coefficients $\mathbf{P}_1, \mathbf{P}_2, \ldots$ are determined from (6.18). Now we take into consideration the block structure of matrices $\mathbf{T}_0, \mathbf{T}_1, \ldots$ and a specific form of the block $\mathbf{Q}_0$. There is only one non-zero row in the latter block. The following identities hold:

$$\mathbf{S}_0\mathbf{Q}_0 = \mathbf{0}, \quad \mathbf{Q}_0\mathbf{S}_0 = \mathbf{0}.$$

The first identity is evident: the s-th column of matrix $\mathbf{S}_0$ is zero while in $\mathbf{Q}_0$ all the elements, besides probably some elements of the s-th line, are zeros. Note that

$$\langle Q_s Q_k\rangle \left\langle \frac{Q_k Q_m}{c_A^4} \right\rangle = \left\langle Q_s \left\langle \frac{Q_m}{c_A^2}\frac{Q_k}{c_A^2} \right\rangle Q_k \right\rangle = \left\langle \frac{Q_s Q_m}{c_A^2} \right\rangle = \delta_{sm},$$

then the other identity is

$$\begin{aligned} (\mathbf{Q}_0\mathbf{S}_0)_{nm} &= -\left(\omega^2 - \omega_m^2(x_0)\right) k_y^2 \left[\frac{\mathrm{d}}{\mathrm{d}x}\omega_s^2(x_0)\right]^{-1} \delta_{ns} \langle \mathbf{Q}_s\mathbf{Q}_k\rangle \left\langle \frac{\mathbf{Q}_k\mathbf{Q}_m}{c_A^4} \right\rangle \\ &= -\left(\omega^2 - \omega_m(x_0)\right) k_y^2 \left[\frac{\mathrm{d}}{\mathrm{d}x}\omega_s^2(x_0)\right]^{-1} \delta_{ns}\delta_{sm} = 0. \end{aligned}$$

Again, here we make summation over repeated indexes. Thus

$$\mathbf{T}_0\mathbf{T}_1 = \begin{bmatrix} \mathbf{0} & \mathbf{0} \\ \mathbf{R}_0\mathbf{Q}_0 & \mathbf{0} \end{bmatrix}, \quad \mathbf{T}_1\mathbf{T}_0 = \begin{bmatrix} \mathbf{0} & \mathbf{0} \\ \mathbf{R}_0\mathbf{Q}_0 & \mathbf{0} \end{bmatrix},$$

$$\mathbf{T}_0\mathbf{T}_1\mathbf{T}_0 = \mathbf{0}, \quad (\mathbf{T}_0\mathbf{T}_1)^2 = (\mathbf{T}_1\mathbf{T}_0)^2 = \mathbf{0}, \quad \mathbf{T}_0\mathbf{T}_1^2\mathbf{T}_0 = \mathbf{0}.$$

Hence, we obtain for $\mathbf{P}_1, \mathbf{P}_2$ the considerably simplified expressions

$$\begin{aligned}
\mathbf{P}_1 &= \mathbf{T}_1 + [\mathbf{T}_0\mathbf{T}_1]\,, \\
\mathbf{P}_2 &= \frac{1}{2}\left(\mathbf{T}_1\mathbf{T}_0\mathbf{T}_1 - \mathbf{T}_1^2\mathbf{T}_0 + \mathbf{C}_2\right) + \frac{1}{4}\left[\mathbf{T}_0\mathbf{C}_2\right] - \frac{1}{4}\mathbf{T}_0\mathbf{T}_2\mathbf{T}_0, \\
\mathbf{C}_2 &= \mathbf{T}_2 + \mathbf{T}_1^2.
\end{aligned}$$

The coefficients in the singular part of (6.17) are even more simple:

$$\begin{aligned}
&\mathbf{P}_0\mathbf{T}_0 = \mathbf{T}_0, \qquad \mathbf{P}_1\mathbf{T}_0 = \mathbf{T}_1\mathbf{T}_0, \\
&\mathbf{P}_2\mathbf{T}_0 = \frac{1}{2}(\mathbf{T}_2 + \mathbf{T}_1^2)\mathbf{T}_0 + \frac{1}{4}\mathbf{T}_0\mathbf{T}_2\mathbf{T}_0.
\end{aligned}$$

Finally, we obtain in block form

$$\begin{aligned}
&\mathbf{P}_1\mathbf{T}_0 = \begin{pmatrix} \mathbf{0} & \mathbf{0} \\ \mathbf{R}_0\mathbf{Q}_0 & \mathbf{0} \end{pmatrix}, \qquad \mathbf{P}_1 = \begin{pmatrix} \mathbf{R}_0 & \mathbf{S}_0 \\ \mathbf{Q}_1 + [Q_0 R_0] & \mathbf{R}_0 \end{pmatrix}, \\
&\mathbf{P}_2\mathbf{T}_0 = \frac{1}{2}\begin{pmatrix} (\mathbf{S}_1 + \mathbf{S}_0\mathbf{R}_0)\,\mathbf{Q}_0 & \mathbf{0} \\ \left(\mathbf{R}_1 + \mathbf{R}_0^2 + \frac{1}{2}\mathbf{Q}_0\mathbf{S}_1\right)\mathbf{Q}_0 & \mathbf{R}_0 \end{pmatrix}.
\end{aligned}$$

Now we obtain the exact expressions describing plasma behavior near a resonance point x_0. It is convenient to present the matrix solution (6.5) in a block form

$$\mathbf{U} \equiv \begin{pmatrix} \mathbf{U}^{(1)} & \mathbf{U}^{(3)} \\ \mathbf{U}^{(2)} & \mathbf{U}^{(4)} \end{pmatrix} = \begin{pmatrix} \mathbf{P}^{(3)}\mathbf{Q}_0 & \mathbf{0} \\ \mathbf{P}^{(4)}\mathbf{Q}_0 & \mathbf{0} \end{pmatrix} \ln \bar{x} + \begin{pmatrix} \mathbf{P}^{(1)} & \mathbf{P}^{(3)} \\ \mathbf{P}^{(2)} & \mathbf{P}^{(4)} \end{pmatrix},$$

where $\bar{x} = k_y(x - x_0)$. It is evident from this expression that the singularity $\ln \bar{x}$ exists in many solutions of the fundamental system (namely in half of the columns of matrix $\mathbf{U}(x)$). However, it turns out that it is possible to find a fundamental matrix $\mathbf{U}(x)$ in which only one column has a logarithmic singularity, but all other columns have no singularity at $x = x_0$. Due to a special form of matrix $\mathbf{Q}$, all the columns of block $\mathbf{P}^{(i)}\mathbf{Q}_0$ $(i = 3, 4)$ in the matrix of coefficients at $\ln \bar{x}$ are collinear to the s-th column of block $\mathbf{P}^{(i)}$:

$$(\mathbf{P}^{(i)}\mathbf{Q}_0)_{nm} = \mathbf{P}^{(i)}_{n\nu}\delta_{\nu s} q_m = q_m \mathbf{P}^{(i)}_{ns}.$$

Hence, subtracting from all other columns of the left-hand part of matrix $\mathbf{U}$, the s-th column of $\mathbf{U}$ multiplied by q_m/q_s $(q_s \neq 0)$, see (6.14)) we eliminate a singularities from all columns of $\mathbf{U}$ except of the s-th column. Then, we get

$$\begin{aligned}
&q_m \mathbf{P}^{(1)}_{ns} \ln \bar{x} + \mathbf{P}^{(i-2)}_{nm} - \frac{q_m}{q_s}(q_s \mathbf{P}^{(i)}_{ns} \ln \bar{x} + \mathbf{P}^{(i-2)}_{ns}) \\
&= \mathbf{P}^{(i-2)}_{nm} - \frac{q_m}{q_s}\mathbf{P}^{(i-2)}_{ns}, \qquad i = 3, 4, m \neq s. \qquad (6.19)
\end{aligned}$$

The relevant singular solution of system (6.1)–(6.3) occurring as a result of the summation of series (6.7) is

$$\begin{pmatrix} \mathbf{b}_s^{(L)}(x,z) \\ \xi_{xs}^{(L)}(x,z) \end{pmatrix} = \left\{ \begin{pmatrix} \mathbf{P}_{ns}^{(1)}(x) \\ \mathbf{P}_{ns}^{(2)}(x) \end{pmatrix} + q_s \ln \bar{x} \begin{pmatrix} \mathbf{P}_{ns}^{(3)}(x) \\ \mathbf{P}_{ns}^{(4)}(x) \end{pmatrix} \right\} \mathbf{Q}_n(x,z). \tag{6.20}$$

The set of regular solutions corresponding to columns of the left-hand part of matrix $\mathbf{U}(x)$ is (see (6.19))

$$\begin{pmatrix} \mathbf{b}_m^{(L)}(x,z) \\ \xi_{xm}^{(L)}(x,z) \end{pmatrix} = \begin{pmatrix} \mathbf{P}_{nm}^{(1)}(x) - \dfrac{q_m}{q_s}\mathbf{P}_{ns}^{(1)}(x) \\ \mathbf{P}_{nm}^{(2)}(x) - \dfrac{q_m}{q_s}\mathbf{P}_{ns}^{(2)}(x) \end{pmatrix} \mathbf{Q}_n(x,z), \quad m \neq s. \tag{6.21}$$

Regular solutions, obtained from columns of the right-hand part of the matrix $\mathbf{U}$, are

$$\begin{pmatrix} \mathbf{b}_m^{(R)}(x,z) \\ \xi_{xm}^{(R)}(x,z) \end{pmatrix} = \begin{pmatrix} \mathbf{P}_{nm}^{(3)}(x) \\ \mathbf{P}_{nm}^{(4)}(x) \end{pmatrix} \mathbf{Q}_n(x,z), \quad m = 1,2,\ldots. \tag{6.22}$$

From (6.7) and (6.11) we find the y-component of plasma displacement $\xi_y(x,z)$ in the form

$$\xi_y(x,z) = -ik_y \frac{\langle Q_n \mathbf{b}\rangle Q_n(x,z)}{\omega^2 - \omega_n^2(x)},$$

where $\mathbf{b} = \mathbf{b}_m^{(L)}$ or $\mathbf{b} = \mathbf{b}_m^{(R)}$.

The behavior of plasma disturbances near a resonance point is determined by the singular solution (6.20). Using the expressions for Taylor series coefficients of matrix functions $\mathbf{P}(x)$ and $\mathbf{P}(x)\mathbf{T}_0$ and the decomposition

$$Q_n(x,z) = Q_n(x_0,z) + Q_n'(x_0,z)(x - x_0) + \cdots,$$

with $u = x - x_0$, we obtain

$$b_s^{(L)}(x,z) = A_s \left\{ \begin{array}{l} Q_s + \mathrm{O}\left(u^2\right) + \\ \left(q_s \dfrac{\omega^2\left(c_a^2\left(x_0\right)\right)}{2c_a^4\left(x_0\right)} Q_s u^2 + \mathrm{O}\left(\mathrm{u}^3\right) \right) \times \ln\left[k_y u\right] \end{array} \right\}, \tag{6.23}$$

$$\xi_{xs}^{(L)}(x,z) = A_s \left\{ \begin{array}{l} \left(Q_s + R(z)\right) u + \mathrm{O}\left(\mathrm{u}^2\right) + \\ \left(q_s Q_s \left(1 + \dfrac{k_y^2 u^2}{4}\right) + \mathrm{O}\left(\mathrm{u}^3\right) \right) \times \ln\left[k_y u\right] \end{array} \right\}, \tag{6.24}$$

$$\xi_{ys}^{(L)}(x,z) = A_s \left\{ \begin{array}{l} -\dfrac{q_s Q_s}{ik_y u} - \dfrac{R(z)}{ik_y} + \mathrm{O}(u) + \\ \left(\dfrac{1}{2} ik_y q_s Q_s u + \mathrm{O}\left(u^2\right) \right) \times \ln\left[k_y u\right] \end{array} \right\}. \tag{6.25}$$

Here $Q_s \equiv Q_s(x_0, z)$, and

$$R(z) = \frac{k_y^2}{2}\left(\frac{\langle Q_s Q_s\rangle}{(\omega_s^2)'}\right)'_{x=x_0} Q_s + k_y^2 \frac{\langle Q_s Q_s\rangle}{(\omega_s^2)'} Q_s|_{x=x_0}$$
$$-k_y^2 \sum_{n\neq 3} \frac{\langle Q_n Q_s\rangle|_{x=x_0}}{\omega^2 - \omega_n^2(x_0)} Q_n(x_0, z).$$

The quantity A is an arbitrary constant which determines the amplitude of resonance Alfvén oscillations.

Singular terms in expressions (6.24) and (6.25) for displacements ξ_x and ξ_y are determined (with an accuracy of up to a normalization factor). However, any regular solution can be added to the regular terms in (6.23) and (6.25) for b_s and ξ_y.

Dissipation

Singular solution (6.20) includes terms with $\ln[k_y(x-x_0)]$, which has a branch point at $x = x_0$. When the integration path goes below (above) x_0, then $i\pi(-i\pi)$ is added to the logarithm. Proper rules for integration over the singular point necessary for making the solution single-valued, can be derived, for example, by taking account of the Joule dissipation.

The obtained results can be generalized to the case of finite ionospheric Pedersen conductivity. For this purpose, the method used should be modified. The necessity for the modification is caused by the fact that when we use (6.1)–(6.3), we come to the generalized nonselfconjugated eigenvalue problem. In this case, the eigenfrequency ω occurs to second power in Dungey's problem. On the other hand, in the case of an infinitely thin ionosphere, the frequency ω occurs in boundary conditions. Eigenfrequencies become complex and corresponding eigenfunctions do not satisfy the orthogonality condition (6.6).

However, an examination of the singularity at a resonance point by the Frobenius method for partial differential equations still remains valid inclusive of the Joule dissipation as well. The main result of such investigation is the following. Taking into account the Joule losses, logarithmic and power singularities are conserved, but real x_0 should be replaced by some complex value w_0. When losses are weak, w_0 can be found from the approximate relationship (5.38):

$$w_0 = x_0 + i\delta_s, \qquad \delta_s \approx \frac{\gamma_s(x_0)}{\omega_s'(x_0)}. \tag{6.26}$$

Here δ_s is the half-width of a resonance region; $\omega_s(x)$ is the frequency of the s-th FLR-harmonic, determined from Dungey's problem without losses; γ_s is the damping rate of the s-th harmonic.

A dissipation results in the damping of oscillations, so

$$\gamma_s > 0.$$

Hence, as it follows from (6.26), when

$$\omega_s'(x_0) > 0,$$

the parameter

$$\delta_s > 0.$$

Then the integration path goes below a resonance point and we have

$$\ln(x - x_0)|_{x=x_0+0} = \ln(x - x_0)|_{x=x_0-0} + i\pi$$

for weak losses. In the opposite case, when

$$\omega_s'(x_0) < 0,$$

and

$$\delta_s < 0,$$

the integration path goes above the resonance point x_0, and

$$\ln(x - x_0)|_{x=x_0+0} = \ln(x - x_0)|_{x=x_0-0} - i\pi.$$

Above, we assumed that for each z, there exists a complex function $c_A(w = x + iv, z)$ regular in O_μ such that $c_A(w \to x, z) \to c_A(x, z)$. The developed theory is, therefore, not valid when function $c_A(x, z)$ contains a discontinuity or a steep slope near x_0. One of example of such problems is a problem of surface waves on discontinuities of Alfvén velocity which requires a particular consideration (see, e.g., [61]).

6.3 MHD-Waves in a Curvilinear Magnetic Field

In the previous section, Alfvén wave resonators have been described using models with straight magnetic field-lines. In the present section, we will show that in the curved magnetic field, not only the basic features of FLR are conserved but also some new features associated with the field curvature occur.

First, we shall write the ideal MHD-equations in the curvilinear coordinates connected with the magnetic field and obtain the FLR-equations from simple qualitative considerations. Then the scheme of calculations described in detail in Section 6.2, will be generalized to the plasma with Hall conductivity in a curve magnetic field.

Coordinate System Potentials

General Case

In the cold magnetized plasma, a convenient description of MHD-oscillations can be obtained in a curvilinear system of coordinates x^1, x^2, x^3 satisfying two conditions:

1. Two coordinates (say x^1, x^2) set the field line of the external magnetic field $\mathbf{B}_0$ and the third (x^3) sets the location of a point on this field-line.
2. Boundary surfaces where boundary conditions are set must coincide with some coordinate surfaces.

Usually, there is no orthogonal coordinate system satisfying these two conditions and field-lines are non-orthogonal to the boundary surfaces. In the Earth's magnetosphere, the geomagnetic field-lines can be considered as orthogonal to the lower ionosphere boundary only in the polar regions, while in the middle and low latitudes geomagnetic field inclination $I \neq \pi/2$ must be taken into account. In this situation, it is reasonable to abandon an orthogonal system of coordinates in favour of a non-orthogonal one which results only in an insignificant complication in writing plasma oscillation equations.

We use an orthogonal curvilinear coordinate system related to the geometry of the background magnetic field $\mathbf{B}_0(\mathbf{r})$. Assuming that $\mathbf{B}_0(\mathbf{r})$ can be expressed through a scalar potential $\Psi(\mathbf{r})$: $\mathbf{B}_0 = -\boldsymbol{\nabla}\, \Psi$, we introduce the following coordinates:

$$y^1 = \Phi^1(x, y, z), \quad y^2 = \Phi^2(x, y, z), \quad y^3 = \Psi(x, y, z), \tag{6.27}$$

where $\{x, y, z\}$ are Cartesian coordinates. The coordinate y^1 marks magnetic shells (for example, y^1 may be proportional to the magnetic flux inside the corresponding shell), coordinate y^2 specifies a field line on a chosen shell, and y^3 is a coordinate along a field-line. In the axially symmetric case $y^2 = \varphi$, where φ is an azimuthal coordinate. The element length is

$$\mathrm{d}s^2 = h_1^2(\mathrm{d}y^1)^2 + h_2^2(\mathrm{d}y^2)^2 + h_3^2(\mathrm{d}y^3)^2,$$

where h_1, h_2, and h_3 are Lamé coefficients.

In this system, the line of intersection of the surfaces $\Phi^1(x, y, z) = \text{const}$ and $\Phi^2(x, y, z) = \text{const}$ with the equipotential surface S_3 ($y^3 = \Psi(x, y, z) = \text{const}$) are the lines of curvature on the surface S_3. These are the lines tangents to which in each point of S_3 belong to one of two orthogonal planes of the principal normal sections. The Lamé coefficients of this coordinate system are

$$h_1 = \left|\boldsymbol{\nabla}\Phi^1\right|^{-1}, \quad h_2 = \left|\boldsymbol{\nabla}\Phi^2\right|^{-1}, \quad h_3 = \left|\boldsymbol{\nabla}\Psi^3\right|^{-1}.$$

Thus, the orthonormal local basis is

$$\mathbf{i}_1 = h_1\boldsymbol{\nabla}\Phi^1, \quad \mathbf{i}_2 = h_2\boldsymbol{\nabla}\Phi^2, \quad \mathbf{i}_3 = h_3\boldsymbol{\nabla}\Psi.$$

Denote by S_a the interface between the ionosphere and the atmosphere and by S_m the magnetopause surface. Write the equation of the surface S_a in the form

$$y^3 = \Gamma_{N(S)}(y^1, y^2), \tag{6.28}$$

where index $N(S)$ denotes the Northern (Southern) Hemisphere. Set the external boundary of the magnetosphere by

$$y^1 = \text{const}\,. \tag{6.29}$$

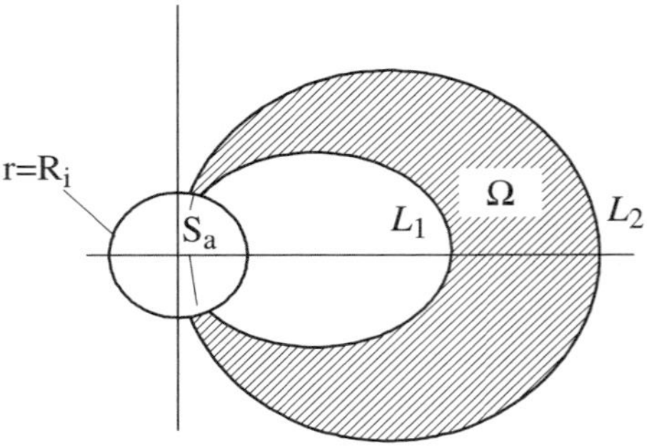

Fig. 6.1. A sketch of the $\boldsymbol{\Omega}$ volume between two magnetic shells

The region $\boldsymbol{\Omega}$ between two magnetic shells (see Fig. 6.1), transforms in the coordinates $\{y_i\}$ into the region $\mathbf{T}$ (see Fig. 6.2a). As the field-lines are not orthogonal to the ionospheres, the meridional cross-section $\mathbf{T}$ is a curvilinear trapezium.

In the non-orthogonal coordinate system, it is possible to achieve the coincidence of boundary S_a with the coordinate surface by replacing variables $\{y^1, y^2, y^3\}$ by $\{x^1, x^2, x^3\}$ as

$$x^1 = y^1, \quad x^2 = y^2, \quad x^3 = K_1(y^1, y^2) + K_2(y^1, y^2)y^3, \tag{6.30}$$

where

$$K_1(y^1, y^2) = \frac{d}{2}\frac{\Gamma_N(y^1, y^2) + \Gamma_S(y^1, y^2)}{\Gamma_N(y^1, y^2) - \Gamma_S(y^1, y^2)},$$

$$K_2(y^1, y^2) = \frac{d}{\Gamma_N(y^1, y^2) - \Gamma_S(y^1, y^2)},$$

and $\Gamma_{N,S}$ is defined in (6.28). By substituting $y^k(x, y, z)$ from (6.27) into (6.30), we obtain the transformation equations from the Cartesian coordinate

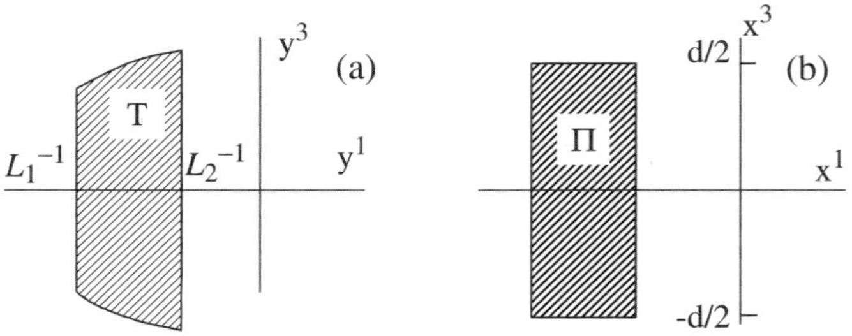

Fig. 6.2. The $\boldsymbol{\Omega}$ volume of Fig. 6.1 is transformed into (a) the curvilinear trapezium $\mathbf{T}$ in the orthogonal system $\{y^i\}$ and (b) the rectangle $\boldsymbol{\Pi}$ in the curvilinear system $\{x^i\}$

to $\{x^1, x^2, x^3\}$:

$$\begin{aligned} x^1 &= \Phi^1(x,y,z), \quad x^2 = \Phi^2(x,y,z), \\ x^3 &= K_1[\Phi^1(x,y,z), \Phi^2(x,y,z)] \\ &\quad + K_2[\Phi^1(x,y,z), \Phi^2(x,y,z)^2]\Psi(x,y,z). \end{aligned} \tag{6.31}$$

With the summation convention, the element length is

$$\mathrm{d}s^2 = g_{ik} dx^i \mathrm{d}x^k.$$

Components of the metric tensor g_{ik} expressed in terms of Lamé coefficients $\{h_1, h_2, h_3\}$ and transformation coefficients (6.30) K_1, K_2 are given in Appendix 6.A.

The vectors of the local basis

$$\mathbf{e}^1 = \boldsymbol{\nabla} x^1, \quad \mathbf{e}^2 = \boldsymbol{\nabla} x^2, \quad \mathbf{e}^3 = \boldsymbol{\nabla} x^3$$

are orthogonal to the coordinate surfaces and the vectors of the local basis $\mathbf{e}_i = g_{ik}\mathbf{e}^k$ $(i = 1,2,3)$ are tangential to the coordinate lines. These two basis are biorthogonal, as it follows immediately from their definitions, that is $\mathbf{e}_i \cdot \mathbf{e}^k = \delta_i^k$.

For an arbitrary vector written in the first basis

$$\mathbf{F} = F_1\mathbf{e}^1 + F_2\mathbf{e}^2 + F_3\mathbf{e}^3,$$

where the covariant components by virtue of the biorthogonal condition are determined by the equality $F_k = \mathbf{F}\cdot\, \mathbf{e}_k$. The vector $\mathbf{F}$ can be written as a sum of two vectors $\mathbf{F}_\perp$ and $\mathbf{F}_\nu$: $\mathbf{F}_\perp$ normal to the field-line, $\mathbf{F}_\perp = F_1\mathbf{e}^1 + F_2\mathbf{e}^2$ and the vector $\mathbf{F}_\nu = F_3\mathbf{e}^3$, orthogonal to the coordinate surface $x^3 = \text{const}$.

A vector $\mathbf{F}$ expanded in the second basis is

$$\mathbf{F} = F^1\mathbf{e}_1 + F^2\mathbf{e}_2 + F^3\mathbf{e}_3,$$

where the contravariant components $F^k = \mathbf{F}\cdot\mathbf{e}^k$. A vector $\mathbf{F}$ can be presented as a sum of the vector $\mathbf{F}_\parallel$ parallel to the field-line, $\mathbf{F}_\parallel = F^3\mathbf{e}_3$, and the vector $\mathbf{F}_\tau = F^1\mathbf{e}_1 + F^2\mathbf{e}_2$ tangential to the coordinate surface $x^3 = \text{const}$.

Equations for the boundary surfaces (6.28) and (6.29) in new coordinates have the form

$$x^3 = \pm\frac{d}{2}, \qquad x^1 = \text{const}\,. \tag{6.32}$$

The region $\boldsymbol{\Omega}$ is a parallelepiped II in the coordinates $\{x_i\}$ (see Fig. 6.2b).

Dipole Coordinates

In the important special case of a dipole magnetic field, special notation is used for the orthogonal coordinates $y^1 = \nu$, $y^2 = \varphi$, $y^3 = \mu$. The potential Ψ of the geomagnetic dipole field in the spherical coordinates $\{r,\ \theta,\ \varphi\}$ with the polar axes directed along the dipole moment is $\Psi \propto \cos\theta / r^2$. One can introduce the dipole coordinates $\{\nu,\ \mu,\ \varphi\}$ connected with the spherical coordinates as

$$\nu = -\frac{\sin^2\theta}{r}, \quad \varphi = \varphi, \quad \mu = \frac{\cos\theta}{r^2}. \tag{6.33}$$

A field line lying within a magnetic shell $\nu = \text{const}$, crosses the equatorial plane $\theta = \pi/2$ at $r = r_e = LR_E$, where $L = -1/(\nu R_E)$ is the McIllwain parameter.

Basic vectors of the dipole coordinate system $(e^\nu,\ e^\varphi,\ e^\mu)$ expressed in terms of orthonormal basis $(\mathbf{i}_r,\ \mathbf{i}_\theta,\ \mathbf{i}_\varphi)$ of the spherical coordinate are

$$\mathbf{e}^\nu = \frac{\sin^2\theta}{r^2}\mathbf{i}_r - \frac{\sin 2\theta}{r^2}\mathbf{i}_\theta\,, \quad \mathbf{e}^\varphi = \frac{1}{r\sin\theta}\mathbf{i}_\varphi\,, \quad \mathbf{e}^\mu = -\frac{2\cos\theta}{r^3}\mathbf{i}_r - \frac{\sin\theta}{r^3}\mathbf{i}_\theta. \tag{6.34}$$

Substitution of (6.34) into $h_n = |\mathbf{e}^n|^{-1}$ yields the Lamé coefficients for the dipole system:

$$h_\nu = \frac{r^2}{\sin\theta(1+3\cos^2\theta)^{1/2}}, \quad h_\varphi = r\sin\theta,$$
$$h_\mu = h_\nu h_\varphi = \frac{r^3}{(1+3\cos^2\theta)^{1/2}}. \tag{6.35}$$

Suppose that the ionosphere-atmosphere interface S_a is a sphere of the radius $R_I \approx R_E$, where R_E is the Earth's radius. Then the equation for S_a is $r = R_E$ and in the dipole coordinates it becomes

$$\mu = \Gamma_{N,S}(\nu) = \pm R_E^{-2}(1+\nu R_E)^{1/2},$$

where plus (minus) corresponds to the northern (southern) ionosphere, respectively.

For the coefficients K_1 and K_2 from (6.30) at $d = 2$ we get

$$K_1 = 0, \quad \text{and} \quad K_2 = R_E^2(1-\nu R_E)^{-1/2}. \tag{6.36}$$

And for new coordinates $\xi = x^1,\ \eta = x^2,\ \zeta = x^3$ we obtain

$$\xi = \nu, \quad \eta = \varphi, \quad \zeta = \mu R_E^2(1+\nu R_E)^{-1/2}, \tag{6.37}$$

and from (6.37)

$$\nu = \xi, \quad \varphi = \eta, \quad \mu = \zeta R_E^{-2}(1+\nu R_E)^{1/2}.$$

In the coordinates ξ, η, ζ, the region $\boldsymbol{\Omega}$ between two magnetic shells L_1 and L_2, is transformed into a parallelepiped $\boldsymbol{\Pi}$: $(-L_1^{-1} < \xi < -L_2^{-1},\ 0 \le \eta < 2\pi, -1 < \zeta < 1)$. The boundary S_a is transformed at these coordinates into rectangles S_+ and S_- for the northern and southern ionospheres. The components of the metrical tensor for this coordinate system are given in Appendix 6.A.

Equations for Alfvén and FMS-Waves

Let us write Maxwell's equations in a curvilinear magnetic field in a coordinate system $\{x^1, x^2, x^3\}$, which is generally not-orthogonal. The contravariant components of $\nabla\times$ are easily expressed through the covariant components. For example, $\nabla\times$ of the electric field is given by

$$\nabla \times \mathbf{E} = \frac{1}{g^{1/2}} \begin{vmatrix} \mathbf{e}_1 & \mathbf{e}_2 & \mathbf{e}_3 \\ \partial_1 & \partial_2 & \partial_3 \\ E_1 & E_2 & E_3 \end{vmatrix}, \quad \partial_k = \frac{\partial}{\partial x_k}, \tag{6.38}$$

where $g = \det(g_{ik})$.

For the perfect longitudinal conductivity, $\sigma_\parallel \to \infty$, the component of the electric field parallel to the geomagnetic field $E_3 = \mathbf{E}\cdot\mathbf{e}_3$ vanishes and $\mathbf{E} = \mathbf{E}_\perp = E_1\mathbf{e}^1 + E_2\mathbf{e}^2$. The other components may be written as a 5-component 1-column matrix

$$\mathbf{U} = (b_2, E_1, b_1, b_3, E_2)^{\mathrm{tr}},$$

where E_k, b_k are the covariant components of the electric and magnetic fields; the tr marks the transpose operation, $\mathbf{A}^{\mathrm{tr}}$ is transposed to $\mathbf{A}$.

Combining Maxwell's equations with Ohm's law and using (6.38), we obtain the equation for $\mathbf{U}$:

$$\left(\mathbf{A_1}\frac{\partial}{\partial \mathbf{x_1}} + \mathbf{A_2}\frac{\partial}{\partial \mathbf{x_2}} + \mathbf{C}\right)\mathbf{U} = 0, \tag{6.39}$$

where

$$\mathbf{A}_1 = \begin{pmatrix} 0 & 0 & 0 & 0 & 0 \\ 0 & 0 & 0 & 0 & 0 \\ 0 & 0 & 0 & 0 & 0 \\ 0 & 0 & 0 & 0 & -1 \\ 0 & 0 & 0 & -1 & 0 \end{pmatrix}, \qquad \mathbf{A}_2 = \begin{pmatrix} 0 & 0 & 0 & 0 & 0 \\ 0 & 0 & 0 & 1 & 0 \\ 0 & 0 & 0 & 0 & 0 \\ 0 & 1 & 0 & 0 & 0 \\ 0 & 0 & 0 & 0 & 0 \end{pmatrix},$$

$$\mathbf{C} = \begin{pmatrix} 0 & -1 & 0 & 0 & 0 \\ -1 & 0 & 0 & 0 & 0 \\ 0 & 0 & 0 & 0 & 1 \\ 0 & 0 & 0 & 0 & 0 \\ 0 & 0 & 1 & 0 & 0 \end{pmatrix} \frac{\partial}{\partial x^3} + ik_0\sqrt{g} \begin{pmatrix} g^{22} & 0 & 0 & g^{23} & 0 \\ 0 & \varepsilon^{11} & 0 & 0 & \varepsilon^{12} \\ 0 & 0 & g^{11} & g^{13} & 0 \\ g^{32} & 0 & g^{31} & g^{33} & 0 \\ 0 & \varepsilon^{21} & 0 & 0 & \varepsilon^{22} \end{pmatrix},$$

$$\varepsilon^{11} = \varepsilon_\perp g^{11}, \qquad \varepsilon^{22} = \varepsilon_\perp g^{22},$$
$$\varepsilon^{12} = \varepsilon_\perp g^{12} + \varepsilon_\top \sqrt{g_{33}/g}, \qquad \varepsilon^{21} = \varepsilon_\perp g^{21} - \varepsilon_\top \sqrt{g_{33}/g}.$$

In the magnetosphere $\varepsilon_\perp = \varepsilon_m$ and in the ionosphere $\varepsilon_\perp = i4\pi\sigma_P/\omega$. Respectively, $\varepsilon_\top = 0$ in the magnetosphere and $\varepsilon_\top = i4\pi\sigma_H/\omega$ in the ionosphere.

Equations (6.39) are written for the covariant components E_k, b_k of the electromagnetic field in an arbitrary coordinate system. Usually, the orthogonal coordinate system is used and electric and magnetic fields are expanded over the local orthonormal basis

$$\mathbf{E} = \hat{E}_1\mathbf{i}_1 + \hat{E}_2\mathbf{i}_2 + \hat{E}_3\mathbf{i}_3, \quad \mathbf{b} = \hat{b}_1\mathbf{i}_1 + \hat{b}_2\mathbf{i}_2 + \hat{b}_3\mathbf{i}_3,$$
$$\hat{E}_k = \mathbf{E}\cdot\mathbf{i}_k, \quad \hat{b}_k = \mathbf{b}\cdot\mathbf{i}_k,$$

where $\mathbf{i}_j = h_j\mathbf{e}_j$, $\mathbf{i}_j\cdot\mathbf{i}_k = \delta_{jk}$, $\hat{E}_k$ and $\hat{b}_k$ are the physical components of the $\mathbf{E}$ and $\mathbf{b}$ fields. The relation between physical and covariant components is especially simple in the orthogonal coordinates:

$$\hat{E}_k = \frac{E_k}{h_k}, \quad \hat{b}_k = \frac{b_k}{h_k}.$$

Since the ULF-frequencies are less than the ion cyclotron frequency, the Hall conductivity can be neglected in the regions above the E-layer. Therefore (6.39) has the form

$$\frac{1}{h_3}\frac{\partial E_1}{\partial x^3} - ik_0\frac{h_1}{h_2}b_2 = 0, \tag{6.40}$$

$$\frac{1}{h_3}\frac{\partial b_2}{\partial x^3} - ik_0\varepsilon_\perp\frac{h_2}{h_1}E_1 = \frac{1}{h_3}\frac{\partial b_3}{\partial x^2}, \tag{6.41}$$

$$ik_0\frac{h_1h_2}{h_3}b_3 = \frac{\partial E_2}{\partial x^1} - \frac{\partial E_1}{\partial x^2}, \tag{6.42}$$

$$\frac{1}{h_3}\frac{\partial E_2}{\partial x^3} + ik_0\frac{h_2}{h_1}b_1 = 0, \tag{6.43}$$

$$\frac{1}{h_3}\frac{\partial b_1}{\partial x^3} + ik_0\varepsilon_\perp\frac{h_1}{h_2}E_2 = \frac{1}{h_3}\frac{\partial b_3}{\partial x^1}. \tag{6.44}$$

If the coupling of the Alfvén and FMS modes is weak, then quazi-Alfvén and quazi-FMS-waves can propagate in plasmas. In the quazi-Alfvén waves, the longitudinal electric current is finite but the longitudinal magnetic field is small. Equation set (6.40)–(6.44) is split into two groups: (6.40)–(6.41) for the toroidal Alfvén mode and (6.43)–(6.44) for the poloidal mode. In quazi-FMS modes, on the contrary the longitudinal magnetic field is not small but the field-aligned current is small.

Some Features of the Alfvén Waves

Large Transversal Wavenumbers

Coupling of Alfvén and FMS-waves is yielding with a decreasing the scale transversal to the main magnetic field. If the scale approaches zero, MHD-wave equations split into the separate equations for Alfvén and FMS-modes.

Consider the disturbances of **E** and $\mathbf{b} \propto \exp[i\Phi(x_1, x_2)]$, where $\Phi(x_1, x_2)$ is the wave phase varying quickly in the transverse direction. Let also the cross-components k_1 and k_2 of the wave vector **k**:

$$k_1 = \frac{\partial \Phi}{\partial x^1}, \qquad k_2 = \frac{\partial \Phi}{\partial x^2},$$

such that $\partial/\partial x^1 \approx ik_1, \partial/\partial x^2 \approx ik_2$. Assume, as is typical in the magnetosphere, that the Alfvén velocity varies more rapidly along, for example, the coordinate x^1 as compared to x^2. In this case, one can extract two wave classes.

The first class corresponds to the case in which the wave field varies along x^1 much more rapidly than along x^2. Since the element lengths in the direction x^1, x^2 are dx^1/h_1, dx^2/h_2, respectively, then the wavelength along x^1 is shorter than along x^2 that results in $|h_1 k_1| \gg |h_2 k_2|$.

In the opposite case, in which the wavelength is shorter along x^2, we have $|h_2 k_2| \gg |h_1 k_1|$.

In these important limiting cases, a set of equations describing Alfvén modes with $b_3 \to 0$ can be separated out from the complete set (6.40)–(6.44) and from compressional FMS oscillations, in which the field-aligned magnetic component b_3 is finite. Assuming that $k_1 \to \infty$, we find from (6.42) and (6.44) that the components $E_2 \to 0$, $b_3 \to 0$ (while $k_1 E_2$, $k_1 b_3$ should not necessarily tend to zero).

Thus, by virtue of (6.43), we also have $b_1 \to 0$. As a result, (6.40) and (6.41) yield a closed set of equations for the components E_1 and b_2:

$$\frac{1}{h_3}\frac{\partial E_1}{\partial x^3} - ik_0 \frac{h_1}{h_2} b_2 = 0, \tag{6.45}$$

$$\frac{1}{h_3}\frac{\partial b_2}{\partial x^3} - ik_0 \varepsilon_\perp \frac{h_2}{h_1} E_1 = 0, \tag{6.46}$$

which, after eliminating b_2, reduces to one second-order equation

$$\frac{h_1}{h_2 h_3}\frac{\partial}{\partial x^3}\frac{h_2}{h_1 h_3}\frac{\partial E_1}{\partial x^3} + \frac{\omega^2}{c^2}\varepsilon_\perp E_1 = 0. \tag{6.47}$$

Alfvén oscillations described by (6.45), (6.46) are polarized so that the **E**-field is directed along the coordinate x^1, while the **b**-field perturbations and plasma displacements are along the coordinate x^2. These oscillations are primarily excited by large-scale sources (with small k_2). This mode corresponds to the toroidal mode, according to geophysical terminology.

For the small-scale perturbations along the coordinate x^2 ($k_2 \to \infty$), components E_1 and b_3 vanish in the system (6.40)–(6.44) while $k_2 E_1$, $k_2 b_3$ should not necessarily tend to zero. In that case, by virtue of the second equation of the set (6.40), we also have $b_2 \to 0$. As a result, (6.43) and (6.44) yield the closed set of equations:

$$\frac{1}{h_3}\frac{\partial E_2}{\partial x^3} + ik_0\frac{h_2}{h_1}b_1 = 0, \tag{6.48}$$

$$\frac{1}{h_3}\frac{\partial b_1}{\partial x^3} + ik_0\varepsilon_\perp\frac{h_1}{h_2}E_2 = 0, \tag{6.49}$$

which reduces to the second-order equation

$$\frac{h_2}{h_1h_3}\frac{\partial}{\partial x^3}\frac{h_1}{h_2h_3}\frac{\partial E_2}{\partial x^3} + \frac{\omega^2}{c^2}\varepsilon_\perp E_2 = 0. \tag{6.50}$$

The polarization of Alfvén oscillations described by (6.48), (6.49) differs from that of the toroidal mode: the electric field is directed along the coordinate x^2, while the magnetic field perturbations and plasma displacements are directed along the coordinate x^1. According to geophysical terminology, we will refer to these oscillations as poloidal mode. These oscillations are mainly excited by localized sources with large k_2.

Equations (6.47) and (6.50) can be rewritten in an invariant form, in which the influence of the geometrical factor – the magnetic field curvature – on the propagation of Alfvén waves is expressed explicitly. Let us introduce the field-aligned coordinate s equal to the distance along the line, such that $ds = h_3dx^3$. Then, (6.47) and (6.50) reduce to

$$\left[\frac{\partial^2}{\partial s^2} + (\kappa_1 - \kappa_2)\frac{\partial}{\partial s} + \frac{\omega^2}{c^2}\varepsilon_\perp\right]E_1 = 0, \tag{6.51}$$

$$\left[\frac{\partial^2}{\partial s^2} - (\kappa_1 - \kappa_2)\frac{\partial}{\partial s} + \frac{\omega^2}{c^2}\varepsilon_\perp\right]E_2 = 0. \tag{6.52}$$

Here

$$\kappa_{1,2} = -\frac{\partial}{\partial s}\log h_{1,2}$$

are the principal curvatures of the equipotential surface $x^3 = \text{const}$ and characterize the rate of convergence/divergence of the field-lines.

Equations (6.51) and (6.52) are supplemented with the boundary condition of the non-current flux from the ionosphere into the atmosphere. A detailed treatment of the interaction of Alfvén waves with the ionosphere will be given in Chapters 7, 8. Here we only note that the atmospheric conductivity is very small. It is several orders of magnitude less than the ionospheric conductivity. That is why the current of an Alfvén wave does not penetrate into the atmosphere and there is no component orthogonal to the atmosphere, bounded by the surface S_a. Thus,

$$j^3\big|_{\mathbf{r}\in S_a} = \mathbf{j}\cdot\mathbf{e}_3\big|_{\mathbf{r}\in S_a} = 0.$$

We get from this equality and from Ampere's law

$$(\boldsymbol{\nabla}\times\mathbf{b})\cdot\mathbf{e}_3 = (4\pi/c)\,j^3$$

that

$$\left(\frac{\partial b_2}{\partial x^1} - \frac{\partial b_1}{\partial x^2}\right)_{\mathbf{r}\in S_a} = 0 \tag{6.53}$$

both for the toroidal mode $b_2|_{\mathbf{r}\in S_a} = 0$, and for the poloidal mode $b_1|_{\mathbf{r}\in S_a} = 0$. Let, for instance, $x^3 = -d/2$ correspond to the Southern ionosphere and $x^3 = d/2$ to the Northern ionosphere. Then by considering (6.45) and (6.48) the boundary conditions for (6.45), (6.46) reduce to

$$\frac{\partial E_2}{\partial x^3} = 0 \quad \text{at} \quad \mathrm{x}^3 = \pm\frac{\mathrm{d}}{2}.$$

and the same for one equation of the second order (6.50) or (6.52). For the system (6.48)–(6.49) (or for one equation (6.50) (or (6.52))), the conditions are given by

$$\frac{\partial E_1}{\partial x^3} = 0 \quad \text{at} \quad x^3 = \pm\frac{d}{2}. \tag{6.54}$$

Equation (6.51) and (6.52) explicitly demonstrate, that the eigenfrequencies of Alfvén oscillations of the geomagnetic shell are somewhat different for modes with different polarizations. This polarization spectrum splitting is caused by a different influence of the magnetic field curvature on the spectrum of oscillations with the plasma displaced along or across the shell. More precisely, it is related to the difference between the principal curvatures of equipotential surfaces or, in other words, to the difference in the rates of the field-lines convergence/divergence in two orthogonal surfaces containing a given field-line. If the principal curvatures coincide at the intersection points of the equipotential surfaces with the given field-line, $\kappa_1 = \kappa_2$ (the field-lines converge/diverge at the same rate, i.e. the cross-sectional shape of a flux tube does not change along its length), then there is no polarization splitting of the spectrum.

It is convenient to rewrite (6.47) and (6.50) in a form similar to the equation for the Alfvén waves propagating in a uniform external magnetic field [44]. The replacement $\mathrm{d}\xi = h_1 h_3 h_2^{-1} dx^3$ reduces (6.47) to

$$\left[\frac{\partial^2}{\partial \xi^2} + \left(\frac{\omega}{c_A^T}\right)^2\right] E_1 = 0, \tag{6.55}$$

whereas (6.50), after the substitution $d\eta = h_2 h_3 h_1^{-1} dx^3$, becomes

$$\left[\frac{\partial^2}{\partial\eta^2} + \left(\frac{\omega}{c_A^P}\right)^2\right] E_2 = 0. \tag{6.56}$$

The modified Alfvén velocities c_A^T ('toroidal') and c_A^P ('poloidal') in (6.55) and (6.56)) are

$$c_A^T = \frac{h_1}{h_2} c_A, \qquad c_A^P = \frac{h_2}{h_1} c_A.$$

Thus, the propagation of both modes of Alfvén waves in a curvilinear magnetic field can be described by (6.55) and (6.56), similar to the equation in a straight field, but with c_A replaced with the modified Alfvén velocities c_A^T and c_A^P.

A simplified derivation of equations for Alfvén waves in the present chapter does not give information about the transversal distribution of the amplitude of Alfvén waves. To obtain them on approximation over the small transversal scale is required. This will be done in the next section with the aid of a generalization of the method used in order to study FLR in a straight magnetic field (see Section 6.2).

FLR Rigorous Solution

In this section the problem of excitation of forced low frequency oscillations of the magnetosphere-ionosphere resonator is considered within the axially symmetric model with the plasma parameters independent on the azimuth. Let $x^2 = \varphi$. Then $g^{12} = g^{21} = 0$, $g_{12} = g_{21} = 0$, and $g^{23} = g^{32} = 0$. We shall restrict ourselves to the analysis of the oscillations $\propto \exp(im\varphi)$ with a fixed azimuthal wavenumber m. Then from (6.39) we obtain the 2D equations:

$$\left(\mathbf{A}\frac{\partial}{\partial x_1} + \mathbf{C}(x^1, x^3)\right)\mathbf{U} = 0, \tag{6.57}$$

where $A = A_1$ from (6.39), $C(x^1, x^3)$ at the fixed x^1 is a differential operator with respect to the coordinate x^3:

$$\mathbf{C}(x^1, x^3) = \begin{pmatrix} ag^{22} & -\dfrac{\partial}{\partial x^3} & 0 & 0 & 0 \\ -\dfrac{\partial}{\partial x^3} & a\varepsilon^{11} & 0 & im & a\varepsilon^{12} \\ 0 & 0 & ag^{11} & ag^{13} & \dfrac{\partial}{\partial x^3} \\ 0 & im & ag^{31} & ag^{33} & 0 \\ 0 & a\varepsilon^{21} & \dfrac{\partial}{\partial x^3} & 0 & a\varepsilon^{22} \end{pmatrix}$$

with $a = ik_0\sqrt{g}$. Equation (6.57) describes monochromatic small oscillations of an axially symmetrical magnetosphere-ionosphere resonator with an arbitrary angle between the field-lines and ionosphere.

Equation (6.57) should be supplemented with the appropriate boundary conditions. A principal difficulty stems from the fact that the Alfvén and FMS-waves are coupled because of the field-lines curvature and plasma inhomogeneity. As this takes place, these modes respond differently to the non-conductive atmosphere and highly conductive Earth. The boundary condition for the FMS is rather complicated. However, as it will be shown below, the only condition of the current non-penetration into the atmosphere is sufficient for the determination of the field in the FLR-region.

The non-penetration condition for the electric current

$$\mathbf{j} = \frac{c}{4\pi}\,\boldsymbol{\nabla}\times\,\mathbf{b} + \frac{i\omega}{c}\mathbf{E}$$

is written as

$$j^3\big|_{\mathbf{x}^3=\pm d/2} = \mathbf{j}\cdot\mathbf{e}_3\big|_{\mathbf{x}^3=\pm d/2} = 0.$$

Neglecting the displacement current, we rewrite it as

$$\frac{\partial b_2}{\partial x^1} + imb_1 = 0 \quad \text{at} \quad \mathbf{x}^3 = \pm d/2. \tag{6.58}$$

The second boundary condition is non-local but it is not necessary to know its exact form to get the resonance structure of the field.

Let us describe the solution of (6.57) with the boundary conditions (6.58) in a rectangle $\boldsymbol{\Pi}$ (see Fig. 6.2(b)), corresponding to the region between two magnetic shells L_1 and L_2 and the ionospheres. It is appropriate to our problem to seek solutions, by analogy with the straight field problem, in the form

$$\mathbf{U}(x^1, x^3) = \frac{\boldsymbol{\alpha}(x^1, x^3)}{x^1 - z_0} + \boldsymbol{\beta}(x^1, x^3)\log(x^1 - z_0), \tag{6.59}$$

where $\boldsymbol{\alpha}(x^1, x^3)$, $\boldsymbol{\beta}(x^1, x^3)$ are analytical vector-functions of x^1:

$$\boldsymbol{\alpha}(x^1, x^3) = \sum_{j=0}^{\infty} \boldsymbol{\alpha}_j(x^3)(x^1 - z_0)^j, \tag{6.60}$$

$$\boldsymbol{\beta}(x^1, x^3) = \sum_{j=0}^{\infty} \boldsymbol{\beta}_j(x^3)(x^1 - z_0)^j, \tag{6.61}$$

and $z_0 = x_0^1 + i\delta$ is a free parameter.

By substituting (6.59) into (6.57) we obtain the chain recurrent boundary problems

$$\mathbf{A}\boldsymbol{\alpha}_0 = 0, \tag{6.62}$$

$$(j+1)\mathbf{A}\boldsymbol{\beta}_{j+1} + \sum_{k=0}^{j} \mathbf{C}_k\boldsymbol{\beta}_{j-k} = 0, \tag{6.63}$$

$$j\mathbf{A}\boldsymbol{\alpha}_{j+1} + \sum_{k=0}^{j} \mathbf{C}_k\boldsymbol{\alpha}_{j-k} + \mathbf{A}\boldsymbol{\beta}_j = 0, \tag{6.64}$$

with boundary condition

$$\boldsymbol{\alpha}_0^{(1)} = 0 \quad \text{at} \quad x^3 = \pm d/2, \tag{6.65}$$
$$(j+1)\boldsymbol{\beta}_{j+1}^{(1)} - im\boldsymbol{\beta}_j^{(3)} = 0 \quad \text{at} \quad x^3 = \pm d/2, \tag{6.66}$$
$$j\boldsymbol{\alpha}_{j+1}^{(1)} + \boldsymbol{\beta}_j^{(3)} - im\boldsymbol{\alpha}_j^{(3)} = 0 \quad \text{at} \quad x^3 = \pm d/2, \tag{6.67}$$

where $j = 0, 1, 2, \ldots$; $\mathbf{C}_k$ are the $(x^1 - z_0)$ - power expansion coefficients of the operator function $\mathbf{C}(x^1, x^3)$. The notations $\boldsymbol{\alpha}_j^{(1)}$ and $\boldsymbol{\beta}_j^{(1)}$, $\boldsymbol{\alpha}_j^{(2)}$ and $\boldsymbol{\beta}_j^{(2)}$, $\boldsymbol{\alpha}_j^{(3)}$ and $\boldsymbol{\beta}_j^{(3)}$ are used for the first three, the fourth and the last two elements of $\boldsymbol{\alpha}_j$ and $\boldsymbol{\beta}_j$, respectively.

A simple but rather a cumbersome analysis of the chain recurrent boundary problems (6.62)–(6.67) shows, that $\alpha_0 \neq 0$, $\beta_0 \neq 0$ (i.e. the solution is really resonance) only when the boundary problem

$$\begin{pmatrix} ik_0\sqrt{g}g^{22} & -\dfrac{\mathrm{d}}{\mathrm{d}x^3} \\ -\dfrac{\mathrm{d}}{\mathrm{d}x^3} & ik_0\sqrt{g}g^{11}\varepsilon_\perp \end{pmatrix} \begin{pmatrix} e_1(x^3) \\ e_2(x^3) \end{pmatrix} = 0, \tag{6.68}$$

$$e_1(-d/2) = 0, e_1(d/2) = 0, \tag{6.69}$$

has a non-trivial solution. Here $g_0 = g(z_0)$, $g_0^{ik} = g^{ik}(z_0)$, $\varepsilon_0^{ik} = \varepsilon^{ik}(z_0)$.

Assume that such solution exists at some z_0 and $\omega = \omega_0$. Using the notation $\mathbf{e} = (e_1, e_2)^{\mathrm{tr}}$ for this solution, we obtain $\boldsymbol{\alpha}_0(x^3)$, $\boldsymbol{\beta}_0(x^3)$ from (6.63), (6.64):

$$\boldsymbol{\alpha}_0 = (e_1, e_2, 0, 0, 0)^{\mathrm{tr}}, \tag{6.70}$$
$$\boldsymbol{\beta}_0 = (C_0e_1,\ C_0e_2,\ -im\frac{g_0^{22}}{g_0^{11}}e_1 - ik_0\sqrt{g_0}\varepsilon_0^{21}\frac{g_0^{13}}{g_0^{11}}e_2, ik_0\sqrt{g_0}\varepsilon_0^{21}e_2,\ ime_2)^{\mathrm{tr}}, \tag{6.71}$$

where C_0 is a constant determined from the condition of solvability of the problem in the next approximation.

Let for each ω in a vicinity of ω_0 there is $z = z(\omega)$ at which the boundary problem (6.68), (6.69) has a solution. Taking the inverse function to $z(\omega)$, we find resonance frequencies $\omega = \omega_r(z)$ corresponding to the "complex magnetic shell" z. Let at $z = z_0$ the derivative

$$\left.\frac{\mathrm{d}\omega_r}{\mathrm{d}z}\right|_{z=z_0} \neq 0, \tag{6.72}$$

The analysis of the chain of recurrent boundary problems (6.62)–(6.67) (not given here because of its awkwardness) shows that they are subsequently resolvable at the condition (6.72) and determine unambiguously the coefficient C_0:

$$C_0 = i\frac{cm^2}{2\omega_r'(z_0)}\left(\int\limits_{-d/2}^{d/2}\sqrt{g_0}g_0^{22}e_1^2\mathrm{d}x^3\right)^{-1}\int\limits_{-d/2}^{d/2}e_1e_2\frac{\mathrm{d}}{\mathrm{d}x^3}\frac{g_0^{22}}{g_0^{11}}\mathrm{d}x^3. \tag{6.73}$$

Here the small terms relate to Hall conductivity are neglected. These terms are small if the length of the field line in the ionospheric E-layer is small in comparison with the length of the field line between the conjugated ionospheres.

It is seen from (6.59) with taking into account (6.70), (6.71) and (6.73) that near the resonance magnetic shell $x^1 = x_0^1$ the distributions of the components $b_2(x^1)$ and $E_1(x^1)$ are the resonance curves with the half-width δ. If to suppose that an external source is given with a real frequency ω, then the boundary problem (6.68), (6.69) determines the corresponding magnetic shell $x_0^1 = \mathrm{Re}\, z_0$ and the resonance half-width $\delta = \mathrm{Im}\, z_0$. At $\delta = 0$, from the boundary problem (6.68), (6.69) we can find resonance frequencies ω and decrement γ.

In (6.68) the boundary conditions (6.69) correspond to one field line and it is not necessary to write it in the non-orthogonal coordinates. Return to the initial orthogonal coordinate system. From the change of variables (6.30) we have

$$\frac{\partial}{\partial x^3} = \frac{1}{K_2}\frac{\partial}{\partial y^3}$$

and (6.68) reduces to

$$\begin{pmatrix} ik_0\dfrac{h_1}{h_2} & -\dfrac{1}{h_3}\dfrac{\mathrm{d}}{\mathrm{d}y^3} \\ -\dfrac{1}{h_3}\dfrac{\mathrm{d}}{\mathrm{d}y^3} & ik_0\dfrac{h_2}{h_1}\varepsilon_\perp \end{pmatrix}\begin{pmatrix} e_1(y^3) \\ e_2(y^3)\end{pmatrix} = 0 \tag{6.74}$$

with $e_1 = 0$ at the field line pierce point between ionosphere and atmosphere.

The obtained formulae yield a key to the principal features of Alfvén resonance oscillations. In the next sections of this chapter, numerical methods for the realization of these theoretical results and their application to the Earth's magnetosphere are given.

6.4 FLR in the Dipole Geomagnetic Field

Before we turn to a closer examination of the wave-field structure in the vicinity of a resonant line in the dipole field, we shall describe the simplified models of the magnetospheric plasma distributions and the ionosphere which are used in numerical calculations.

Model Magnetospheric Plasma Distribution

The models of a plasma distribution in the Earth's magnetosphere are based on equations of hydrostatic equilibrium with taking into account the electrostatic forces, Earth's magnetic field and Coriolis force ([3], [18]). The ion and electron temperature variation along the field line and plasma diffusion also have an effect on the results of the model calculations. In some regions the anisotropy of the electron temperature is also significant.

For our purposes it is enough to use a simplified model of the cold plasma distribution in the form

$$\rho(r, L, \varphi) = \rho_e(L, \varphi)\bar{\rho}(r, L, \varphi),$$

where $\rho_e(L, \varphi)$ is the distribution in the equatorial plane and $\bar{\rho}(r, L, \varphi) = \rho(r, L, \varphi)/\rho_e(L, \varphi)$ is the normalized distribution along a field line with the McIllwain parameter L. $\bar{\rho}(r, L, \varphi) = 1$ at the top of a field-line; φ is the geomagnetic longitude; the location of a point on the field line is determined by the geocentric distance r. Neglecting the azimuthal asymmetry, we shall consider $\rho(r, L, \varphi)$ independent on φ, i.e.

$$\rho = \rho_e(L)\bar{\rho}(r, L). \tag{6.75}$$

Such a model is valid only for the MHD-disturbances localized either at the dayside or at the nightside. Set a normalized field line distribution of plasma density as a power function of the geocentric distance:

$$\bar{\rho}(r, L) = \left(\frac{LR_E}{r_L}\right)^{p(L)}, \tag{6.76}$$

where r_L is the geocentric distance on a given field line L. Let us approximate the equatorial distribution ρ_e with the expression

$$\rho_e(L) = \frac{\rho_1(L)}{1 + \exp\left(2\dfrac{L - L_{pp}}{\Delta L}\right)} + \frac{\rho_2(L)}{1 + \exp\left(-2\dfrac{L - L_{pp}}{\Delta L}\right)},$$

$$\rho_1(L) = m_p n_1 \left(\frac{L}{L_1}\right)^{-s_1(L)}, \quad \rho_2(L) = m_p n_2 \left(\frac{L}{L_1}\right)^{-s_2(L)}. \tag{6.77}$$

Here L_{pp} and ΔL are the position and the half-width of the plasmapause; m_p is the proton mass. We shall use two models for the plasma density distribution in the numerical calculations. In the first model $L_{pp} = 4.9$, in the second one $L_{pp} = 4.4$ (see Fig. 6.3).

The Dipole Dispersion Equation General Case

Let us write (6.74) for the dipole magnetic field. It is more convenient to use a variable w defined by

$$\frac{w}{(1 - w^2)^2} = \frac{\mu}{\nu^2} \tag{6.78}$$

instead of the coordinate $\mu = \cos\theta/r^2$. Substituting $\mu = \cos\theta/r^2 = (1 + \nu r)^{1/2}/r^2$ into (6.78) with the branch of the root is selected according to the

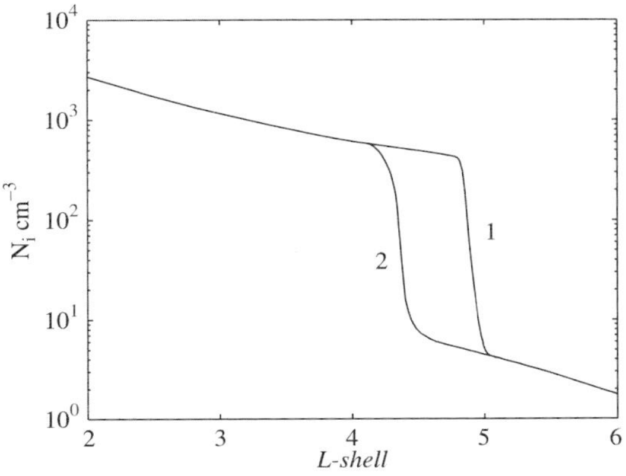

Fig. 6.3. Number density of the cold plasma versus L-shell is given by (6.77). Curve 1 represents a model with the plasmapause at $L_{pp} = 4.9$, and curve 2 is the same but $L_{pp} = 4.4$

sign of $\cos\theta$, we have

$$\frac{w}{(1-w^2)^2} = \frac{(1+\nu r)^{1/2}}{\nu^2 r^2}.$$

and for w, we obtain

$$w = (1+\nu r)^{1/2}.$$

Substitution of $\mu(w)$ in Lamé coefficients (6.35) yields

$$\frac{h_\nu h_\mu}{h_\varphi} = h_\nu^2 = \frac{(1-w^2)^3}{\nu^4(1+3w^2)^2}, \quad \frac{h_\varphi h_\mu}{h_\nu} = h_\varphi^2 = \frac{(1-w^2)^3}{\nu^2}.$$

Since

$$\frac{\mathrm{d}}{\mathrm{d}\mu} = \frac{1}{\mu^2}\frac{(1-w^2)^3}{1+3w^2}\frac{\mathrm{d}}{\mathrm{d}w},$$

changing of the variable $\mu = \mu(w)$ reduces (6.74) to

$$\begin{aligned} \frac{\mathrm{d}e_1}{\mathrm{d}w} &= ik_0(1+3w^2)\varepsilon_\perp e_2, \\ \frac{\mathrm{d}e_2}{\mathrm{d}w} &= \frac{ik_0}{\nu^2}e_1, \end{aligned} \tag{6.79}$$

with conditions

$$e_1(w = \pm w_0) = 0. \tag{6.80}$$

Here $w_0 = (1+\nu R_I)^{1/2}$ is the boundary between the ionosphere and atmosphere, $R_I = R_E + h_a$, and h_a is the atmosphere thickness.

The boundary problem (6.79)–(6.80) at real ν has a complex spectrum of resonance frequencies $\omega_j - i\gamma_j$ for a magnetic shell with the coordinate ν. The variable $w = (1 + \nu r)^{1/2} = x = \cos\theta$ is, with that, real. If the oscillations are excited by an external driver at a real frequency ω, then (6.79)–(6.80) leads to the complex $\nu_r = \nu_r' + i\nu_r''$. In this case, $\nu = \nu_r$ provides a non-trivial solution of this boundary problem.

The variable w is then complex and (6.79) are integrated over the curve $w = (1 + \nu r)^{1/2}$ in the complex plane. Let $x = (1 + \nu' r)^{1/2}$, then

$$w = \left[\left(1 + i\frac{\nu''}{\nu'}\right) x^2 - i\frac{\nu''}{\nu'}\right]^{1/2} .$$

It is convenient to use only the second-order equation instead of the system (6.79). Let us eliminate, for example e_1 from (6.79). From the second equation of (6.79), $e_1 = \left(\nu^2/ik_0\right)\left(de_2/dw\right)$. Substituting it into the first equation and into the condition (6.80), we have

$$\frac{\mathrm{d}^2 e_2}{\mathrm{d}w^2} + \frac{k_0^2}{\nu^2}(1 + 3w^2)\varepsilon_\perp e_2 = 0, \tag{6.81}$$

$$\left.\frac{\mathrm{d}e_2}{\mathrm{d}w}\right|_{w=w_0} = 0. \tag{6.82}$$

At real ν, from the boundary problem (6.79), (6.80), or similar problem (6.81), (6.82), it is possible to find the spectrum of the resonance frequencies $\omega_j(\nu)$ and decrements $\gamma_j(\nu), j = 1, 2, 3, \ldots$. But, if we put real frequency ω, then one can define the location of the resonance magnetic shell $(\mathrm{Re}\,\nu_j)$ and its half-width $(\mathrm{Im}\,\nu_j)$.

Similarly to (6.26)

$$\nu_j'' \approx \gamma_j \left(\frac{\mathrm{d}\omega_j(\nu = \nu_j')}{\mathrm{d}\nu}\right)^{-1}, \tag{6.83}$$

which defines the half-width of the resonance shell with a beforehand known decrement of the resonance oscillations.

'Thin' Ionosphere

Because the skin depth in the ionosphere at frequencies $f \lesssim 0.1\,\mathrm{Hz}$ exceeds the thickness of the ionospheric conductive layer, it is possible to integrate (6.74) or (6.79) over the height and use the 'thin' ionosphere approximation which is valid if

$$\omega \ll \frac{c^2}{4\pi \Sigma_P l_P}, \qquad \Sigma_P = \int_0^{l_P} \sigma_P \sin I_0 \mathrm{d}s, \tag{6.84}$$

where integration is to be performed from the pierce point of the field line with the boundary S_a up to the upper boundary of the conductive layer, I_0 is the inclination of the geomagnetic field, l_P is the length of the field line segment within the Pedersen layer. Certainly, the boundaries of the layer cannot be determined exactly, but the decrease of conductivity is very steep and this inaccuracy is of no practical importance.

By going to the thin ionosphere approximation, integration of (6.79) within each ionosphere with the conditions (6.80) gives

$$e_1(w) = ik_0 \int_{\pm w_0}^{w} (1+3u^2)\varepsilon_\perp(u)e_2(u)\mathrm{d}u, \tag{6.85}$$

$$e_2(w) = e_2(w_0) + \frac{ik_0}{\nu^2} \int_{\pm w_0}^{w} e_1(u)\mathrm{d}u. \tag{6.86}$$

Hereafter, the upper (lower) sign corresponds to the northern (southern) ionosphere. Solve (6.85)–(6.86) with iterations and put the values of e_1 and e_2 for $\omega = 0$ as a zero approximation. That is,

$$e_1^{(0)}(w) = 0 \quad e_2^{(0)}(w) = e_2(\pm w_0).$$

Substitution of the zero approximation into the right-hand side of (6.85)–(6.86) and $\varepsilon_\perp = i4\pi\sigma_P/\omega$ yields the first approximation

$$e_1(w) = \frac{4\pi}{c} e_2(\pm w_0) \int_{\pm w_0}^{w} \sigma_P(u)(1+3u^2)\mathrm{d}u,$$
$$e_2(w) = e_2(\pm w_0).$$

In the same manner we obtain the second, third, etc., approximations. With the condition (6.84) holding, the second-order corrections are small in comparison with the first-order ones over the whole ionosphere. Omitting these terms, we obtain from the second equation the boundary condition at the upper boundary of the Pedersen layer:

$$e_1(w) = -\nu(1+3w^2)\frac{X_\pm}{\sin I_0} e_2(w) \quad \text{at} \quad w = \pm w_0, \tag{6.87}$$

where $X_\pm = 4\pi\Sigma_P^\pm/c$, $\tan I_0 = 2\cot\theta_0$, θ_0 is co-latitude of the field line pierce point on the ionosphere.

Resonance Fields

The resonance fields are written in (6.59) for the covariant components. Rewrite these equations for the physical components which are found from the covariant components with the aid of Lamé coefficients. For the dipole coordinates we have

$$\hat{b}_\nu = b_\nu/h_\nu, \quad \hat{b}_\varphi = b_\varphi/h_\varphi, \quad \hat{b}_\mu = b_\mu/h_\mu,$$
$$\hat{E}_\nu = E_\nu/h_\nu, \quad \hat{E}_\varphi = E_\varphi/h_\varphi, \quad \hat{E}_\mu = E_\mu/h_\mu.$$

Then (6.59) can be rewritten

$$\left(\hat{b}_\varphi, \hat{E}_\nu, \hat{b}_\nu, \hat{b}_\mu, \hat{E}_\varphi\right)^{\mathrm{tr}} = \frac{\hat{\boldsymbol{\alpha}}_0}{\nu - \nu_r} + \hat{\boldsymbol{\alpha}}_1 + \cdots + \log\left[im\frac{h_\nu}{h_\varphi}(\nu - \nu_r)\right]\left[\hat{\boldsymbol{\beta}}_0 + \hat{\boldsymbol{\beta}}_1(\nu - \nu_r) + \cdots\right]. \tag{6.88}$$

Equations (6.70)–(6.71) for the vector-functions $\hat{\boldsymbol{\alpha}}_0$ and $\hat{\boldsymbol{\beta}}_0$ in the dipole coordinates become

$$\hat{\boldsymbol{\alpha}}_0(w) = \left(h_\varphi^{-1} e_1(w), h_\nu^{-1} e_2(w), 0, 0, 0\right)^{\mathrm{tr}},$$
$$\hat{\boldsymbol{\beta}}_0(w) = \left(C_0 e_1(w), C_0 e_2(w), -im\frac{h_\nu}{h_\varphi^2} e_1(w), 0, \frac{im}{h_\varphi} e_2(w)\right)^{\mathrm{tr}}.$$

Substitution of Lamé coefficients of the dipole coordinates (6.35) gives

$$\hat{\boldsymbol{\alpha}}_0(w) = \frac{\nu_r}{(1 - w^2)^{3/2}}\left(-e_1(w), \nu_r(1 + 3w^2)^{1/2} e_2(w), 0, 0, 0\right)^{\mathrm{tr}}, \tag{6.89}$$

$$\hat{\boldsymbol{\beta}}_0(w) = C_0\hat{\boldsymbol{\alpha}}_0(w) + \frac{im}{(1 - w^2)^{3/2}}\left(0, 0, -\frac{e_1(w)}{(1 + 3w^2)^{1/2}}, 0, \nu_r e_2(w)\right)^{\mathrm{tr}}. \tag{6.90}$$

The boundary problem (6.79)–(6.80) determines $e_1(w)$, and $e_2(w)$ accurate to an arbitrary constant factor. It can be found from the distribution of the longitudinal magnetic field $b_\parallel$ along a resonance field-line. We obtain from the conditions of solvability for (6.63) and (6.64):

$$-\frac{\nu_r^2 c^2}{\Omega^2}\int_{-w_0}^{w_0}(1 - w)^{6-p} e_2^2 \mathrm{d}w = -\frac{mc}{2\nu_r}\left(\frac{\mathrm{d}\omega_r}{\mathrm{d}\nu}\right)^{-1}_{\nu=\nu_r}\int_{-w_0}^{w_0}(1 + 3w^2) b_\parallel e_2 \mathrm{d}w. \tag{6.91}$$

Here the small terms relate to the Pedersen and Hall conductivities are neglected. Equation (6.91) defines the transformation of the FMS-mode into FLR-oscillations. If the longitudinal magnetic field $b_\parallel$ at the resonance field line is known, then the amplitude of the resonance Alfvén oscillations can be found from (6.91).

The Dipole Dispersion Equation Special Case

For the power distribution of the plasma density defined by (6.76) and dipole magnetic field, the boundary problem (6.81) can be solved explicitly. In the magnetosphere, $\Sigma_H = 0$ and the transversal dielectric permeability is

$$\varepsilon_\perp = \varepsilon_m = \frac{c^2}{c_A^2} = \frac{4\pi\rho c^2}{B_0^2}. \tag{6.92}$$

For the dipole geomagnetic field $B_0^2 = M^2(1+3w^2)/r^6$, with the dipole moment M, and plasma density (6.75), then (6.92) becomes

$$\varepsilon_m = \frac{\nu^2 c^2}{\Omega^2}\bar{\rho}\frac{(1-w^2)^6}{1+3w^2}, \tag{6.93}$$

where $\Omega = M\nu^4 \Big/ \left(2\pi^{1/2}\rho_e^{1/2}\right)$. Then (6.81) reduces to

$$\frac{\mathrm{d}^2 e_2}{\mathrm{d}w^2} + \frac{\omega^2}{\Omega^2}\bar{\rho}(1-w^2)^6 e_2 = 0. \tag{6.94}$$

For the power distribution of the plasma density (6.76), the boundary problem (6.94) and (6.87) becomes

$$\frac{\mathrm{d}^2 e_2}{\mathrm{d}w^2} + \frac{\omega^2}{\Omega^2}(1-w^2)^{6-p} e_2 = 0, \tag{6.95}$$

$$\left.\frac{\mathrm{d}e_2}{\mathrm{d}w}\right|_{w=\pm w_0} = -i\frac{\omega}{c\nu}(1+3w_0^2)\frac{X_\pm}{\sin I_0}\, e_2(w)|_{w=\pm w_0}\,. \tag{6.96}$$

Here the altitude dependency of the Lamé coefficients in the ionosphere is neglected.

Let us also write the equations for calculating the coefficient C_0 at the logarithmic term in the expansion of the field (6.59). From (6.73) we obtain

$$C_0 = \frac{m^2\nu^2c^2}{2\Omega^2(1+3w^2)}\omega_j \frac{\int\limits_{-w_0}^{w_0}\left[(1-w^2)^{6-p}e_2^2(w) - \left(\frac{\Omega}{\omega}\frac{\mathrm{d}e_2}{\mathrm{d}w}\right)^2\right]\mathrm{d}w}{\frac{\mathrm{d}\omega_j}{\mathrm{d}\nu}\int\limits_{-w_0}^{w_0} e_1^2(w)\mathrm{d}w}. \tag{6.97}$$

The formulae obtained in the present section allow us to get the explicit expressions for the leading terms describing the resonance structure of the field. Corresponding formulae are given in the next section. For the high frequency harmonics, we present the expressions for the resonance frequencies obtained in the WKB approximation. Neglecting the dissipation and assuming a power distribution of the plasma concentration, we find that the resonance frequency is

$$\omega_j = \frac{j\pi\Omega}{2\int\limits_0^{w_0}(1-w^2)^{6-p}\,\mathrm{d}w}. \tag{6.98}$$

Explicit Solutions

Dispersion Equation

In the general case, the boundary problem (6.94), (6.95) can be solved numerically. In this section we shall consider a special case of the plasma density distribution allowing us to solve explicitly the boundary problem. Consider the power distribution of the plasma density (6.63) along a field-line. Then the general solution of (6.94) at $p = 6$ is the simplest and is given by

$$e_2 = A\cos(\omega w/\Omega) + B\sin(\omega w/\Omega).$$

If the Pedersen conductivities of the conjugate ionospheres are equal, that is $X_+ = X_- = X$, then the symmetric solution with respect to the equatorial plane is

$$e_2 = A\cos(\omega w/\Omega), \tag{6.99}$$

and the anti-symmetric one is

$$e_2 = B\sin(\omega w/\Omega). \tag{6.100}$$

Substituting (6.99) and (6.100) into the boundary conditions (6.96), we obtain the dispersion equations for the symmetrical and anti-symmetrical modes:

$$\begin{aligned} \cot(w_0\frac{\omega}{\Omega}) &= iq(\nu) \quad \text{symmetrical mode,} \\ \tan(w_0\frac{\omega}{\Omega}) &= -iq(\nu) \quad \text{anti-symmetrical mode,} \end{aligned}$$

with

$$q(\nu) = -\frac{c\nu}{\Omega(1+3w_0^2)^{1/2}}\frac{\sin I_0}{X_\pm}.$$

The FLR-frequencies are determined by

$$\omega_j(\nu) = j\omega_A(\nu) - i\gamma_j(\nu), \tag{6.101}$$

where

$$\omega_A(\nu) = \frac{\pi}{2}\frac{\Omega(\nu)}{w_0(\nu)}, \quad \bar{\gamma}_j(\nu) = \frac{\gamma_j(\nu)}{\omega_A(\nu)} = \frac{2}{j\pi}\operatorname{atanh} q(\nu). \tag{6.102}$$

Symmetrical (anti-symmetrical) modes correspond to odd (even) j.

Let us estimate the FLR-frequencies $\omega_A(\nu)$ and decrements $\bar{\gamma}_j(\nu)$ for the dipole field. The parameter $\nu = -1/LR_E$, the geomagnetic dipole moment $M = 8\times10^{25}\,\text{G}\cdot\text{cm}^3$, the Earth's radius $R_E = 6.37\times10^8\,\text{cm}$, the proton mass $m_p = 1.673 \times 10^{-24}$ g/ cm^3. Then from (6.102) we have

$$\omega_A(L) = \frac{166.5}{n_e^{1/2} L^4 (1 - 1.015/L)^{1/2}}, \quad \Omega = \frac{106}{n_e^{1/2} L^4},$$
$$\bar{\gamma}_j(L) = \frac{0.637}{j} \operatorname{atanh}\left(0.889 n_e^{1/2} \frac{(1 - 1.015/L)^{1/2} L^3}{4 - 3.047/L} \frac{1}{X}\right),$$
$$q = \frac{94.19}{L\Omega X} \frac{w_0}{1 + 3w_0^2}. \tag{6.103}$$

If the frequencies $\omega_A(L)$ and decrements γ_L are found, one can determine the FLR-half-width $\delta_L^{(j)}$ in the magnetosphere if we take hold of (6.83) in the form

$$\delta_L^{(j)} \approx \frac{\gamma_j}{\omega_j'(L)}, \tag{6.104}$$

where $\omega_j'(L) = d\omega_j/dL$. Mapping $\delta_L^{(j)}$ onto the ionosphere, one can express the half-width of the resonance region $\delta_i^{(j)}$ at the ionospheric height in terms of $\delta_L^{(j)}$

$$\delta_i^{(j)} \approx \frac{6523}{L(4L - 3)} \delta_L^{(j)} \text{ (km)}. \tag{6.105}$$

Since the decrement is small, then from (6.103) follows that

$$\frac{d \ln(\omega_j(L))}{dL} \approx -\frac{4}{L} + \frac{1}{2L(L-1)} + \frac{p}{2L}. \tag{6.106}$$

From (6.105) and (6.106), we obtain for the half-width $\delta_L^{(j)}$ outside the plasmasphere:

$$\delta_L^{(j)} \approx \frac{\bar{\gamma}_j L}{-4 + 1/2(L-1) + p/2} \sim \frac{\bar{\gamma}_j L}{3}.$$

Distribution of the Wave Fields

A field line distribution of the magnetic disturbances is given by $\mathbf{e}_j(w)$. Let $\bar{\mathbf{e}}_j(w)$ be normalized by conditions

$$\nu_j^{-2} \int_{-w_0}^{w_0} \bar{e}_1^2(w) \mathrm{d}w - \frac{\nu_j^2 c^2}{\Omega^2} \int_{-w_0}^{w_0} \bar{e}_2^2(w) \mathrm{d}w = 1. \tag{6.107}$$

Then (6.101) and (6.102) become

$$\bar{\mathbf{e}}_j(w) = a_j \begin{pmatrix} i\dfrac{\nu^2 c}{\Omega}\sin\left[j\dfrac{\pi}{2}(1-i\bar{\gamma}_j)\dfrac{w}{w_0}\right] \\ \cos\left[j\dfrac{\pi}{2}(1-i\bar{\gamma}_j)\dfrac{w}{w_0}\right] \end{pmatrix} \quad \text{at} \quad j = 2k+1, \tag{6.108}$$

$$\bar{\mathbf{e}}_j(w) = a_j \begin{pmatrix} -i\dfrac{\nu^2 c}{\Omega}\cos\left[j\dfrac{\pi}{2}(1-i\bar{\gamma}_j)\dfrac{w}{w_0}\right] \\ \sin\left[j\dfrac{\pi}{2}(1-i\bar{\gamma}_j)\dfrac{w}{w_0}\right] \end{pmatrix} \quad \text{at} \quad j = 2k. \tag{6.109}$$

The coefficients a_j are determined from the normalization condition (6.107) and are given by

$$a_j = \frac{i\Omega}{c\nu(2w_0)^{1/2}}. \tag{6.110}$$

Substitution of numerical values of parameters in (6.110) yields

$$a_j = i\,\frac{1.59}{n_e^{1/2} L^3 (2w_0)^{1/2}}.$$

The coefficient C_{0j} is obtained from (6.97). With normalization condition (6.107), we have at $p = 6$

$$C_{0j} = (-1)^{j+1} m^2 a_j^2 \frac{2c^2 w_0}{\Omega^2 \dfrac{\mathrm{d}\ln\omega_j}{\mathrm{d}\nu}} \frac{I_1}{\pi L^2 R_E}, \tag{6.111}$$

where

$$I_1 = \int_0^\pi \frac{\cos\left(j(1-i\bar{\gamma}_j)\,t\right.}{1+3w_0^2 t^2/\pi^2}\mathrm{d}t.$$

Then

$$C_{0j} = (-1)^{j+1} \frac{2\times 10^8 m^2}{\dfrac{\mathrm{d}\ln\omega_j}{\mathrm{d}\nu}} I_1.$$

For calculation of the wave field we use lengths in kilometers. After the normalization we have

$$L\bar{C}_{0j} = \frac{2m^2(L-1)^{1/2}}{R_E^2} C_{0j} = (-1)^{j+1} \frac{10^{-4}(L-1)^{1/2}}{\dfrac{\mathrm{d}\ln\omega_j}{\mathrm{d}L}} I_1(\mathrm{km}^{-1}).$$

Introduce the coefficient of excitation of the j-mode as

$$e_j = n_j \bar{e}_j.$$

Equation (6.91) allows us to express n_j in terms of the longitudinal magnetic component. Denote the physical component of the longitudinal magnetic field as $\hat{b}_\parallel$. Then field line distribution of the cold plasma expressed in terms of $\hat{b}_\parallel$ can be written as

$$n_j = \frac{mc}{L\dfrac{\mathrm{d}\omega_j}{\mathrm{d}L}} \int_{-w_0}^{w_0} (1+3u^2)^{1/2} \hat{b}_\parallel(u) \bar{e}_{2j}(u) \mathrm{d}u, \tag{6.112}$$

where the integral is calculated over a contour in the complex plane.

From (6.89), (6.107), and (6.112) we have

$$\hat{b}_{\varphi j} = \frac{\Lambda_{Bj} U_j}{(L-L_j)(1-w^2)^{3/2}} \begin{cases} \sin W_j(w), & j = 2n+1 \\ -\cos W_j(w), & j = 2n \end{cases}, \tag{6.113}$$

$$\hat{E}_{\nu j} = \frac{\Lambda_{Ej} U_j (1+3w^2)^{1/2}}{(L-L_j)(1-w^2)^{3/2}} \begin{cases} \cos W_j(w), & j = 2n+1 \\ \sin W_j(w), & j = 2n \end{cases}, \tag{6.114}$$

where $n = 0, 1, 2, 3, \ldots$, $W_j(w) = j\frac{\pi}{2}(1 - i\bar{\gamma}_j)\frac{w}{w_0}$,

$$U_j = \int_{-w_0}^{w_0} \mathrm{d}u \, \hat{b}_\parallel \left(1+3u^2\right)^{1/2} \begin{cases} \cos W_j(u), & j = 2n+1 \\ \sin W_j(u), & j = 2n \end{cases}, \tag{6.115}$$

$$\Lambda_{Bj} = im \frac{a_j^2 c^2}{L^2 R_E^2 \Omega \dfrac{\mathrm{d}\omega_j}{\mathrm{d}L}}, \qquad \Lambda_{Ej} = m \frac{a_j^2 c}{L R_E \dfrac{\mathrm{d}\omega_j}{\mathrm{d}L}}. \tag{6.116}$$

6.5 Numerical Simulation

Dispersion Equation

Let us describe a method of derivation of the dispersion equation based on the impedance successive sweep method. Introduce a partial admittance $y = e_1/e_2$. Note that this admittance is equal to the ratio of covariant field components contrary to the usual admittance determined in terms of physical components. From (6.81) obtain a Riccati-type equation for y:

$$y' + i\frac{\omega}{\nu^2 c} y^2 - i\frac{\omega}{c}\left(1+3w^2\right)\varepsilon_\perp = 0, \tag{6.117}$$

with the boundary conditions

$$y|_{w=\pm w_0} = 0. \tag{6.118}$$

For the power plasma distribution (6.75) within the thin ionosphere approximation we get

$$y' + i\frac{\omega}{\nu^2 c} y^2 - i\frac{\nu^2 c}{\Omega^2}\omega \left(1 - w^2\right)^{6-p} = 0, \tag{6.119}$$

$$y|_{w=\pm w_0} = -\nu \left(1 + 3w_0^2\right)^{1/2} \eta_+. \tag{6.120}$$

Take the integral of (6.117) or (6.119) numerically beginning with $w = -w_0$ with the initial values determined by (6.118) or (6.120) to $w = w_0$. Then we equate the obtained boundary values of $y_+ = y(w_0)$ to the boundary value of the admittance at $w = w_0$. In doing so we construct the admittance equation written for the boundary conditions (6.120) in the form

$$F(\omega, \nu) = y_+ + \nu(1 + 3w_0^2)^{1/2}\eta_+. \tag{6.121}$$

Zeroes of the function $F(\omega, \nu)$ determine the Alfvén resonance frequencies.

Consider now the ionosphere of a finite thickness. The coefficients of (6.117) have different characteristic scales in the ionosphere and magnetosphere. Therefore, split the interval of integrating into two subintervals. The first corresponds to the ionosphere and the second to the magnetosphere. Introduce new variables for the ionospheric interval:

$$y = \frac{2X}{LR_E}u, \qquad \tau = \frac{(r - R_I)}{l_P},$$

with

$$X = \frac{4\pi \Sigma_P}{c}, \qquad l_P = \frac{\Sigma_P}{\langle \sigma_P \rangle}.$$

Here $\langle \sigma_P \rangle$ is the height averaged Pedersen conductivity, l_P is the thickness of the ionospheric conductive layer. Then from (6.117) we have

$$\frac{\mathrm{d}u}{\mathrm{d}\tau} = ik_0 l_P X \frac{u^2}{\left(1 - \mu\tau\right)^{1/2}} \mp \frac{\sigma_P}{\langle \sigma_P \rangle} \frac{1 - \mu\tau/4 - 1/L}{\left(1 - \mu\tau - 1/L\right)^{1/2}}, \tag{6.122}$$

where the upper (lower) sign corresponds to the northern (southern) ionosphere; $\mu = l_P/(LR_I)$. The boundary conditions (6.118) are given by

$$u|_{\tau=0} = 0. \tag{6.123}$$

Integrating (6.122) with the boundary conditions (6.123) for both hemispheres, we find u at $\tau = 1$. Calculating the admittance y and returning to (6.117) we construct the function $F(\omega, L)$ and obtain the dispersion equation (6.121).

Roots of (6.121) can be found using, for instance, Newton's method. Let $\omega_j^{(0)}(L)$ be an zero approximation of the j-th FLR frequency $\omega_j(L)$. Then the next approximation is

$$\omega_j^{(n+1)}(L) = \omega_j^{(n)}(L) - \frac{F(\omega_j^{(n)}, L)}{\dfrac{\partial F(\omega_j^{(n)}, L)}{\partial L}}.$$

A zero approximation can be calculated from (6.99).

So, the numerical algorithms enabled us

- to find frequencies and decrements of the FLR for a given L-shell and
- to find the resonance L-shell and half-width of the FLR for a prescribed value of the frequency of the external source.

FLR-Frequencies and Decrements

We explored two ionospheric models. In the first model, we used height dependencies of the ionospheric conductivities (see Chapter 2) computed with IRI 2000 [5]. We call this model a 'thick' ionosphere in contrary to the 'thin' ionosphere (the 2-nd model). We used also two models of equatorial distribution of cold plasma (see Fig. 6.3). Dependencies of the fundamental FLR-period $T_1 = 2\pi/\omega_A$ and relative decrement γ_1/ω_A on McIllwain parameter L for the 'thin' and 'thick' ionospheres are given in Figure 6.4. The curves 1 and 3 correspond to the 1-st model; 2 and 4 are for the 2-nd model. The curves 1, 2 and 3, 4 show T_1 and γ_1/ω_A, respectively, of the first harmonics calculated for the 'thin' ionosphere with $\Sigma_P = 1.55 \times 10^8$ km/s.

The curves 5 (T_1) and 6 (γ_1/ω_A) are for a 'thick' (i.e. of finite thickness) ionosphere of the same Σ_P with an account taken of the steep growth of the ion concentration in the F-layer. The curves 5 and 6 are shown only for $L < 3.5$.

Outside this region the curves for T_1 and γ_1/ω_A merge with the same curves calculated for the thin ionospheric model. It can be seen from Fig. 6.4 that resonance periods at $L \gtrsim 2$ calculated for thin and thick ionospheric models are almost the same. At $L \lesssim 2$ resonance period T_1 for the thin ionosphere decreases monotonously with L while the thick ionosphere gives a minimum of $T_1 \approx 10$ s at $L \approx 1.5$.

In order to find the decrement γ_1 and the resonance half-width δ_L at $L \gtrsim 2.5$, it is sufficient to use the 'thin' ionospheric approximation which gives the same dependencies as the real ionospheric model. As to the inner regions with $L \lesssim 2$, γ_1 and δ_1 are found in the real ionospheric model, to go up rather steeply, whereas for the 'thin' ionosphere they go down monotonically. Thus at $L \lesssim 2$, the Q-factor of the FLR decreases severely and the resonance effects become very weak.

One can see that in the $T_1(L)$ dependency there is a minimum at $L = 1.6$ in which $T_1 \approx 10$ s. The FLR-period goes down in the low latitudes to $L < 1.6$ and then goes up to the plasmapause at $L = L_{pp}$. Behind the plasmapause $T_1(L)$ decreases steeply and increases smoothly to the large L.

The low latitudinal minimum is caused with the fact that the total traveling time between conjugate ionospheres consists of the ionospheric and

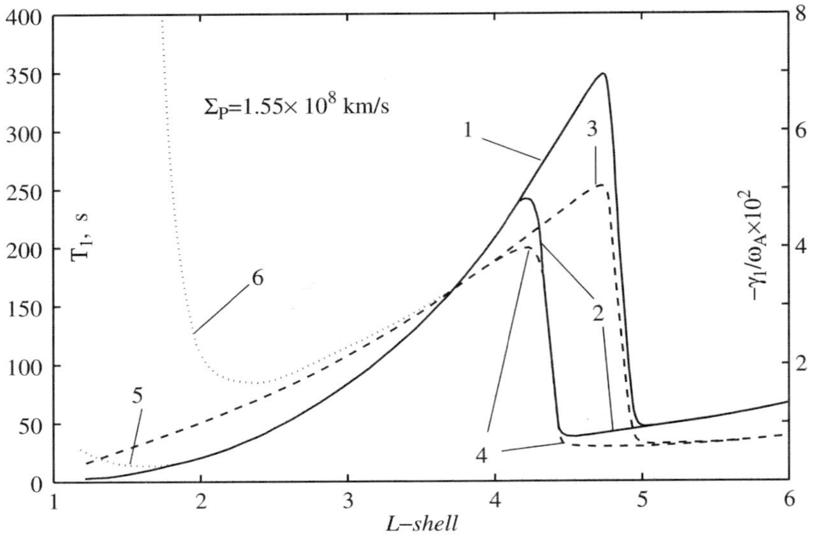

Fig. 6.4. The FLR-period $T_1 = 2\pi/\omega_A$ and relative decrements γ_1/ω_A versus L-shell. Lines $1, 2, 3, 4$ are plotted for the 'thin' model, while $5, 6$ for the 'thick' ionospheric model. Two models of the equatorial distribution of the cold plasma were used (see Fig 6.3): the curves 1 and 3 refer to the 1-st magnetospheric model; the curves 2 and 4 are for the 2-nd model. The 1-st, 2-nd, and 3-rd, 4-th show T_1 and γ_1/ω_A, respectively, of the first harmonics calculated for the 'thin' ionosphere with the integral Pedersen conductivity $\Sigma_P = 1.55 \times 10^8$ km/s. The curves 5 (T_1) and 6 (γ_1/ω_A) are for a 'real' (i.e. of finite thickness) ionosphere of the same Σ_P

magnetospheric parts. The FLR-period can be estimated roughly as

$$\frac{T_1}{2} \sim 2\frac{l_I}{c_{AI}} + \frac{l_m}{c_{Am}}$$

where c_{AI} and c_{Am} are typical Alfvén velocities in the ionosphere (I) and magnetosphere (m); respectively, l_I and l_m are lengths of the ionospheric and magnetospheric segments of the field-line. Since, $c_{AI} < c_{Am}$ ($c_{AI} \approx 300\,\text{km/s}$ and $c_{Am} \approx 1000\,\text{km/s}$), then the FLR-period $T_1 \sim 2l_I/c_{AI}$ at the low latitudes where $l_I > l_m\, c_{AI}/c_{Am}$. It means that at these latitudes the MHD-wave travels mostly within the ionosphere. In the near equatorial region, the field line is almost totally immersed into the ionosphere. The resonance period vanishes near the equator. This latitudinal range is not shown in Fig. 6.4. The length of the ionospheric field line segment decreases with latitude, therefore the FLR-period, in turn, decreases.

Approaching the plasmapause, the length of the magnetospheric segments becomes so large that the inequality $l_m > l_I\, c_{Am}/c_{AI}$ is satisfied. So, in this region the traveling time is determined by the magnetospheric part. The field line length increases and the T_1 FLR-period increases with latitude. Behind

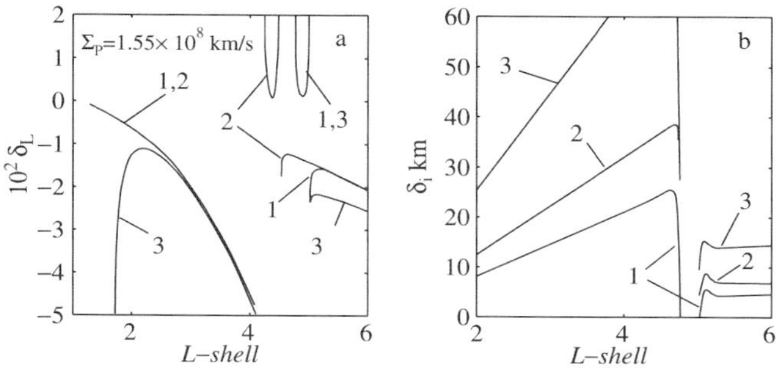

Fig. 6.5. Half-width of the FLR-shell in the magnetosphere δ_L (panel (a)) and its mapping onto the ionosphere δ_i (b) versus the $L-$ shell. The curves 1 and 2 at panel (a) refer to the thin ionosphere and to the I ($L_{pp} = 4.9$) and II ($L_{pp} = 4.4$) magnetospheric models, respectively. The 3-rd shows δ_L for the thick ionosphere and I magnetosphere. Curves at (b) refer, respectively, to $\Sigma_P = (1.55, 1, 0.5) \times 10^8$ km/s and to the I model

the plasmapause, the cold plasma density falls, and as a consequence of this, c_{Am} increases sharply and the FLR-period is shortened.

An error in δ_L caused by the use of (6.104) instead of (6.121) is $\sim$0.1% at $L < L_{pp}$ and $\sim$1% at $L > L_{pp}$. At the L corresponding to the plasmapause location, the error in δ_L increases but remains less than 10%. Thus, the approximate formula (6.104) is valid for calculation of δ_L in the whole range of L and will be exploited in further numerical calculations. The half-width of the resonance region on the ground at the low latitudes is about $\delta_i + h$ (see Chapter 7) and is controlled predominantly by the ionospheric losses. Equation (6.105) gives $\delta_i \approx 200 - 300\,\text{km}$ at $L \approx 1.5$. The height of the high conductive ionospheric layer $h \approx 100\,\text{km}$.

Close study of the distribution of geomagnetic pulsations at $L \lesssim 2$ could give information about the boundary of the region where the FLR contributes mainly to the spatial pulsations' distribution. Note that at low latitudes the resonance half-width can be found from the observations with the accuracy of $\sim$10 km. At $L \gtrsim 2$ the relative error of δ_i obtained from the ground-based data is essentially higher because δ_i itself is about several tens of kilometers.

The half-width of the FLR-shell in the magnetosphere δ_L (frame (a)) and its mapping onto the ionosphere δ_i (frame (b)) as a function of the L-shell are shown in Figures 6.5a and 6.5b. The curves 1 and 2 at frame (a) refer to the thin ionosphere and to the I ($L_{pp} = 4.9$) and II ($L_{pp} = 4.4$) magnetospheric models, respectively. The 3-rd curve represents δ_L for the thick ionosphere and I magnetosphere. The dependencies $\delta_i\,(L)$ found from (6.105) for the I model at $\Sigma_P = 1.55\times10^8$, 1×10^8, 0.5×10^8 km/s (curves 1, 2, and 3, respectively) are shown in Fig. 6.5b. The half-width δ_i increases monotonously from $\approx$10 km at $L = 2$ to several tens of kilometers closer to L_{pp} and decreases steeply to

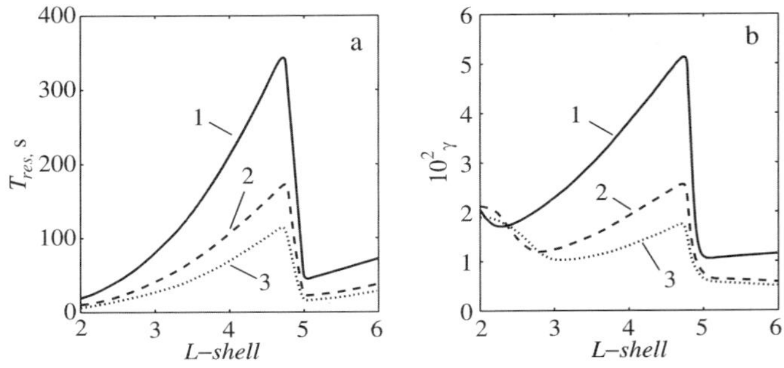

Fig. 6.6. (a) The FLR-period and (b) relative decrement versus L. Exponent $p = 6$ in (6.76) describing the field-aligned plasma distribution. Indexes refer to the 1-st, 2-nd and 3-rd harmonics

about $\sim$10 km at $L \gtrsim L_{pp}$. These results for δ_i do not contradict the ones obtained from the experimental data.

The resonance periods T_{res} (left frame) and decrements γ (right frame) as functions of the L-shell for the first three harmonics are shown in Fig. 6.6 for the cold plasma distribution (6.76) with $p = 6$. Figure 6.7 represents the half-widths $\delta_L(L)$ of three harmonics for the distributions with exponent $p = 6$ (left frame) and $p = 3$ (right frame). At $p = 3$ the harmonics are not equidistant and it is possible to use this fact in sounding the magnetosphere by estimating the parameter p. The decrement and the resonance half-width in the regions $2 \lesssim L \lesssim 4.5$ for the 2-nd harmonic and $3 \lesssim L \lesssim 4.5$ for the 3-rd harmonic are almost constant and the resulting shape of the resonance curve changes weakly in these ranges of L.

Note also that γ decreases in the thin ionosphere with harmonic number. Contrary, the finite ionospheric thickness leads to the growth of the relative

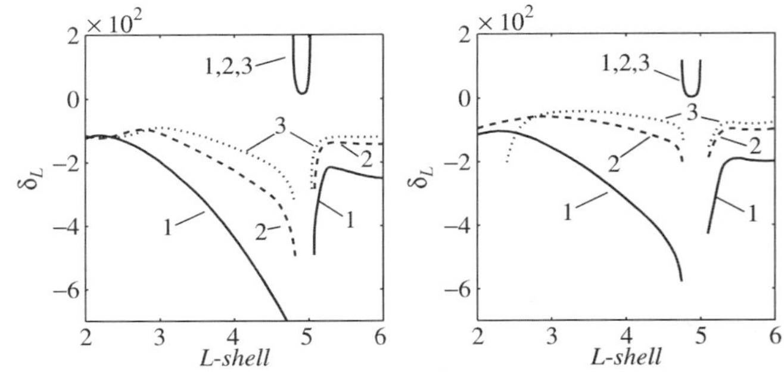

Fig. 6.7. The half-widths δ_L in the magnetosphere for the first three resonance harmonics functioning L. The curves are calculated for the field-aligned plasma distributions with exponent $p = 6$ (left panel) and $p = 3$ (right panel) of (6.76)

decrement with the number beginning with the resonance period of about 20 s.

Consider the dependence of the decrement on Pedersen conductivity. At large Σ_P (dayside ionosphere), when $\bar{\Sigma}_P = 4\pi\Sigma_P c_A/c^2 \gg 1$, the decrement $\gamma_1 \propto 1/\Sigma_P$ vanishes ($\Sigma_P \to \infty$). At $\bar{\Sigma}_P \ll 1$ when the Σ_P is small (night ionosphere), the decrement $\gamma_1 \propto \Sigma_P$. At $\bar{\Sigma}_P \to 1$, the coefficient of reflection of the Alfvén wave vanishes and $\gamma_1 \to \infty$. This effect is connected with the 'thin' ionosphere approximation and for the 'thick' ionosphere γ is finite.

Resonance Electric and Magnetic Fields Along a Field-Line

Distribution of the electric and magnetic wave components along a resonance field line is determined by $e_j(x)$ (6.79). In the dipole magnetic field the physical components of the azimuthal magnetic field $\hat{b}_\varphi$ and orthogonal to the magnetic shells electric field $\hat{E}_\nu$ are determined by (6.89), (6.90). Keeping only the main terms in (6.89), (6.90), we obtain the distribution of $\hat{b}_\varphi$ and $\hat{E}_\nu$ along the field line at the shell $\nu = \nu_r$:

$$\begin{aligned} \hat{b}_\varphi &\approx \nu_r(1-x^2)^{-3/2} e_1^{(j)}(x), \\ \hat{E}_\nu &\approx \nu_r^2(1-x^2)^{-3/2}(1+3x^2)^{1/2} e_2^{(j)}(x), \end{aligned} \tag{6.124}$$

where $x = \cos\theta$ is the coordinate of a point on a field-line.

An example of the distribution of the FLR-fields for two harmonics along a field line $L = 2$ is shown in Fig. 6.8. Figures 6.8a and 6.8b represent, respectively, the fields of the 2-nd harmonic with resonance frequency $\omega_2 = 0.6136 - 0.1186i$ and the 3-rd harmonics with $\omega_3 = 0.6336 - 0.1323i$. One can see that ionospheric fields are larger than magnetospheric ones. The amplitude of the magnetic field decreases steeply in the ionospheric F-layer. The electric

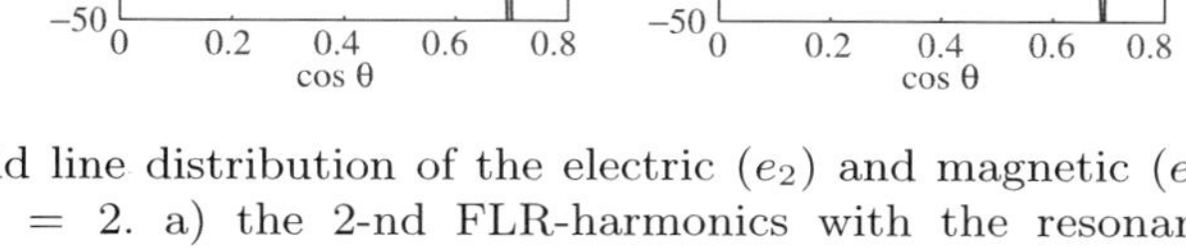

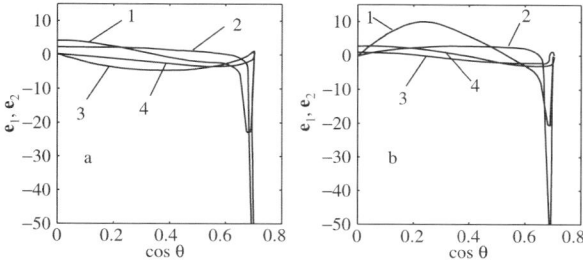

Fig. 6.8. Field line distribution of the electric (e_2) and magnetic (e_1) wave components at $L = 2$. a) the 2-nd FLR-harmonics with the resonance frequency $\omega_2 = 0.6136 - 0.1186i$; (b) the 3-rd harmonics $\omega_3 = 0.6336 - 0.1323i$. Indexes 1 and 3 refer, respectively, to real parts of e_1 and e_2 and $2, 4$ – to their imaginary parts. The azimuthal magnetic component b_φ and the electric component orthogonal to the E_ν magnetic shell are connected with e_1 and e_2 by (6.124)

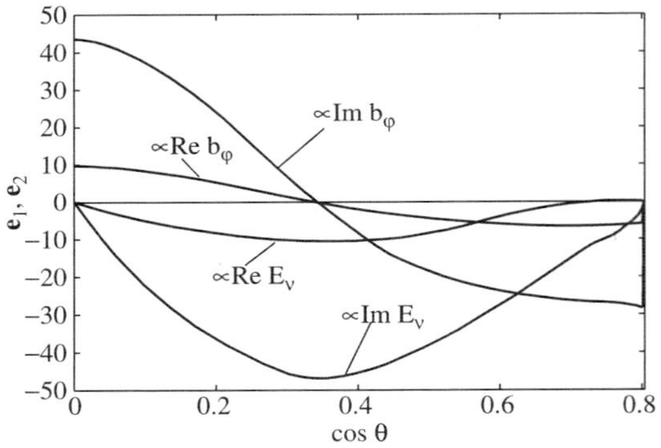

Fig. 6.9. Field line distribution of the real and imaginary parts of the electric and magnetic components of the 2-nd FLR-harmonics at field line $L = 3$ as a function of colatitude θ. Point $\cos\theta = 0$ is the equatorial point and $\cos\theta = 0.8$ is the pierce point of the field line in the ionosphere

field varies significantly at heights about 200 km and it changes its sign at $h \approx 300$ km. The relative growth of the electric field in the ionosphere results in increasing Joule dissipation and decrement.

The same dependencies for the 2-nd harmonic at the shell $L = 3$ are shown in Fig. 6.9. Note that the azimuthal component of the magnetic field $\hat{b}_\varphi(x) \propto e_1(x)$ varies rapidly in the ionospheric E-layer. It grows from zero under the ionosphere to the maximal value above the ionosphere. On the other hand, the component $\hat{E}_\nu(x) \propto e_2(x)$ is almost constant in the E-layer.

In the above calculations the ionospheric parameters typical for the solar maximum are used.

Meridional Distribution of the FLR-Amplitudes and Phases

The meridional distribution of the FLR-amplitude and phase above the ionosphere are determined by (6.88), (6.89) and (6.90). In order to utilize their we need to define FLR-periods, half-widths, and coefficients C_0 and Λ_0. Resonance period and half-width of the FLR are determined from the dispersion equation (6.121). C_0 and Λ_0 are given by (6.111)–(6.116). The calculations have been carried out at $\Sigma_P = 0.8 \times 10^8$ km/s. The equatorial distribution of the ion concentration has been taken from Fig. 6.3 (curve 2) with plasmapause at $L_{pp} = 3.9$ that corresponds to a mean value of the plasmapause position at $K_p = 5$. The calculations are performed for $m = 1$. Results of the calculations are shown in Fig. 6.10. Panels (a) and (b) show dependencies of the FLR-period and half-width against L-shell, and (c) and (d) show abs $C_0(L)$ and abs $\Lambda_0(L)$.

Let us now find the meridional distribution of the wave field produced by a driver of period T. From 6.10a follows that there are two fundamentally

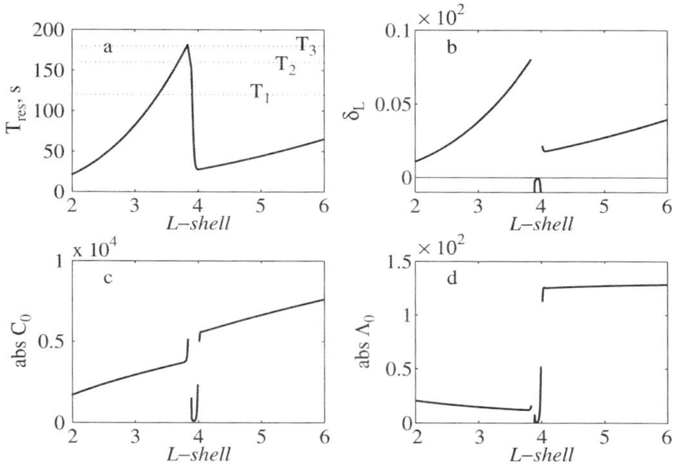

Fig. 6.10. Plots of (a) FLR-period T_1 of the 1-st harmonics and (b) the half-width δ_L of the FLR-shells versus L. Panels (c) and (d) demonstrate, correspondingly, dependencies of coefficients C_0 of (6.73), and coefficient Λ_0 of (6.116) on L

different situations. In the first one, T is out of the FLR-band and in the other one, T is within the FLR-band. It is obvious, that in this case the driver excites FLR-oscillations of at least one FLR-shell.

Figure 6.11a ($|b_\varphi|$) and Figure 6.11b ($\arg b_\varphi$) show azimuthal component b_φ in the equatorial magnetosphere plane as a function of the L-shell. The solid lines correspond to the period of the driver $T_1 = 120\,$s, the dashed line – $T_2 = 160\,$s, and the dotted line – $T_3 = 180.2\,$s. Three horizontal dot-lines in Fig. 6.10a show the L-shell dependencies of these periods. It is clear from Fig. 6.10a that each of periods T_1 and T_2 correspond two magnetic shells. The shells $L = 3.35$ and 3.90 most closely correspond to T_1 and $L = 3.72$ and 3.88 correspond to T_2. The period T_3 exceeds slightly the maximal FLR-period 180 s, at $L = 3.83$.

In $|b_\varphi(L)|$, there are two maxima between $L = 3$ and $L = 4$ for each of T_1 and T_2 (solid and dashed lines in Fig. 6.10a). There is a minimum between the FLR-shells. The ratio of minimal amplitude to maximal amplitude is significantly lower for T_1 compared to the same ratio for T_2. The amplitude minima are located near the southern edge of the plasmapause for both periods. From the conservation of the energy flux, it follows that the amplitude $\propto c_A^{1/2} \propto T_1(L)^{-1/2}$. Hence, the minimal amplitude should be located near the maximal FLR-period, i.e. at the southern edge of the plasmapause (see Fig. 6.10a).

Let the driver period be a little less than the maximal FLR-period. Then two FLR-shells approach and only one peak of $|b_\varphi(L)|$ remains in the meridional distribution instead of two resonance maxima. In the opposite case, if

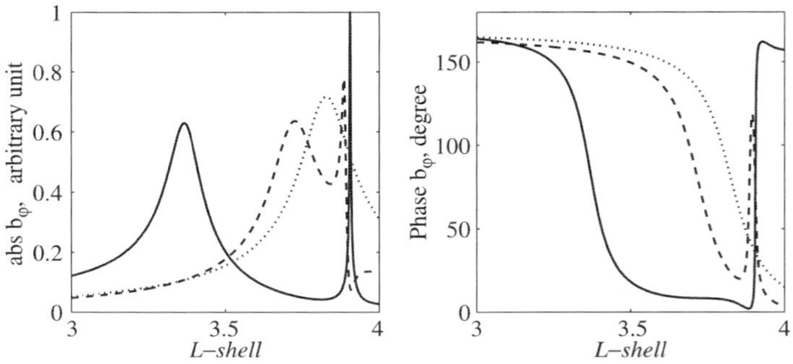

Fig. 6.11. The calculated amplitude (a) $|b_\varphi|$ and phase (b) $\arg(b_\varphi)$ of the azimuthal magnetic component vs. L (see (6.88) and (6.90)) for the 1-st harmonics in the vicinity of the resonance magnetic shell. The solid lines represent $T_1 = 120\,\mathrm{s}$, the dashed lines – $T_2 = 160\mathrm{s}$, and the dotted lines – $T_3 = 180.2\mathrm{s}$

the driver period exceeds maximal FLR-period, the amplitude distribution also has one maximum gradually flattened with increase of T. Such amplitude distribution for the driver period $T_3 = \max T_1(L)$ is shown by the dotted line in Fig. 6.11a.

If the FLR-shells are distant from each other, the phase jumps at $\approx \pm 180°$ as each of the FLR-shell transections. The sign '+' is taken for the FLR-period growth in passage of the resonance shell, and '−' is for the FLR-period diminution. The solid line in Fig. 6.11b demonstrates such L-phase dependency. The phase decreases at $\approx 180°$ with an increase of L from 3 to 3.5 and increases at 180° for a change of L from 3.8 to 4. As a result the total phase change at the consequent passage of two resonance shells is $\approx 0°$.

The mutual contribution of two close resonances with different excitation coefficients Λ_{Bj} to the total magnetic field. The dashed line in Fig. 6.10b presents an example of the phase L-distribution in the vicinity of the two FLR-shells. Close to the first shell with $L_1 = 3.73$ the phase rapidly goes down almost at 180° and near the second shell, $L_2 = 3.88$, the phase initially increases. An impact of L_1 prevails with moving away from L_2 to the north. As a consequence, the total phase change $\approx -180°$ as L changes from 3 to 4. Approaching the FLR-shells, the phase change becomes monotonous and has the form shown by the dotted line in Fig. 6.11a for the drive period T_3.

Meridional Distribution of the FLR Polarization Ellipses

Let us note some features of the FLR-polarization above the ionosphere. A polarization ellipse is defined by the ratio of the meridional component $\hat{b}_\nu$ to the azimuthal component $\hat{b}_\varphi$. From (6.89), (6.90) we get

$$\frac{\hat{b}_\nu}{\hat{b}_\varphi} \approx im\xi \ln(-im\xi),$$

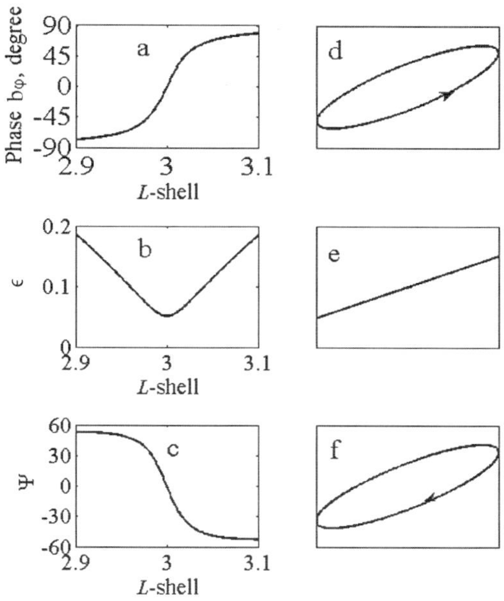

Fig. 6.12. (a) Dependency of the $b_\varphi\,(L)$-phase; (b) ellipticity $\epsilon\,(L)$ and (c) phase shift $\Psi\,(L)$ between the normal to magnetic shell component b_ν and azimuthal component b_φ. The plots are calculated close to the resonance shell. The panels (d), (e) and (f) demonstrate sequential changes of the rotation sign of the polarization ellipse passing through the FLR-shell. (d) shows elliptical polarization and counter-clockwise rotation of the transversal magnetic field within the FLR-shell at $T < T_r$, (e) linear polarization at the shell, $T = T_r$ and (f) the elliptical polarization with clockwise rotation of the transversal magnetic component outside the shell, $T > T_r$

where $\xi = (L_r/L - 1)\,(4 - 3/L_r)^{-1/2}$ and m is the azimuthal wavenumber.

The rotation of the polarization ellipse is determined by the phase shift Ψ between the transversal magnetic components:

$$\Psi = \arg\left(\frac{\hat{b}_\nu}{\hat{b}_\varphi}\right) = \pi/2 + \arg(\xi) + \arg[\ln|m\xi| + i(-\pi/2 + \arg(\xi))].$$

Figure 6.12 demonstrates the dependencies of the magnetic field polarization close to the resonance magnetic shell $L = L_r$ as functions of L. In this example $L_r = 3$. Spatial distribution of the b_φ-phase is shown in Fig. 6.12a. The ellipticity $\epsilon\,(L)$ and the phase shift $\Psi\,(L)$ are shown in the frames (b) and (c) respectively. The frames (d), (e) and (f) demonstrate sequential changes of the rotation sign in the polarization ellipse at the transition through the FLR-shell. Herewith, (d) shows elliptical polarization and counter-clockwise rotation of the transversal magnetic field within the FLR-shell at $T < T_r$, (e) linear polarization at the shell, $T = T_r$ and (f) the elliptical polarization with clockwise rotation outside the shell, $T > T_r$.

6.6 Summary

In this chapter we treated both free and excited oscillations and the structure of the MHD-waves in the axially symmetric magnetospheric models as well.

1. We found that spectrums of Alfvén oscillations in 2D case can be reduced to two 1D boundary problems:
   - determination of the toroidal oscillation spectrum from (6.51), (6.53) and
   - determination of the poloidal oscillation spectrum from (6.52), (6.54).
2. We examined also the excitement of the oscillations by a harmonic source of frequency ω close to the frequency $\omega_j^{(T)}(L_r)$ of the toroidal resonance mode j of one of the FLR-shells with $L = L_r$. It was shown that the spatial distribution of the azimuthal magnetic component b_φ and of the electric component E_ν orthogonal to the FLR-shellin the toroidal mode are Lorentz resonance curves:
$$\hat{b}_\varphi \approx \frac{\hat{b}_\varphi^{(j)}}{L - (L_r + i\delta_L^{(j)})}, \qquad \hat{E}_\nu \approx \frac{\hat{E}_\varphi^{(j)}}{L - (L_r + i\delta_L^{(j)})},$$
where $\hat{b}_\varphi^{(j)}$ and $\hat{E}_\nu^{(j)}$ can be found from the longitudinal magnetic component $\hat{b}_\|$ (see (6.113), (6.114)).
3. The decrement $\gamma_j^{(T)}(L)$ of the toroidal oscillations is connected with the half-width $\delta_L^{(j)}$ and the spatial scale of the FLR-frequency $\omega_j^{(T)}$:
$$\delta_L^{(j)} \approx \gamma_j^{(T)} \left(\frac{\mathrm{d}\omega_j^{(T)}}{\mathrm{d}L} \right)_{L=L_r}^{-1}.$$
4. Besides of toroidal modes, there are also poloidal modes. Contrary to toroidal oscillations which manifest themselves in azimuthal magnetic components and in electrical components orthogonal to the shell, poloidal oscillations of the shell are resonance oscillations of the magnetic component orthogonal to the shell and of the azimuthal electric component.
5. Polarization splitting of the FLR-frequencies for the toroidal and poloidal oscillations is determined by the difference of the two main curvatures of the surface orthogonal to geomagnetic field-lines. In other words, the split is determined by the difference of the convergency/divergency rate of the field-lines within the meridional and equatorial plains.

Appendix A

The metric tensors g_{ik} and g^{ik} of the non-orthogonal system coordinates $\{x^1, x^2, x^3\}$ are

$$
(g_{ik}) = \begin{pmatrix} h_1^2 + \frac{p_1^2}{K_2^2} h_3^2 & \frac{p_1 p_2}{K_2^2} h_3^2 & -\frac{p_1}{K_2^2} h_3^2 \\ \frac{p_1 p_2}{K_2^2} h_3^2 & h_2^2 + \frac{\gamma_2^2}{K_2^2} h_3^2 & -\frac{p_2}{K_2^2} h_3^2 \\ -\frac{p_1}{K_2^2} h_3^2 & -\frac{p_2}{K_2^2} h_3^2 & \frac{h_3^2}{K_2^2} \end{pmatrix}, \tag{6.A.1}
$$

$$
(g^{ik}) = \begin{pmatrix} \frac{1}{h_1^2} & \frac{p_1 p_2}{K_2^2} h_3^2 & \frac{p_1}{h_1^2} \\ \frac{p_1 p_2}{K_2^2} h_3^2 & \frac{1}{h_2^2} & \frac{p_2}{h_2^2} \\ \frac{p_1}{h_1^2} & \frac{p_2}{h_2^2} & \frac{K_2^2}{h_3^2} + \frac{p_1^2}{h_1^2} + \frac{p_2^2}{h_2^2} \end{pmatrix}, \tag{6.A.2}
$$

where $p_\alpha = \partial x^{(3)}/\partial y^{(\alpha)} = \partial K_1/\partial y^{(\alpha)} + y^{(3)} \partial K_2/\partial y^{(\alpha)}$. The metric tensors of the non-orthogonal dipole coordinates ξ, η, ζ, obtained from (6.36), (6.A.1) and (6.A.2), are

$$
\begin{aligned}
g_{11} &= h_\nu^2 + \frac{R_E^2 h_\mu^2}{4r^4} \frac{\cos^2\theta}{(1 - R_E \sin^2\theta/r)^2}, \quad g_{22} = h_\varphi^{-2}, \\
g_{33} &= h_\mu^2 (1 - R_E \sin^2\theta/r)/R_E^4, \\
g_{13} &= g_{31} = \frac{h_\mu^2}{2r^2 R_E} \frac{\cos\theta}{(1 - R_E \sin^2\theta/r)^{1/2}},
\end{aligned} \tag{6.A.3}
$$

$$
\begin{aligned}
g^{11} &= h_\nu^{-2}, \quad g^{22} = h_\varphi^{-2}, \\
g^{33} &= \left[\frac{1}{h_\mu^2} + \frac{R_E^2 h_\mu^2}{4r^4} \frac{\cos^2\theta}{(1 - R_E \sin^2\theta/r)^2}\right] \frac{R_E^4}{1 - R_E \sin^2\theta/r}, \\
g^{13} &= g^{31} = -\frac{R_E^3 \cos\theta}{2r^2 h_\nu^2 (1 - R_E \sin^2\theta/r)^{3/2}}.
\end{aligned} \tag{6.A.4}
$$

References

1. Allan, W., S. P. White, and E. M. Poulter, pulse-excited hydromagnetic cavity and field resonances in the magnetosphere, *Planet. Space. Sci.*, **34**, 371, 1986.
2. Allan, W. and D. R. McDiarmid, Magnetospheric cavity modes: numerical model of a possible case, *J. Geophys. Res.*, **94**, 309, 1989.
3. Angerami, J. J. and J. O. Thomas, The distribution of ions and electrons in the Earth's exosphere, *J. Geophys. Res.*, **69**, 4537, 1964.
4. Bhatacharjee, A., C. A. Kletzing, Z. W. Ma, C. S. Ng, N. F. Otani, and X. Wang, Fourfield model for dispersive fieldline resonances: Effects of coupling between shear Alfvén and slow modes, *Geophys. Res. Lett.*, **26**, 3281, 1999.
5. Bilitza, D., IRI 2000, *Radio Science*, **36**, 261, 2001.
6. Boyce, W. E. and R. C. DiPrima, *Elementary Differential Equations and Boundary Value Problems*, 7th edn., John Wiley & Sons, Inc., New York, 2001.

7. Chen, L., and A. Hasegawa, Plasma heating by spatial resonance of Alfvén wave, *Phys. Fluids*, **17**, 1399, 1974.
8. Chen, L., and A. Hasegawa, A theory of long period magnetic pulsations, 1, steady state excitation of field line resonance, *J. Geophys. Res.*, **79**, 1024, 1974.
9. Chen, L., and S. C. Cowley, On field line resonances of hydromagnetic Alfvén waves in dipole magnetic field, *Geophys. Res. Lett.*, **16**, 895, 1989.
10. Cheng, C. Z., T . C. Chang, C. A. Lin, and W. H. Tsai, Magnetohydrodynamics theory of field line resonances in the magnetosphere, *J. Geophys. Res.*, **98**, 11339, 1993.
11. Cheng, C. Z., Three-dimensional magnetospheric equilibrium with isotropic pressure, *Geophys. Res. Lett.*, **22**, 2401, 1995.
12. Cheng, C. Z., MHD field line resonances and global modes in three-dimensional magnetic fields, *J. Geophys. Res.,* **108**(A1), 1002, doi:10.1029/2002JA009470, 2003.
13. Cheng, C. Z. and S. Zaharia, field line resonances in quiet and disturbed time three-dimensional magnetospheres, *J. Geophys. Res.,* **108**(A1), 1001, 2003.
14. Cummings, W. D., R. J. O'Sullivan, and P. J. Coleman, Standing Alfvén waves in the magnetosphere, *J. Geophys. Res.*, **74**, 778, 1969.
15. Denton, R. E., K. Takahashi, R. R. Anderson, and M. P. Wuest , Magnetospheric toroidal Alfvén wave harmonics and the field line distribution of mass density, *J. Geophys. Res.*, **109**, doi: 10. 1029/ 2003JA010201, 2004.
16. Denton, R. E., J. D. Menietti, J. Goldstein, S. L. Young, and R. R. Anderson, Electron density in the magnetosphere, *J. Geophys. Res.*, **109**, A09215, doi: 10. 1029/ 2003JA010245, 2004.
17. Dungey, J. W., The propagation of AlfvCn waves through the ionosphere. Ionospheric Res. Sci. Rpt. No. 57, Pennsylvania State Univ., 19, 1954.
18. Eviatar, A., A. M. Lenchek, and S. F. Singer, Distribution of density in an ion-exosphere of a non-rotating planet, *Phys. Fluids*, **7**, 1775, 1964.
19. Fedorov, E. N. , B. N. Belen'kaya, M. B. Gokhberg, S. P. Belokris, L. N. Baransky, and C. A. Green, *Planet. Space Sci.*, **38**, 269, 1990.
20. Fedorov, E. N., N. G. Mazur, V. A. Pilipenko, and K. Yumoto, On the theory of field line resonances in plasma configurations, *Phys. Plasmas*, **2**, 527, 1995.
21. Fedorov, E., V. Pilipenko, and M. J. Engebretson, ULF wave damping in the auroral acceleration region, *J. Geophys. Res.*, **106**, 6203, 2001.
22. Fedorov, E., V. Pilipenko, M. J. Engebretson, and T. J. Rosenberg, Alfvén wave modulation of the auroral acceleration region, *Earth Planets Space*, **56**, 649, 2004.
23. Grossmann, W., and J. Tataronis, Decay of MHD waves by phase mixing–The Theta-Pinch in cylindrical geometry, *Z. Physik.*, **261**, 217, 1973.
24. Hansen, P. J., and C. K. Goertz, Validity of the field line resonance expansion, *Phys. Fluids* , **B4**, 2713, 1992.
25. Inhester, B., Resonance absorption of Alfvén oscillations in a nonaxixymmetric magnetosphere, *J. Geophys. Res.*, **91**, 1509, 1986.
26. Kivelson, M. G. and D. J. Southwood, Coupling of global maqgnetospheric MHD eigenmodes to field line resonances, *J. Geophys. Res.*, **91**, 4345, 1986.
27. Klimushkin, D. Yu., A. S. Leonovich, and V. A. Mazur, On the propagation of transversally small scale standing Alfvén waves in a three dimensionally inhomogeneous magnetosphere, *J. Geophys. Res.*, **100**, 9527, 1995.
28. Klimushkin, D. Yu., Resonators for hydromagnetic waves in the magnetosphere, *J. Geophys. Res.*, **103**, 2369, 1998.

29. Knudsen, D. J., M. C. Kelley, and J. F. Vickrey, Alfvén waves in the auroral ionosphere: A numerical model compared with measurements, *J. Geophys. Res.*, **97**, 77, 1992.
30. Krylov, A. L. and E. N. Fedorov, Concerning eigen oscillations of bounded volume of a cold magnetzed plasma, *Doklady AN SSSR*, **231**, 68, 1976.
31. Krylov, A. L., A. E. Lifshitz, and E. N. Fedorov, On the plasma resonances in a curvelinear magnetic field, *Doklady AN SSSR*, **247**, 1095, 1979.
32. Krylov, A. L., A. E. Lifshitz, and E. N. Fedorov, About rezonant properties of the magnetospheric field-lines, *Geomagn. and Aeronomy*, **20**, 689, 1980.
33. Krylov, A. L., A. E. Lifshitz, and E. N. Fedorov, About rezonant properties of the magnetosphere, *Izv. Akad. Nauk SSSR, Fiz. Zemli*, **6**, 49, 1981.
34. Krylov, A. L. and A. E. Lifshitz, Quasi-Alfvén oscillations of magnetic surfaces, *Planet. Space Sci.*, **32**, 481, 1984.
35. Leonovich, A. S. and V. A. Mazur, Resonance eacitation of standing Alfvén waves in an axisymmetytric magnetosphere (monochromatic oscillations), *Planet. Space Sci.*, **37**, 1095, 1989.
36. Leonovich, A. S. and V. A. Mazur, On the spectrum the magnetocpheric eigenoscillations of an axisymmetytric magnetosphere, *J. Geophys. Res.*, **106**, 3919, 2001.
37. Leonovich, A. S. and V. A. Mazur, Self-consistent model of a dipole-like magnetosphere with an azimutal solar wind flow, *J. Plasma Physics.*, **70**, 99, 2004.
38. Lee, D. H. and R. L. Lysak, ULF wave coupling in the dipole model: the pulse exitation, *J. Geophys. Res.*, **94**, 17097, 1989.
39. Lee, D. H. and R. L. Lysak, Effects of azimuthal asymmetry on ULF waves in the dipole magnetosphere, *Geophys. Res. Lett.*, **17**, 53, 1990.
40. Lifshitz, A. E. and E. N. Fedorov, *Doklady AN SSSR*, **287**, 90, 1986.
41. Lysak, R. L., Magnetosphere-ionosphere coupling by Alfvén waves at midlatitudes, *J. Geophys. Res.*, **109**, A07201, doi:10.1029/2004JA010454, 2004.
42. Mann, I. R., On the internalradial structure of field line resonances, *J. Geophys. Res.*, **102**, 27, 1997.
43. Mann, I. R., A. N. Wright, K. Mills, and V. M. Nakariakov, Excitation of magnetospheric waveguide modes by magnetosheath flows, *J. Geophys. Res.*, **104**, 333, 1999.
44. Mazur, N. G., E. N. Fedorov, and V. A. Pilipenko, On the Possibility of reflection of Alfvén waves in a curvilinear magnetic field , *Plasma Physics Reports*, **30**, 413, 2004.
45. Mond, M., E. Hameiri, and P. N. Hu, *J. Geophys. Res.*, **95**, 89, 1990.
46. Ozeke, L. G. and I. R. Mann, Modeling the properties of guided poloidal Alfvén waves with finite asymmetric, ionospheric conductivities in a dipole field, *J. Geophys. Res.*, **109**, A05205, doi:10.1029/2003JA010151, 2004.
47. Pilipenko, V., E. Fedorov, M. J. Engebretson, and K. Yumoto, Energy budget of Alfvén wave interactions with the auroral acceleration region, *J. Geophys. Res.*, **109**, A10204, doi:10.1029/2004JA010440, 2004.
48. Rankin, R., P. Frycz, and V. T. Tikhonchuk, Shear Alfvén waves on stretched field-lines near midnight in the Earth's magnetosphere, *Geophys. Res. Lett.*, **27**, 3265, 2000.
49. Singer, H. J., D. J. Southwood, R. J. Walker, and M. G. Kivelson, Alfvén wave resonances in a realistic magnetospheric magnetic field geometry, *J. Geophys. Res.*, **86**, 4589, 1981.

50. Southwood, D. J., Some features of field line resonances in the magnetosphere, *Planet. Space Sci.*, **22**, 483, 1974.
51. Southwood, D. J. and M.G. Kivelson, The effect of parallel inhomogeneity on magnetospheric wave coupling, *J. Geophys. Res.*, **91**, 6871, 1986.
52. Southwood, D. J. and M. A. Saunders, Curvature coupling of slow and Alfvén MHD waves in a magnetotail field configuration, *Planet. Space Sci.*, **33**, 127, 1985.
53. Streltsov, A. and W. Lotko, Dispersive field line resonances on auroral field-lines, *J. Geophys. Res.*, **100**, 19, 457, 1995.
54. Schindler, K. and J. Birn, MHD stability of magnetotail. equilibria including a background pressure, *J. Geophys. Res.*, **109**, A10208, doi: 10. 1029/ 2004JA010537, 2004.
55. Schulze-Berge, S., S. Cowley, and L. Chen, Theory of field line resonances of standing sheatr Alfvén waves in three-dimensional inhomogeneous plasmas, *J. Geophys. Res.*, **97**, 3219, 1992.
56. Takahashi, K., C. Z. Cheng, R. W. McEntire, and L. M. Kistler, Observation and theory of $Pc5$ waves with harmonically related transverse and compressional components, *J. Geophys. Res.*, **95**, 977, 1990.
57. Takahashi, K., R. E. Denton, and D. Gallagher, Toroidal wave frequency at $L = 6 - 10$: Active Magnetospheric Particle Tracer Explorers/CCE observations and comparison with theoretical model, *J. Geophys. Res.*, **107** (A2), 2002.
58. Tataronis, J., and W. Grossmann, Decay of MHD waves by phase mixing, I, The sheet pinch in plane geometry, *Z. Phys.*, **261**, 203, 1973.
59. Thompson, M. J. and A. N. Wright, Resonant Alfvén wave excitation in two-dimensional systems: Singularities in partial differential equations, *J. Geophys. Res.*, **98**, 15541, 1993.
60. Timofeev, A. V., To the theory of Alfven oscillations in an inhomogeneous plasma, in *Reviews of Plasma Physics*, (Ed. by M. A. Leontovich), **9**, 205, Plenum Publishing Corp., 1979.
61. Uberoi, C., The Alfvén wave equation for magnetospheric plasmas, *J. Geophys. Res.*, **93**, 295 1988.
62. Uberoi, C., Resonant Alfvén wave exitation in two-dimensional systems, *J. Geophys. Res.*, **101**, 13345 (1996).
63. Wasow, W., Asymptotic Expansions for Ordinary Differential Equations, J. Wiley, New York, 1965.
64. Wright, A. N., MHD wave coupling in inhomogeneous media, *Geophys. Res. Lett.*, **18**, 1951, 1991.
65. Wright, A. N. and M. J. Thompson, Analytical treatment of Alfvén resonances and singularities in nonuniform magnetoplasmas, *Phys. Plasmas*, **1**, 691, 1994.
66. Wright, A. N., K. J. Mills, A. W. Longbottom, and M. S. Ruderman, The nature of convectively unstable waveguide mode disturbances on the magnetospheric flanks, *J. Geophys. Res*, **107**, 1242, 2002.

7

MHD-Waves in Layered Media

7.1 Introduction

The main properties of MHD-waves in the magnetosphere have been discussed in the previous chapters wherein the general features of these waves traveling from a magnetospheric source towards the ionosphere have been established. Moreover, we have constructed the shape of the MHD-wave caused by the FLR not only in the magnetosphere but also on the upper boundary of the ionosphere. In the present and forthcoming chapters, we will consider penetration of the MHD-waves incident to the ionosphere from the magnetosphere to the ground observer. We shall find the ionospheric frequency response function in the wide-scale size of the ULF-range.

We begin by constructing models for the wave and the system ground-atmosphere-ionosphere-magnetosphere. Consider a layered model of the ionosphere. Here, the horizontal scale size of change of the ionospheric conductivity is much more than the horizontal scale size of the magnetospheric wave coming to the ionosphere. An Alfvén wave activated close to the equatorial magnetospheric plane is trapped by the geomagnetic field and guided along the field-lines. It results in the mapping of the place of generation onto the ionosphere. Due to the convergency of the field-lines, the horizontal scale of the wave above the ionosphere becomes essentially less than the Earth's radius. It is valid for the latitudinal dependency of the wave.

Conductivities of the dayside and nightside ionospheres depend weakly on the latitude and longitude. And only within specific regions like the boundary between dayside and nightside ionospheres and in the equatorial and polar regions, the horizontal scale of the ionospheric change becomes comparable with the horizontal scale of the Alfvén wave.

The FMS-waves undergo a strong refraction in the magnetosphere and usually their reflection points are located far away from the ionosphere. The amplitude of these waves decreases exponentially as they propagate towards the ionosphere and their intensity becomes smaller compared to that of Alfvén waves. In this way, the polarization of the waves reaching the ionosphere can be considered to be basically corresponding to that of Alfvén type waves.

Two wave ionospheric transformations are:

- a direct transformation of the Alfvén/FMS waves into Alfvén/FMS waves, and
- a cross transformation Alfvén waves are transformed into FMS waves and vice-versa.
  In order to explore these wave transformations, consider the ULF wave propagation in the whole system; the ground-atmosphere-ionosphere-magnetosphere modeled with a layered medium (see Chapter 2 and Fig. 2.10).
- In so doing, the ground is presented as a conductor with the conductivity depending only on depth.
- The atmosphere can be conceived as a perfect insulator in the most of the wave phenomena in the ULF range with rare exception such as penetration of vertical currents from the ionospheric heights to the ground and vice versa and influence of the vertical electric field produced by lightning discharges, surface and sub-surface high-energy sources (e.g. volcano eruptions, earthquakes, tsunami, etc.) on the ionosphere.
- We consider the ionosphere as an anisotropic conductor with the perfect longitudinal conductivity.

The equipotentiality of the ionospheric field-lines allows us to change from the real ionosphere to the model of a thin ionosphere with integral anisotropic conductivity. Thus the ionospheric currents produced by the MHD-waves can be replaced by the total current spreading over the whole ionospheric thickness. In spite of the fact that this is an idealization, it enables us to reveal the basic physical effects which are conserved in the more complicated models that can be studied numerically.

Contrary to the previous chapters, where the MHD-wave features have been studied in rather complicated magnetospheric models, consider here the magnetosphere as a homogeneous half-space filled with collisionless magnetized cold plasma. The magnetosphere in this case can be approximated as an anisotropic dielectric with infinite longitudinal dielectric permeability, whereas the transversal component of the permeability is the ratio c^2/c_A^2 (see Chapter 2).

In addition to the ionospheric penetration of the MHD-wave, a valid demonstration of the applied ionospheric boundary conditions for the magnetospheric resonances will also be presented.

7.2 Model and Basic Equations

Plain Stratified Model

As a model of the ionosphere-magnetosphere, we consider a plasma in a homogeneous straight magnetic field $\mathbf{B}_0$. The inclination or dip of $\mathbf{B}_0$ is I (see

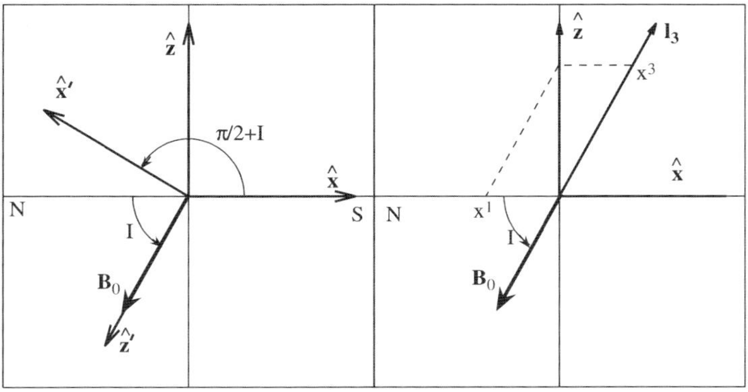

Fig. 7.1. Two chosen Cartesian systems for the Northern Hemisphere. The system (x, y, z) is Cartesian-altitude coordinate system and $(x', y'z')$ is connected with the geomagnetic field. I is the inclination angle

Figure 7.1). Dip angle $I < 0$ in the Southern Hemisphere and $I > 0$ in the Northern Hemisphere. Input two Cartesian coordinate systems: the first one $\{x', y', z'\}$ related to the main magnetic field $\mathbf{B}_0$ and another one $\{x, y, z\}$ is a Cartesian-altitude system. Let $\{\hat{\mathbf{x}}', \hat{\mathbf{y}}', \hat{\mathbf{z}}' = \mathbf{B}_0/B_0\}$ and $\{\hat{\mathbf{x}}, \hat{\mathbf{y}}, \hat{\mathbf{z}}\}$ be unit vectors of the x'-, y'-, z'- and x-, y-, z- axes, respectively. Let the $z = 0$ plane be the boundary between the conducting ionosphere and the atmosphere. Although this boundary is not exactly determined, this inaccuracy does not lead to any practical difficulties. We neglect the curvature of the ground surface and match it with the plane $z = -h$. Coordinate system $\{x', y', z'\}$ turns out from $\{x, y, z\}$ by the rotation with respect to axis y at an angle $\pi/2 + I$. The x and y axes are parallel to the ionosphere and ground surface. The x axis is southwards, y is eastwards and z axis is vertically upwards. Let us direct axis z' along $\mathbf{B}_0$ and axis y' eastwards. Set also an oblique coordinate system $\{x^1, x^2, x^3\}$ with horizontal surfaces $x^3 = \text{const}$ and coordinate lines x^3 coincident with the field-lines.

The orthogonal coordinate system related to the magnetic field for the Northern Hemisphere is shown in the left frame of Fig. 7.1 and the oblique coordinate system is shown in the right frame. The coordinate systems are related as

$$\begin{aligned} x &= -x' \sin I - z' \cos I, & x' &= -x \sin I + z \cos I, \\ y &= y', & y' &= y, \\ z &= x' \cos I - z' \sin I, & z' &= -x \cos I - z \sin I, \end{aligned} \tag{7.1}$$

$$\begin{aligned} x &= x^1 + x^3 \cot I, & x^1 &= x - z \cot I, \\ y &= x^2, & x^2 &= y, \\ z &= x^3, & x^3 &= z, \end{aligned} \tag{7.2}$$

$$
\begin{aligned}
x^1 &= -\frac{x'}{\sin I}, & x' &= -x^1 \sin I, \\
x^2 &= y', & y' &= x^2, \\
x^3 &= x' \cos I - z' \sin I, & z' &= -x^1 \cos I - \frac{x^3}{\sin I}.
\end{aligned} \tag{7.3}
$$

The contravariant basis

$$
\mathbf{l}^n = \nabla x^n = \frac{\partial x^n}{\partial x}\hat{\mathbf{x}} + \frac{\partial x^n}{\partial y}\hat{\mathbf{y}} + \frac{\partial x^n}{\partial z}\hat{\mathbf{z}}
$$

of the oblique coordinate system is

$$
\mathbf{l}^1 = \hat{\mathbf{x}} - \cot I\,\hat{\mathbf{z}} = -\frac{1}{\sin I}\hat{\mathbf{x}}', \quad \mathbf{l}^2 = \hat{\mathbf{y}} = \hat{\mathbf{y}}', \quad \mathbf{l}^3 = \hat{\mathbf{z}} = \cos I\,\hat{\mathbf{x}}' - \sin I\,\hat{\mathbf{z}}'.
$$

The covariant basis

$$
\mathbf{l}_n = \frac{\partial \mathbf{r}}{\partial x^n}
$$

is

$$
\mathbf{l}_1 = \hat{\mathbf{x}} = -\sin I\hat{\mathbf{x}}' - \cos I\,\hat{\mathbf{z}}', \quad \mathbf{l}_2 = \hat{\mathbf{y}} = \hat{\mathbf{y}}', \quad \mathbf{l}_3 = \cot I\hat{\mathbf{x}} + \hat{\mathbf{z}} = -\frac{1}{\sin I}\hat{\mathbf{z}}'.
$$

These two bases are mutually orthonormal:

$$
(\mathbf{l}^i \cdot \mathbf{l}_k) = \delta^i_k.
$$

Here $\delta^i_k = 1$ if $i = k$ and $\delta^i_k = 0$ if $i \neq k$. The metric tensor of the oblique coordinate system is

$$
g_{ik} = (\mathbf{l}_i \cdot \mathbf{l}_k) = \begin{pmatrix} 1 & 0 & \cot I \\ 0 & 1 & 0 \\ \cot I & 0 & \sin^{-2} I \end{pmatrix}
$$

and

$$
g^{ik} = (\mathbf{l}^i \cdot \mathbf{l}^k) = \begin{pmatrix} \sin^{-2} I & 0 & -\cot I \\ 0 & 1 & 0 \\ -\cot I & 0 & 1 \end{pmatrix}.
$$

Determinant of the metric tensor $\det(g_{ik}) = 1$, so that

$$
\mathbf{l}_1 = \mathbf{l}^2 \times \mathbf{l}^3, \quad \mathbf{l}_2 = \mathbf{l}^3 \times \mathbf{l}^1, \quad \mathbf{l}_3 = \mathbf{l}^1 \times \mathbf{l}^2.
$$

Then, for operator ∇ in $\{\mathbf{l}^1, \mathbf{l}^2, \mathbf{l}^3\}$ basis, we get

$$
\nabla = \mathbf{l}^1 \frac{\partial}{\partial x^1} + \mathbf{l}^2 \frac{\partial}{\partial x^2} + \mathbf{l}^3 \frac{\partial}{\partial x^3}.
$$

Hence the contravariant components of $\mathbf{a} = \nabla \times \mathbf{c}$ are

$$a^1 = \frac{\partial c_3}{\partial x^2} - \frac{\partial c_2}{\partial x^3}, \quad a^2 = \frac{\partial c_1}{\partial x^3} - \frac{\partial c_3}{\partial x^1}, \quad a^3 = \frac{\partial c_2}{\partial x^1} - \frac{\partial c_1}{\partial x^2}.$$

and

$$\nabla \cdot \mathbf{c} = \frac{\partial c^1}{\partial x^1} + \frac{\partial c^2}{\partial x^2} + \frac{\partial c^3}{\partial x^3}$$

The covariant components of **a** are $a_i = g_{ik}a^k$, $i,\ k\ = 1, 2, 3$. Summation here is over repeated subscripts.

Electric (**E**) and magnetic (**b**) fields in the oblique coordinate system can be presented as

$$\mathbf{E} = E_1\mathbf{l}^1 + E_2\mathbf{l}^2 + E_3\mathbf{l}^3, \qquad \mathbf{b} = b_1\mathbf{l}^1 + b_2\mathbf{l}^2 + b_3\mathbf{l}^3.$$

Then for the horizontal components of **E** and **b** in the Cartesian and oblique coordinate systems we have

$$E_x = E_1, \qquad E_y = E_2,$$
$$b_x = b_1, \qquad b_y = b_2.$$

and for the vertical z-component:

$$b_z = -\cot I\, b_1 + b_3,$$
$$j_z = -\cot I\, j_1 + j_3. \tag{7.4}$$

We begin the study of MHD-wave interaction with ionospheric plasma by investigating the plane-layered model shown in Fig. 7.2. The complex

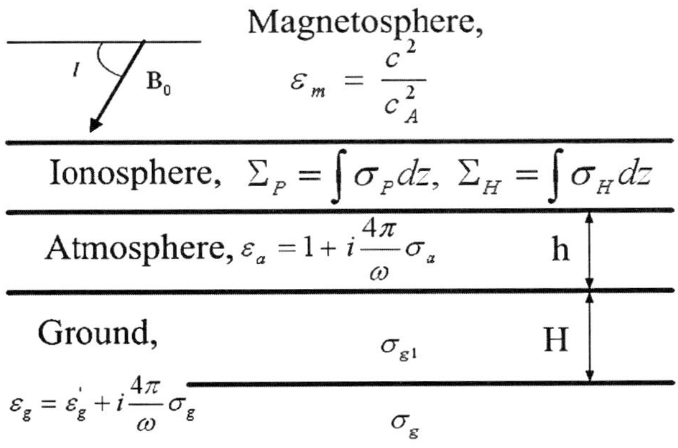

Fig. 7.2. Plane model of the Earth-magnetosphere system. The magnetosphere is a dielectric with the transversal dielectric permeability ε_m. The ionosphere is taken as a thin layer with the Pedersen Σ_P and Hall Σ_H components of the tensor height integrated conductivity. The atmosphere is a conductor with complex dielectric permeability. The ground is a conductive layer of thickness H and the specific conductivity σ_{g1} underlain by the half-space of conductivity σ_g

dielectric permeability tensor $\boldsymbol{\varepsilon}$ is a function only of z. First consider the magnetosphere to be a homogeneous half-space filled with collisionless magnetized cold plasma. Such a medium becomes strongly anisotropic. Across the magnetic field the medium can be approximated as a dielectric with transverse dielectric permeability

$$\varepsilon_m = \frac{c^2}{c_A^2},$$

where c is the light velocity and c_A is the Alfvén velocity

$$c_A = \frac{B_0}{\sqrt{4\pi\rho_0}}.$$

B_0 is the geomagnetic field strength, ρ_0 is the cold plasma density. Along the magnetic field this medium can be considered as a perfect conductor.

Ohm's law for the ionospheric plasma has the form (see Chapter 1)

$$\mathbf{j}_\perp = \boldsymbol{\sigma}_\perp \mathbf{E}_\perp = \sigma_P \mathbf{E}_\perp + \sigma_H \frac{\mathbf{B}_0 \times \mathbf{E}_\perp}{B_0}, \quad \mathbf{j}_\parallel = \sigma_\parallel \mathbf{E}_\parallel,$$

where $\mathbf{j}_\perp$, $\mathbf{E}_\perp$ are transversal current density and the electric field, $\mathbf{j}_\parallel$, $\mathbf{E}_\parallel$ are the field-aligned current density and electric field. The tensor of conductivity $\boldsymbol{\sigma}_\perp$ in the coordinate system connected with $\mathbf{B}_0$ is

$$\boldsymbol{\sigma}_\perp = \begin{pmatrix} \sigma_P & -\sigma_H \\ \sigma_H & \sigma_P \end{pmatrix}.$$

Here σ_P and σ_H are the Pedersen and Hall specific conductivities, respectively. The relationship of σ_P and σ_H with ionospheric parameters was obtained in Chapter 1. Note that $\sigma_H > 0$. Components of $\mathbf{j}$ in the coordinates $\{x', y', z'\}$ are

$$\begin{aligned} j_{x'} &= \sigma_P E_{x'} - \sigma_H E_{y'}, \\ j_{y'} &= \sigma_H E_{x'} + \sigma_P E_{y'}, \\ j_{z'} &= \sigma_\parallel E_{z'}. \end{aligned}$$

The atmosphere and the ground are considered as isotropic conductors with conductivities $\sigma_a(z)$ and $\sigma_g(z)$. In this case, the dielectric permeability tensor is reduced to scalar

$$\varepsilon_{a,g}(z) = \varepsilon'_{a,g} + i\frac{4\pi\sigma_{a,g}(z)}{\omega}.$$

In the lower part of the atmosphere (the first 10 kilometers), at frequencies $\omega \gtrsim 1 - 10^{-1}\mathrm{s}^{-1}$, the magnitude of the displacement currents is much larger than the conduction currents. The extremely low value of atmospheric conductivity enables us to regard it as a perfect insulator. Only one exception is known when the atmospheric conductivity must be taken into account: that is the excitation of the so-called TEM wave mode (see Chapter 8).

Basic Equations and Boundary Conditions

The Maxwell's equations for a monochromatic wave with time dependence $\exp(-i\omega t)$ are

$$\nabla \times \mathbf{b}(\mathbf{r}) = -ik_0\varepsilon\,\mathbf{E}(\mathbf{r}), \tag{7.5}$$

$$\nabla \times \mathbf{E}(\mathbf{r}) = ik_0\mathbf{b}(\mathbf{r}). \tag{7.6}$$

Here $k_0 = \omega/c$ is a wavenumber in the vacuum, ε is the complex permeability tensor.

By virtue of horizontal homogeneity of the medium, solutions of these equations can be found for Fourier harmonics. Then a solution is obtained for an arbitrary distribution with the help of the inverse Fourier transformation. Substitute into (7.5)–(7.6) harmonics in the form

$$\mathbf{E}(x^1, x^2, x^3) = \mathbf{E}(x^3)\exp(ik_1x^1 + ik_2x^2),$$
$$\mathbf{b}(x^1, x^2, x^3) = \mathbf{b}(x^3)\exp(ik_1x^1 + ik_2x^2).$$

Denote the amplitudes of spatial Fourier harmonics in the same way as the initial fields, explicitly indicating whenever necessary the arguments (x^1, x^2) or $\mathbf{k}_\tau = (k_1, k_2)$. For instance, the Fourier transformation with respect to x^1, x^2 from $E_1(x^1, x^2, x^3)$ will be written as $E_1(x^3; \mathbf{k}_\tau)$ or simply E_1, etc.

Now, consider the Maxwell's equations (7.5), (7.6) in the coordinates $\{x^1, x^2, x^3\}$. Let the longitudinal conductivities of the ionosphere and the magnetosphere tend to infinity. In this case, the longitudinal electric field vanishes, thus $E_3 = 0$. If the variable b_3 is eliminated from (7.5) and (7.6), we obtain the equations for b_1, b_2, E_1, E_2 in the ionosphere and magnetosphere:

$$\frac{\partial b_1}{\partial x^3} = \left(-\frac{4\pi\sigma_H}{c\sin I} - i\frac{k_1k_2}{k_0}\right)E_1 + \left(\frac{4\pi\sigma_P}{c} + i\frac{k_1^2}{k_0}\right)E_2 + ik_1b_1\cot I, \tag{7.7}$$

$$\frac{\partial b_2}{\partial x^3} = \left(-\frac{4\pi\sigma_P}{c\sin^2 I} - i\frac{k_2^2}{k_0}\right)E_1 - \left(\frac{4\pi\sigma_H}{c\sin I} - i\frac{k_1k_2}{k_0}\right)E_2 + ik_2b_1\cot I, \tag{7.8}$$

$$\frac{\partial E_1}{\partial x^3} = ik_0b_2, \tag{7.9}$$

$$\frac{\partial E_2}{\partial x^3} = -ik_2E_1\cot I + ik_1E_2\cot I - ik_0b_1. \tag{7.10}$$

Equations (7.7)–(7.10) are similar to the equations of the radio wave in the ionosphere [5]. Equations in the atmosphere and on the ground can be obtained in the same manner.

On the boundary between the atmosphere and the ionosphere ($x^3 = 0$) it is necessary to demand continuity of horizontal components of $\mathbf{E}$ and $\mathbf{b}$. It is more convenient to split the problem solution into two parts. First, let us find the fields reflected from the ionosphere back to the magnetosphere. Then, from the known values of the initial incident field and the reflected field, we can find the field penetrated to the ground surface.

Reflection R and Transformation T Matrices

Let us introduce the admittance matrix $\mathbf{Y}$

$$\mathbf{Y}_a = \begin{pmatrix} Y_{11}^a & Y_{12}^a \\ Y_{21}^a & Y_{22}^a \end{pmatrix},$$

which relates horizontal components of the electric $\mathbf{E}$ and magnetic $\mathbf{b}$ fields:

$$\begin{pmatrix} b_1 \\ b_2 \end{pmatrix}\Bigg|_{z=-0} = \mathbf{Y}_a \begin{pmatrix} E_1 \\ E_2 \end{pmatrix}\Bigg|_{z=-0}. \tag{7.11}$$

The total field above the ionosphere is represented as

$$\mathbf{b}(\mathbf{r}) = \mathbf{b}^{(i)}(\mathbf{r}) + \mathbf{b}^{(r)}(\mathbf{r}),$$

where $\mathbf{b}^{(i)}(\mathbf{r})$ is the magnetic component of the incident wave; $\mathbf{b}^{(r)}(\mathbf{r})$ is the reflected field.

In order to find the relations between $\mathbf{b}^{(i)}(\mathbf{r})$ and $\mathbf{b}^{(r)}(\mathbf{r})$ as well as between $\mathbf{b}^{(i)}(\mathbf{r})$ and the field $\mathbf{b}^{(g)}(\mathbf{r})$ on the ground surface, we shall represent the fields being expressed by their Fourier transforms. Then the relation between the corresponding Fourier harmonics of the magnetic field on the ground and above the ionosphere in the general case may be written

$$\mathbf{b}^{(g)}(\mathbf{k}) = \mathbf{T}'(\mathbf{k})\,\mathbf{b}_\tau^{(i)}(+0, \mathbf{k}), \qquad \mathbf{b}_\tau^{(r)}(+0, \mathbf{k}) = \mathbf{R}'(\mathbf{k})\,\mathbf{b}_\tau^{(i)}(+0, \mathbf{k}).$$

Here $\mathbf{R}'(\mathbf{k})$ and $\mathbf{T}'(\mathbf{k})$ are reflection and transformation matrices for harmonics with horizontal wavenumber $\mathbf{k}$. Vectors $\mathbf{b}_\tau^{(i)}(+0, k)$ and $\mathbf{b}_\tau^{(r)}(+0, k)$ are the amplitudes of the horizontal magnetic field in the incident and the reflected wave above the ionosphere. The transformation matrix $\mathbf{T}'_\Sigma$ defined as a ratio of the horizontal ground magnetic field to the total horizontal wave magnetic component is

$$\mathbf{b}^{(g)} = \mathbf{T}'_\Sigma\,\mathbf{b}_\tau.$$

One can find the matrix $\mathbf{T}'$ by relating the amplitude of the horizontal magnetic fields on the ground to the amplitude of the initial wave $\mathbf{b}_\tau^{(i)}(+0)$ by

$$\mathbf{b}^{(g)} = \mathbf{T}'_\Sigma\,(\mathbf{1} + \mathbf{R}')\,\mathbf{b}_\tau^{(i)}.$$
$$\mathbf{T}' = \mathbf{T}'_\Sigma\,(\mathbf{1} + \mathbf{R}').$$

7.3 Atmospheric and Ground Fields

Here we shall obtain and analyze expressions for coefficients $\mathbf{R}'(\mathbf{k})$ and $\mathbf{T}'(\mathbf{k})$. We begin with the investigation of fields in the atmosphere and on the ground, which enables us to connect the electric and magnetic fields observed on the ground surface and to obtain specific expressions for admittances included in boundary condition (7.11).

Potentials

Let us present the total field as a sum of electric (e) and magnetic (m) modes

$$\mathbf{b} = \mathbf{b}^{(\mathbf{e})} + \mathbf{b}^{(\mathbf{m})}, \qquad \mathbf{E} = \mathbf{E}^{(\mathbf{e})} + \mathbf{E}^{(\mathbf{m})}. \tag{7.12}$$

Electric $\mathbf{E}^{(e)}$ and magnetic $\mathbf{b}^{(e)}$ components of the electric mode (e-mode) is also called the transverse magnetic mode (TM-mode). The vertical magnetic component $b_z^{(e)}$ of the TM-mode is

$$b_z^{(e)} = 0.$$

and the horizontal magnetic components are 2D solenoidal, i.e.

$$\frac{\partial b_x^{(e)}}{\partial x} + \frac{\partial b_y^{(e)}}{\partial y} = 0. \tag{7.13}$$

Magnetic $\mathbf{b}^{(m)}$ and electric $\mathbf{E}^{(m)}$ components relate to the magnetic mode (H-mode) or transverse electric mode (TE-mode). In the TE-mode the vertical component of the electric field vanishes, while the horizontal magnetic field is 2D-potential

$$\frac{\partial b_y^{(m)}}{\partial x} - \frac{\partial b_x^{(m)}}{\partial y} = 0. \tag{7.14}$$

For finding the fields in the atmosphere let us take advantage of Debye's potentials [6]. For example, the fields of the electric mode according to (7.13) can be presented as

$$\begin{aligned}
E_x^{(e)}(\mathbf{r}) &= \frac{1}{\varepsilon(z)}\frac{\partial^2 \Phi(\mathbf{r})}{\partial x\,\partial z}, & b_x^{(e)}(\mathbf{r}) &= -ik_0\frac{\partial \Phi(\mathbf{r})}{\partial y},\\
E_y^{(e)}(\mathbf{r}) &= \frac{1}{\varepsilon(z)}\frac{\partial^2 \Phi(\mathbf{r})}{\partial y\,\partial z}, & b_y^{(e)}(\mathbf{r}) &= ik_0\frac{\partial \Phi(\mathbf{r})}{\partial x},\\
E_z^{(e)}(\mathbf{r}) &= -\frac{1}{\varepsilon(z)}\boldsymbol{\nabla}_\perp^2 \Phi(\mathbf{r}), & b_z^{(e)}(\mathbf{r}) &= 0,
\end{aligned} \tag{7.15}$$

and for the magnetic mode from (7.14)

$$\begin{aligned}
E_x^{(m)}(\mathbf{r}) &= ik_0\frac{\partial \Psi(\mathbf{r})}{\partial y}, & b_x^{(m)}(\mathbf{r}) &= \frac{\partial^2 \Psi(\mathbf{r})}{\partial x\partial z},\\
E_y^{(m)}(\mathbf{r}) &= -ik_0\frac{\partial \Psi(\mathbf{r})}{\partial x}, & b_y^{(m)}(\mathbf{r}) &= \frac{\partial^2 \Psi(\mathbf{r})}{\partial y\partial z},\\
E_z^{(m)}(\mathbf{r}) &= 0, & b_z^{(m)}(\mathbf{r}) &= -\boldsymbol{\nabla}_\perp^2 \Psi(\mathbf{r}),
\end{aligned} \tag{7.16}$$

where $\boldsymbol{\nabla}_\perp^2 = \dfrac{\partial^2}{\partial x^2} + \dfrac{\partial^2}{\partial y^2}$.

Combining (7.15), (7.16) and the Maxwell's equations (7.5) gives

$$\nabla^2\Phi(\mathbf{r}) - \frac{1}{\varepsilon(z)}\frac{\partial\varepsilon(z)}{\partial z}\frac{\partial\Phi(\mathbf{r})}{\partial z} + k_0^2(z)\varepsilon(z)\,\Phi(r) = 0, \tag{7.17}$$

$$\nabla^2\Psi(\mathbf{r}) + k_0^2(z)\,\varepsilon(z)\,\Psi(r) = 0. \tag{7.18}$$

For a harmonics we get

$$\varepsilon(z)\frac{\partial}{\partial z}\left(\frac{1}{\varepsilon(z)}\frac{\partial\Phi(z;\mathbf{k}_\tau)}{\partial z}\right) + k_z^2(z)\,\Phi(z;\,\mathbf{k}_\tau) = 0, \tag{7.19}$$

$$\frac{\partial^2\Psi(z;\mathbf{k}_\tau)}{\partial z^2} + k_z^2(z)\,\Psi(z;\,\mathbf{k}_\tau) = 0, \tag{7.20}$$

where $k_z^2 = k_0^2\varepsilon(z) - k_\perp^2$, $k_\tau^2 = k_x^2 + k_y^2$.

The horizontal electric and magnetic components should be continuous on the discontinuity $z = z_i$ of dielectric permeability $\varepsilon(z)$. Hence, we find boundary conditions for potentials

$$\{\Phi\}_{z=z_i} = 0, \qquad \left\{\frac{1}{\varepsilon(z)}\frac{\partial\Phi}{\partial z}\right\}_{z=z_i} = 0, \tag{7.21}$$

$$\{\Psi\}_{z=z_i} = 0, \qquad \left\{\frac{\partial\Psi}{\partial z}\right\}_{z=z_i} = 0, \tag{7.22}$$

where $\{\Phi\}_{z=z_i} = \Phi(z_i+0) - \Phi(z_i-0)$, etc.

Admittance Matrix

Electric and Magnetic Admittances

Let us define spectral admittances $Y^{(e,m)}$ and impedances $Z^{(e,m)}$ for the electric and magnetic modes as

$$Y^{(e)}(z;\mathbf{k}_\tau) = \frac{1}{Z^{(e)}(z;\mathbf{k}_\tau)} = -\frac{b_y^{(e)}(z;\mathbf{k}_\tau)}{E_x^{(e)}(z;\mathbf{k}_\tau)} = \frac{b_x^{(e)}(z;\mathbf{k}_\tau)}{E_y^{(e)}(z;\mathbf{k}_\tau)}, \tag{7.23}$$

$$Y^{(m)}(z;\mathbf{k}_\tau) = \frac{1}{Z^{(m)}(z;\mathbf{k}_\tau)} = -\frac{b_y^{(m)}(z;\mathbf{k}_\tau)}{E_x^{(m)}(z;\mathbf{k}_\tau)} = \frac{b_x^{(m)}(z;\mathbf{k}_\tau)}{E_y^{(m)}(z;\mathbf{k}_\tau)}. \tag{7.24}$$

They can be expressed in terms of potentials:

$$Y^{(e)}(z) = -ik_0\varepsilon(z)\Phi(z)\left[\frac{\partial\Phi}{\partial z}\right]^{-1}, \tag{7.25}$$

$$Y^{(m)}(z) = -\frac{1}{ik_0}\Psi^{-1}(z)\frac{\partial\Psi}{\partial z}. \tag{7.26}$$

Relations (7.23) and (7.24) for $z = -h$ give a definition of the spectral impedance (or of the surface admittances) on the ground surface

$$\left[Z_g^{(e)}(\mathbf{k}_\tau)\right]^{-1} = Y_g^{(e)}(\mathbf{k}_\tau) = Y^{(e)}(-h;\mathbf{k}_\tau),$$
$$\left[Z_g^{(m)}(\mathbf{k}_\tau)\right]^{-1} = Y_g^{(m)}(\mathbf{k}_\tau) = Y^{(m)}(-h;\mathbf{k}_\tau). \tag{7.27}$$

Let the ground be approximated by a half-space with constant geoelectric permeability ε_g. Then the solution for Φ and Ψ in the ground ($z < -h$) can be written as

$$\begin{pmatrix} \Phi \\ \Psi \end{pmatrix} \propto \exp(-i\kappa_g z)$$

and electric and magnetic admittances become

$$Y_g^{(e)} = \frac{1}{Z_g^{(e)}} = \frac{k_0\varepsilon_g}{\kappa_g}, \qquad Y_g^{(m)} = \frac{1}{Z_g^{(m)}} = \frac{\kappa_g}{k_0}, \tag{7.28}$$

where

$$\kappa_g = \sqrt{k_0^2\varepsilon_g - k_\tau^2}.$$

Assume also that σ_g is so large that the inequality holds $\left|k_0\sqrt{\varepsilon_g}\right| \gg k_\tau$. Then $\kappa_g^2 \approx k_0^2\varepsilon_g$ and (7.28) yields

$$Y_g^{(e)} = \frac{1}{Z_g^{(e)}} \approx Y_g^{(m)} = \frac{1}{Z_g^{(m)}} \approx \sqrt{\varepsilon_g}.$$

Spectral Admittance Matrix

In the general case, the horizontal components of total electric and magnetic fields are connected by an admittance matrix

$$\mathbf{b}_\tau(z) = \mathbf{Y}(z)\,\mathbf{E}_\tau(z), \tag{7.29}$$

where $\mathbf{E}_\tau(z)$ and $\mathbf{b}_\tau(z)$ are horizontal magnetic and electric fields. We find an expression for $\mathbf{Y}(z)$ in the atmosphere from (7.12), (7.15), (7.16) and (7.23), (7.24):

$$\mathbf{Y}(z) = \begin{pmatrix} Y_{11} & Y_{12} \\ Y_{21} & Y_{22} \end{pmatrix}, \tag{7.30}$$

where

$$Y_{11} = -Y_{22} = -\frac{k_x k_y}{k_\tau^2}\left(Y^{(m)}(z) - Y^{(e)}(z)\right),$$
$$Y_{12} = \frac{1}{k_\tau^2}\left[k_x^2 Y^{(m)}(z) + k_y^2 Y^{(e)}(z)\right],$$
$$Y_{21} = -\frac{1}{k_\tau^2}\left[k_y^2 Y^{(m)}(z) + k_x^2 Y^{(e)}(z)\right].$$

Atmospheric Potentials and Admittances

Let us denote the solution of (7.19) by $\zeta_1(z)$ and $\partial\zeta_1(z)/\partial z$ by $ik_0\varepsilon_a(z)\zeta_2(z)$. They obey the ground conditions:

$$\zeta_1(-h) = -Y_g^{(e)}, \qquad \zeta_2(-h) = \frac{1}{ik_0\varepsilon_a(-h)} \left.\frac{\partial\zeta_1(z)}{\partial z}\right|_{z=-h} = 1. \tag{7.31}$$

Denote also the solution of (7.20) by $\zeta_3(z)$ and $\partial\zeta_3(z)/\partial z$ by $ik_0\zeta_4(z)$ defined by the boundary conditions

$$\zeta_3(-h) = 1, \qquad \zeta_4(z) = \frac{1}{ik_0} \left.\frac{\partial\zeta_3(z)}{\partial z}\right|_{z=-h} = -Y_g^{(m)}. \tag{7.32}$$

Here

$$Y_g^{(e)} = Y^{(e)}(-h), \qquad Y_g^{(m)} = Y^{(m)}(-h).$$

Express solutions of (7.19), (7.20) and their derivatives in terms of $\zeta_i(z)$

$$\Phi(z) = -\Phi_0\frac{\zeta_1(z)}{Y_g^{(e)}}, \qquad \frac{\partial\Phi(z)}{\partial z} = -\frac{ik_0\,\varepsilon_a(z)}{Y_g^{(e)}}\Phi_0\,\zeta_2(z), \tag{7.33}$$

$$\Psi(z) = \Psi_0\,\zeta_3(z), \qquad \frac{\partial\Psi(z)}{\partial z} = ik_0\,\Psi_0\,\zeta_4(z). \tag{7.34}$$

These formulae are general and are also applicable to an inhomogeneous atmosphere. Substituting (7.27) and (7.34) into (7.25) and (7.26), we obtain

$$Y^{(e)}(z) = -\frac{\zeta_1(z)}{\zeta_2(z)}, \qquad Y^{(m)}(z) = -\frac{\zeta_4(z)}{\zeta_3(z)}. \tag{7.35}$$

Then components of the admittance matrix under the ionosphere are given by

$$Y_{22} = -Y_{11} = \frac{k_xk_y}{k_\tau^2}\left(\frac{\zeta_1}{\zeta_2} - \frac{\zeta_4}{\zeta_3}\right), \tag{7.36}$$

$$Y_{12} = -\frac{1}{k_\tau^2}\left(k_x^2\frac{\zeta_4}{\zeta_3} + k_y^2\frac{\zeta_1}{\zeta_2}\right), \tag{7.37}$$

$$Y_{21} = \frac{1}{k_\tau^2}\left(k_y^2\frac{\zeta_4}{\zeta_3} + k_x^2\frac{\zeta_1}{\zeta_2}\right). \tag{7.38}$$

Let

$$\zeta_i = \zeta_i(-0). \tag{7.39}$$

Substituting (7.33) and (7.34) into (7.15) and (7.16), we obtain the horizontal components $b_\tau^{(e),(m)} = (b_x^{(e),(m)}, b_y^{(e),(m)})$ in terms of ζ_i:

$$\mathbf{b}_\tau^{(e)}(z) = -k_0\Phi_0\frac{\zeta_1(z)}{Y_g^{(e)}}\begin{pmatrix} k_y \\ -k_x \end{pmatrix},$$

$$\mathbf{b}_\tau^{(m)}(z) = -k_0\Psi_0\zeta_4(z)\begin{pmatrix} k_x \\ k_y \end{pmatrix}. \tag{7.40}$$

By substituting (7.40) for $z = -h$ and $z = -0$ into (7.31)–(7.32) we can express the ground magnetic components and vertical electric field in terms of admittances:

$$\mathbf{b}^{(g)(e)} \equiv \mathbf{b}_{\tau}^{(e)}(-h) = -\frac{Y_g^{(e)}}{\zeta_1}\mathbf{b}_{\tau}^{(e)}(-0), \tag{7.41}$$

$$\mathbf{b}^{(g)(m)} \equiv \mathbf{b}_{\tau}^{(m)}(-h) = -\frac{Y_g^{(m)}}{\zeta_4}\mathbf{b}_{\tau}^{(m)}(-0), \tag{7.42}$$

$$E_z^{(g)} = E_z(-h) = -\frac{\varepsilon_a(-0)}{\varepsilon_a(-h)}\frac{Y_g^{(e)}}{\zeta_1}E_z(-0), \tag{7.43}$$

$$b_z^{(g)} = b_z(-h) = \frac{b_z(-0)}{\zeta_3}. \tag{7.44}$$

For the homogeneous atmosphere $\varepsilon_a(z) = \varepsilon_a = \text{const}$, we have

$$\begin{aligned}
\zeta_1(z) &= -Y_g^{(e)}\cos\kappa_a(z+h) + \frac{ik_0\varepsilon_a}{\kappa_a}\sin\kappa_a(z+h), \\
\zeta_2(z) &= \frac{\kappa_a}{ik_0\varepsilon_a}Y_g^{(e)}\sin\kappa_a(z+h) + \cos\kappa_a(z+h), \\
\zeta_3(z) &= \cos\kappa_a(z+h) - \frac{ik_0}{\kappa_a}Y_g^{(m)}\sin\kappa_a(z+h), \\
\zeta_4(z) &= -\frac{\kappa_a}{ik_0}\sin\kappa_a(z+h) - Y_g^{(m)}\cos\kappa_a(z+h),
\end{aligned} \tag{7.45}$$

where

$$\kappa_a = \sqrt{k_0^2\varepsilon_a - k_\tau^2}.$$

Equations (7.45) are obtained by taking into account the vertical displacement currents and of atmospheric conductivity. It can be assumed that $\sigma_a = 0$. Then only the condition for the magnetic mode needs to be retained in boundary condition (7.11) while the boundary condition for the electric mode is replaced by the condition of zero electric current from the ionosphere to the atmosphere

$$\left.\frac{4\pi}{c}j_z\right|_{z=-0} = \frac{\partial b_y}{\partial x} - \frac{\partial b_x}{\partial y} = 0, \tag{7.46}$$

where j_z is the vertical current component normal to the boundary between the atmosphere and the ionosphere.

7.4 'Thin' Ionosphere

In the general case of ionospheric MHD-wave propagation, it is necessary to solve the full wave equations (7.7)–(7.10) with the permeability tensor dependent on altitude. Obviously, that problem can be solved only numerically.

Its solution requires rather complicated computations to be carried out for a wide range of wavenumbers $\mathbf{k}_\tau$. However, these equations can be simplified in the case of a 'thin' ionosphere. This approach holds when the 'optical' thickness of the ionosphere for the transversal Alfvén waves is small, and the skin depth scale $d_P = c/(2\pi\omega\sigma_P)^{1/2}$ is larger than the thickness l_I of the highly conductive ionosphere region. We shall restrict ourselves to the consideration of oscillations with periods $T \geq 10\,\mathrm{s}$, for which the $d_P \gg l_I$.

Two limiting cases must be distinguished here:

- Small horizontal wavenumbers k_τ, when $k_\tau l_I \ll 1$.
  Then the electric field variation in the ionosphere is a small constant with an accuracy of an order of d_P/l_I and $k_\tau l_I$.
- Large k_τ. In this case $k_\tau l_I \gg 1$. It is seen from (7.7)–(7.10) that the electric field can vary significantly at the ionospheric thickness. But at large $\mathbf{k}_\tau$, wave mode coupling is weak, FMS-waves scarcely affect the Alfvén waves and it is possible to use the thin ionosphere model for the Alfvén waves in this case as well. The precise condition for the applicability of this approximation will be given in Section 7.7. The numerical calculations have shown that the application area of these two simple limiting cases overlap and the interaction of the MHD-waves with the ionosphere is described within these simple approximations.

Large Horizontal Scales

Let the scales $k_\tau l \ll 1$. Outside the equatorial region, integrating (7.7)–(7.10) with a condition $k_\tau l_I \ll 1$ along the field-lines, yields

$$\{\mathbf{E}_\tau\} = 0, \qquad \{\mathbf{b}_\tau\} = \frac{4\pi}{c}\mathbf{I}\times\hat{\mathbf{z}}, \qquad \mathbf{I} = \boldsymbol{\Sigma}\mathbf{E}, \tag{7.47}$$

$$\mathbf{b}_\tau = \begin{pmatrix} b_1 \\ b_2 \end{pmatrix}, \qquad \mathbf{E}_\tau = \begin{pmatrix} E_1 \\ E_2 \end{pmatrix}, \qquad \boldsymbol{\Sigma} = \begin{pmatrix} \boldsymbol{\Sigma}_{11} & \boldsymbol{\Sigma}_{12} \\ \boldsymbol{\Sigma}_{21} & \boldsymbol{\Sigma}_{22} \end{pmatrix}, \tag{7.48}$$

$$\boldsymbol{\Sigma}_{11} = \frac{\boldsymbol{\Sigma}_P}{\sin^2 I}, \qquad \boldsymbol{\Sigma}_{21} = -\boldsymbol{\Sigma}_{12} = -\frac{\boldsymbol{\Sigma}_H}{\sin I}, \qquad \boldsymbol{\Sigma}_{22} = \boldsymbol{\Sigma}_P. \tag{7.49}$$

Here $\{A\} = A_+ - A_-$ is the discontinuity of 'A' on the ionosphere; $\boldsymbol{\Sigma}_P$ and $\boldsymbol{\Sigma}_H$ are the integral Pedersen and Hall conductivities, respectively, given by

$$\boldsymbol{\Sigma}_P = \int_0^{l_I} \sigma_P(z)dz, \qquad \boldsymbol{\Sigma}_H = \int_0^{l_I} \sigma_H(z)dz.$$

Then the boundary conditions at the ionosphere for the tangential components of the magnetic field $\mathbf{b}_\tau$ and electric field $\mathbf{E}_\tau$ become

$$\mathbf{b}_\tau(+0) = \mathbf{Y}_I\,\mathbf{E}_\tau(+0)\,, \tag{7.50}$$

where $\mathbf{Y}_I = \mathbf{Y}(+0)$ is the admittance matrix above the ionosphere.

Let us relate the admittance matrix above the ionosphere with the admittance matrix below the ionosphere. It follows from (7.47) that

$$\mathbf{Y}_I \equiv \mathbf{Y}\,(+0) = \mathbf{Y}_a + \mathbf{G}\,,$$

where

$$\mathbf{G} = \begin{pmatrix} -Y/\sin I & X \\ -X/\sin^2 I & -Y/\sin I \end{pmatrix},$$

$$X = \frac{4\pi}{c}\boldsymbol{\Sigma}_P, \quad Y = \frac{4\pi}{c}\boldsymbol{\Sigma}_H .$$

This results in

$$\mathbf{Y}_I = \begin{pmatrix} -\dfrac{k_x k_y}{k_\tau^2}\left(\dfrac{\zeta_1}{\zeta_2} - \dfrac{\zeta_4}{\zeta_3}\right) - \dfrac{Y}{\sin I} & -\left(\dfrac{k_x^2}{k_\tau^2}\dfrac{\zeta_4}{\zeta_3} + \dfrac{k_y^2}{k_\tau^2}\dfrac{\zeta_1}{\zeta_2}\right) + X \\ \dfrac{k_y^2}{k_\tau^2}\dfrac{\zeta_4}{\zeta_3} + \dfrac{k_x^2}{k_\tau^2}\dfrac{\zeta_1}{\zeta_2} - \dfrac{X}{\sin^2 I} & \dfrac{k_x k_y}{k_\tau^2}\left(\dfrac{\zeta_1}{\zeta_2} - \dfrac{\zeta_4}{\zeta_3}\right) - \dfrac{Y}{\sin I} \end{pmatrix}.$$

In the next sections will be found both the matrix of transformation $\mathbf{T}'_\Sigma$ of the total horizontal magnetospheric fields $b_\tau^{(m)}\,(+0)$ above the ionosphere into the ground magnetic fields $\mathbf{b}^{(g)}$

$$\mathbf{b}^{(g)} = \mathbf{T}'_\Sigma\,\mathbf{b}_\tau^{(m)}\,(+0) \qquad (7.51)$$

and the matrix $\mathbf{T}'$ relating $\mathbf{b}^{(g)}$ and the amplitude of an initial wave $\mathbf{b}_\tau^{(i)}\,(+0)$ by

$$\mathbf{b}^{(g)} = \mathbf{T}'\,\mathbf{b}_\tau^{(i)}\,(+0)\,. \qquad (7.52)$$

Since

$$\mathbf{b}_\tau^{(m)}\,(+0) = \left(\mathbf{1} + \mathbf{R}'\right)\,\mathbf{b}_\tau^{(i)}\,(+0)\,,$$

these two matrices are related with one another as

$$\mathbf{T}' = \mathbf{T}'_\Sigma\,\left(\mathbf{1} + \mathbf{R}'\right). \qquad (7.53)$$

7.5 Homogeneous Magnetosphere

The Alfvén and FMS-Wave Potentials

The magnetospheric electric and magnetic wave fields we present as a sum of the Alfvén $(\mathbf{E}_A\,(\mathbf{r})\,, \mathbf{b}_A\,(\mathbf{r}))$ and FMS $(\mathbf{E}_S\,(\mathbf{r})\,, \mathbf{b}_S\,(\mathbf{r}))$ fields:

$$\mathbf{E}\,(\mathbf{r}) = \mathbf{E}_A\,(\mathbf{r}) + \mathbf{E}_S\,(\mathbf{r})\,, \qquad \mathbf{b}\,(\mathbf{r}) = \mathbf{b}_A\,(\mathbf{r}) + \mathbf{b}_S\,(\mathbf{r})\,. \qquad (7.54)$$

Let us first obtain expressions for the fields in the coordinate system (x', y', z') connected with the external magnetic field. In the oblique coordinate system

(x^1, x^2, x^3), the electric components will be found with the help of formulae for transformation vector components from one coordinate system to another. Let us express $\mathbf{E}(\mathbf{r})$ and $b(\mathbf{r})$ in terms of the scalar φ and vector $\mathbf{A}$ potentials

$$\mathbf{E}(\mathbf{r}) = -\nabla\varphi + ik_0\mathbf{A}, \tag{7.55}$$

$$\mathbf{b}(\mathbf{r}) = \nabla \times \mathbf{A}. \tag{7.56}$$

However, the $\mathbf{E}(\mathbf{r})$ and $\mathbf{b}(\mathbf{r})$ fields are conserved for transform of kind: $A \to A + \nabla\Psi$, $\varphi \to \varphi + ik_0\Psi$. A particular choice of φ and $\mathbf{A}$ potentials is a gauge, and a Ψ scalar function used to change a gauge is called a gauge function ([7], [14]).

In Alfvén waves $b_{Az'} = b_{\|} = 0$, so the condition $\nabla \cdot \mathbf{b} = \mathbf{0}$ becomes

$$\frac{\partial b_{Ax'}}{\partial x'} + \frac{\partial b_{Ay'}}{\partial y'} = 0. \tag{7.57}$$

Therefore, a gauge function can be chosen so that only one z'-component of the vector potential $\mathbf{A}$ is nonvanishing: $b_A(\mathbf{r}) = \nabla \times \mathbf{A}$, where $\mathbf{A} = A\mathbf{z}'$. Then the magnetic field in the Alfvén wave is given by

$$b_{Ax'} = \frac{\partial A}{\partial y'}, \qquad b_{Ay'} = -\frac{\partial A}{\partial x'}. \tag{7.58}$$

Components of the electric field $\mathbf{E}_A = -\nabla\varphi + ik_0\mathbf{A}$ and longitudinal current $j_{\|}$ are

$$E_{Ax'} = -\frac{\partial \varphi}{\partial x'}, \qquad E_{Ay'} = -\frac{\partial \varphi}{\partial y'}, \quad E_{Az'} = -\frac{\partial \varphi}{\partial z'} + ik_0 A, \tag{7.59}$$

$$j_{\|} = j_{Az'} = -\frac{c}{4\pi}\left(\frac{\partial^2 A}{\partial x'^2} + \frac{\partial^2 A}{\partial y'^2}\right). \tag{7.60}$$

Substituting (7.58), (7.59) into (7.5) and taking into account vanishing of the longitudinal electric component $E_{Az'} = 0$, we obtain

$$\frac{\partial A}{\partial z'} = ik_0\varepsilon_m\,\varphi, \tag{7.61}$$

$$\frac{\partial \varphi}{\partial z'} = ik_0 A. \tag{7.62}$$

Eliminating A from this system we get the 1D-wave equation for φ :

$$\frac{\partial^2 \varphi}{\partial z'^2} + \frac{\omega^2}{c_A^2}\,\varphi = 0.$$

In the FMS-wave the field-aligned component of the electric current vanishes

$$j_{Sz'} = j_{\|} = 0.$$

Therefore, it is possible to write the equation for transverse electric components:

$$\frac{\partial E_{Sx'}}{\partial x'} + \frac{\partial E_{Sy'}}{\partial y'} = 0. \tag{7.63}$$

Henceforth, index S marks the FMS-wave components. Let the gauge potentials be so that the scalar potential of the FMS-wave vanishes ($\varphi_S = 0$) [14]. Then, (7.55) gives

$$A_{S\|} = 0$$

and the S-transversal electric field is

$$\mathbf{E}_{S\perp} = ik_0 \mathbf{A}_{S\perp}.$$

From (7.63) a condition for the potential can be written as $\mathbf{A}_{S\perp}$

$$\frac{\partial A_{Sx'}}{\partial x'} + \frac{\partial A_{Sy'}}{\partial y'} = 0.$$

Thus, the transversal components of the vector-potential $\mathbf{A}_{S\perp}$ can be expressed by using a scalar potential ψ:

$$A_{Sx'} = \frac{\partial \psi}{\partial y'} \quad A_{Sy'} = -\frac{\partial \psi}{\partial x'}$$

and all components in the FMS-wave can be expressed in terms of $\psi\left(x', y', z'\right)$ as

$$\begin{aligned} b_{Sx'} &= \frac{\partial^2 \psi}{\partial x' \partial z'}, \quad b_{Sy'} = \frac{\partial^2 \psi}{\partial y' \partial z'}, \quad b_{\|} = b_{Sz'} = -\frac{\partial^2 \psi}{\partial x'^2} - \frac{\partial^2 \psi}{\partial y'^2}, \\ E_{Sx'} &= ik_0 \frac{\partial \psi}{\partial y'}, \quad E_{Sy'} = -ik_0 \frac{\partial \psi}{\partial x'}, \quad E_{Sz'} = 0. \end{aligned} \tag{7.64}$$

Substituting $b_{Sx'}$ and $b_{Sy'}$ from (7.64) into (7.5), we obtain

$$\nabla^2 \psi + \frac{\omega^2}{c_A^2} \psi = 0. \tag{7.65}$$

R and T Matrices

Now we want to find the relation between the amplitude of the initial incident wave of arbitrary polarization with reflected and transmitted waves. For this purpose we represent Fourier harmonics $\exp\left(ik_1 x^1 + ik_2 x^2\right)$ of the electric and magnetic fields as the sum of four waves: two Alfvén waves (incident, denoted by index 'i', and reflected 'r') and two FMS-waves ('i' and 'r'). Let $\mathbf{E}_{A\perp}^{(i,r)}$, $\mathbf{b}_{A\perp}^{(i,r)}$, $\mathbf{E}_{S\perp}^{(i,r)}$, $\mathbf{b}_{S\perp}^{(i,r)}$ be vectors ('polarization vectors') directed along corresponding wave electric and magnetic components. Polarization vectors in the coordinate system $\{x', y', z'\}$, associated with $\mathbf{B}_0$, are found directly from (7.58)–(7.65):

Alfvén Waves

$$\mathbf{E}_{A\perp}^{(i,r)} = \begin{pmatrix} E_{Ax'}^{(i,r)} \\ E_{Ay'}^{(i,r)} \end{pmatrix} = \mp \frac{c_A}{ck_A^{(i,r)}} \begin{pmatrix} k_{Ax'}^{(i,r)} \\ k_{Ay'}^{(i,r)} \end{pmatrix}, \qquad E_{Az'} = 0,$$

$$\mathbf{b}_{A\perp}^{(i,r)} = \begin{pmatrix} b_{Ax'}^{(i,r)} \\ b_{Ay'}^{(i,r)} \end{pmatrix} = \frac{1}{k_A^{(i,r)}} \begin{pmatrix} -k_{Ay'}^{(i,r)} \\ k_{Ax'}^{(i,r)} \end{pmatrix}, \qquad b_{Az'} = 0, \tag{7.66}$$

Now the upper sign '−' refers to the 'i'-wave, while '+' refers to the 'r'-wave.

FMS-Waves

$$\mathbf{E}_{S\perp}^{(i,r)} = \begin{pmatrix} E_{Sx'}^{(i,r)} \\ E_{Sy'}^{(i,r)} \end{pmatrix} = \frac{k_0}{k_{S,z'}^{(i,r)} k_S^{(i,r)}} \begin{pmatrix} k_{Sy'}^{(i,r)} \\ -k_{Sx'}^{(i,r)} \end{pmatrix}, \qquad E_{Sz'} = 0,$$

$$\mathbf{b}_{S\perp}^{(i,r)} = \begin{pmatrix} b_{Sx'}^{(i,r)} \\ b_{Sy'}^{(i,r)} \end{pmatrix} = \frac{1}{k_S^{(i,r)}} \begin{pmatrix} k_{Sx'}^{(i,r)} \\ k_{Sy'}^{(i,r)} \end{pmatrix}, \qquad b_{Sz'} = -\frac{k_{Sx'}^2 + k_{Sy'}^2}{k_{Sz'}^{(i,r)} k_S^{(i,r)}}. \tag{7.67}$$

In (7.66)–(7.67)

$$k_\alpha^{(i,r)} = \left(\left| k_{\alpha,x'}^{(i,r)} \right|^2 + \left| k_{\alpha,y'}^{(i,r)} \right|^2 \right)^{1/2}, \quad \text{and} \quad \alpha = \text{`}A\text{'} \quad \text{or} \quad \alpha = \text{`}S\text{'}.$$

Vectors $\mathbf{b}_{A\perp}^{(i,r)}$ and $\mathbf{b}_{S\perp}^{(i,r)}$ are orthogonal to $\mathbf{B}_0$ and normalized to unity. The scalar products are

$$\left| \mathbf{b}_{A\perp}^{(i,r)} \right| = \left(\mathbf{b}_{A\perp}^{(i,r)} \cdot \mathbf{b}_{A\perp}^{(i,r)} \right)^{1/2} = 1,$$

$$\left| \mathbf{b}_{S\perp}^{(i,r)} \right| = \left(\mathbf{b}_{S\perp}^{(i,r)} \cdot \mathbf{b}_{S\perp}^{(i,r)} \right)^{1/2} = 1. \tag{7.68}$$

Normalization of $\mathbf{E}_{A\perp}^{(i,r)}$ and $\mathbf{E}_{S\perp}^{(i,r)}$ are given by

$$\left| \mathbf{E}_{A\perp}^{(i,r)} \right| = \left(\mathbf{E}_{A\perp}^{(i,r)} \cdot \mathbf{E}_{A\perp}^{(i,r)} \right)^{1/2} = \frac{c_A}{c},$$

$$\left| \mathbf{E}_{S\perp}^{(i,r)} \right| = \left(\mathbf{E}_{S\perp}^{(i,r)} \cdot \mathbf{E}_{S\perp}^{(i,r)} \right)^{1/2} = \frac{k_0}{\left| k_{Sz'}^{(i,r)} \right|}.$$

Here the dot denotes the scalar product which for the example for two vectors $\mathbf{A} = (A_{x'}, A_{y'}, A_{z'})$ and $\mathbf{B} = (B_{x'}, B_{y'}, B_{z'})$ is

$$(\mathbf{A} \cdot \mathbf{B}) = A_{x'} B_{x'}^* + A_{y'} B_{y'}^* + A_{z'} B_{z'}^*.$$

Wave-vector components are determined by covariant components k_1, k_2, k_{A3} and k_1, k_2, k_{S3}. Transformation of the covariant components of oblique coordinates to Cartesian coordinates has the form

$$\begin{aligned} k_{\alpha x'} &= -\frac{k_1}{\sin I} + k_{\alpha 3}\cos I, \\ k_{\alpha y'} &= k_2, \\ k_{\alpha z'} &= -k_{\alpha 3}\sin I, \end{aligned} \tag{7.69}$$

where $\alpha = A$ or $\alpha = S$.

The transformation inverse to (7.69), applied to components $\mathbf{E}$ or $\mathbf{b}$, permits the polarization vector to be found in oblique coordinates. For instance, for $\mathbf{b}$ we get:

$$\begin{aligned} b_1 &= -b_{x'}\sin I - b_{z'}\cos I, \\ b_2 &= b_{y'}, \\ b_3 &= -\frac{b_{z'}}{\sin I}. \end{aligned} \tag{7.70}$$

Using transformation (7.70), we find:

Alfvén Waves

$$\begin{aligned} \mathbf{E}_{A\tau}^{(i,r)} &= \begin{pmatrix} E_{A1}^{(i,r)} \\ E_{A2}^{(i,r)} \end{pmatrix} = \mp\frac{c_A}{k_A^{(i,r)}}\begin{pmatrix} -k_{Ax'}^{(i,r)}\sin I \\ k_{Ay'}^{(i,r)} \end{pmatrix}, \\ \mathbf{b}_{A\tau}^{(i,r)} &= \begin{pmatrix} b_{A1}^{(i,r)} \\ b_{A2}^{(i,r)} \end{pmatrix} = \frac{1}{k_A^{(i,r)}}\begin{pmatrix} k_{Ay'}^{(i,r)}\sin I \\ k_{Ax'}^{(i,r)} \end{pmatrix}, \\ E_{A3}^{(i,r)} &= 0, \quad b_{A3}^{(i,r)} = 0. \end{aligned} \tag{7.71}$$

FMS-Waves

$$\begin{aligned} \mathbf{E}_{S\tau}^{(i,r)} &= \begin{pmatrix} E_{S1}^{(i,r)} \\ E_{S2}^{(i,r)} \end{pmatrix} = \frac{k_0}{k_{z'}^{(i,r)}k_S^{(i,r)}}\begin{pmatrix} -k_{Sy'}^{(i,r)}\sin I \\ -k_{Sx'}^{(i,r)} \end{pmatrix}, \\ \mathbf{b}_{S\tau}^{(i,r)} &= \begin{pmatrix} b_{S1}^{(i,r)} \\ b_{S2}^{(i,r)} \end{pmatrix} = \frac{1}{k_S^{(i,r)}}\begin{pmatrix} -k_{Sx'}^{(i,r)}\sin I + \dfrac{k_{Sx'}^2 + k_{Sy'}^2}{k_{Sz'}}\cos I \\ k_{Sy'} \end{pmatrix}, \\ E_{S3}^{(i,r)} &= 0, \quad b_{S3}^{(i,r)} = -\frac{b_{Sz'}}{\sin I} = \frac{1}{\sin I}\,\frac{k_{Sx'}^2 + k_{Sy'}^2}{k_{Sz'}^{(i,r)}k_S^{(i,r)}}. \end{aligned} \tag{7.72}$$

It now remains to express k_{A3} and k_{S3} in terms of k_1 and k_2 from the dispersion equation. It is possible to substitute fields $\propto \exp\left(ik_3x^3\right)$ into the set (7.7)–(7.10). It is, however, more easy to substitute (7.69) into the dispersion equations

$$k_{Az'}^2 = \frac{\omega^2}{c_A^2} \qquad \text{and} \qquad k_{Sx'}^2 + k_{Sy'}^2 + k_{Sz'}^2 = \frac{\omega^2}{c_A^2}.$$

Then

$$k_{A3}^{(i,r)} = \mp \frac{\omega}{c_A \sin I}, \qquad k_{S3}^{(i,r)} = k_1 \cot I \mp \sqrt{\frac{\omega^2}{c_A^2} - k_1^2 - k_2^2}, \tag{7.73}$$

where the upper sign refers to the incident waves and the lower to the reflected waves. Thus, the electric and magnetic components of the Alfvén and FMS-waves above the ionosphere can be expressed as

$$\begin{aligned} \begin{pmatrix} \mathbf{E}_\tau \\ \mathbf{b}_\tau \end{pmatrix} &= b_A^{(i)} \begin{pmatrix} \mathbf{E}_{A\tau}^{(i)} \\ \mathbf{b}_{A\tau}^{(i)} \end{pmatrix} \exp\left(ik_A^{(i)}x^3\right) + b_S^{(i)} \begin{pmatrix} \mathbf{E}_{S\tau}^{(i)} \\ \mathbf{b}_{S\tau}^{(i)} \end{pmatrix} \exp\left(ik_S^{(i)}x^3\right) \\ &+ b_A^{(r)} \begin{pmatrix} \mathbf{E}_{A\tau}^{(r)} \\ \mathbf{b}_{A\tau}^{(r)} \end{pmatrix} \exp\left(ik_A^{(r)}x^3\right) + b_S^{(r)} \begin{pmatrix} \mathbf{E}_{S\tau}^{(r)} \\ \mathbf{b}_{S\tau}^{(r)} \end{pmatrix} \exp\left(ik_S^{(r)}x^3\right), \end{aligned} \tag{7.74}$$

where

$$\mathbf{E}_\tau = \begin{pmatrix} E_1 \\ E_2 \end{pmatrix} = \begin{pmatrix} E_x \\ E_y \end{pmatrix}, \qquad \mathbf{b}_\tau = \begin{pmatrix} b_1 \\ b_2 \end{pmatrix} = \begin{pmatrix} b_x \\ b_y \end{pmatrix}.$$

By virtue of normalization (7.68), the coefficients $b_A^{(i)}$, $b_A^{(r)}$, $b_S^{(i)}$ and $b_S^{(r)}$ are the amplitude of the corresponding wave magnetic components transverse to $\mathbf{B}_0$.

R Matrix

Introduce the (2×2) reflection coefficients matrix $\mathbf{R}(\mathbf{k})$ as

$$\begin{pmatrix} b_A^{(r)} \\ b_S^{(r)} \end{pmatrix} = \mathbf{R}(\mathbf{k}) \begin{pmatrix} b_A^{(i)} \\ b_S^{(i)} \end{pmatrix}, \qquad \mathbf{R} = \begin{pmatrix} R_{AA} & R_{AS} \\ R_{SA} & R_{SS} \end{pmatrix}. \tag{7.75}$$

The sense of elements R_{ik} in matrix $\mathbf{R}(\mathbf{k})$ is clear from (7.75). If, for instance, an Alfvén wave with a unit magnetic amplitude is incident on the ionosphere, then the reflected Alfvén wave has the amplitude R_{AA} and the reflected FMS-wave amplitude is R_{SA}. Rewrite (7.74) in matrix form:

$$\begin{pmatrix} \mathbf{E}_\tau \\ \mathbf{b}_\tau \end{pmatrix} = \mathbf{U}_\tau^{(i)} \exp\left(i\mathbf{K}^{(i)}x^3\right) \mathbf{b}^{(i)} + \mathbf{U}_\tau^{(r)} \exp\left(i\mathbf{K}^{(r)}x^3\right) \mathbf{b}^{(r)},$$

where

$$\mathbf{b}^{(i,r)} = \begin{pmatrix} b_A^{(i,r)} \\ b_S^{(i,r)}, \end{pmatrix},$$

and the (2×2) block matrix $\mathbf{U}_\tau^{(i,r)}$ and $\mathbf{K}^{(i,r)}$ are:

$$\mathbf{U}_\tau^{(i,r)} = \begin{pmatrix} \mathbf{E}_A^{(i,r)} & \mathbf{E}_S^{(i,r)} \\ \mathbf{b}_A^{(i,r)} & \mathbf{b}_S^{(i,r)} \end{pmatrix}, \quad \mathbf{K}^{(i,r)} = \begin{pmatrix} k_A^{(i,r)} & 0 \\ 0 & k_S^{(i,r)} \end{pmatrix}.$$

With(7.75)

$$\begin{pmatrix} \mathbf{E}_\tau \\ \mathbf{b}_\tau \end{pmatrix} = \left[\mathbf{U}_\tau^{(i)} \exp\left(i\mathbf{K}^{(i)} x^3\right) + \mathbf{U}_\tau^{(r)} \exp\left(i\mathbf{K}^{(r)} x^3\right) \mathbf{R}(\mathbf{k}) \right] \mathbf{b}^{(i)}. \tag{7.76}$$

The problem of finding the electromagnetic fields is thus reduced to determining the reflection coefficients matrix $\mathbf{R}(\mathbf{k})$. To find $\mathbf{R}$ substitute (7.76) into boundary condition (7.50). After a simple algebra, we get

$$\mathbf{R} = -\mathbf{D}_r^{-1} \mathbf{D}_i, \tag{7.77}$$

where the 2×2 matrices $\mathbf{D}_i$ and $\mathbf{D}_r$ are

$$\mathbf{D}_\alpha = \left(\mathbf{b}_A^{(\alpha)} - \mathbf{Y}_I \mathbf{E}_A^{(\alpha)},\ \mathbf{b}_S^{(\alpha)} - \mathbf{Y}_I \mathbf{E}_S^{(\alpha)} \right), \quad \alpha = i \ \text{ or } \ r. \tag{7.78}$$

Equation (7.74) allows the horizontal components of $\mathbf{E}$ and $\mathbf{b}$ to be found for a known incident wave. The component $b_3 = -b_\parallel / \sin I$ above the ionosphere can be found directly from Maxwell's equations. The substitution of $b^1 = g^{1n} b_n = b_1 / \sin^2 I - b_3 \cot I$ in $b_3 = g_{3k} b^k = b^1 \cot I + b^3 / \sin^2 I$ gives

$$b_3 = b^3 + b_1 \cot I$$

On substituting $b^3 = (k_1 E_2 - k_2 E_1)/k_0$ in the last equation we have

$$b_3 = \frac{k_1 E_2 - k_2 E_1}{k_0} + b_1 \cot I. \tag{7.79}$$

Finally, the longitudinal magnetic component is

$$b_\parallel = -b_3 \sin I = -\sin I \left(\frac{k_1 E_2 - k_2 E_1}{k_0} + b_1 \cot I \right). \tag{7.80}$$

T Matrix

Fields under the ionosphere are found from (7.47):

$$\mathbf{E}_\tau\,(0) = \mathbf{E}_\tau\,(-0) = \mathbf{E}_\tau\,(+0)\,, \qquad \mathbf{b}_\tau\,(-0) = \mathbf{b}_\tau\,(+0) - \mathbf{G}\,\mathbf{E}_\tau\,(+0)\,.$$

Substituting the solution (7.76), we obtain

$$\mathbf{b}_\tau(-0) = \left(\mathbf{D}_i^{(0)} + \mathbf{D}_r^{(0)}\,\mathbf{R}\right)\mathbf{b}^{(i)}, \tag{7.81}$$

where

$$\mathbf{D}_\alpha^{(0)} = \left(\mathbf{b}_A^{(\alpha)} - \mathbf{G}\mathbf{E}_A^{(\alpha)},\ \mathbf{b}_S^{(\alpha)} - \mathbf{G}\mathbf{E}_S^{(\alpha)}\right), \quad \alpha = i \text{ or } r.$$

The covariant components b_1, b_2 and E_1, E_2 coincide with the horizontal magnetic b_x, b_y and electric E_x, E_y, that is

$$b_x = b_1, \quad b_y = b_2, \quad E_x = E_1, \quad E_y = E_2,$$

while for b_z we get

$$b_z = -b_1 \cot I + b_3.$$

The formulae obtained relate the total MHD-wave field above and under the thin ionosphere with an Alfvén or FMS-wave incident upon it from a homogeneous half-space filled with cold plasma.

Components of $\mathbf{E}(z)$ and $\mathbf{b}(z)$ in the atmosphere can be easily calculated from (7.81). Explicit expressions will be given here permitting $\mathbf{E}(z)$ and $\mathbf{b}(z)$ to be found in the most important case when displacement currents and conductivity in the atmosphere can be neglected.

It is convenient to write the field in the atmosphere and on the ground surface in the Cartesian-altitude system $\{x, y, z\}$. For horizontal wavenumbers $|k| \gg k_0 |\varepsilon_a|^{1/2}$ from (7.41), (7.42), and (7.45), it follows that the atmospheric fields are given by

$$\mathbf{b}_{\boldsymbol{\tau}}(z) = \frac{1 + R_g e^{-2k(z+h)}}{1 + R_g e^{-2kh}} e^{kz}\,\mathbf{b}_{\boldsymbol{\tau}}(0), \tag{7.82}$$

$$\mathbf{E}_{\boldsymbol{\tau}}(z) = -i\frac{k_0}{k}\frac{1 - R_g e^{-2k(z+h)}}{1 + R_g e^{-2kh}} e^{kz}\,\hat{\mathbf{z}} \times \mathbf{b}_{\boldsymbol{\tau}}(0),$$

$$b_z(z) = \frac{k_x E_y(z) - k_y E_x(z)}{k_0}, \tag{7.83}$$

where

$$R_g = \frac{1 - ikZ_g^{(m)}/k_0}{1 + ikZ_g^{(m)}/k_0}.$$

The expressions obtained for fields are rather cumbersome and their application requires numerical calculations. Nevertheless, the utilization of these general equations permits several useful results to be obtained. In particular, no special difficulties are involved in carrying out the inverse numerical Fourier transformation, which allows field distribution to be obtained both above the ionosphere and on the ground from MHD-wave beam.

The approximation used in this section is applicable only to waves with small horizontal wavenumbers k_1 and k_2. The condition $k_1 l_I \ll 1$ and $k_2 l_I \ll 1$

must be satisfied, where l_I is the thickness of the conductive region of the ionosphere. Since $l_I \approx 30\,\text{km}$, it means that rather strict conditions are obtained on the horizontal scale $L = \min\left(k_1^{-1}, k_2^{-1}\right) \gg 30\,\text{km}$.

This condition may not hold, for instance, for the fields near the FLR. In that case, as is known from Chapter 6, small-scale approximation is valid. Since uncoupled equations for Alfvén waves do not include horizontal wavenumbers k_1, k_2 , the thin ionosphere approximation is valid for them even at $k_1 l_I \gtrsim 1$ and $k_2 l_I \gtrsim 1$.

It it worth noting that the large $\mathbf{k}$ approximation and the thin ionosphere approximation taken together cover almost all the virtually important cases.

7.6 Propagation Along a Meridian

R and T Matrices

Consider MHD-waves propagating in a meridional plane, $k_2 = 0$. In that case, the reflection and transmission matrices have rather simple analytical expressions. In this section, expressions will be given for matrix $\mathbf{R}'$ connecting horizontal magnetic components in reflected $\mathbf{b}_\tau^{(r)}$ and incident waves $\mathbf{b}_\tau^{(i)}$ above the ionosphere as well as for matrix $\mathbf{T}'$ connecting the ground $\mathbf{b}^{(g)}$ and $\mathbf{b}_\tau^{(i)}$. Equations (7.41) and (7.42) are simplified for $k_2 = 0$ and become

$$b_1^{(g)} = -\frac{Y_g^{(m)}}{\zeta_4} b_1(-0), \quad b_2^{(g)} = -\frac{Y_g^{(e)}}{\zeta_1} b_2(-0). \tag{7.84}$$

From (7.36)–(7.38) with (7.47), it follows that

$$b_1(-0) = -\frac{\zeta_4}{\zeta_3} E_2(0), \quad b_2(-0) = \frac{\zeta_1}{\zeta_2} E_1(0). \tag{7.85}$$

Combining (7.84) and (7.85), we obtain

$$b_1^{(g)} = \frac{Y_g^{(m)}}{\zeta_3} E_2(0), \quad b_2^{(g)} = -\frac{Y_g^{(e)}}{\zeta_2} E_1(0). \tag{7.86}$$

The magnetic field $\mathbf{b}^{(i)}$ in incident and $\mathbf{b}^{(r)}$ in the reflected waves and the horizontal components $\mathbf{b}^{(g)}$ on the ground can be written as

$$\mathbf{b}_\tau^{(r)} = \mathbf{R}'\,\mathbf{b}_\tau^{(i)}, \qquad \text{and} \qquad \mathbf{b}^{(g)} = \mathbf{T}'\,\mathbf{b}_\tau^{(i)} = \mathbf{T}\begin{pmatrix} b_S^{(i)} \\ b_A^{(i)} \end{pmatrix}. \tag{7.87}$$

Substituting (7.76) into (7.87), at $x^3 = 0$ we find

$$\begin{aligned} \mathbf{R}' &= \left(\mathbf{b}_S^{(r)}, \mathbf{b}_A^{(r)}\right) \mathbf{R} \left(\mathbf{b}_S^{(i)}, \mathbf{b}_A^{(i)}\right)^{-1}, \\ \mathbf{T}' &= \left(\mathbf{b}_S^{(i)}, \mathbf{b}_A^{(i)}\right) \mathbf{T}. \end{aligned} \tag{7.88}$$

Sometimes matrix $\mathbf{T}'_\Sigma$ is useful:

$$\mathbf{b}^{(g)} = \mathbf{T}'_\Sigma \, \mathbf{b}_\tau , \qquad \mathbf{T}' = \mathbf{T}'_\Sigma \left(\mathbf{1} + \mathbf{R}'\right) . \tag{7.89}$$

Matrices $\mathbf{R}'$ and $\mathbf{T}'$ could be found after cumbersome but simple enough algebra from the equalities (7.71)–(7.78) for $k_2 = 0$. However, it is more simple to proceed from (7.7)–(7.10) and the boundary condition (7.50).

Without considering the Hall conductivity and for $k_2 = 0$, equations (7.7)–(7.10) become

$$\left(\frac{\partial}{\partial x^3} - ik_1 \cot I\right) b_1 = \frac{\kappa_s^2}{ik_0} E_2, \tag{7.90}$$

$$\left(\frac{\partial}{\partial x^3} - ik_1 \cot I\right) E_2 = -ik_0 b_1, \tag{7.91}$$

for the FMS-waves and

$$\frac{\partial b_2}{\partial x^3} = -\frac{\kappa_A^2}{ik_0} E_1, \tag{7.92}$$

$$\frac{\partial E_1}{\partial x^3} = ik_0 b_2, \tag{7.93}$$

for Alfvén waves. Here

$$\kappa_s = \left(k_0^2 \varepsilon_m - k_1^2\right)^{1/2}, \qquad \mathrm{Re}\, \kappa_S > 0$$

and

$$\kappa_A = k_A / \left|\sin I\right| , \quad k_A = k_0 \varepsilon_m^{1/2} = \omega / c_A .$$

The admittance matrix in the boundary condition (7.50) for $k_2 = 0$ has the form

$$\mathbf{Y}_I = \begin{pmatrix} -\dfrac{Y}{\sin I} & X - \dfrac{\zeta_4}{\zeta_3} \\ \dfrac{\zeta_1}{\zeta_2} - \dfrac{X}{\sin^2 I} & -\dfrac{Y}{\sin I} \end{pmatrix} . \tag{7.94}$$

Components b_2 and E_1 (Alfvén waves) and components b_1 and E_2 (FMS-waves) from (7.90)–(7.93) are given by

$$b_1 = e^{ik_1 x^3 \cot I} \left[\left(e^{-i\kappa_s x^3} + R'_{SS} e^{i\kappa_s x^3}\right) b_1^{(i)} + R'_{SA} e^{i\kappa_s x^3} b_2^{(i)}\right], \tag{7.95}$$

$$E_2 = \frac{k_0}{\kappa_s} e^{ik_1 x^3 \cot I} \left[\left(e^{-i\kappa_s x^3} - R'_{SS} e^{i\kappa_s x^3}\right) b_1^{(i)} - R'_{SA} e^{i\kappa_s x^3} b_2^{(i)}\right], \tag{7.96}$$

$$b_2 = R'_{AS} e^{i\kappa_A x^3} b_1^{(i)} + \left(e^{-i\kappa_A x^3} + R_{AA} e^{i\kappa_A x^3}\right) b_2^{(i)}, \tag{7.97}$$

$$E_1 = -\frac{k_0}{\kappa_A} \left[-R'_{AS} e^{\,i\kappa_A x^3} b_1^{(i)} + \left(e^{-i\kappa_A x^3} - R_{AA} e^{i\kappa_A x^3}\right) b_2^{(i)}\right], \tag{7.98}$$

where $b_1^{(i)}$ and $b_2^{(i)}$ are the amplitudes of the incident Alfvén and FMS-waves. By substituting electric and magnetic fields into boundary conditions (7.50), we find the reflection matrix

$$\mathbf{R}' = \begin{pmatrix} R'_{SS} & R'_{SA} \\ R'_{AS} & R'_{AA} \end{pmatrix} = \mathbf{1} + \frac{2}{\Delta} \begin{pmatrix} \dfrac{\kappa_s \Delta_A}{k_0} & \dfrac{\kappa_s}{k_0} Y \operatorname{sign} I \\ -\varepsilon_m^{1/2} \dfrac{Y}{\sin I} & \varepsilon_m^{1/2} \Delta_S \end{pmatrix} . \tag{7.99}$$

Let us obtain a transformation matrix of the magnetospheric fields into ground fields. We have from (7.50) on the upper ionospheric boundary, relation $E_\tau(0) = \mathbf{Y}_I^{-1} \mathbf{b}_\tau(+0)$. Then, from (7.86) and (7.94),

$$\mathbf{T}'_\Sigma = \frac{1}{|\mathbf{Y}_I|} \begin{pmatrix} -\dfrac{Y_g^m}{\zeta_3} \left(\dfrac{\zeta_1}{\zeta_2} - \dfrac{X}{\sin^2 I} \right) & -\dfrac{Y_g^m}{\zeta_3} \dfrac{Y}{\sin I} \\ \dfrac{Y_g^e}{\zeta_2} \dfrac{Y}{\sin I} & -\dfrac{Y_g^e}{\zeta_2} \left(\dfrac{\zeta_4}{\zeta_3} - X \right) \end{pmatrix} . \tag{7.100}$$

The following denotations are used in (7.99) and (7.100):

$$\Delta = \Delta_S \Delta_A + \frac{Y^2}{|\sin I|}, \tag{7.101}$$

$$\Delta_A = \frac{\zeta_1}{\zeta_2} |\sin I| - \frac{\tilde{X}}{|\sin I|}, \qquad \Delta_S = \frac{\zeta_4}{\zeta_3} - \frac{\kappa_S}{k_0} - X, \tag{7.102}$$

$$|\mathbf{Y}_I| = \frac{Y^2}{\sin^2 I} + \left(\frac{\zeta_4}{\zeta_3} - X \right) \left(\frac{\zeta_1}{\zeta_2} - \frac{X}{\sin^2 I} \right), \tag{7.103}$$

$$\tilde{X} = X + \varepsilon_m^{1/2} |\sin I|, \quad X = \frac{4\pi \Sigma_P}{c}, \qquad Y = \frac{4\pi \Sigma_H}{c}. \tag{7.104}$$

Inclination $I < 0$ in the Southern Hemisphere and $I > 0$ in the Northern Hemisphere. $\zeta_i \equiv \zeta_i(-0)$.

One can find, in a similar manner from (7.95)–(7.98), the transmission matrix $\mathbf{T}'$:

$$\mathbf{T}' = -\frac{2}{\Delta} \begin{pmatrix} \dfrac{Y_g^m \Delta_A}{\zeta_3} & \dfrac{Y_g^m}{\zeta_3} Y \operatorname{sign} I \\ -\dfrac{Y_g^e}{\zeta_2} Y \operatorname{sign} I & \dfrac{Y_g^e}{\zeta_2} \Delta_S |\sin I| \end{pmatrix} \tag{7.105}$$

Characteristic Parameters

In the present section the transmission and reflection matrices are studied within a more realistic model of the distribution of the Earth's conductivity.

Table 7.1. Characteristic wave and medium parameters

τ (s)	k_0 (km^{-1})	h (km)	c_A (km/s)	$\sqrt{\varepsilon_m}$	$X \approx Y$
$\gg 1$	$\ll 2 \times 10^{-5}$	$\approx 10^2$	$\approx 10^3$	3×10^2	5×10^1 (night) 5×10^3 (day)
σ_g (s^{-1})	$\sqrt{\varepsilon_g}$	H(km)	Σ_g (km/s)	$\sqrt{\varepsilon_a}$	
$10^5 - 10^8$	$\gg 3 \times 10^2$	5×10^2	$\approx 10^8$	< 5	

Assume that a layer with a constant conductivity σ_g lies on a perfect conductor at a depth H. The thin conductive ionosphere is located, as before, at the height h. The magnetosphere is modeled by a uniform half-space with a constant Alfvén velocity. The model contains several characteristic spatial scales: h, H, d_g, l_m, $(k_0\sqrt{\varepsilon_a})^{-1}$ and k^{-1}. The sense of the first two parameters is evident. $d_g = (\mathrm{Re}\, k_0\sqrt{\varepsilon_g})^{-1}$ is the ground skin depth. The wavelength in the magnetosphere $l_m = 2\pi(k_0\sqrt{\varepsilon_m})^{-1} = 2\pi c_A/\omega$. $2\pi(k_0\sqrt{\varepsilon_a})^{-1}$ is the wavelength in the atmosphere and k^{-1} is the horizontal wave-scale.

Table 7.1 shows the characteristic scales of the wave and the medium. One can see that

$$|k_0\sqrt{\varepsilon_a}| \ll |k_0\sqrt{\varepsilon_g}| \quad \text{and} \quad k_0\sqrt{\varepsilon_m} \ll H^{-1} < h^{-1}.$$

The skin depth and thickness of the underlying half-space is compared on periods

$$T = 10^{-9} H^2 \sigma_g,$$

where H is in km and σ_g in s^{-1}.

Analytical Properties of R and T Matrices

The behavior of the ground fields and signals reflected into space is determined by the analytical properties of $\mathbf{R}'(k)$ and $\mathbf{T}'(k)$ in the complex plane of wavenumbers $k = k_x$.

The first thing to pay attention to is the presence of branch points (see Fig. 7.3):

- In a model 'open' at $H = \infty$ (see Fig. 7.2) there are four branch points $k_{1,2} = \pm k_0\sqrt{\varepsilon_m}$ and $k_{3,4} = \pm k_0\sqrt{\varepsilon_g}$;
- In the two-layer model with a layer of the finite thickness underlain by a perfect conductor (Fig. 7.2), only two points remain: $k_{1,2} = \pm k_0\sqrt{\varepsilon_m}$.

In order to isolate univalent branches of the functions under investigation their behavior will be considered on a many-sheeted Riemann surface with the number of sheets determined by the number of branch points.

Construct a Riemannian surface for the 1-st model of the Earth as a half-space with finite conductivity σ_g . Draw four cuts on plane k from the branch points to infinity. $\mathrm{Im}\,\kappa_s = \mathrm{Im}\,\kappa_g = 0$ on cuts (see Fig. 7.3). Numeration the sheets is shown in Table 7.2.

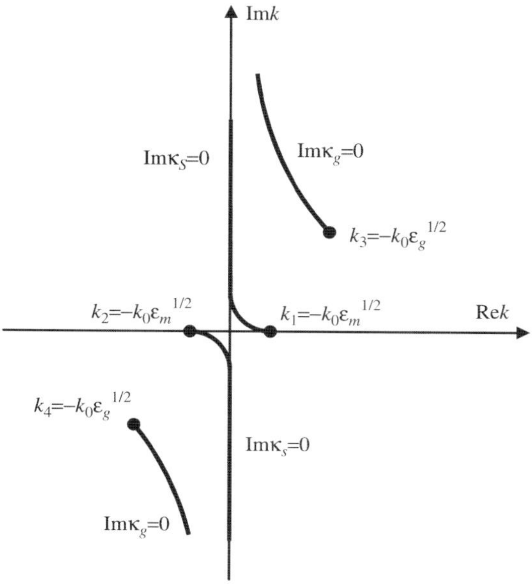

Fig. 7.3. Location of cuts $\operatorname{Im}\kappa_g = \operatorname{Im}\kappa_s = 0$ and branch points $k_{1,2} = \pm k_0\sqrt{\varepsilon_m}$, $k_{3,4} = \pm k_0\sqrt{\varepsilon_g}$ on the complex plane of wave numbers k

In the case of the finite conductivity layer underlain by a perfect conductor, only two sheets remain:

- the physical sheet: $\operatorname{Im}\kappa_s > 0$ and
- the non-physical sheet: $\operatorname{Im}\kappa_s < 0$.

Apart from branch points, the poles of $\mathbf{R}'$ and $\mathbf{T}'$ can appear on the complex plane, and their location is determined by the roots of $\varDelta(k)$ (7.101):

$$\varDelta(k) = 0. \tag{7.106}$$

The Role of a Magnetic Inclination

Before proceeding to the investigation of $\mathbf{R}'(k)$ and $\mathbf{T}'(k)$ in various wavenumber ranges, note that the character of the reflection of MHD-waves does not

Table 7.2. Numeration of the sheets of the Riemann surface

Physical sheet	Non-physical sheets
1. $\operatorname{Im}\kappa_g > 0$ and $\operatorname{Im}\kappa_s > 0$	2. $\operatorname{Im}\kappa_s < 0$, $\operatorname{Im}\kappa_g > 0$
	3. $\operatorname{Im}\kappa_s < 0$, $\operatorname{Im}\kappa_g < 0$
	4. $\operatorname{Im}\kappa_s > 0$, $\operatorname{Im}\kappa_g < 0$

change qualitatively in a broad range of ionospheric and magnetospheric parameters at the transition from $I = \pi/2$ to arbitrary magnetic inclinations. In order to clarify the validity of that statement, we rewrite (7.99) and (7.100), introducing parameter X_K:

$$X_K = X - \frac{Y^2}{\Delta_A\,|\sin I|}. \tag{7.107}$$

Then

$$\Delta = \Delta_A \Delta_{SK},$$

where

$$\Delta_{SK} = \frac{\zeta_4}{\zeta_3} - \frac{\kappa_s}{k_0} - X_K.$$

Equation (7.99) for the reflection coefficients matrix can be then easily written as

$$\mathbf{R}' = \mathbf{1} + \frac{2}{\Delta_A \Delta_{SK}} \begin{pmatrix} \dfrac{\kappa_S \Delta_A}{k_0} & Y\dfrac{\kappa_S}{k_0}\,\mathrm{sign}\, I \\ -\dfrac{Y\sqrt{\varepsilon_m}}{\sin I} & \Delta_S\sqrt{\varepsilon_m} \end{pmatrix}. \tag{7.108}$$

A little later we shall show that, except of some exotic and practically uninteresting cases, dependency of Δ_A on the wavenumber k is insignificant and can be replaced with $\Delta_A^{(0)} = -(X/|\sin I| + \sqrt{\varepsilon_m})$, and X_K, in turn, with $X_K^{(0)} = X - Y^2/(X + \sqrt{\varepsilon_m}\,|\sin I|)$. Equation (7.108) for matrix $\mathbf{R}'$ shows the results of calculating for $\mathbf{R}'$ with a vertical $\mathbf{B}_0$ to be transferred to an arbitrary inclination of $\mathbf{B}_0$. It is necessary that we only replace Y by $Y/\sin I$ and X by X_K.

Likewise, matrix $\mathbf{T}'$ with account being taken of inclination I becomes

$$\mathbf{T}' = \frac{2}{\Delta_A \Delta_{SK}} \begin{pmatrix} -\dfrac{Y_g^{(m)}\Delta_A}{\zeta_3} & -\dfrac{Y_g^{(m)}}{\zeta_3} Y\,\mathrm{sign}\, I \\ \dfrac{Y_g^{(e)}}{\zeta_2} Y\,\mathrm{sign}\, I & -\Delta_S \dfrac{Y_g^{(e)}}{\zeta_2}\,|\sin I| \end{pmatrix}. \tag{7.109}$$

The Role of the Ground Conductivity

We obtain approximate expressions for transformation matrices for the horizontal wavenumbers

$$k \gg k_0\sqrt{\varepsilon_a} \tag{7.110}$$

for which

$$\frac{\zeta_1}{Y_g^{(e)}} = \cosh kh \left(-1 + \frac{ik_0\varepsilon_a}{kY_g^{(e)}} \tanh kh \right), \tag{7.111}$$

$$\frac{\zeta_2}{Y_g^{(e)}} = i\frac{k}{k_0\varepsilon_a} \sinh kh \left(1 - \frac{ik_0\varepsilon_a}{kY_g^{(e)}} \coth kh \right), \tag{7.112}$$

$$\frac{\zeta_3}{Y_g^{(m)}} = -i\frac{k_0}{k} \sinh kh \left(1 + \frac{ik}{k_0 Y_g^{(m)}} \coth kh \right), \tag{7.113}$$

$$\frac{\zeta_4}{Y_g^{(m)}} = -\cosh kh \left(1 + \frac{ik}{k_0 Y_g^{(m)}} \tanh kh \right). \tag{7.114}$$

For the two-layer model

$$Y_g^{(e)} = \frac{1}{Z_g^{(e)}} = \frac{ik_0\varepsilon_g}{\kappa_g} \cot \kappa_g H, \qquad Y_g^{(m)} = \frac{1}{Z_g^{(m)}} = \frac{i\kappa_g}{k_0} \cot \kappa_g H, \tag{7.115}$$

where

$$\kappa_g = \sqrt{k_0^2\varepsilon_g - k^2} = \sqrt{2} d_g^{-1} \sqrt{i - \frac{(kd_g)^2}{2}}.$$

Admittances $Y_g^{(e)}$ and $Y_g^{(m)}$ for $H \to \infty$, $\cot \kappa_g H \to -i$, tend to the electric and magnetic admittances of the conductive half-space.

Let us find the conditions when the ground can be considered as a perfect conductive medium. The second term in (7.111)–(7.114) can be neglected for high enough conductivity if the inequalities

$$\left| \frac{\sigma_a}{\sigma_g} \frac{\kappa_g}{k} \tan \kappa_g H \right| \ll \left| \coth kh \right| + \left| \tanh kh \right|,$$

$$\left| \frac{k}{\kappa_g} \tan \kappa_g H \right| \ll \left| \coth kh \right| + \left| \tanh kh \right|$$

hold. Then

$$\frac{\zeta_1}{Y_g^{(e)}} = -\cosh kh, \qquad \frac{\zeta_2}{Y_g^{(e)}} = i\frac{k}{k_0\varepsilon_a} \sinh kh, \tag{7.116}$$

$$\frac{\zeta_3}{Y_g^{(m)}} = -i\frac{k_0}{k} \sinh kh, \qquad \frac{\zeta_4}{Y_g^{(m)}} = -\cosh kh. \tag{7.117}$$

Let us consider first an influence of the conductivity on the electric mode. The corrective term in (7.111) can be ignored if

$$\left| \frac{\varepsilon_a}{\varepsilon_g} \frac{\kappa_g}{k} \tan \kappa_g H \tanh kh \right| \ll 1. \tag{7.118}$$

The inequality is invalid near the branch cut $\mathrm{Im}\, \kappa_g = 0$ (Fig.7.3) and imaginary axes of the complex plane of wavenumbers k since $\tan \kappa_g H$ and $\tanh kh$

have there poles. Equation (7.118) is fulfilled for the high conductivity over the whole k plane except a narrow zone near the branch cuts. Close to the cuts $\tan\kappa_g H \approx 1$ and $\tanh kh \approx 1$. In this case the inequality (7.118) in the wavenumber range (7.110) holds for $|\varepsilon_a/\varepsilon_g|^{1/2} \ll 1$. Use for the magnitude of tangent the estimation

$$\max|\tan\kappa_g H|_{\mathrm{Im}\,z=\mathrm{const}} = \frac{1+\exp\left(-2\left|\mathrm{Im}\,\kappa_g H\right|\right)}{1-\exp\left(-2\left|\mathrm{Im}\,\kappa_g H\right|\right)} \approx 1 + \frac{1}{\mathrm{Im}\,\kappa_g H},$$

$$\max|\tanh kh|_{\mathrm{Im}\,k=\mathrm{const}} = \frac{1+\exp\left(-2\left|\mathrm{Im}\,k\right|h\right)}{1-\exp\left(-2\left|\mathrm{Im}\,k\right|h\right)} \approx 1 + \frac{1}{\mathrm{Im}\,kh}.$$

Thus, close to the branch cuts we have

$$\tan\kappa_g H \lesssim \frac{1}{\mathrm{Im}\,\kappa_g H}, \qquad \tan kH \lesssim \frac{1}{\mathrm{Im}\,kh}$$

and for

$$\mathrm{Im}\kappa_g H\,\mathrm{Im}\,kh \gg \left|\frac{\sigma_a}{\sigma_g}\right|$$

the influence of finite conductivity on $\zeta_1/Y_g^{(e)}$is insignificant. The same estimate is obtained for $\zeta_2/Y_g^{(e)}$. A ratio of the atmospheric conductivity to the ground conductivity is extremely small. For typical values of the atmospheric $\sigma_a = 10^{-4}\,\mathrm{s}^{-1}$ and crust $\sigma_g = 10^7\,\mathrm{s}^{-1}$ conductivities, we have

$$\left|\frac{\sigma_a}{\sigma_g}\right| \sim \frac{10^{-4}}{10^7} = 10^{-11}.$$

Thus, the finite-ground conductivity has no effect on the electric mode. Note, that this conclusion is well known in the theory of radio wave propagation.

On the other hand, ground conductivity can essentially effect the magnetic mode. For the magnetic mode, the ground can be replaced with a perfect conductor only for an horizontal scale larger than the skin depth in the ground. Equation (7.117) is true if

$$kd_g \ll 1, \quad d_g = \frac{c}{(2\pi\sigma_g\omega)^{1/2}}.$$

Let us now define the range of the wavenumbers for which one can neglect penetration of the current from the ionosphere to the ground. First, let us consider the function Δ_A. It follows at once from (7.102) and (7.116) that at

$$\mathrm{Re}k \gg |k_*|, \quad k_* = \left(\frac{k_0\varepsilon_a}{h\widetilde{X}}\,|\sin I|\right)^{1/2} \tag{7.119}$$

the expression for

$$\Delta_A = \frac{\zeta_1}{\zeta_2}\,|\sin I| - \frac{\widetilde{X}}{|\sin I|}$$

is simplified and has the form

$$\Delta_A = \Delta_A^{(0)} \left(1 + 0\left(\frac{k_*}{\mathrm{Re}\, k}\right)\right), \qquad \Delta_A^{(0)} = -\frac{\widetilde{X}}{|\sin I|}, \tag{7.120}$$

where $\widetilde{X} = X + \sqrt{\varepsilon_m}\,|\sin I|$. The estimation of k_* is

$$|k_*| \approx \left(\frac{4\pi}{ch\widetilde{X}}\right)^{1/2} \begin{cases} (2T)^{-1/2} \\ \sigma_a^{1/2} \end{cases} \text{for} \begin{cases} \sigma_a T \ll 1 \\ \sigma_a T \gg 1 \end{cases}. \tag{7.121}$$

Then, if

$$\begin{cases} \sigma_a T \ll 1 \\ \sigma_a T \gg 1 \end{cases}, \quad |k_*|\left(\mathrm{km}^{-1}\right) \sim \begin{cases} (2T)^{-1/2} \\ \sigma_a^{1/2} \end{cases} \times \begin{matrix} 10^{-5} \ (\text{day}) \\ 10^{-4} \ (\text{night}) \end{matrix}.$$

The simplified expression (7.120) for Δ_A, independent of the wavenumber, also permits us to simplify X_K which, for Re$k \gg k_*$, no longer depends on k:

$$X_K^{(0)} = X + \frac{Y^2}{\widetilde{X}}. \tag{7.122}$$

As a result, the problem of long-period MHD-wave reflection and transformation by ionospheric plasma with an inclined magnetic field is reduced to the study of MHD-wave propagation through a thin isotropic ionosphere with conductivity $\Sigma_K = X_K c/4\pi$. The transformation coefficients (of 'A' into 'S' and of 'S' into 'A') are then found by simple multiplication of the solutions obtained by coefficients proportional to $\Sigma_H = Yc/4\pi$.

Denote by $\mathbf{R}^{(0)}$, $\mathbf{T}^{(0)}$ and $T_\Sigma^{(0)}$ matrices $\mathbf{R}, \mathbf{T}$ and $\mathbf{T}_\Sigma$ at $\Delta_A = \Delta_A^{(0)}$ and $X_K = X_K^{(0)}$ determined by (7.120), (7.122). The introduced matrices will differ from exact only at $|\mathrm{Re}k| \le k_*$.

Features of the R and T Matrices

The values of the transmitted and reflected fields depend on nine dimensional parameters:

$$c_A, \quad \Sigma_P, \quad \Sigma_H, \quad \sigma_g, \quad \sigma_a, \quad \omega, \quad x, \quad h, \quad H.$$

The consideration can be simplified by transferring to a dimensionless description. One can introduce six dimensionless parameters:

$$\bar{k}_A = k_0 h\sqrt{\varepsilon_m} = \frac{\omega}{c_A}h, \quad \bar{\tau}_K = k_0 h X_K,$$
$$\bar{d}_g = \frac{d_g}{h}, \quad \bar{H} = \frac{H}{d_g}, \quad \bar{x} = \frac{x}{h}, \quad \bar{k} = kh. \tag{7.123}$$

The matrix $\mathbf{R}^{(0)}$ and $\mathbf{T}^{(0)}$ in dimensionless variables may be written

$$\mathbf{R}^{(0)} = \begin{pmatrix} 1 + \dfrac{2\sqrt{\bar{k}_A^2 - \bar{k}^2}}{\bar{\Delta}_{SK}^{(0)}} & \dfrac{2Y\sqrt{\bar{k}_A^2 - \bar{k}^2}}{\Delta_A^{(0)}\bar{\Delta}_{SK}^{(0)}}\mathrm{sign}I \\ -\dfrac{2Y\bar{k}_A}{\Delta_A^{(0)}\bar{\Delta}_{SK}^{(0)}\sin I} & R_{AA}^{(0)} + \dfrac{2\bar{k}_A Y^2}{|\sin I|\left(\Delta_A^{(0)}\right)^2\bar{\Delta}_{SK}^{(0)}} \end{pmatrix}, \tag{7.124}$$

$$\mathbf{T}^{(0)} = \begin{pmatrix} T_{SS}^{(0)} & \dfrac{Y\,\mathrm{sign}\,I}{\Delta_A^{(0)}}T_{SS}^{(0)} \\ -\dfrac{\bar{k}_0 Y}{\bar{\Delta}_S^{(0)}\sin I}T_{AA}^{(0)} & T_{AA}^{(0)} \end{pmatrix}, \tag{7.125}$$

where

$$R_{AA}^{(0)} = \frac{X - \sqrt{\varepsilon_m}\,|\sin I|}{X + \sqrt{\varepsilon_m}\,|\sin I|}, \quad T_{AA}^{(0)} = \frac{2i\bar{k}_0\varepsilon_a}{\Delta_A^{(0)}\bar{k}\sinh\bar{k}}\,\frac{\bar{\Delta}_S^{(0)}}{\bar{\Delta}_{SK}^{(0)}}\,|\sin I|\,, \tag{7.126}$$

$$T_{SS}^{(0)} = -\frac{2}{\bar{\Delta}_{SK}^{(0)}\cosh\bar{k}\left[(k_0 h)^{-1}Z_g^{(m)} - i\tanh\bar{k}/\bar{k}\right]}, \tag{7.127}$$

$$\Delta_A^{(0)} = -\frac{X}{|\sin I|} - \sqrt{\varepsilon_m}, \qquad \bar{\Delta}_{SK}^{(0)} = k_0 h\frac{\zeta_4}{\zeta_3} - \left(\bar{\tau}_K + \sqrt{\bar{k}_A^2 - \bar{k}^2}\right) \tag{7.128}$$

where

$$\bar{\tau}_K = \bar{\tau}_D + \frac{k_0 h Y^2}{\widetilde{X}}, \quad \bar{\tau}_D = k_0 h X.$$

Equations (7.124), (7.125) were obtained for the right-hand half-plane on the physical sheet ($\mathrm{Re}\,k > 0$). In the left-hand half-plane, it is necessary to replace k by $|k|$. It can be seen that

$$\left|\bar{\Delta}_{SK}\right| \gtrsim 1.$$

This allows a simplified expression to be written immediately for the transformation coefficient of an incident Alfvén wave into the reflected Alfvén wave:

$$R_{AA} = R_{AA}^{(0)} = \frac{X - \sqrt{\varepsilon_m}\,|\sin I|}{X + \sqrt{\varepsilon_m}\,|\sin I|}. \tag{7.129}$$

In (7.129), the neglected term is of the order of

$$\left|\frac{k_0 h Y^2}{\left(X + \sqrt{\varepsilon_m}\right)^2}\,\frac{2\sqrt{\varepsilon_m}\sin I}{\Delta_{SK}}\right| \approx 2\bar{k}_A \ll 1.$$

It is taken into account here that $X \sim Y$, $\Delta_{SK} \sim 1$, $\sqrt{\varepsilon_m} = c/c_A \lesssim X, Y$. Equation (7.129) at $I = \pi/2$ transforms into the known expression for R_{AA} ([20], [21]).

The coefficient of transformation of an FMS-wave into a reflected Alfvén wave (R_{AS}) is small in the entire range of wavenumbers and is of the order of $\bar{k}_A = k_0\sqrt{\varepsilon_m}h$ at small $\bar{k}$ and $\bar{k}_A/\bar{k}$ at large $\bar{k}$.

Further, $R_{SS} = 1$ and $R_{SA} = 0$ at $\bar{k} = \bar{k}_A$. At larger wavenumbers R_{SS} approaches zero as $1/\bar{k}$. The coefficient $R_{SA} \sim \bar{k}_A$ at small $\bar{k}$ ($|\bar{k}| \ll \bar{k}_A$). At large k

$$R_{SA} \approx -\frac{2Y}{\Delta_A^{(0)}} \operatorname{sign} I. \tag{7.130}$$

The transition from small values R_{SA} to (7.130) is determined by the concrete model of the ground. For instance, in the case when the skin depth d_g is larger than the atmosphere thickness h, R_{SA} begins to reach (7.130) at $|kd_g| \approx 1$.

For transmitted Alfvén waves (electric modes) $|T_{AA}| \ll 1$ and $|T_{AS}| \ll 1$. For instance,

$$|T_{AA}| \approx \frac{2k_0 h\, |\varepsilon_a| \sin^2 I}{\widetilde{X} k \sinh \bar{k}} = 2\frac{|k_* h|^2}{\bar{k}\sinh \bar{k}}. \tag{7.131}$$

For estimates of $|k_*|$ see (7.121).

Let the Earth's conductivity be high, so that the skin depth d_g is small in comparison with the atmosphere's height h. Consider the wavenumbers meeting the condition $h^{-1} \ll k \ll d_g^{-1}$ and $k \gg k_A$. Then, $\tanh \bar{k}/\bar{k} \approx 1$, $\sqrt{(\bar{k}_A^2 - \bar{k}^2)} = i\bar{k}$, $Z_g^{(m)} \approx 0$. In this approximation (7.117) gives $\zeta_4/\zeta_3 \approx -ik/k_0$. Substituting last equations into (7.125) we get

$$T_{SS}^{(0)} = \frac{2\bar{k}e^{-\bar{k}}}{\bar{k} - i\bar{\tau}_K/2} \tag{7.132}$$

$T_{SA}(k)$ differs from $T_{SS}(k)$ only by the factor $\operatorname{sign} I\ Y/\Delta_A^{(0)}$. Hence,

$$T_{SA} \approx -\frac{2Y \sin I}{X + \sqrt{\varepsilon_m}|\sin I|}\, \frac{\bar{k}e^{-\bar{k}}}{\bar{k} - i\bar{\tau}_K/2}. \tag{7.133}$$

It can be seen that $|T_{SA}|$ goes down exponentially at large $\bar{k}$.

It follows from (7.131) and (7.133) that an incident Alfvén wave is transformed to a magnetic mode. This means that the initial magnetic field vector turns by $\pi/2$ counterclockwise (as seen along $\mathbf{B}_0$). Recall that this conclusion at $k_y = 0$ is independent on the inclination angle of $\mathbf{B}_0$. One can argue that in the limit $\omega \to 0$, the system of 2D-inclined field-aligned currents does not contribute to the magnetic field under the ionosphere at $\Sigma_H = 0$.

The obtained results are important in interpreting observations of the ULF electromagnetic fields conducted by low-orbit satellites: a longitudinal component appears in the reflected FMS-waves whose amplitude can become commensurate with the incident magnetic field of an Alfvén wave (see also

Chapter 3). The appearing FMS-waves damp exponentially at the characteristic scale of k^{-1}. Also note here that despite $R_{SA} \approx 1$, the energy flux transformed from the incident Alfvén wave into an FMS is small because the electric field in an FMS reflected wave is small and is of the order of $k_0 R_{SA}/k$, and in an Alfvén wave it is $\approx R_{AA}/\sqrt{\varepsilon_m}$.

7.7 Small-Scale Perturbations

One of the conditions for the applicability of the thin ionosphere model, viz smallness of horizontal wavenumbers $k_\tau l_I \ll 1$, is not always satisfied. For instance, FLR-oscillations can have a horizontal scale of about 10 km at the ionosphere level. In that case, the approximation of large wavenumbers is useful. In this section we want to obtain an uncoupled equation for Alfvén waves at $k_1 \to \infty$.

As distinct from Chapter 6, where an uncoupled equation was obtained within the same limit for Alfvén waves propagating in transversely inhomogeneous plasma, we shall regard the plasma as horizontally homogeneous, but Hall conductivity and the change in plasma characteristics with altitude will be taken into account. The obtained equations are not always valid and not applicable for an arbitrarily small horizontal perturbation scale.

The reason is that we did not take into account the interaction of Alfvén and FMS-waves with other oscillation branches remarkable at some small scales, let us say, less than the electron inertial scale $\lambda_e = c/\omega_{pe}$. The sequential derivation of asymptotic equations for small k_1^{-1} is omitted here because it is very cumbersome. Here we shall turn not fully faithfully. We extract a small parameter, writing equations in the zeroth-order approximation with respect to this parameter. And then we shall check that the corrective terms to the obtained equations are small. The introduced earlier designations are

$$k_0 = \frac{\omega}{c}, \quad X = \frac{4\pi}{c}\Sigma_P = \frac{4\pi}{c}\int \sigma_P\, dz, \quad Y = \frac{4\pi}{c}\Sigma_H = \frac{4\pi}{c}\int \sigma_H dz.$$

σ_P, σ_H and Σ_P, Σ_H are the specific and integral Pedersen and Hall conductivities of the ionosphere, respectively. It can be shown that when

$$k_2 \ll k_1 \tag{7.134}$$

and the following condition is satisfied:

$$\frac{k_0 Y^2}{k_1 X} \ll 1 \tag{7.135}$$

in a zero approximation for this parameter, the full wave equation in the ionosphere are split into equations for Alfvén and FMS-oscillations. We shall write these equations in an oblique coordinate system (x^1, x^2, x^3).

Alfvén Waves

In a zero approximation the equations for Alfvén oscillations have the form

$$\frac{dE_A}{dx^3} = ik_0 \frac{b_A}{\sin I}, \qquad \frac{db_A}{dx^3} = -\frac{4\pi\sigma_P}{c}\frac{E_A}{\sin I}, \tag{7.136}$$

where

$$E_A = \frac{1}{\sin I}E_1, \qquad b_A = b_2,$$

and k_i, E_i, b_i are the covariant components in an oblique coordinate system. Equation (7.136) are supplemented at the lower boundary of the ionosphere by the condition of no vertical current ($j_n = 0$) equivalent to the requirement

$$b_A|_{z=-0} = 0. \tag{7.137}$$

FMS-Waves

The FMS-waves are described by

$$\left(\frac{d}{dx^3} - ik_1 \cot I\right) E_S = -ik_0 b_S,$$
$$\left(\frac{d}{dx^3} - ik_1 \cot I\right) b_S = \left(\frac{4\pi\sigma_P}{c} + i\frac{k^2}{k_0}\right) E_S - \frac{4\pi\sigma_H}{c}E_A, \tag{7.138}$$

where

$$E_S = E_2, \qquad b_S = b_1, \qquad k^2 = k_1^2.$$

The procedure for determining the ionospheric distribution of electric and magnetic fields produced by the incident Alfvén waves is as follows. Firstly, we define the ionospheric fields caused by the Alfvén wave itself making use of (7.136) with the boundary condition (7.137). Then we define the Hall currents initiated by the ionospheric Alfvén electric field. The term proportional to E_A in (7.138) is the Hall current produced by the electric field of an Alfvén wave. In turn, the induced Hall current generates an FMS-wave (7.138) with the electric field $\mathbf{E_S}$ perpendicular to $\mathbf{E_A}$. In the next approximation, $\mathbf{E_S}$ gives a correction to $\mathbf{E}_A = \mathbf{E}_A^{(0)}$ of the order of

$$E_A^{(1)} \approx \frac{k_0}{k_1}\frac{Y^2}{X}E_A. \tag{7.139}$$

The approximation can be used if this a correction is small. We shall restrict our consideration to a zero approximation for the parameter (7.135). The same equation is correct both in the atmosphere and within the ground, but with a corresponding replacement of σ_P either by atmospheric conductivity σ_a or by the ground conductivity σ_g.

The fields at the ground surface are found after the altitude-dependence of $\mathbf{E}$ and $\mathbf{b}$ in the ionosphere is determined. This requires that numerical integration of (7.136) and (7.138) be carried out twice for each harmonic. Converting the boundary conditions from $-\infty$ to the upper surface of the ionosphere, we find the reflection coefficients and the field over the ionosphere. Using the determined fields as the initial data, by numerical integration of (7.136), (7.137), we determine the fields on the ground for fixed k_1.

If

$$T > 20 - 30\,\mathrm{s} \tag{7.140}$$

it is possible to avoid mentioning the tedious computations. We will use the thin ionosphere approximation. For (7.136) this presents no difficulties with the single limitation (7.140) on the periods of oscillations because the coefficients of (7.136) do not depend on the horizontal wavenumber k_1. It immediately follows from (7.136) that the electric field is conserved along a field-line. In the computation of Alfvén waves this makes it possible to replace the whole ionosphere by a thin layer with the integral Pedersen conductivity Σ_P and use the boundary conditions for such a layer in the form

$$\mathbf{b}_A = -X\,\mathbf{E}_A|_{x=+0}\,. \tag{7.141}$$

For arbitrary T, it is necessary to take into account the dependence of $\mathbf{B}_0$ and plasma density on altitude. For example, let c_A in the F-layer be 300 km/s and, therefore, the wavelength of the $T \approx 1$ s oscillations becomes comparable with the thickness of the F-layer. However, for the low-frequency range features of the height distribution of c_A within the ionosphere are of little importance.

Henceforth, we will not take into account changes in c_A in the upper ionosphere and magnetosphere and return to the model of the homogeneous half-space. Omitting intermediate computations, we obtain for R_{AA} from (7.136) the same relation as (7.129):

$$R_{AA} \approx R_{AA}^{(0)} = \frac{X - \sqrt{\varepsilon_m}\,|\sin I|}{X + \sqrt{\varepsilon_m}\,|\sin I|}. \tag{7.142}$$

Now we shall return to (7.138) which describes FMS-waves. The main difference between FMS-waves and Alfvén waves is the independence of the phase velocity value on the angle between the direction of wave propagation and $\mathbf{B}_0$, that is, the $\mathbf{B}_0$ does not form a specific direction. The inclined system connected with $\mathbf{B}_0$ has no advantage, therefore, we return to a Cartesian coordinate system.

We express the components of the field of FMS-waves through the potential Ψ. Let us present (7.16) in the form

$$\mathbf{E}_\tau = -k_0\Psi\mathbf{k}\times\hat{\mathbf{z}}, \qquad \mathbf{b}_\tau = i\mathbf{k}\frac{d\Psi}{dz},$$

where

$$\frac{d\Psi}{dz} = \frac{d\Psi}{dx^3} - ik_1\cot I\Psi,$$

$\hat{\mathbf{z}}$ is the unit vector of the z-axis, $\mathbf{E}_\tau$, $\mathbf{b}_\tau$ are the horizontal components of the $\mathbf{E}$ and $\mathbf{b}$ vectors. For the potential Ψ, from (7.138) we have

$$\frac{d^2\Psi}{dz^2} + \left(k_0^2\varepsilon_\perp - k^2\right)\Psi = -\frac{4\pi\sigma_H}{ik_1 c}E_A, \tag{7.143}$$

where $\varphi = -E_A \sin I/ik_1$, $\varepsilon_\perp = i4\pi\sigma_P/\omega$ in the ionosphere. Equation (7.143) is supplemented by the boundary conditions of the potential Ψ at $z \to \infty$ and the impedance condition at the ground surface

$$\left.\left(\frac{d\Psi}{dz} + \frac{ik_0}{Z_g^{(m)}}\Psi\right)\right|_{z=-h} = 0, \tag{7.144}$$

where $Z_g^{(m)}$ is the surface impedance (7.27).

We will examine disturbances with a horizontal scale less than the wavelengths in the magnetosphere (m), ionosphere (i) and the atmosphere (a), that is, with the wavenumbers k satisfying the conditions

$$k \gg k_0|\varepsilon_{m,i,a}|^{1/2}. \tag{7.145}$$

For the ionosphere, it can be rewritten in terms of the Pedersen skin depth $d_P = c/\sqrt{2\pi\omega\sigma_P}$ as

$$kd_P \gg 1.$$

For frequency $\omega < 3\times10^{-1}\,\mathrm{s}^{-1}$ and $\sigma_P = 10^7\,\mathrm{s}^{-1}$ the skin depth $d_P \gtrsim 100\,\mathrm{km}$ and the last inequality is reduced to $k \gtrsim 10^{-2}\,\mathrm{km}^{-1}$. Similar reasonings for the atmosphere lead us to the conclusion that on numerous occasions the magnetostatic approximation is adequate for the analysis of propagation of FMS-waves in the atmosphere-ionosphere-magnetosphere system.

Thus, computations of the field with an explicit indication of the ionospheric source can be carried out in the following way:

$$\frac{d^2\Psi}{dz^2} - k^2\Psi = -\frac{4\pi\sigma_H}{ik_1 c}E_A, \tag{7.146}$$

with the boundary conditions (7.144) on the ground surface. The Green's function $G(z, z_1)$ of (7.146) is

$$G(z, z_1) = \frac{1}{2k}\exp\left(-k\,|z - z_1|\right) - R_g\frac{\exp\left[-k(z + z_1 + 2h)\right]}{2k}, \tag{7.147}$$

where

$$R_g = \frac{1 - ikZ_g^{(m)}/k_0}{1 + ikZ_g^{(m)}/k_0}.$$

Then the solution of (7.146) is given by

$$\Psi(z) = i\frac{2\pi}{c}\frac{E_A}{k_1 k}\int \sigma_H(z_1)\left(e^{-k|z-z_1|} - R_g e^{-k(z+2h+z_1)}\right)dz_1, \tag{7.148}$$

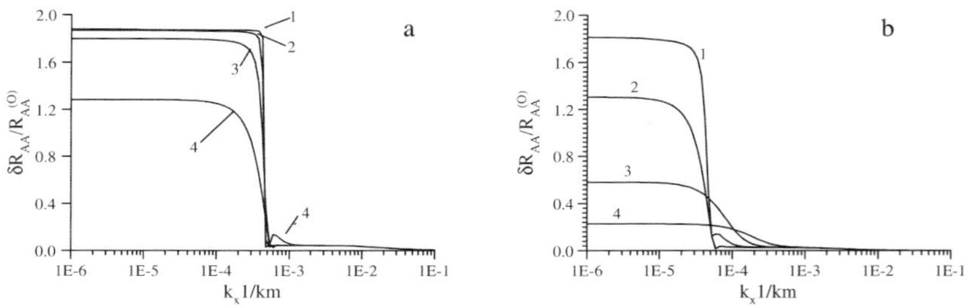

Fig. 7.4. Comparison of the analytical (by (7.129)) and numerical results for $T = 10\,\mathrm{s}$ (a) and $T = 100\,\mathrm{s}$ (b)

Here integration is extended to the ionospheric region occupied by Hall currents. We expand the integrand exponent into a series. Then the magnetic field caused by a height distributed current is equivalent to the field of a thin current layer located at the altitude

$$h = \frac{\int z_1 \sigma_H(z_1)\, dz_1}{\int \sigma_H(z_1)\, dz_1}. \tag{7.149}$$

In another limiting case, the transformation to the thin ionosphere approximation, as in (7.145), is done directly in the initial equations.

Thus, in problems of ionospheric propagation of low-frequency MHD-waves, it is possible to use the thin ionosphere approximation for the range of wavenumbers $|\mathbf{k}| < L^{-1} \approx 10^{-1}\,\mathrm{km}^{-1}$.

7.8 Numerical Examples

Small-Scale Approximation

We begin discussing the results of numerical calculations by estimating the errors $\delta R_{AA} = R_{AA} - R_{AA}^{(0)}$ of approximation (7.129) for various k_x, and k_y. In Figures 7.4a (for $T = 10\,\mathrm{s}$) and Fig. 7.4b (for $T = 100\,\mathrm{s}$) dependencies of ratio $\delta R_{AA}/R_{AA}^{(0)}$ on k_x are shown for various values of k_y. The curve 1 corresponds to $k_y = 10^{-5}\,\mathrm{km}^{-1}$, 2 is to $0.316 \times 10^{-4}\,\mathrm{km}^{-1}$, 3 is to $10^{-4}\,\mathrm{km}^{-1}$, and 4 is to $0.316 \times 10^{-3}\,\mathrm{km}^{-1}$. It can be seen that the error in determining R_{AA} by (7.129) is not more than 2% at $T = 10\,\mathrm{s}$ for $k_x > 4 \times 10^{-4}\,\mathrm{km}^{-1}$ and at $T = 100\,\mathrm{s}$ for $k_x > 2 \times 10^{-4}\,\mathrm{km}^{-1}$.

Figures 7.5a and 7.5b present amplitude (7.5a) and phase (7.5b) dependencies of $b_3 = b_{\|} \sin I$ from k_x above the ionosphere (curves $1, 2$) and b_z on the ground $(3, 4)$ on k_x at $k_y = 0$. The curves in Figures 7.5a,b are computed for the period $T = 30\,\mathrm{s}$, and in Fig. 7.5c for $T = 300\,\mathrm{s}$. Curves $2, 4$ correspond to small-scale approximation and curves $1, 3$ correspond to the exact

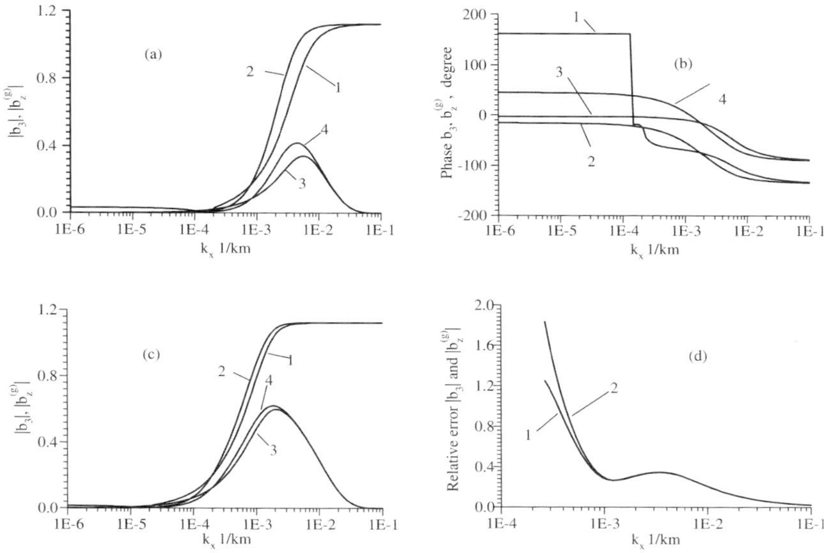

Fig. 7.5. Longitudinal magnetic field above the ionosphere (1, 2) and the vertical on the ground (3, 4) as a function of the horizontal wavenumber. The amplitude (a) and phase (b) for $T = 30$ s. Curves 1, 3 are numerical results, 2, 4 is by the small-scale approximation formula. (c) is the same as (a) but for $T = 300$ s. (d) is the relative error

solution. An error of the small-scale approximation is shown in Fig. 7.5d as well. The other parameters are indicated in Table 7.3. It can be seen that at $k_x \gg k_0 X$ the approximation equation yields good results. For instance, at $k_x > 10^{-2}\,\mathrm{km}^{-1}$, $k_0 X = 3 \times 10^{-2}\,\mathrm{km}^{-1}$ for $T = 30$ s, the error less than a few percents.

The calculations conducted for different models showed that the small-scale approximation can be used in calculating electric and magnetic fields in the ULF for wavenumbers $k_x \gtrsim 10^{-2}\,\mathrm{km}^{-1}$.

Table 7.3. Magnetospheric wavelength $l_m = (k_0\sqrt{\varepsilon_m})^{-1}$ km, ground skin depth d_g and wavenumber k_* for periods $T = 30, 120$ and 600 s

Σ_P km/s	0.116×10^9	$\sigma_g\ \mathrm{s}^{-1}$	1×10^6
Σ_H km/s	0.136×10^9	c_A km/s	1×10^3
h km	100	I	$\pi/2$
T, s	30	120	600
l_m, km	1.4×10^6	5.7×10^6, km	2.9×10^7
d_g, km	260	520	1170
k_*, km^{-1}	1.2×10^{-6}	6×10^{-7}	2.6×10^{-7}

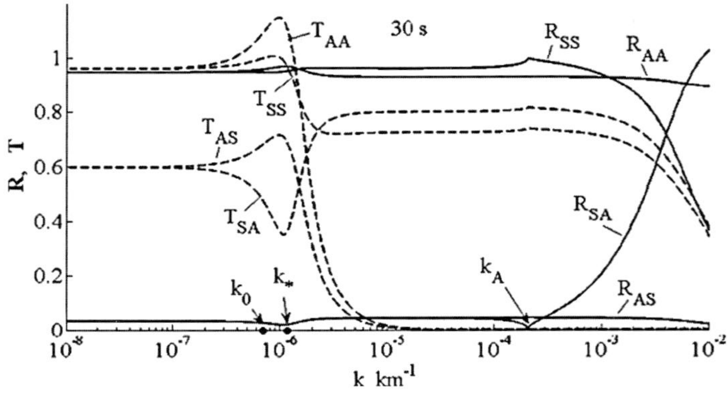

Fig. 7.6. Dependencies of the reflection **R** and transmission **T** matrices on the horizontal wavenumber k for the dayside ionosphere. The second subindex refers to the incident wave (A = Alfvén wave, S = FMS-wave), the first one refers to a converted wave. For instance, R_{SA} is reflection coefficient of the incident Alfvén wave into the reflected FMS-wave. The wave period $T = 30\,\mathrm{s}$, and other parameters are given in Table 7.3

Meridional Propagation

Figure 7.6 demonstrates various elements of **R** and **T** matrices depending on horizontal wavenumber $k = k_x$ for pulsations with periods $T = 30\,\mathrm{s}$. The curves for other longer periods T are not shown but are very similar to those of Fig. 7.6. The parameters for this example are indicated in Table 7.3. The ground is a half-space ($H = \infty$), $X = 4\pi\Sigma_P/c = 4.86 \times 10^3$, and $Y = 4\pi\Sigma_H/c = 5.7 \times 10^3$.

The first thing to pay attention to is the steep decrease of T_{AS} and T_{AA} at $k \gtrsim k_*$. It means that a considerable contribution to the ground field in the electric mode will be made only by spatial harmonics with $k \lesssim k_*$:

$$T_{SA} \approx \text{const}, \quad T_{SS} \approx \text{const}, \text{ at } k_* < k < h^{-1}$$

and damp exponentially at $k > h^{-1}$.

Comparison of the approximation formula (7.129) for $R_{AA}(k)$ and numerical results presented in Fig. 7.6 shows that (7.129) is indeed valid in a wide range of horizontal wavenumbers ($k > k_*$). Minor differences between calculated values and those computed by (7.129) correspond to correction terms discarded in (7.129).

Dependencies of the vertical magnetic ground component $b_z(k)$ for different periods are shown in Fig. 7.7. Panel (a) is the amplitude of b_z, (b) is its phase. The curves $1, 2, 3, 4$ correspond to $30, 100, 300$, and $1000\,\mathrm{s}$ respectively. One can see that there is a maximum in the spatial spectrum shifting with the oscillation period.

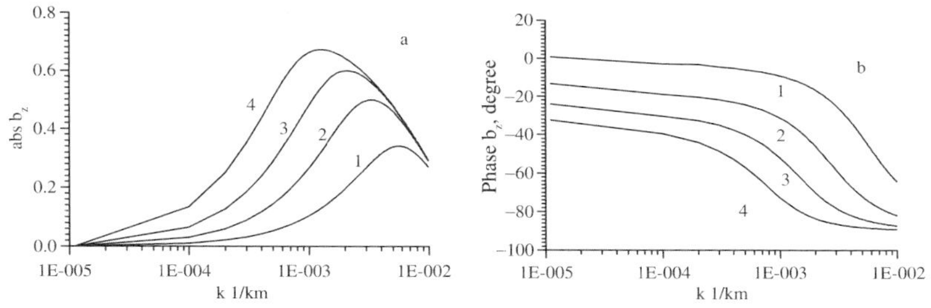

Fig. 7.7. The ground vertical magnetic field. The curves 1, 2, 3, 4 correspond to $T = 30, 100, 300, 1000\,\mathrm{s}$, $k_y = 0$

Non-Meridional Propagation

Inclined Geomagnetic Field

The change in amplitude and phase of R_{AA} depending on k_y for series of fixed values k_x is depicted in Fig. 7.8. Curves 1, 2, 3, 4 refer to $k_x = 10^{-5}, 10^{-4}, 10^{-3}$, $10^{-2}\,\mathrm{km}^{-1}$, respectively. The calculations were carried out for $\Sigma_P = 0.116 \times 10^9\,\mathrm{km/s}$, $\Sigma_H = 0.139 \times 10^9\,\mathrm{km/s}$, $c_A = 10^3\,\mathrm{km/s}$, $h = 100\,\mathrm{km}$, and $I = \pi/4$. The ground is a half-space with $\sigma_g = 9 \times 10^8\,\mathrm{s}^{-1}$. Note a deep minimum in the value of R_{AA} and a sharp change in phase at $k_y = 10^{-4}\,\mathrm{km}^{-1}$. Since perturbations of horizontal scale greater than 10^4 km are not interesting to ionospheric physics, then R_{AA} may be considered as a constant.

Figure 7.9 demonstrates the dependence of the reflection coefficient R_{AA} of the Alfvén wave and the coefficient of its transformation R_{SA} into the FMS-wave on the integral Pedersen conductivity of the ionosphere. The Alfvén velocity $c_A = 1100\,\mathrm{km/s}$ and the high-conductive ground half-space with conductivity $\sigma_g = 9 \times 10^8\,\mathrm{s}^{-1}(0.1\,\mathrm{Ohm}^{-1})$. The inclination $I = 20°$ (dashed lines) and $I = 60°$ (solid lines). $T = 100\,\mathrm{s}$ and meridional and azimuthal wavenumbers $k_x = 10^{-2}\,\mathrm{km}^{-1}$ and $k_y = 10^{-3}\,\mathrm{km}^{-1}$, respectively.

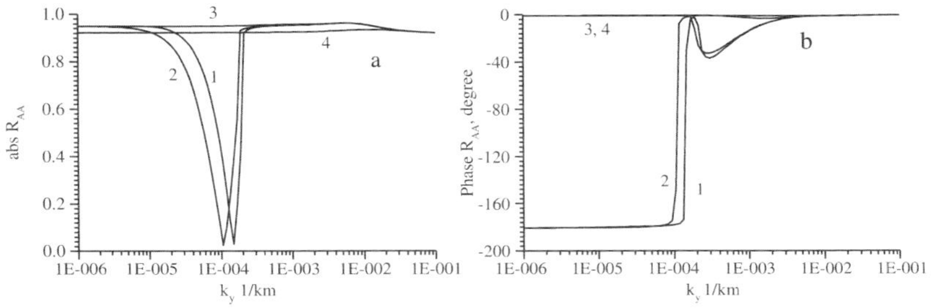

Fig. 7.8. Absolute value (a) and phase (b) of the reflection coefficient R_{AA} as a function of k_y for fixed values $k_x = 10^{-5}$ (1), 10^{-4} (2), 10^{-3} (3), and $10^{-2}\,\mathrm{km}^{-1}$ (4). Dip angle $I = \pi/4$

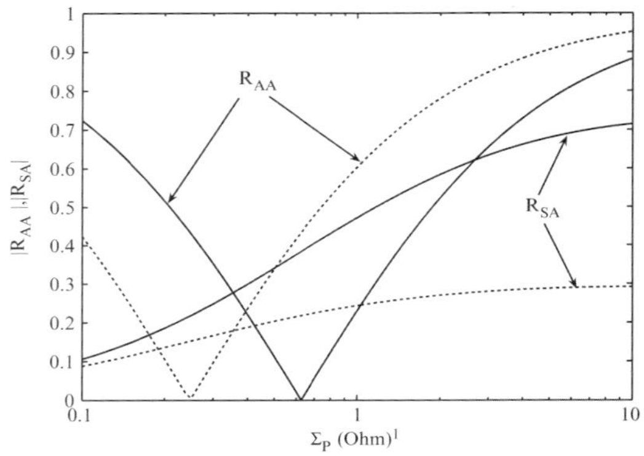

Fig. 7.9. Non-meridional propagation. Coefficients R_{AA} and R_{SA} for the inclined magnetic field $\mathbf{B}_0$ versus the Pedersen conductivity Σ_P. The Hall conductivity $\Sigma_H = \Sigma_P$. Dashed lines and solid lines correspond to inclination $I = 20°$ and $I = 60°$, respectively. The other parameters are $k_x = 0.01\,\mathrm{km}^{-1}$, $k_y = 0.001\,\mathrm{km}^{-1}$, $f = 2\pi/100\,\mathrm{Hz}$, $\sigma_g = 0.1\,\mathrm{Ohm}^{-1}$, $c_A = 1100\,\mathrm{km/s}$, $h = 100\,\mathrm{km}$

The steep decreasing of $R_{AA}\,(\Sigma_P)$ takes place if Σ_P tends to $\Sigma_P^* = \Sigma_A \sin I$ (see (7.129)). Here the Alfvén wave is not reflected from the ionosphere [15]. It follows from Fig. 7.9 that $R_{AA} \approx 0$ when $\Sigma_P = \Sigma_P^* \approx 0.63\,\mathrm{Ohm}^{-1}$ for $I = 60°$. The same value of Σ_P^* follows from (7.129). Transition from the large $\Sigma_P \gg \Sigma_P^*$ to the small $\Sigma_P \ll \Sigma_P^*$ leads to the change of R_{AA} from 1 by day to -1 by night and to the phase change of the reflected wave at the point Σ_P^* from 0 to 180°. Direction of the magnetic vector of the reflected wave changes to the opposite one when going from day ($\Sigma_P > \Sigma_P^*$) to night ($\Sigma_P < \Sigma_P^*$) conditions. The total horizontal magnetic field $\mathbf{b}_\tau$ at the ionosphere then changes from $|\mathbf{b}_\tau| \approx 2b_A^{(i)}$ to $|\mathbf{b}_\tau| \ll b_A^{(i)}$. It means that the dayside ionosphere is similar to a perfect conductor and the nightside ionosphere to a dielectric for the Alfvén wave. Figure 7.9 also points out the important fact that the zero point for the reflection coefficient R_{AA} disappears in the low-latitudes because of Σ_P^* decreasing.

At $\Sigma_P > \Sigma_P^*$, the $\mathbf{b}_\tau$ transformation coefficient R_{SA} of the incident Alfvén wave into the reflected FMS-wave tends to $R_{SA} \propto \sin I$ and at $\Sigma_P < \Sigma_P^*$ the dependence of R_{SA} on inclination I weakens. This result can be found from the analysis of the approximate formula (7.142) for the reflection coefficient. For simplicity, let us consider the case of quasi-meridional propagation $|k_y| \ll |k_x|$. The electric field of the incident Alfvén wave is $E^{(i)} = -b^{(i)}/\sqrt{\varepsilon_m}$, and the electric field of the reflected wave is $E^{(r)} = R_{AA} b^{(i)}/\sqrt{\varepsilon_m}$. Then, the total electric field is $\approx -(1-R_{AA})b^{(i)}/\sqrt{\varepsilon_m}$. Substituting the approximate reflection coefficient $R_{AA}^{(0)}$ from (7.142), we have the electric field of the Alfvén wave:

$$E_A \approx -\frac{2\,|\sin I|}{\widetilde{X}} b_A^{(i)}, \tag{7.150}$$

with $\widetilde{X} = X + \sqrt{\varepsilon_m}\,|\sin I|$. The Hall current caused by this field is

$$I_y \approx \frac{2\Sigma_H \sin I}{\widetilde{X}} b_A^{(i)}.$$

Then, the horizontal magnetic components excited by this current are

$$\begin{aligned}
b_{Sx}^{(r)} &\approx P\left(1 - R_g e^{-2kh}\right) b_A^{(i)} && \text{at } z = +0,\\
b_{Sx}^{a} &\approx -P\left(e^{kz} + R_g e^{-k(z+2h)}\right) b_A^{(i)} && \text{at } -h \le z < 0,\\
b_{Sx}^{(g)} &\approx -Pe^{-kh}\left(1 + R_g\right) b_A^{(i)} && \text{at } z = -h,
\end{aligned} \tag{7.151}$$

where

$$P = \frac{Y \sin I}{\widetilde{X}}, \quad R_g = \frac{k_0/ik - Z_g}{k_0/ik + Z_g}.$$

In the dispersion equation of the FMS-wave, k_A^2 can be neglected and then it is written as $k_x^2 + k_z^2 = 0$ (since y–component of the wavevector is neglected). The substitution of $k_z = \pm i k_x$ into $\nabla \cdot \mathbf{b} = 0$ gives the polarization $b_z/b_x = \pm i$. Hence, the magnetic field of the FMS-wave is circularly polarized in the meridional plane. For upgoing wave the polarization is $+i$. The b_x and b_z are in the quadrature and have the same amplitude. Thus, the component of the magnetic field of the FMS-wave transversal to $\mathbf{B}_0$ is $b_{x\prime} = i\exp(iI)\,b_x$. Then (7.151) may be written (since $R_{SA} = b_{x'}/b_A^{(i)}$) as

$$R_{SA} \approx i\exp(iI)\,P\left(1 - R_g e^{-2kh}\right) \tag{7.152}$$

Let $\Sigma_P > \Sigma_P^*$, then (7.152) and $\widetilde{X} \approx X$ gives the reflection coefficient

$$R_{SA} \approx i\exp(iI)\,\frac{Y \sin I}{X}\left(1 - R_g e^{-2kh}\right).$$

For $\Sigma_P < \Sigma_P^*$, $\widetilde{X} \approx \sqrt{\varepsilon_m}\sin I$ and

$$R_{SA} \approx i\exp(iI)\,\frac{Y}{\sqrt{\varepsilon_m}}\left(1 - R_g e^{-2kh}\right)$$

with a weak dependence of the transformation coefficient on I.

It is seen from Fig. 7.9 that the efficiency of FMS-waves excitation by incident Alfvén waves is higher at big magnetic field inclination I and high conductivities.

The electric field in the Alfvén wave is the most sensitive to ionospheric conductivity changes. Component-wise dependencies of the total ionospheric electric fields on Σ_P are shown in Fig. 7.10a. The solid lines correspond to the initial Alfvén wave, and dashed lines to the FMS-wave. The inclination angle $I = 60°$. The intensity of the total electric field is in the inverse proportion to Σ_A in the Alfvén wave (see (7.150)). The maximum ionospheric

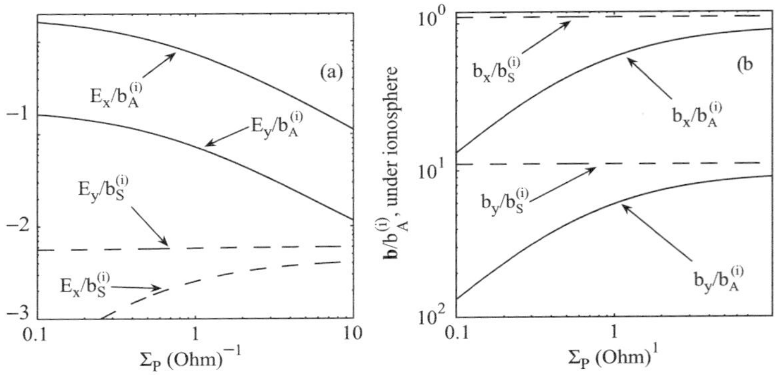

Fig. 7.10. (a) Dependencies of the E_x and E_y components of the electric field in the ionosphere on its integral Pedersen conductivity. (b) b_x and b_y components under the ionosphere as functions of the integral Pedersen conductivity Σ_P. Solid lines correspond to incident Alfvén waves. Dashed lines correspond to FMS-waves. For a description of the medium model, see the caption to Fig. 7.9

electric field caused by such waves should appear on the nightside of the ionosphere.

Figure 7.10b shows the dependence of the horizontal magnetic component in the atmosphere, under the ionosphere, on Σ_P. It can be seen that ionospheric conductivity does not influence the magnitude of the magnetic component of the FMS-wave (dashed lines). When the amplitude of the initial magnetic component $b^i = 1$, the magnetic field b^a under the ionosphere is a little bit more than 1. The atmospheric magnetic component produced by the Alfvén wave b_x (b_y) is generally caused by the Hall current j_y (j_x), which is generated due to the E_x (E_y) component of the incident Alfvén wave. The total magnetic field of the Alfvén wave in the ionosphere depends strongly on the ionospheric conductivity. As it can be seen from Fig. 7.10b, it changes in roughly 3 orders for change Σ_P in 2 orders. It is in the consent with the estimation (7.150) of the total electric field of the Alfvén wave. Transition from the well conductive dayside ionosphere to the nighttime one which has very small conductivity leads to decreasing of b_x and b_y components by a factor of ≈ 0.14 (see (7.151)).

The very low conductivity of the atmosphere means that the currents are very small. Then $(\nabla \times \mathbf{b})_z = 4\pi/c\ j_z \approx 0$ for the magnetic fields under the ionosphere, produced as the Alfvén waves incidence, as and the FMS-waves incidence. Hence, $k_x b_y - k_y b_x \approx 0$ and $b_x/b_y \approx k_x/k_y$. In the example shown in Fig. 7.10, the ratio $b_x/b_y \approx 10$, i.e., orientation of the horizontal magnetic field under ionosphere does not depend on the ionospheric conductivity. However, this result is valid only for the horizontally homogeneous ionosphere. Ionospheric irregularities can result in change of the wave-polarization b_x/b_y .

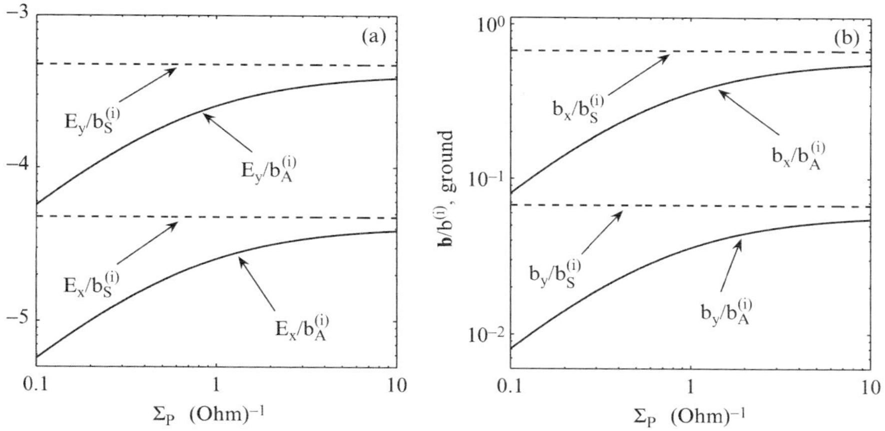

Fig. 7.11. Dependence of the electric field on the ground on Σ_P. Designations are the same as in Fig. 7.9. Dependence of magnetic components of the Alfvén and magnetosonic waves on Σ_P on the ground. See also the caption to Fig. 7.9

Dependency of horizontal electric fields on the ground as a function of Σ_P is shown in Fig. 7.11a. It is the same as the dependence of b_a. Electric fields in the Alfvén wave on the ground and in the ionosphere are turned over $\pi/2$ contrary to the FMS-wave.

By comparing values of the induced electric components in the ionosphere and in the ground and corresponding electric currents it is easy to understand why the ionosphere does not influence the magnitude of the atmospheric and ground magnetic components of the FMS (dashed lines in Fig. 7.10b and Fig. 7.11b). The skin depth in the conductive ground is

$$d_g\,(\text{km}) \approx 0.503\sqrt{\frac{T\,(\text{s})}{\sigma_g\,(\text{Ohm}^{-1}\text{m}^{-1})}}.$$

For $\sigma_g = 0.1\text{Ohm}^{-1}\text{m}^{-1}$ and $T = 100\,\text{s}$, we have $d_g \approx 16\,\text{km}$. A total current inside the ground $|J^{(g)}| = d_g\sigma_g|E^{(g)}|/\sqrt{2}$, and in the ionosphere $|J^{(I)}| = \Sigma_P|E^{(I)}|$. The electric field in the ground $E^{(g)}$ (Fig. 7.11a) induced by the FMS–waves with amplitude of incident wave $b^{(i)} = 1\,\text{nT}$ and $T = 100\,\text{s}$ is $|E^{(g)}| \approx 4.76\times10^{-4}\,\text{mV}/\text{m}$. The ionospheric electric fields (Fig. 7.10a) are $|E^{(I)}| \approx 5.55\times10^{-3}$ and $5.97\times10^{-3}\,\text{mV}/\text{m}$ for $\Sigma_P = 0.1$ and $\Sigma_P = 10\,\text{Ohm}^{-1}\text{m}^{-1}$, respectively. Hence the current in the ground and in the ionosphere are $J^{(g)} \approx 0.54\,\text{mA}/\text{m}$, $J^{(I)} \approx 5.55\times10^{-4}$ and $5.97\times10^{-3}\,\text{mA}/\text{m}$, respectively. The ionospheric current is essentially less than the ground current. Therefore, the change from day to night conditions does not influence the magnitude of the magnetic component in the atmosphere (Fig. 7.10b) and on the ground surface (Fig. 7.11b). The homogeneous ionosphere, is transparent for the FMS–waves and ionospheric current gives inessential contribution into the magnetic field. The reflected magnetic field, and consequently, magnetic field on the ground and under the ionosphere are determined first of all by currents inside the ground.

In the Alfvén wave the magnetic field under the ionosphere appears to be due to the ionospheric Hall currents which are the source of the magnetotelluric field [4]. Comparing the solid curves in Fig. 7.11b and Fig. 7.10b one can see that the magnetic field under the ionosphere in this case is greater approximately by a factor of 1.64 than that on the ground. Estimation by (7.151) gives the same ratio for the field amplitudes.

7.9 Discussion

There are two points that emerge from the theoretical consideration. The first is that fundamental distinction exists between propagation of vertical and inclined MHD-waves. The approximation of the normally incident wave is equivalent to the assumption of independence of the field component on the transversal coordinate, i.e. the infinite transversal scale is assumed. It is obvious that there is a natural horizontal scale, the Earth's radius, which defines the largest transversal wave-scale. A finite horizontal scale or the decline of the wave-vector results in the total reflection because of the strong discontinuity of the phase velocities at the magnetosphere-atmosphere boundary. The wave propagates in the magnetosphere with the Alfvén velocity of $c_A \sim 10^3$ km/s, and in the atmosphere with light velocity $c = 3 \times 10^5$ km/s. As a result, the wave front rotates back to the magnetosphere even for very small inclination of the wave vector.

Under the ionosphere, the wave field decreases exponentially at the distances of order of the horizontal wave-scale size. The wave reflected totally from the ionosphere can be detected at distances from the ionosphere not larger than the transversal scale of the wave. When oscillations are observed on the ground it means that these oscillations are in fact a tail of the MHD-wave exponentially attenuating in the atmosphere.

The second is that only the FMS-mode can penetrate the atmosphere. This follows directly from the polarization structure of the Alfvén wave which contains field-aligned electrical current. This current, meeting on its way the non-conductive atmosphere, spreads over the ionosphere creating the ionospheric current system which is the same as large-scale current systems produced by the magnetospheric processes with the exception of its configuration and intensity.

Taking into account the finite conductivity of the atmosphere has no influence on the general conclusion about ionospheric screening of the Earth for the Alfvén waves with the exception of the rather exotic wave-like TEH mode. For discussion of some features of this wave, regularities of its propagation and generation refer to Chapter 8.

The magnetic component of the wave reached the ground rotated on $\pi/2$ compared to the magnetic component of the initial Alfvén wave. One can see at once that this is defined by the Hall conductivity Σ_H. This phenomena occurs also for stationary influx currents into the ionosphere from the magnetosphere ([8], [9], [10]). The Hall conductivity is of great concern in the penetration of the MHD-wave to the ground. While Σ_H is reduced the amplitude of the magnetic component on the ground vanishes. This effect is a result of the total compensation of the magnetic fields caused by the longitudinal current of the Alfvén wave and the Pedersen currents spreading over the ionosphere.

Three key factors define the efficiency of the transformation of the Alfvén wave into the ground wave of the FMS polarization: (1) atmosphere does not let leak the longitudinal current; (2) The Hall conductivity provides the current system, the magnetic field of which can reach the ground, and (3) a transversal scale-size of the initial wave. One can assert that only the oscillations of the scale-size of the order of the atmosphere thickness or more can be seen on the ground.

The FMS-wave in contrary to the Alfvén wave does not contain the longitudinal current and delivers to the ionosphere the horizontal electric field. The ionospheric, atmospheric and the ground electrical currents connected with this electric field generating corresponding magnetic fields. Magnetic fields on the ground and under the ionosphere are defined by the resultant action of these currents and their relative contribution depends on the ionospheric and ground conductivities. Since the conductivity of the ground is higher in comparison with the ionospheric one, the influence of the ionosphere on the propagation of the FMS-mode is rather weak. For instance, a change from dayside ionosphere to the nightside ionosphere almost has no affects on the transmission coefficient of the FMS-mode.

Comparing propagation mechanisms of the Alfvén and FMS-modes, one can say that penetration to the ground of the incident Alfvén wave in the same mode is defined by the total resistivity of the system: magnetosphere-ionosphere-atmosphere-Earth because of the longitudinal current. This system can be considered as an electrical chain with series resistances and, vise versa for the FMS-mode, this system can be treated as a series which is the same but parallely connected resistances. The effective conductivity of the system is defined mainly by the geoelectrical cross-section at the observation point.

As for reflected waves, the essential point is that the reflection coefficient of the $A \Rightarrow A$ modes is independent on the horizontal wave-scale. Hence, the package of the Alfvén waves incident onto the ionosphere does not disperse and goes back to the magnetosphere conserving the shape.

The expression for R_{AA} is independent of the Σ_H and it is defined only by the relative values of the dimensionless terms associated with Pedersen conductivity Σ_P ($X = 4\pi\Sigma_P/c$) and the Alfvén velocity c_A $\left(\varepsilon_m = c^2/c_A^2\right)$. For the dayside ionosphere $X \gg \varepsilon_m^{1/2}$, the coefficient $R_{AA} \approx 1$, i.e. the total magnetic component above the ionosphere tends to 2 as at the surface of a perfect conductor. At the night ionosphere, relations between X and $\varepsilon_m^{1/2}$ are inverse $\left(X \ll \varepsilon_m^{1/2}\right)$. Then $R_{AA} = -1$, the total wave magnetic field above the ionosphere is nil, as it takes place at the boundary of the perfect dielectric. When passing from day to night, the R_{AA} changes from positive to negative values, so there is a transition region where $R_{AA} \to 0$. The Q-factor of the magnetospheric resonator at this local time steeply goes down.

There is another important point that follows from the calculations. The Alfvén wave incident onto the ionosphere is transformed not only into the same mode but into the FMS as well that follows from the above reasoning. There are two peculiar properties of this mode. The first one is that the mode damps exponentially as moving away from the ionosphere to the magnetosphere with the scales of order of the transversal scale of the initial wave. Other words, it is possible to reveal this wave only on low altitude satellites. An intensity of this reflected mode is controlled not so much by the magnetospheric properties as by the conductivity of the ground.

References

1. Alperovich, L. S. and E. N. Fedorov, Effect of the ionosphere on the propagation of MHD wave beams, *Izv. Vuzov, Radiofizika*, **27**, 1238, 1984a.
2. Alperovich, L. S. and E. N. Fedorov, The role of the finite conductivity of Earth in the spatial distributions of geomagnetic pulsations, *Izv. Ac. Sc. USSR, Physics of the Solid Earth*, **20**, 858, 1984b.
3. Alperovich, L. S. and E. N. Fedorov, The propagation of hydromagnetic waves through the ionospheric plasma and spatial characteristics of the geomagnetic variations, *Geomagn. Aeron.*, **24**, 529–534, 1984c.
4. Alperovich L. S., E. N. Fedorov, and T. B. Osmakova, Characteristics of telluric field near resonance magnetic shell, *Izvestia Ac. Sci. USSR, Physics of the Solid Earth*, **7**, 23, 1991.
5. Badden, K. G., *Radio Waves in the Ionosphere*, Cambridge Univ. Press, London, 1961.
6. Born, M. and E. Wolf, *Principles of Optics. Electromagnetic Theory of Propagation, Interference and Diffraction of Light,* 2nd edn., Pergamon Press, Oxford, London, Edinburgh, New York, Paris, Frankfurt, 1964.
7. Landau, L. D. and E. M. Lifshitz, *Electrodynamics of Continuous Media,* 2nd edn., New York, Pergamon, **8**, 1984.
8. Fukushima, N., Equivalence in ground magnetic effect of Chapman-Vestine's and Birkeland-Alfvén's electric currents-systems for polar magnetic storms. *Rept. Ionos. Space. Res. Japan*, **23**: 219, 1969.
9. Fukushima, N., Electric currents systems for polar magnetic storms and their magnetic effect below and above the ionosphere, *Radio Sci.*, **6**: 269–275 (1971).
10. Fukushima, N., Generalized theorem for no ground magnetic effect of vertical currents connected with Pedersen currents in the uniform-conductivity ionosphere, *Rept. Ionos. Space. Res., Japan*, **30**: 401–409, 1976.
11. Hughes, W. J., The effect of the atmosphere and ionosphere on long period magnetospheric micropulsations, *Plan. and Space Sci.*, **22**, Issue 8, 1157, 1974.
12. Hughes, W. J. and D. J. Southwood, Effect of atmosphere and ionosphere on magnetospheric micropulsation signals, *Nature,* **248**, 493, 1974.

13. Hughes, W. J., and Southwood D. J., An illustration of modification of geomagnetic pulsation structure by the ionosphere, *J. Geophys. Res.*, **81**, 3241, 1976.
14. Jackson J. D., *Classical Electrodynamics*, 3rd edn., Univ. of California, Berkeley, 1999.
15. Newton, R. S., D. J. Southwood, and W. J. Hughes, Damping of geomagnetic pulsations by the ionosphere, *Planet. Space Sci.*, **26**, 201, 1978.
16. Nishida, A., Ionospheric screening effect and storm sudden commencement, *J. Geophys. Res.*, **69**, 1861, 1964.
17. Saka, O., and Alperovich L., Sunrise effect on dayside *Pc* pulsations at dip equator. *J. Geophys. Res.,* **98**, 13779, 1993.
18. Sciffer, M. D., C. L. Waters, and F. W. Menk, Propagation of ULF waves through the ionosphere: Inductive effect for oblique magnetic fields, *Ann. Geophys.*, **22**, 1155, 2004.
19. Sciffer, M. D. and C. L. Waters, Propagation of ULF waves through the ionosphere: Analytic solutions for oblique magnetic fields, *J. Geophys. Res.*, **107**, A10, 1297, doi:1029/2001JA000184, 2002.
20. Scholer, M., On the motion of artificial ion clouds in the magnetosphere, *Planet. Space Sci.*, **18**, 977, 1970.
21. Vanyan, L. L., M. Berdichevsky, and L. Alperovich, *Geomagnetic Pulsations*, Nauka, Moscow, 1973.

8

Propagation of MHD-Beams

8.1 Introduction

The previous chapter discussed field transformation to the Fourier domain due to the homogeneity in the horizontal plane and time. The reflection and transformation matrices were found for the wave periodical in space. The thin ionosphere approximation enabled us to proceed with a minimum computations to a rough estimate of the general features of the MHD-waves. The next step, then, is to compute spatial field distributions produced by specific MHD-wave beams.

When a wave front propagates through the ionospheric plasma, it suffers reflection back to the magnetosphere and an attenuation due to the Joule heating. At some distances away from the pierce point of the wave beam, two waves are mixed on the ground and above the ionosphere. The first one is the transmitted wave and the second one is the wave reflected from the ground and the layers within the ground. The phases and intensities of these partial waves are mixed in different rates and various distances from the beam axis.

The wave can reach the ground observer with various paths. The electromagnetic field can be seen directly from the place where the beam enters the ionosphere. This place can be considered as an ionospheric equivalent source. In the simplest case of the ground modeled by a conductive half-space, the secondary mirror source arises within the ground. In the layered ground, there are infinite chains of the imaginary sources. An efficiency of radiation of the 'ground' (imaginary) sources is defined by the ground conductivity and a distance where the field produced by these underground sources can be seen. This distance is about the skin depth which, in turn, is proportional to the root of the wave period. For instance, the skin depth can reach 1000 km for the oscillations of the $1 - 10$ min period. The phases and intensities of the 'initial' and 'ground' sources are stacked and form a rather complicated interference pattern of the waves on small distances from the beam axis.

The wave can travel also within the magnetosphere along the upper ionospheric boundary and be re-emitted to the ground observer from above far from the beam axis. And at last, the wave passing through the ionosphere and finite conductive atmosphere results in the production of the atmospheric TEM-mode.

All these questions will be treated in the present chapter. We will develop a 2D theory of the horizontal propagation of the MHD-wave beams. Much attention is given to the Alfvén beams. Examples of spatial distributions of various field components are given with particular applications to synchronous bound Alfvén wave beams in which different points of the beam in the plane transversal to the beam axis oscillate in-phase and non-synchronous beams like a resonance Alfvén shell.

8.2 Coordinate Dependencies

Let a beam of the Alfvén waves generated in the magnetospheric equatorial region propagate along the field-lines. The transversal dispersion of the beam is a comparatively slow process and only for small spatial scales does the transversal dispersion and spreading of a beam become significant (decrease of the spatial scale is due to the phase mixing, see Chapter 5).

The spatial distribution of an electromagnetic field can be found by the inverse Fourier transform of the electric and magnetic fields in the magnetosphere (7.74) and on the ground (7.83) over wavenumbers $\mathbf{k}$. In the magnetosphere at $k_y = 0$, only E_x and b_y components remain in an Alfvén wave (see (7.95), (7.96)) and b_x, b_z, and E_y components remain in an FMS-wave (see (7.97), (7.98)). The beam is given by the distribution of incident magnetic wave components $b_x^{(i)}(x, +0)$ and $b_y^{(i)}(x, +0)$ above the ionosphere. The Fourier transform of the above ionosphere incident field is given by

$$\mathbf{b}_\tau^{(i)}(k) = \frac{1}{\sqrt{2\pi}} \int_{-\infty}^{+\infty} \mathbf{b}_\tau^{(i)}(x, +0) \exp(-ikx)\, dx, \tag{8.1}$$

where $b_\tau^{(i)}(x, +0)$ is the horizontal vector of the magnetic field. In this chapter, for notational simplicity, we shall put $k_x = k$. In Chapter 7, coefficients of reflection and transmission have been given at $k \geq 0$. Coefficients $R'(k)$ and $T'(k)$ at arbitrary k can be obtained from $R'(k)$ and $T'(k)$ of Chapter 7 by replacing k with $|k|$.

The fields reflected from the ground-ionosphere system $\mathbf{b}^{(r)}(k)$ and transmitted to the ground surface $\mathbf{b}^{(g)}(k)$ are determined by

$$\mathbf{b}_\tau^{(r)}(k) = \mathbf{R}'(k)\, \mathbf{b}_\tau^{(i)}(k) \qquad \text{and} \qquad \mathbf{b}^{(g)}(k) = \mathbf{R}'(k)\mathbf{b}_\tau^{(i)}(k), \tag{8.2}$$

where matrices $\mathbf{R}'$ and $\mathbf{T}'$ are given by (7.108) and (7.105). For components of the electric field and current, we have

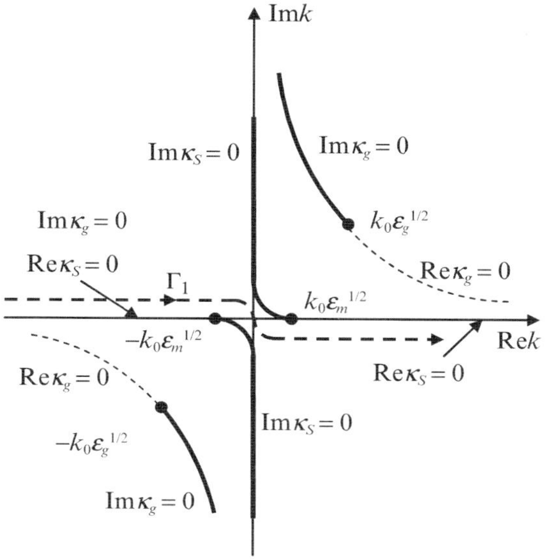

Fig. 8.1. Integration path Γ_1 going along the real axis over the physical sheet of the complex wave number k plane. $\kappa_g = \sqrt{k_0^2\varepsilon_g - k^2}$, $\kappa_s = \sqrt{k_0^2\varepsilon_m - k^2}$, $\varepsilon_g = 2i\sigma_g T$. Four points of branching $\left(\pm k_0\varepsilon_g^{1/2} \text{and} \pm k_0\varepsilon_m^{1/2}\right)$ are denoted by solid dots. Solid lines are the cut lines $\operatorname{Im}\kappa_s = 0$ and $\operatorname{Im}\kappa_g = 0$

$$E_x^{(i,r)}(k) = \mp\frac{|\sin I|}{\sqrt{\varepsilon_m}} b_y^{(i,r)}(k), \quad E_z^{(i,r)}(k) = -E_x^{(i,r)}(k)\cot I, \tag{8.3}$$

$$E_y^{(i,r)}(k) = \pm\frac{k_0}{\kappa_S} b_x^{(i,r)}(k), \qquad b_z^{(i,r)}(k) = \pm\frac{k}{\kappa_S} b_x^{(i,r)}(k), \tag{8.4}$$

$$j_x^{(i,r)}(k) = i\frac{c}{4\pi} b_y^{(i,r)}(k)\frac{\pm k_A + k\cos I}{\sin I}, \qquad j_z^{(i,r)}(k) = i\frac{c}{4\pi} k b_y^{(i,r)}(k), \tag{8.5}$$

with $\kappa_S = (k_A^2 - k^2)^{1/2}$, $k_A = \omega/c_A$. The upper and lower signs refer here, respectively, to the incident and reflected waves. Their coordinate dependencies may be presented as

$$\mathbf{b}_\tau^{(r)}(x, +0) = \frac{1}{\sqrt{2\pi}}\int_{\Gamma_1} \mathbf{R}'(k)\mathbf{b}_\tau^{(i)}(k)\exp(ikx)\, dk,$$

$$\mathbf{b}_\tau^{(g)}(x) = \mathbf{b}_\tau(x, -h) = \frac{1}{\sqrt{2\pi}}\int_{\Gamma_1} \mathbf{T}'(k)\mathbf{b}_\tau^{(i)}(k)\exp(ikx)\, dk. \tag{8.6}$$

Figure 8.1 shows the integration contour Γ_1 running from $-\infty$ to $+\infty$ along the real axis over the physical sheet of the complex plane k where $\operatorname{Im}\kappa_S > 0$, $\operatorname{Im}\kappa_g > 0$ (see Table 7.2).

Replace the exact expressions for $\mathbf{R}'(k)$ (7.108) and $\mathbf{T}'(k)$ (7.105) in (8.6) by approximate $\mathbf{R}^{(0)}(k)$ (7.124) and $\mathbf{T}^{(0)}(k)$ (7.125). With such replacement we get a good approximation for the magnetic mode on the ground. As for the transmitted electric mode, its computation must be conducted by exact formulae for $\mathbf{T}'(k)$, which is necessary to define the field at the large distances.

The reflected electric and magnetic fields above the ionosphere are then

$$\begin{pmatrix} E_{SS}^{(r)} \\ b_{zS}^{(r)} \end{pmatrix} = -\frac{1}{\sqrt{2\pi}} \int\limits_{\Gamma_1} \begin{pmatrix} k_0 h \\ \bar{k} \end{pmatrix} \frac{b_S^{(i)}(\bar{k}) \exp(i\bar{k}\bar{x})}{\sqrt{\bar{k}_A^2 - \bar{k}^2}} d\bar{k} + \sqrt{\frac{2}{\pi}} \begin{pmatrix} -k_0 h \\ i\partial/\partial\bar{x} \end{pmatrix} \Phi_{2S}, \quad (8.7)$$

$$\begin{pmatrix} b_{SS}^{(r)} \\ b_{SA}^{(r)} \end{pmatrix} = \begin{pmatrix} b_S^{(i)} \\ 0 \end{pmatrix} + \sqrt{\frac{2}{\pi}} \begin{pmatrix} \Phi_{1S} \\ -Y \sin I / \widetilde{X}\ \Phi_{1A} \end{pmatrix}, \quad (8.8)$$

$$\begin{pmatrix} b_{AS}^{(r)} \\ b_{AA}^{(r)} \end{pmatrix} = \begin{pmatrix} 0 \\ R_{AA}^{(0)} b_A^{(i)} \end{pmatrix} + \sqrt{\frac{2}{\pi}} \frac{\bar{k}_A Y}{\widetilde{X}} \begin{pmatrix} \operatorname{sign} I\, \Phi_{2S} \\ |\sin I|\, Y/\widetilde{X}\ \Phi_{2A} \end{pmatrix}, \quad (8.9)$$

$$\begin{pmatrix} b_{zA}^{(r)} \\ E_{SA}^{(r)} \end{pmatrix} = \sqrt{\frac{2}{\pi}} \frac{Y \sin I}{\widetilde{X}} \begin{pmatrix} -i\partial/\partial\bar{x} \\ k_0 h \end{pmatrix} \Phi_{2A}, \quad (8.10)$$

where $\widetilde{X} = X + \sqrt{\varepsilon_m}\,|\sin I|$, $\bar{k} = kh$, $\bar{x} = x/h$, h is the height of the ionospheric thin conductive layer, and

$$\begin{pmatrix} \Phi_{1(A,S)} \\ \Phi_{2(A,S)} \end{pmatrix} = \int\limits_{-\infty}^{+\infty} \begin{pmatrix} \sqrt{\bar{k}_A^2 - \bar{k}^2} \\ 1 \end{pmatrix} \frac{\exp(i\bar{k}\bar{x})}{\bar{\Delta}_{SK}^{(0)}} \bar{b}_{(A,S)}^{(i)}(\bar{k})\, d\bar{k},$$

$$\bar{b}_{(A,S)}^{(i)}(\bar{k}) = \frac{b_{(A,S)}^{(i)}(\bar{k})}{h} = \frac{1}{\sqrt{2\pi}} \int\limits_{-\infty}^{+\infty} b_{(y,x)}^{(i)}(\bar{x}, +0)\ \exp(i\bar{k}\bar{x})\ d\bar{x}.$$

Low indexes in the integrals denote either Alfvén (A) or FMS-wave (S).

Similarly, the dependence of ground electric and magnetic fields on the horizontal coordinate are given by

$$\begin{pmatrix} b_{SS}^{(g)} \\ E_{SS}^{(g)} \end{pmatrix} = -\sqrt{\frac{2}{\pi}} \begin{pmatrix} \Phi_{3S} \\ \Phi_{4S} \end{pmatrix}, \quad (8.11)$$

$$\begin{pmatrix} b_{SA}^{(g)} \\ E_{SA}^{(g)} \end{pmatrix} = \sqrt{\frac{2}{\pi}} \frac{Y}{\widetilde{X}} \sin I \begin{pmatrix} \Phi_{3A} \\ \Phi_{4A} \end{pmatrix}, \quad (8.12)$$

$$\begin{pmatrix} b_{zS}^{(g)} \\ b_{zA}^{(g)} \end{pmatrix} = \frac{1}{ik_0 h} \sqrt{\frac{2}{\pi}} \frac{\partial}{\partial\bar{x}} \begin{pmatrix} -\Phi_{4S} \\ (Y/\widetilde{X}) \sin I\ \Phi_{4A} \end{pmatrix}, \quad (8.13)$$

where

$$\begin{pmatrix} \Phi_{3(A,S)} \\ \Phi_{4(A,S)} \end{pmatrix} = \int_{-\infty}^{+\infty} \frac{\left(\bar{\Delta}_{SK}^{(0)}\right)^{-1} \bar{b}_{(A,S)}^{(i)}\left(\bar{k}\right) \exp\left(i\bar{k}\bar{x}\right)}{\left[Z_g^{(m)}/(k_0 h) - i \tanh|\bar{k}|/|\bar{k}|\right] \cosh|\bar{k}|} \begin{pmatrix} 1 \\ Z_g^{(m)} \end{pmatrix} d\bar{k}. \quad (8.14)$$

For a definition of the impedance $Z_g^{(m)}$, see (7.27) and (7.28).

Spatial dependencies of the total fields above the ionosphere and on the ground surface can be obtained from (8.7)–(8.10) and (8.11)–(8.13) employing the fact that

$$b_x^{(i,r,g)}(\bar{x}) = b_{SS}^{(i,r,g)}(\bar{x}) + b_{SA}^{(i,r,g)}(\bar{x}), \qquad b_y^{(i,r,g)}(\bar{x}) = b_{AS}^{(i,r,g)}(\bar{x}) + b_{AA}^{(i,r,g)}(\bar{x}),$$
$$b_z^{(i,r,g)}(\bar{x}) = b_{zA}^{(i,r,g)}(\bar{x}) + b_{zS}^{(i,r,g)}(\bar{x}).$$

Equations (8.7)–(8.10) and (8.11)–(8.13) can be conveniently used in numerical calculations at small distances less than 1000–2000 km. At large distances, the values of the integrals are determined by the behavior of integrand expressions in the vicinity of the singular points. Therefore, direct use of the numerical Fourier transformation can lead to wrong results.

8.3 Small Distances

Resonance Magnetic Shell

Let us suppose that an Alfvén wave beam excited by the FLR is incident on the ionosphere. Resonance disturbances above the ionosphere for fixed azimuthal harmonics $\propto \exp(ik_y y)$ are given by (6.88) which can be rewritten

$$b_y^{(i)}(x) = b_0^{(i)} \left(\frac{\delta_i}{x + i\delta_i} + Ck_y^2 \ln k_y(x + i\delta_i) + \cdots \right). \quad (8.15)$$

Here x is the horizontal coordinate directed southward from the base of the resonance shell, $b_y^{(i)}$ is the azimuthal component above the ionosphere, $b_0^{(i)}$ is the intensity of the magnetic component in the maximum of the incident wave, and C is a constant determined by the field line geometry and field line distribution of the cold plasma. δ_i is the half-width of the FLR above the ionosphere. The azimuthal wavenumber $|k_y| \ll |k_x| = k$, and, thus, approximate estimates for the reflected and transmitted magnetic and electric fields are found from the equations (8.7)–(8.14) for the meridional propagation.

The spatial spectrum of $b_y^{(i)}(x)$ is

$$b_A^{(i)}(k) = \begin{cases} A_0 \exp(-k\delta_i) + \cdots & \text{if} \quad k > 0, \\ 0 & \text{if} \quad k < 0. \end{cases} \tag{8.16}$$

where $A_0 = -i\sqrt{2\pi} b_0^{(i)} \delta_i$. The Fourier transform of the logarithmic and subsequent terms in series (8.15) is expressed by generalized functions vanishing at $k < 0$. We do not present here these spectrums, restricting ourselves just to the final results in the coordinate representation for the logarithmic term. Since the function $S(k) = 0$ at $k < 0$, then the coordinate dependencies of the field components is found using the inverse Fourier transformations along the positive semi-axis of wavenumbers k.

High Conductive Ground

Ratio ζ_4/ζ_3 (see (7.113), (7.114)) reduces to $\zeta_4/\zeta_3 \approx -i|\bar{k}| \coth|\bar{k}|$ for the high conductive ground. For the wavenumber $|\bar{k}| \gg \bar{k}_A$ and $|\bar{k}| \gg \bar{\tau}_K = k_0 h X_K$ from (7.128), we have

$$\bar{\Delta}_{SK}^{(0)}(\bar{k}) \approx -i\,|\bar{k}|\,\frac{\exp|\bar{k}|}{\sinh|\bar{k}|}. \tag{8.17}$$

By neglecting $Z_g^{(m)}$ in (8.14) for the high conductivity $(\sigma_g \to \infty)$, we obtain

$$\Phi_{3A} = -\int\limits_{-\infty}^{+\infty} \bar{b}_A^{(i)}(\bar{k}) \exp\left[\bar{k}\,(i\bar{x} - 1)\right] d\bar{k}. \tag{8.18}$$

Finally, substituting (8.16) into (8.18), we have

$$\Phi_{3A} = -(2\pi)^{1/2} b_0^{(i)} \frac{\delta_i}{x + i(h + \delta_i)}. \tag{8.19}$$

Let us consider the situation of the high enough conductivity of the near-surface ground layer in which the wavelength λ is much more than the skin depth $\left(\bar{k} \ll \bar{d}_g^{\;-1} = h/d_g\right)$. Here d_g is the skin depth in the ground. Substituting the expressions (8.19) for Φ_{3A} into (8.12), (8.13), we find

$$b_x^{(g)} \approx -2\frac{\Sigma_H}{\Sigma_P} b_0^{(i)} \sin I \left\{ \frac{\delta_i}{x + i(\delta_i + h)} + C k_y^2 \ln\left[k_y x + i k_y(\delta_i + h)\right] + \cdots \right\}, \tag{8.20}$$

$$b_y^{(g)} \approx 2ik_y \frac{\Sigma_H}{\Sigma_P} b_0^{(i)} \delta_i \sin I \ln\left[k_y x + i k_y(\delta_i + h)\right] + \cdots, \tag{8.21}$$

$$b_z^{(g)} \approx -2\frac{iZ_h}{k_0}\frac{\Sigma_H}{\Sigma_P} b_0^{(i)} \sin I \left[\frac{\delta_i}{[x + i(\delta_i + h)]^2} - \frac{C k_y^2}{x + i(\delta_i + h)} \right], \tag{8.22}$$

$$E_x^{(g)} \approx -Z_h b_y^{(g)}, \qquad E_y^{(g)} \approx Z_h b_x^{(g)}, \tag{8.23}$$

where $Z_h = Z_g^{(m)}(\bar{k} \to 0)$ is the ground spectral impedance for the vertically incident plane wave. Relations (8.20)–(8.23) were obtained for the dayside ionosphere, i.e. it was assumed that

$$\frac{4\pi\Sigma_P}{c} \equiv X >> \frac{c}{c_A}.$$

Low Conductive Ground

The scale of FLR on the ground surface is about $100 - 200$ km. There are regions with the effective skin depth d_g in the $Pc3,4$ range comparable to the horizontal scale L of the field $L \sim |k|^{-1}$. In this case, the simple impedance conditions cease to be true. However, a simple explicit expression can be obtained in the contrary case of the low-conductive ground when $|kd_g| \gg 1$ and $\kappa_g = (k_0^2\varepsilon_g - k^2)^{1/2} \approx i|k|$. At so small horizontal scales, the ground conductivity scarcely affects the field distribution. The surface admittance $Z_g^{(m)} \approx i|k|/k_0$ and $\zeta_4/\zeta_3 \approx -i|k|/k_0$ and we have (see (7.128)) $\Delta_{SK}^{(0)}(\bar{k}) \approx -2i(|k| - i\tau_K/2)$. Substitution of these expressions into the integrals Φ_{3A} and Φ_{4A} gives

$$\Phi_{3A} = -\frac{1}{2}\int\limits_0^\infty \bar{b}_A^{(i)} \frac{\bar{k}\exp[\bar{k}(i\bar{x} - 1)]}{\bar{k} - i\bar{\tau}_K/2} d\bar{k},$$

$$\Phi_{4A} = \frac{1}{2} i k_0 h \int\limits_0^\infty \bar{b}_A^{(i)} \frac{\exp[\bar{k}(i\bar{x} - 1)]}{\bar{k} - i\bar{\tau}_K/2} d\bar{k}.$$

Let us estimate Φ_{3A} for $\bar{\tau}_K \ll 1$ and $\bar{\delta}_i = \delta_i/h \lesssim 1$. In this case, one can neglect the small term $i\bar{\tau}_K/2$ in the denominator of the integrand. Integrating, we obtain

$$\Phi_{3A} \approx -i\frac{A_0}{2h}\frac{1}{\bar{x} + i(1 + \bar{\delta}_i)}. \tag{8.24}$$

In estimating Φ_{4A}, it is impossible to ignore the small corrected term in the denominator of the sub-integral expression. We must take into account the pole at $\bar{k} = i\bar{\tau}_K/2$. Then Φ_{4A} can be written as

$$\Phi_{4A} \approx -i\frac{A_0 k_0}{2} \exp(-Z_0)\,\mathrm{Ei}\,(Z_0),$$

where

$$Z_0 = \frac{\bar{\tau}_K}{2}\left[\bar{x} + i(1 + \bar{\delta}_i)\right].$$

$\mathrm{Ei}\,(Z_0)$ is the exponential integral which can be presented as the series

$$\mathrm{Ei}\,(Z_0) = C + \ln(-Z_0) + \sum_{n=1}^\infty \frac{Z_0^n}{n!n},$$

$C \approx 0.577$ is the Euler constant. At $|Z_0| \ll 1$ we restrict our consideration to the first two terms in the expansion of $\mathrm{Ei}\,(Z_0)$, and $\exp(-\bar{\tau}_K\bar{x}/2) \approx 1$. Write the final expression for Φ_{4A} :

$$\Phi_{4A} \approx -i\frac{A_0 k_0}{2}\left[C + \ln(-Z_0)\right]. \tag{8.25}$$

Substituting (8.24) and (8.25) into (8.12) and (8.13) for components of the magnetic mode at the ground, we find

$$\begin{pmatrix} b_{SA}^{(g)} \\ b_{zA}^{(g)} \end{pmatrix} = -\frac{A_0}{\sqrt{2\pi}}\frac{Y\sin I}{\widetilde{X}}\frac{1}{x + i\left(\delta_i + h\right)}\begin{pmatrix} i \\ 1 \end{pmatrix},$$

$$E_{SA}^{(g)}\left(x\right) = E_y^{(g)}\left(x\right) = -ik_0\frac{A_0}{\sqrt{2\pi}}\frac{Y\sin I}{\widetilde{X}}\left[C + \ln\left(-Z_0\right)\right]. \tag{8.26}$$

For $|kd_g| \gg 1$ the vertical magnetic component $b_z^{(g)}$ carries the same information about above ionosphere wave structure as the horizontal component $b_x^{(g)}$. Ratio $b_z^{(g)}/E_y^{(g)}$ is completely determined by the structure of the field inside the FLR-region:

$$ik_0\frac{b_z^{(g)}}{E_y^{(g)}} = \frac{1}{\left[x + i(\delta_i + h)\right]\left[C + \ln(-Z_0)\right]}, \qquad \frac{b_x}{b_z} = i.$$

Numerical Examples

Let the ground be a half-space with conductivity σ_g, and the ionospheric model corresponds to the middle latitude dayside ionosphere in the maximum of solar activity. The half-width δ_i and the resonance periods for $L = 2,\ 3,\ 4$ are found from Fig. 6.6. Chosen parameters are shown in Table 8.1.

Let the total magnetic wave-field above the ionosphere be given by (8.15), in which instead of $b_0^{(i)}$ is substituted by the total amplitude of the incident

Table 8.1. A ground-ionosphere model used in the numerical calculations. σ_g is a specific conductivity of the ground half-space. Σ_P and Σ_P are the Pedersen and Hall integral conductivities. The ionospheric model corresponds to the middle latitude dayside ionosphere in the maximum of solar activity. The half-width δ_i of a resonance shell corresponds to the resonance period for $L = 2,3,4$. h is the height of the conductive thin ionospheric layer

Parameter	Value
Σ_P	1.6×10^8 km/s
Σ_H	1.75×10^8 km/s
σ_g	$3 \times 10^6 s^{-1}$
h	130 km
δ_i	8 km ($T = 20$ s, $L = 2$)
δ_i	13.5 km ($T = 80$ s, $L = 3$)
δ_i	25 km, ($T = 210$ s, $L = 4$)

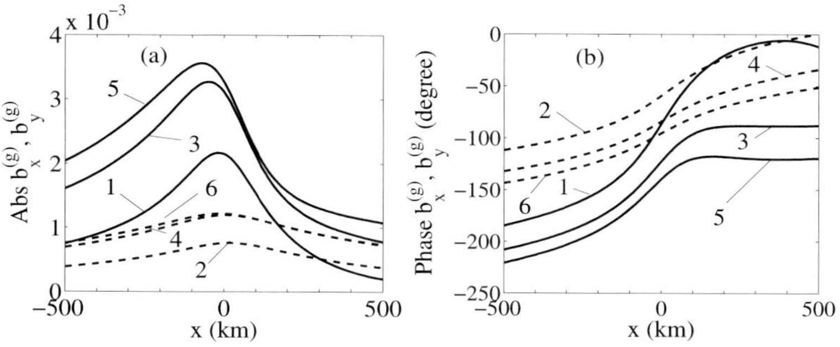

Fig. 8.2. The ground mapping of the FLR magnetic field. Panel (**a**) is the amplitude and panel (**b**) is the phase of the magnetic field. Solid line is a $b_x^{(g)}(x)$ (meridional) – component, dotted line is $b_y^{(g)}(x)$ (azimuthal)-component. Lines 1 and 2 refer to the oscillation period $T = 20\,\mathrm{s}$, lines 3 and 4 – $T = 80\,\mathrm{s}$, lines 5 and 6 – $T = 210\,\mathrm{s}$

and reflected Alfvén waves b_0 with normalization $b_0\delta_i = 1$. For the dayside ionosphere, the respective amplitude of the incident wave $b_0^{(i)} \approx b_0/2$. For the perturbations $\propto \exp(im\varphi)$, the wave numbers k_y are $k_y = mL^{1/2}/(R_E + h)$ where $R_E = 6370\,\mathrm{km}$. For $L = 2,\ 3,\ 4$, we have $k_y = 1.1 \times 10^{-3}$, 1.3×10^{-3} and $1.5 \times 10^{-3}\ \mathrm{km}^{-1}$ for $m = 6$.

The spatial distribution of ground magnetic field calculated according to (7.51), (7.81) and (7.83) is shown in Fig. 8.2. The panel (a) shows the spatial distribution of amplitudes $b_x^{(g)}(x)$ and $b_y^{(g)}(x)$, and panel (b) – phases. Lines 1 and 2 correspond to $T = 20\,\mathrm{s}$, lines 3 and 4 – $T = 80\,\mathrm{s}$, lines 5 and 6 – $T = 210\,\mathrm{s}$. $b_x^{(g)}$ is shown by solid lines and $b_y^{(g)}$ is shown by broken lines.

Maximum values of the field are 2.2×10^{-3}, 3.3×10^{-3} and 3.6×10^{-3} for $L = 2$, 3, and 4, respectively. Estimates, from (8.26) for the low conductive ground, are, correspondingly, 3.5×10^{-3}, 3.6×10^{-3} and 3.4×10^{-3}. One can see that they coincide roughly with the results of the numerical integration.

The maximum of field is shifted relative to the base of the FLR. For example, the maximum of $b_x^{(g)}$, are about $x = x_m = -20\,\mathrm{km}$ for $T = 20\,\mathrm{s}$, $x_m = -50\,\mathrm{km}$ for $T = 80\,\mathrm{s}$ and $x_m = -70\,\mathrm{km}$ for $T = 210\,\mathrm{s}$. The half-width δ_i found from Fig. 8.2 with (8.26) is correspondingly equal to $10\,\mathrm{km}$ ($T = 20\,\mathrm{s}$), $47\,\mathrm{km}$ ($T = 80\,\mathrm{s}$) and $77\,\mathrm{km}$ ($T = 210\,\mathrm{s}$).

The $b_x^{(g)}(x)$-phase (Fig. 8.2b) increases with x and reaches a maximum for $x > x_m$. With further increasing x, the phase slowly goes down. The maximum difference of phases $\Delta\varphi = (\arg b_x^{(g)})_{\max} - (\arg b_x^{(g)})_{\min}$ is inversely proportional to T. For instance, $\Delta\varphi = 175°$ for $T = 20\,\mathrm{s}$ and $\Delta\varphi = 110°$ for $T = 210\,\mathrm{s}$. The ratio of the major to minor axes of the horizontal ellipse of polarization changes from 2 for $x = 500\,\mathrm{km}$ up to 15 for $x = 0$. Beneath FLR the polarization becomes almost linear. The sense of rotation of the magnetic vector can be both the same and opposite on the different sides of FLR. For example, the sense of rotation changes for oscillations with $T = 20\,\mathrm{s}$ and does not change for $T = 80\,\mathrm{s}$.

Now that we have obtained spatial patterns of the magnetic field on the Earth for various periods of the FLR oscillations we are in the position to solve an inverse problem and define the half-width of the resonance shells using computed ground fields. In order to do this we fit the curves in Figures 8.2a by various curves with different δ_i varying them so that the fitting is to be best. We explored the range $\pm 50\,\mathrm{km}$ from the spatial maximum for the fitting and found follow reconstructed values for δ_i: 5.4 km for $T = 20\,\mathrm{s}$, 33.6 km for $t = 80\,\mathrm{s}$ and 55.6 km for $T = 210\,\mathrm{s}$. It follows by inspection of the respective initial values of $\delta_i = 8$, 13.5, and 25 km with the found δ_i that this approach indeed yields rather proper results in the considered case only for the short-period (low-latitude) FLR oscillations.

Synchronous Alfvén Wave Beams

Point Source

Suppose an Alfvén 2D wave beam is incident on the ionosphere. We assume that the beam is narrow and describe the distribution of the magnetic field in it as a δ-function. Let the latitudinal distribution of the magnetic component in the initial incident wave be given by $b_y^{(i)}(x) = b_0^{(i)} L_0 \delta(x)$, where $b_0^{(i)}$ is a characteristic magnetic amplitude in the beam and L_0 is its scale-width. The spectrum of $b_0^{(i)} L_0 \delta(x)$ is

$$b_A^{(i)}(k) = b_0^{(i)} L_0 / \sqrt{2\pi}. \tag{8.27}$$

For the high ground conductivity ($\bar{\tau}_K \ll 1$) similar to (8.18), we find

$$\Phi_{3A} = -\frac{b_0^{(i)} L_0}{\sqrt{2\pi} h} \int_{-\infty}^{+\infty} \exp\left[\left(i\bar{k}\bar{x} - |\bar{k}|\right)\right] d\bar{k}. \tag{8.28}$$

Substituting it into (8.12) and integrating, we obtain

$$b_{SA}^{(g)}(x) = b_x^{(g)}(x) = -\frac{2}{\pi} \frac{Y \sin I}{\widetilde{X}} \frac{h L_0}{h^2 + x^2} b_0^{(i)}. \tag{8.29}$$

According to (7.28) valid for the high conductive ground, the surface impedance $Z_g^{(m)}$ depends weakly on k. As a result, for the vertical magnetic component we have

$$b_z^{(g)}(x) = -\frac{1}{ik_0} \frac{\partial E_y^{(g)}}{\partial x} = -\frac{Z_g^{(m)}}{ik_0} \frac{\partial b_x^{(g)}(x)}{\partial x}.$$

This is combined with (8.29) to give

$$b_z^{(g)} = -\frac{4}{\pi} \frac{Y \sin I}{\widetilde{X}} \frac{Z_g^{(m)}}{ik_0} \frac{h L_0 x}{(h^2 + x^2)^2} b_0^{(i)}. \tag{8.30}$$

A similar calculation can be made for the low conductive ground. Then the horizontal and magnetic fields are given by

$$b_x^{(g)}(x) = -\frac{1}{\pi}\frac{Y \sin I}{\widetilde{X}}\frac{hL_0}{h^2+x^2}b_0^{(i)},$$
$$b_z^{(g)}(x) = -\frac{1}{\pi}\frac{Y \sin I}{\widetilde{X}}\frac{xL_0}{h^2+x^2}b_0^{(i)}. \tag{8.31}$$

The meridional magnetic component on the ground $b_x^{(g)}(x)$ for $x \gg h$ is seen from (8.30), (8.31), diminishes as x^{-2}, and the vertical component $b_z^{(g)}(x)$ as x^{-3} for the high conductive ground and as x^{-1} for the low conductive ground. At $x = 0$ under the source, the vertical magnetic component vanishes ($b_z^{(g)} = 0$) independently on the ground conductivity value contrary to a resonance field line in which the vertical component $b_z^{(g)}$ has a maximum under the field line axis.

Note that (8.29), (8.30), (8.31) can be obtained from the Bio–Savart law by the Hall current which is in turn produced by the horizontal electric field E_x of the Alfvén wave in the ionosphere

$$j_y L_0 \approx -\Sigma_H E_x L_0 \operatorname{sign} I = 2\Sigma_H L_0 b_0^{(i)}\frac{\sin I}{\widetilde{X}}.$$

Gaussian Beam

Consider another model of an incident beam: synchronous beam of Alfvén waves comes to the ionosphere with bell-coordinate dependence:

$$b_y(x) = b_0 \exp\left(-\frac{x^2}{2l_0^2}\right), \tag{8.32}$$

where $b_0 = b_0^{(i)} + b_0^{(r)}$ is the magnitude of the magnetic wave field above the ionosphere, l_0 is the beam half-width. Here, we assume that in contrast to FLR (8.15), various parts of the incident wave come simultaneously onto the ionosphere. We shall show at a later stage, that presetting the initial wave in the form (8.32) radically alters the spatial distribution as below the source as far from it. The spectral expression of $b_y(x)$ has the form

$$b_A(k) = b_y(k) = b_0 l_0 \exp\left(-\frac{k^2 l_0^2}{2}\right).$$

If to consider a high conductive ground, then using (8.14), (8.17) and (8.12) we obtain

$$b_{SA}^{(g)}(x) = -\sqrt{\frac{2}{\pi}}\frac{Y \sin I}{\widetilde{X}} b_0 l_0 \int_{-\infty}^{+\infty} \exp\left(-\frac{(kl_0)^2}{2} + ikx - h|k|\right) dk.$$

After a change of variables

$$p = \frac{i}{\sqrt{2}}\left(kl_0 - \frac{ix - h\,|k|\,/k}{l_0}\right),$$

the integral reduces to

$$-i\frac{\sqrt{2}}{l_0}\exp\left(-\frac{(x+ih)^2}{2l_0^2}\right)\int\limits_{\frac{x+ih}{\sqrt{2}l_0}}^{i\infty}\exp\left(z^2\right)dz + \text{c.c.}$$

and the magnetic field on the ground may be written as

$$b_{SA}^{(g)}(x) = b_x^{(g)}(x) = -\frac{Yb_0\sin I}{\widetilde{X}}\operatorname{Re} w(p_0), \tag{8.33}$$

where $w(p_0)$ is

$$w(p_0) = \exp(-p_0^2)\operatorname{erfc}(-ip_0), \qquad p_0 = \frac{x+ih}{\sqrt{2}l_0}. \tag{8.34}$$

The complementary error function

$$\operatorname{erfc}(-ip_0) = 1 + \frac{2i}{\sqrt{\pi}}\int\limits_0^{p_0}\exp(p^2)dp.$$

Approximation (8.33) can be used up to a few thousands kilometers from the beam axes for ground conductivities larger than $10^7\,\mathrm{s}^{-1}$. Figure 8.3 shows the $b_x^{(g)}(x)$-component for different values of periods and ground conductivities calculated by approximation (8.12), (8.19), and (8.33). The computations are carried out for the Alfvén velocity $c_A = 1000\,\mathrm{km}/\mathrm{s}$. Respectively, the wave Alfvén conductivity Σ_A is $\Sigma_A = c^2/(4\pi c_A) = 7.16\times10^6\,\mathrm{km/s}$. Then for the normalized dayside Pedersen $\bar{\Sigma}_P$ and Hall $\bar{\Sigma}_H$ conductivities, we have $\bar{\Sigma}_P = \Sigma_P/\Sigma_A = 10$, $\bar{\Sigma}_H = \Sigma_H/\Sigma_A = 14$, and for the nightside $\bar{\Sigma}_P = 0.1$, $\bar{\Sigma}_H = 0.14$. The corresponding dimensional conductivities are $\Sigma_P = 7.16\times10^7\,\mathrm{km/s}$, $\Sigma_H = 1\times10^8\,\mathrm{km/s}$ and $\Sigma_P = 7.16\times10^5\,\mathrm{km/s}$, $\Sigma_H = 1\times10^6\,\mathrm{km/s}$, for the day and night ionospheres, respectively. The last parameters are given in Table 8.2. The dashed line in Fig. 8.3 shows the initial field (8.32). The thick solid line is the spatial distribution for the high conductive ground is given by (8.33). The other curves are the result of numerical integration for the parameters shown in the Table 8.2.

One can see that approximation (8.33) adequately describes the decreasing of the field for the day conditions and ground conductivities $\sigma_g = 10^6 - 10^7\,\mathrm{s}^{-1}$ up to 2000 km. However, for the Pedersen conductivities typical for the night ionosphere, the approximation (8.33) is applicable till ≈500 km for $\sigma_g = 10^6\,\mathrm{s}^{-1}$ and till ≈1000 km for $\sigma_g = 10^7\,\mathrm{s}^{-1}$.

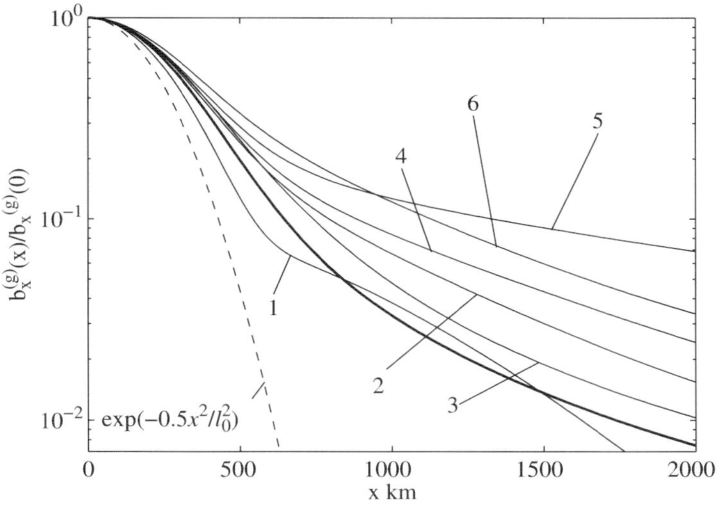

Fig. 8.3. Plots of the relative amplitude of the ground horizontal magnetic component $b_x^{(g)}(x)/b_x^{(g)}(0)$ as a function of horizontal coordinate. The curves are calculated for different values of the periods, ground and ionospheric conductivities are given in the Table 8.2. Numbering of the curves correspond to indexes of the 1st column of the Table 8.2

Phase incursion at propagation a wave perturbation is not described in the static approximation, in which the equation is obtained. In this approximation magnetic oscillations are in-phase at all distances if the phase in the incident beam does not depend on the horizontal coordinate x. Note that the phase changes found above for the resonance beam are connected with, that the different parts of the initial FLR-beam are out of phase. This is effect of a source but not propagation of perturbations from the axis.

For an in-phase beam-bell we shall define phase incursion by (8.11)–(8.14). Figures 8.4 and 8.5 show phase φ and phase velocity V_{phase} as functions of

Table 8.2. Ground and ionospheric conductivities chosen in numerical modeling

No.	$\sigma_g(\text{s}^{-1})$	T(s)	$b_x^{(g)}(0)$	Ionosphere
1	10^5	60	0.67	Day
2	10^6	60	0.88	Day
3	10^7	60	1.23	Day
4	10^6	100	0.92	Day
5	10^5	60	0.10	Night
6	10^6	60	0.11	Night

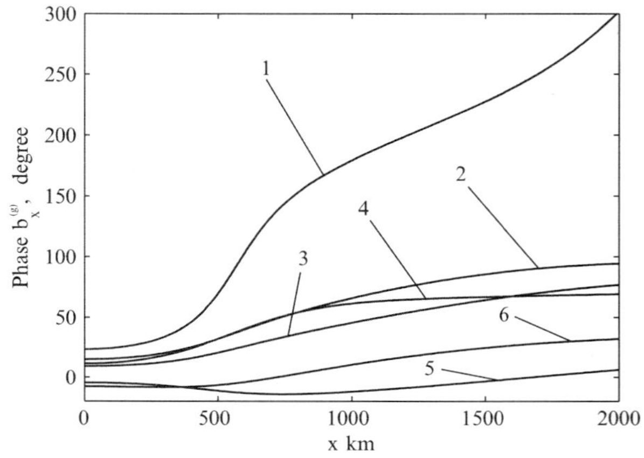

Fig. 8.4. Plots of phases (in degrees) versus horizontal distance from the bell-beam axis

a distance from the bell-beam for the ground and ionospheric conductivities from Table 8.2. It is important to note that $\varphi(x)$ has an extremum in the case of the low conductive ground and night ionosphere. Correspondingly, in the meridional distributions of the phase velocity $V_{phase}(x)$ can appear regions with the opposite sign of $V_{phase}(x)$. Hence, when the ground is low conductive, regions can appear with inverse phase velocity where the wave is seeming to propagates towards the source.

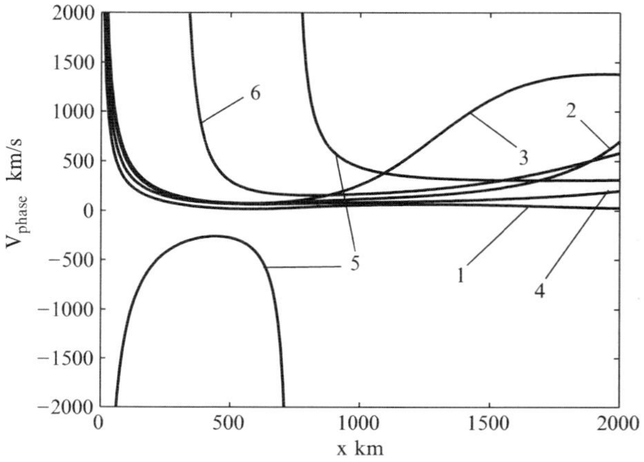

Fig. 8.5. Plots of the apparent phase velocities in km/s versus horizontal distance from the bell-beam axis

8.4 Large Distances

Sommerfeld–Watson Transformation

For investigating spatial distributions at large distances from the beam axis as well as electric type wave propagation, let us return to exact expressions (7.108) and (7.105) for matrices $\mathbf{R}'(k)$ and $\mathbf{T}'(k)$ and to the integral presentation (8.6). It is convenient to study the amplitude and phase dependencies with the field expanded by normal modes propagating along the layers. We shall use the Sommerfeld–Watson transformation [5]. Deform the initial contour of integration Γ_1 in the complex plane of horizontal wavenumbers k to the contour Γ_m (see Fig. 8.1). If the ground is modeled by a half-space with a finite conductivity σ_g, the integration contour must be supplemented by the contour Γ_g.

The transition from Γ_1 to Γ_m requires knowledge of the analytical properties of $\mathbf{R}'(k)$ and $\mathbf{T}'(k)$ on the entire complex plane k (see Section 7.6). The poles of $\mathbf{R}'(k)$ and $\mathbf{T}'(k)$ determine the wavenumbers of normal waveguide modes propagating in the ground-lower ionosphere waveguide. They can be found from

$$\Delta_{SK}\Delta_A = 0. \tag{8.35}$$

Two types of waveguide modes, magnetic and electric, prove to be coupled with one another because of the anisotropy of the ionosphere, and normal modes are of mixed character. However, the separation of the waves into two types remains useful. Divide the roots of (8.35) into two groups. The wavenumbers of the first group are determined by

$$\Delta_{SK}(k) = \frac{\zeta_4}{\zeta_3} - X_K - \frac{\kappa_S}{k_0} = 0, \tag{8.36}$$

and of the other group by

$$\Delta_A(k)/|\sin I| = \frac{\zeta_1}{\zeta_2} - \frac{\widetilde{X}}{\sin^2 I} = 0, \tag{8.37}$$

where $X_K = X + Y^2/\widetilde{X}$, ζ_i $(i = 1 - 4)$ see (7.39) and (7.45).

At $\Sigma_H \to 0$, when the magnetic and electric modes do not couple, find from (8.36) the wavenumbers of magnetic-type normal modes and from (8.37) those for the electric type. In the general case, $\Sigma_H \neq 0$, we preserve the designations of magnetic and electric modes for the branches turning by continuity into the respective modes at $\Sigma_H \to 0$.

Make a general note concerning the behavior of Δ_A and Δ_{SK}: one can separate two regions on the complex plane k with essentially different dependencies $\Delta_A(k)$ and $\Delta_{SK}(k)$. One of them is far from the contours Γ_m and Γ_g where $\Delta_A \approx -\widetilde{X}/|\sin I|$ (see (7.120)) and the other one is close to Γ_m and Γ_g. Here a special study is necessary.

Roots of a Dispersion Equation

Magnetic Mode

Let us study (8.35) far from Γ_m and Γ_g which is reduced to (8.36). The wavenumbers denoted as $k_{(n)S}$ are roots of (8.36) and determine normal magnetic-mode waves. As shown below, waveguide modes of this type do not always exist. In the general case, for arbitrary conductivities σ_g, the analysis of (8.36) is rather complicated. Therefore, consider first the case of the high conductive ground.

Ground Conductivity $\sigma_g \to \infty$

In this case, the surface admittance $|Y_g^{(m)}| \to \infty$ and (8.36) reduce to

$$i\bar{k}\coth\bar{k} + \sqrt{\bar{k}_A^2 - \bar{k}^2} + \bar{\tau}_K = 0. \tag{8.38}$$

It is proved in Appendix 8.A that (8.38) has no roots on the physical sheet. A root can be found just in the 2-nd and 3-rd quadrants of the non-physical sheet. The estimate for the damping of these virtual modes shows that the damping is very strong with $\operatorname{Im} k > 3\pi/4h$, where h is an atmospheric thickness.

Finite Conductive Ground

The last statement is valid for a perfect conductive ground. The following question is of interest: are there roots of (8.36) in the band $0 < \operatorname{Im} k < 3\pi/4$ if $\sigma_g \neq \infty$? If such roots exist then it can strongly affect the results of numerical integration. If it appears on a non-physical sheet close to the integration path, then the root becomes apparent as a poor integrable singularity. If the root is found on the physical sheet, it manifests itself as a pole corresponding to the horizontally propagating mode. Analysis of (8.36) for a finite conductive ground presented in Appendix 8.B proves that the root in fact exists for moderate conductivities.

Figure 8.6 charts the dependence of the root $k_{(1)S}$ on the ground conductivity obtained by a numerical solution of (8.36) for the period $T = 100\,\mathrm{s}$, $\Sigma_P = 0.116 \times 10^9\,\mathrm{km/s}$. The ground was replaced in these computations by a homogeneous half-space. Comparison of roots found by (8.B.7) and by the numerical solution (the curve in Fig. 8.6) shows that (8.B.7) is valid at $\sigma_g \leq 10^6\,\mathrm{s}^{-1}$.

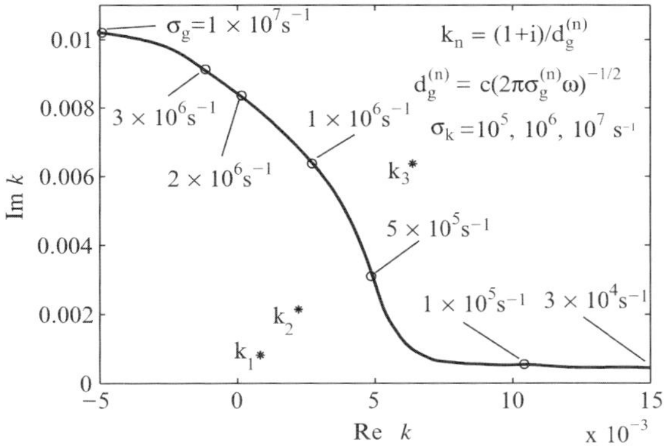

Fig. 8.6. The plot of the root $k_{(1)S}$ versus ground conductivity obtained by a numerical solution of (8.36) for the 100 s wave period and the Pedersen conductivity $\Sigma_P = 0.116 \times 10^9$km/s

Electric Mode

It has been shown in Chapter 7 that, far from cuts Γ_m and Γ_g (see Fig. 8.1), function

$$\Delta_A \approx -\frac{\widetilde{X}}{\sin I} = \text{const},$$

i.e. all singularities of an electric mode are located close to cuts Γ_m and Γ_g. Analysis carried out in Appendix 8.C demonstrates that the damping coefficients of these waves, except the basic one, are greater than h^{-1} (h is the thickness of the atmosphere) and increase in proportion to the mode number. The expression for the basic mode $\kappa^2_{(0)}$ (type TEM-wave in the terminology adopted in waveguide theory) for

$$\left|\sqrt{\varepsilon_g}\right| >> \{X, Y\} \qquad \text{and} \qquad k_0 h X \ll 1,$$

is

$$k_{(0)} = \exp\left(\frac{i\pi}{4}\right)\frac{\sqrt{\beta}}{h}, \qquad \beta = k_0 h \varepsilon_a \frac{\sin^2 I}{\widetilde{X}}, \tag{8.39}$$

where I is the inclination of the geomagnetic field, ε_a is atmospheric dielectric permeability, h is the atmosphere thickness (in km). β can be estimated for the typical values of the ionospheric conductivities (see Table 7.1) as

$$\beta \propto \begin{cases} 10^{-7}T^{-1} & \text{day,} \\ 10^{-5}T^{-1} & \text{night,} \end{cases}$$

where T is the period of the wave (in s).

The practical significance of the TEM-mode in the ULF-range is limited because it is hard to excite it by incident MHD-waves, its amplitude usually being small. The matter is that an electric mode must have a large vertical electric component which cannot be effectively generated by magnetospheric Alfvén waves, since the atmosphere's conductivity is small as compared to that of the ionosphere.

The TEM-mode can propagate to large distances over the ground surface with practically no damping and at velocities close to the velocity of light [11]. Therefore, despite the extremely low effectiveness of its excitation, it may contribute to pulsation field far from the MHD-wave beam incident on the ionosphere. Note, besides, that this mode can be excited by strong atmospheric sources like lightning discharge. The wave caused by the lightning propagates in the TEM-mode producing Shumann resonance in the atmospheric waveguide [12].

Spatial Distributions

Magnetic Mode

Let us return to the analysis of field behavior at large distances from the beam axis. Assuming the incident beam amplitude is spatially localized, set the magnetic field distribution in the form of δ-function:

$$\mathbf{b}^{(i)}(x) = \mathbf{b}_0^{(i)}\delta(x), \quad \mathbf{b}_0^{(i)} = \begin{pmatrix} b_{0x}^{(i)} \\ b_{0y}^{(i)} \end{pmatrix} = \text{const}\,.$$

where $\mathbf{b}^{(i)}$ is the horizontal magnetic field of the incident wave above the ionosphere. The incident field spectrum is

$$\mathbf{b}^{(i)}(k) = \frac{\mathbf{b}_0^{(i)}}{\sqrt{2\pi}}.$$

The magnetic field on the ground surface $\mathbf{b}^{(g)}(x)$ and above the ionosphere $\mathbf{b}^{(r)}(x)$ can be written as

$$\mathbf{b}^{(r,g)}(x) = \mathbf{G}^{(r,g)}(x)\mathbf{b}_0^{(i)}, \tag{8.40}$$

where $\mathbf{G}^{(r,g)}$ are the Green matrices for the reflected (r) from the ionosphere and transmitted (t) through it to the ground (g) wave, respectively. Equation (8.40) can be rewritten as

$$\begin{pmatrix} b_x^{(r,g)} \\ b_y^{(r,g)} \end{pmatrix} = \begin{pmatrix} G_{xx}^{(r,g)} & G_{xy}^{(r,g)} \\ G_{yx}^{(r,g)} & G_{yy}^{(r,g)} \end{pmatrix} \begin{pmatrix} b_{0x}^{(i)} \\ b_{0y}^{(i)} \end{pmatrix}. \tag{8.41}$$

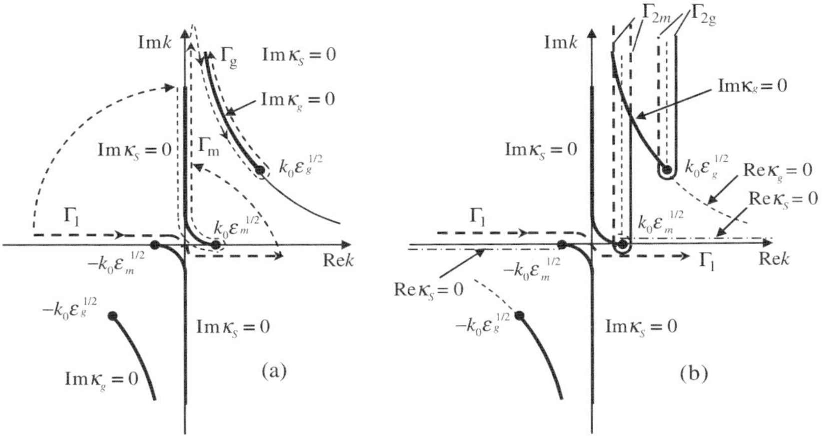

Fig. 8.7. The integration path Γ_1 of the integral (8.1) and its deformation to two branch cuts. Deformation of the primary contour Γ_1 to contours Γ_{2m} and Γ_{2g}

Equations (8.6), combined with (8.40), gives

$$\mathbf{G}^{(r)} = \frac{1}{2\pi}\int_{\Gamma_1} \mathbf{R}'(k)e^{ikx}dk, \quad \mathbf{G}^{(g)} = \frac{1}{2\pi}\int_{\Gamma_1} \mathbf{T}'(k)e^{ikx}dk. \tag{8.42}$$

The dependence of $\mathbf{b}^{(g)}(x)$ for arbitrary distribution $\mathbf{b}^{(i)}(x)$ in the incident beam can be found in the form of Green integrals.

It follows from the consideration of a sequence of extending contours passing between the poles $\mathbf{T}'(k)$ ($\mathbf{R}'(k)$) that at $x > 0$ the integration path Γ_1 can be transformed to the contour $\Gamma_2 = \Gamma_{2m} + \Gamma_{2g}$ (see Fig. 8.7). This procedure is formally non-applicable to the integral of $\mathbf{R}'(k)$, but it is valid for

$$\bar{\mathbf{R}}'(k) = \mathbf{R}'(k) - \mathbf{R}'(\infty).$$

$\mathbf{R}'(\infty)$ is determined in (7.142) and (8.B.7), while the integral of $\mathbf{R}'(\infty)$ evidently yields either the reflected Alfvén wave (7.142) or FMS-wave (8.B.7) caused by an incident Alfvén wave in the form of delta function $\delta(x)$.

A magnetic field on the ground surface is determined by a matrix $\mathbf{G}^{(g)}(x)$ which may be written in the form

$$\mathbf{G}^{(g)} = \mathbf{G}_m + \mathbf{G}_g + \sum_n \mathbf{G}_n e^{ik_{(n)}x} + \sum_n \mathbf{G}_{nS} e^{ik_{(n)S}x}, \tag{8.43}$$

where

$$\mathbf{G}_n = i\,\mathrm{Res}\,\left(\mathbf{T}'(k)\right)\big|_{k=k_{(n)}}, \qquad \mathbf{G}_{nS} = i\,\mathrm{Res}\,\left(\mathbf{T}'(k)\right)\big|_{k=k_{(n)S}}, \tag{8.44}$$

$$\mathbf{G}_{m,g} = \frac{1}{2\pi}\int\limits_{k_0\sqrt{\varepsilon_{m,g}}}^{k_0\sqrt{\varepsilon_{m,g}}+i\infty} \Delta\mathbf{T}'_{m,g}\exp\left(ikx\right)\,dk. \tag{8.45}$$

$\Delta \mathbf{T}'_{m,g}$ is the difference between the values of the transmission matrix taken from the corresponding sheets on contours Γ_{2m} and Γ_{2g}:

$$\Delta \mathbf{T}'_m = \mathbf{T}'_1 - \mathbf{T}'_2, \qquad \text{or} \qquad \Delta \mathbf{T}'_m = \mathbf{T}'_4 - \mathbf{T}'_3,$$
$$\Delta \mathbf{T}'_g = \mathbf{T}'_1 - \mathbf{T}'_4. \tag{8.46}$$

In (8.46) the subscripts attached to $\mathbf{T}'$ indicate the index of the sheet from which the value of $\mathbf{T}'$ is taken (see Table 7.2). With the chosen integration contours, the residues (see (8.C.2), (8.C.3)) are taken from the 2-nd sheet of the Riemannian surface $\mathbf{T}'(k)$. The poles $k = k_{(n)S}$ are determined by zeroes of Δ_{SK} (see (8.36)) covered by the deformation of the integration contours at the transition from Γ_1 and $\Gamma_2 = \Gamma_{2m} + \Gamma_{2s}$.

The computation of integral terms in (8.43) can be complicated by the presence of a pole near the integration contour. For example, in the case shown in Fig. 8.6, for the ground conductivities $10^5\,\mathrm{s}^{-1} \lesssim \sigma_g \lesssim 10^6\,\mathrm{s}^{-1}$ the pole $k_{(1)S}(\sigma_g)$ is located on the 1-st sheet (the physical sheet, $\operatorname{Im}\kappa_g > 0$ and $\operatorname{Im}\kappa_s > 0$). For $\sigma_g \approx 10^6\,\mathrm{s}^{-1}$ the pole $k_{(1)S}(\sigma_g)$ intersects the right bank of cut $\operatorname{Im}(\kappa_g) = 0$ and for $10^6\,\mathrm{s}^{-1} \lesssim \sigma_g \lesssim 2 \times 10^6\,\mathrm{s}^{-1}$ is located on the 4-th sheet ($\operatorname{Im}\kappa_s > 0$, $\operatorname{Im}\kappa_g < 0$). Thus, the pole can occur close to the contour of integration affecting the accuracy of the numerical computation. When the pole is located between the contours Γ_{2m} and Γ_{2g}, it is necessary in computing $G^{(g)}$ to take into account the residue in the pole. However, one can select a singular part $\mathbf{T}'(k)$ and calculate explicitly the contribution caused by the singularities in the integrand with the help of the residue.

Let us change the variable of integration in (8.45). Substitution

$$k = k_0\sqrt{\varepsilon_{m,g}} + i\frac{s}{x} = k_{m,g}(s) \tag{8.47}$$

reduces (8.45) to

$$\mathbf{G}_{m,g} = \frac{1}{2\pi}\frac{i\exp\left(ik_0\sqrt{\varepsilon_{m,g}}x\right)}{x}\int\limits_0^\infty \Delta \mathbf{T}'_{m,g}\left(k_{m,g}(s)\right)\exp\left(-s\right)\,ds. \tag{8.48}$$

In the neighborhood of $k = k_{(1)S}$, the corresponding branch of the matrix function $\mathbf{T}'(k)$ can be presented in the form of the Laurent series:

$$\mathbf{T}'(k) = \frac{C_{-1}}{k - k_{(1)S}} + \mathbf{C}_0 + \mathbf{C}_1(k - k_{(1)S}) + \cdots, \tag{8.49}$$

where C_{-1} is the residue matrix $\mathbf{T}'$ at $k = k_{(1)S}$. Taking into account that near $s = 0$ the functions $\Delta \mathbf{T}'_{m,g}(s) \propto \sqrt{s}$, we write

$$\frac{\Delta \mathbf{T}'_{m,g}}{\sqrt{s}} = \frac{\Delta \mathbf{T}'_{m,g}}{\sqrt{s}} \pm \frac{\mathbf{C}_{-1}}{\sqrt{s_{m,g}}\left[k_{m,g}(s) - k_{(1)S}\right]} \mp \frac{\mathbf{C}_{-1}}{\sqrt{s_{m,g}}\left[k_{m,g}(s) - k_{(1)S}\right]},$$

where

$$s_{m,g} = ix(k_0\sqrt{\varepsilon_{m,g}} - k_{(1)S}).$$

Here and below the upper sign refers to 'm', the lower to 'g'. For $s = s_{m,g}$ we have $k_{m,g}(s_{m,g}) = k_{(1)S}$. The branch of $\sqrt{s}$ $\sqrt{s_{m,g}}$) being determined by the condition $\operatorname{Re}\sqrt{s} > 0$ ($\operatorname{Re}\sqrt{s_{m,g}} > 0$). If $\operatorname{Re} k_0\sqrt{\varepsilon_{m,g}} > \operatorname{Re} k_{(1)S}$, then $\operatorname{Im}\sqrt{s_{m,g}} > 0$, and if $\operatorname{Re} k_0\sqrt{\varepsilon_{m,g}} < \operatorname{Re} k_{(1)S}$, then $\operatorname{Im}\sqrt{s_{m,g}} < 0$. functions

$$\frac{\Delta \mathbf{T}'_{m,g}}{\sqrt{s}} \pm \frac{\mathbf{C}_{-1}}{\sqrt{s_{m,g}}\left[k_{m,g}(s) - k_{(1)S}\right]}$$

are regular in the vicinity of $s = 0$ and $s = s_{m,g}$

Consider the integral

$$I_{m,g} = \pm\frac{\mathbf{C}_{-1}}{\sqrt{s_{m,g}}}\int\limits_0^\infty \frac{\sqrt{s}\exp(-s)}{k_{m,g}(s) - k_{(1)S}}ds.$$

Use new variable $s = t^2$ and, with (8.47),

$$I_{m,g} = \mp\frac{2ix\mathbf{C}_{-1}}{\sqrt{s_{m,g}}}\int\limits_0^\infty \frac{t^2\exp\left(-t^2\right)}{t^2 - s_{m,g}}dt.$$

Let us rewrite it in the form

$$I_{m,g} = \mp\frac{2ix\mathbf{C}_{-1}}{\sqrt{s_{m,g}}}\frac{\sqrt{\pi}}{2}\left(1 - \frac{2s_{m,g}}{\sqrt{\pi}}\int\limits_0^\infty \frac{\exp\left(-t^2\right)}{s_{m,g} - t^2}dt\right). \tag{8.50}$$

The integral can be expressed in terms of the Kramp function $w(V)$ ([1], [8]) which for $\operatorname{Im} V > 0$ is defined by the integral

$$w(V) = \frac{2iV}{\pi}\int\limits_0^\infty \frac{\exp\left(-t^2\right)}{V^2 - t^2}dt, \tag{8.51}$$

and for $\operatorname{Im} V < 0$ it is found by the analytic continuation of (8.51) into the lower half-plane. As a result,

$$\int\limits_0^\infty \frac{\exp\left(-t^2\right)}{V^2 - t^2}dt = -\frac{i\pi}{2V}w(V) + \begin{cases} 0 & \text{at} \quad \operatorname{Im} V > 0, \\ i\pi\exp\left(-V^2\right)/V & \text{at} \quad \operatorname{Im} V < 0. \end{cases} \tag{8.52}$$

From (8.50) and (8.52) we get

$$I_{m,g} = \mp i\sqrt{\pi}\frac{x}{\sqrt{s_{m,g}}}\mathbf{C}_{-1}\left[1 + i\sqrt{\pi s_{m,g}}w(\sqrt{s_{m,g}})\right]$$
$$\mp\begin{cases} 0 & \text{at} \quad \operatorname{Im}\sqrt{s_{m,g}} > 0, \\ 2\pi x\mathbf{C}_{-1}\exp(-s_{m,g}) & \text{at} \quad \operatorname{Im}\sqrt{s_{m,g}} < 0. \end{cases}$$

Turning back to the expressions for the Green's matrices $\mathbf{G}_{m,g}$ (8.45), we obtain

$$G_{m,g} = \frac{i\exp\left(ik_0\sqrt{\varepsilon_{m,g}}x\right)}{2\pi x}$$
$$\times\left\{\int_0^\infty\left(\frac{\Delta\mathbf{T}'_{m,g}}{\sqrt{s}} \pm \frac{\mathbf{C}_{-1}}{\sqrt{s_{m,g}}\left[k_{m,g}(s)-k_{(1)S}\right]}\right)\sqrt{s}\exp\left(-s\right)ds\right.$$
$$\left.\pm\, i\sqrt{\pi}\frac{x}{\sqrt{s_{m,g}}}\mathbf{C}_{-1}\left[1+i\sqrt{\pi s_{m,g}}w(\sqrt{s_{m,g}})\right]\pm Q\right\}, \qquad (8.53)$$

where

$$Q = \begin{cases} 0 & \text{at} \quad \operatorname{Im}\sqrt{s_{m,g}} > 0, \\ 2\pi x\mathbf{C}_{-1}\exp\left(-s_{m,g}\right) & \text{at} \quad \operatorname{Im}\sqrt{s_{m,g}} < 0. \end{cases}$$

The last term in the braces is either zero when the singularity of $\mathbf{T}'(k)$ is the right of the integration contour or equal to res $\mathbf{T}'(k_{(1)S})$.

Substituting this expression into (8.43), we write the Green matrix

$$\mathbf{G}^{(g)} = \sum_{n=m,g} G_n + \sum_{n=0}^{\infty}\mathbf{G}_n\exp\left(ik_{(n)}\,x\right) + \sum_{n\neq 1}^{\infty}\mathbf{G}_{nS}\exp\left(ik_{(n)S}\,x\right). \qquad (8.54)$$

In the last sum, term $n=1$ is off because it was already taken into account in (8.53). The integrals which define $G_{m,g}$ in (8.54), now have no singularities. The influence of the singularities is accounted by the Kramp function $w(V_{m,g})$. The impact of waveguide modes with numbers $n \geq 2$ on field behavior is insignificant because of the strong damping of these modes.

Electric Mode

The contribution of electric modes into the ground magnetic field is given in (8.54) by

$$\sum_{n=0}^{\infty}\mathbf{G}_n e^{ik_{(n)}x},$$

where only the term with $n=0$ is significant. The higher modes, like in the case of magnetic modes, damp rapidly because

$$\operatorname{Im} k_{(n)} \approx \frac{n\pi}{h}.$$

Besides, the amplitude of the basic mode (determined by the residue) is much larger than that of higher modes. Write ζ_i, κ_S, Δ_S and Δ_A near

$$k = k_{(0)} = e^{i\pi/4}\left(\frac{k_0\varepsilon_a}{h}\frac{\alpha}{1-\gamma^{(0)}}\right)^{1/2},$$

where

$$\alpha = \frac{\sin^2 I}{\widetilde{X}}, \qquad \gamma^{(0)} = \frac{Y^2}{\widetilde{X}\Delta_S^{(0)}}.$$

Let $|\kappa_a h| \ll 1$. Then

$$\begin{aligned}
\zeta_1 &\approx \zeta_1^{(0)} = -\sqrt{\varepsilon_g}, & \zeta_2 &\approx 1 + \frac{\kappa_a^2 h \sqrt{\varepsilon_g}}{ik_0\varepsilon_a}, \\
\zeta_3 &\approx \zeta_3^{(0)} = 1 - ik_0 h\sqrt{\varepsilon_g}, & \zeta_4 &\approx \zeta_4^{(0)} = -\sqrt{\varepsilon_g}, \\
\kappa_S^{(0)} &\approx k_0\sqrt{\varepsilon_m}, & \Delta_S &\approx \Delta_S^{(0)} = -\sqrt{\varepsilon_m} - X + \frac{\zeta_4^{(0)}}{\zeta_3^{(0)}}.
\end{aligned}$$

The approximate expression for κ_S is obtained under the assumption that $\left|k_{(0)}^2/(k_0^2\varepsilon_m)\right| \ll 1$, where $k_{(0)}$ is determined by (8.39). This inequality holds at $T < 10^4$ s. Function

$$\Delta(k) = \Delta_{SK}(k)\Delta_A(k) \approx -i\Delta_S^{(0)} \frac{k_0\varepsilon_a |\sin I|}{k_{(0)}^4 h} \left(k^2 - k_{(0)}^2\right).$$

near $k = k_{(0)}$

For the contribution of the electric mode to the ground field, we get

$$\mathbf{G}_0 = \frac{k_0\varepsilon_a}{k_{(0)} h \Delta_S^{(0)} \sin I} \frac{\alpha^2}{(1-\gamma^{(0)})^2} \times \begin{pmatrix} \dfrac{\sqrt{\varepsilon_g} Y^2}{\zeta_3^{(0)}\Delta_S^{(0)} \sin I} & -\dfrac{\sqrt{\varepsilon_g} Y}{\zeta_3^{(0)}} \\ -\dfrac{Y(1-\gamma^{(0)})}{\alpha} & \dfrac{1-\gamma^{(0)}}{\alpha}\Delta_S^{(0)} \sin I \end{pmatrix}. \qquad (8.55)$$

If the Hall conductivity is nil ($Y = 4\pi\Sigma_H/c = 0$), then the wave propagates in the atmospheric waveguide as a TEM-mode ($E_z \neq 0, \quad b_z = 0$). In the general case ($Y \neq 0$), we call it a TEM-type wave, keeping in mind, of course, that the wave has a longitudinal component b_x, as well.

The matrix (8.55) is sufficiently bulky, therefore, let us take into account that for a wide range of frequencies and conductivities the ground conductivity is much more than the ionospheric conductivity, i.e. $|\varepsilon_g| > X \sim Y$ and $|k_0 h X| \ll 1$. Then (8.55) becomes (see (8.39))

$$\mathbf{G}_0 \exp\left(ik_{(0)}x\right) = -ik_{(0)} \begin{pmatrix} -ik_0 h \dfrac{Y^2}{\widetilde{X}} & \dfrac{Y}{\widetilde{X}} \sin I \\ -ik_0 h \dfrac{Y}{\sin I} & 1 \end{pmatrix} \exp\left(ik_{(0)}x\right). \qquad (8.56)$$

This solution corresponds to a wave propagating with very weak damping at large distances with phase velocity

$$V_{ph} = \frac{\omega}{\operatorname{Re} k_{(0)}} \approx c\sqrt{2k_0 h\widetilde{X}}, \tag{8.57}$$

and linear polarization

$$\frac{b_x}{b_y} = \frac{\alpha Y}{\sin I} = \frac{Y \sin I}{\widetilde{X}}. \tag{8.58}$$

One can see from (8.58) that at $Y \neq 0$ no waves with purely transverse magnetic polarization exist, i.e. in the general case the magnetic component in the direction of propagation is finite.

Comparison of amplitudes of TEM-mode and the integral part in the ground magnetic field in components T_{AS} and T_{AA} shows that the ground magnetic field in the electric mode is determined basically by the TEM-mode. The validity of this assertion follows both from consideration of diverse limiting cases and from the results of numerical computation.

Fields in a Far Zone

Obtain asymptotic expressions for fields on the ground at distances much more than the value of the skin depth d_g in the ground ($x \gg d_g$). It enables us to neglect the integral term $\mathbf{G}_g$ (see (8.53)) in (8.43), since (8.48) gives

$$|\mathbf{G}_g| \propto \frac{\exp\left(-x/d_g\right)}{x}.$$

In order to calculate $\Delta\mathbf{T}'_m$, one can use approximate (7.120), because (7.119) is true on the integration contour Γ_{2m}. Indeed, on Γ_{2m} we get $\operatorname{Re} k = k_0\sqrt{\varepsilon_m}$, therefore, the applicability condition for (7.120) can be written in the form

$$k_0 \gg \frac{|\varepsilon_a| \sin^2 I}{\varepsilon_m \widetilde{X} h},$$

which holds for periods

$$\begin{array}{ll} T \ll 10^4\,\mathrm{s} & \text{day}, \\ T \ll 2\times 10^3\,\mathrm{s} & \text{night}. \end{array}$$

In order to obtain $\Delta\mathbf{T}'_m$, we choose the branches of the root at the sides of the branch cut according to the rule

$$\kappa_S = \begin{cases} +\sqrt{\bar{k}_A^2 - \bar{k}^2} & \text{left-hand cut} \\ -\sqrt{\bar{k}_A^2 - \bar{k}^2} & \text{right-hand cut} \end{cases}, \tag{8.59}$$

where assuming that $\operatorname{Re}\sqrt{\bar{k}_A^2 - \bar{k}^2} > 0$.

The other variables in matrices $\mathbf{R}'$ and $\mathbf{T}'$ remain continuous in transition from the left branch cut of contour Γ_{2m} to the right. Therefore

$$\begin{aligned} \frac{1}{\Delta_{SK}^{R}} - \frac{1}{\Delta_{SK}^{L}} &= -\frac{2\sqrt{\bar{k}_A^2 - \bar{k}^2}}{\left(\bar{\zeta}_4/\bar{\zeta}_3 - i\bar{\tau}_K\right)^2 - \bar{k}_A^2 + \bar{k}^2}, \\ \frac{\Delta_S^R}{\Delta_{SK}^{R}} - \frac{\Delta_S^R}{\Delta_{SK}^{L}} &= -\frac{k_0 h Y^2}{\Delta_A^{(0)}|\sin I|}\left(\frac{1}{\Delta_{SK}^{R}} - \frac{1}{\Delta_{SK}^{L}}\right), \end{aligned} \tag{8.60}$$

where $\Delta_{SK}^R, \Delta_S^R$ and Δ_{SK}^L , Δ_S^L are the values of Δ_{SK} , Δ_S on the right-hand and left-hand parts of the integration contour, respectively. It follows from (8.48) that the main contribution to the integral is from $s \le 1$. Since

$$\bar{k} = \bar{k}_A + i\frac{sh}{x},$$

then at $x \gg h$ with $\bar{k}_A \ll 1$ we get $|\bar{k}| \ll 1$. From (7.125) and (8.60) we obtain

$$\Delta\mathbf{T}'_m \approx \Delta\mathbf{T}_m^{(0)} \approx 4i\frac{\sqrt{\bar{k}_A^2 - \bar{k}^2}}{\left(k_0 h\zeta_4^{(0)}/\zeta_3^{(0)} - \bar{\tau}_K\right)^2} \times \begin{pmatrix} -i\dfrac{k_0 h Y_g^{(m)}}{\zeta_3^{(0)}} & -i\,\mathrm{sign}\, I\dfrac{Y}{\Delta_A^{(0)}}\dfrac{k_0 h Y_g^{(m)}}{\zeta_3^{(0)}} \\ \mathrm{sign}\, I\dfrac{Y}{\Delta_A^{(0)}}\dfrac{k_0^2 h^2 \varepsilon_a}{\bar{k}^2} & \dfrac{Y^2}{\left(\Delta_A^{(0)}\right)^2}\dfrac{k_0^2 h^2\varepsilon_a}{\bar{k}^2} \end{pmatrix}. \tag{8.61}$$

Let now the conductivity be a layer with $\sigma_g = \mathrm{const}$ and perfectly conducting underlying half-space at depth H . Let also $|kh| \ll 1$ and $|kd_g| \ll 1$. Then (7.113)–(7.115) and (7.125)–(7.128) give

$$\begin{aligned} \zeta_3^{(0)} &= 1 - ik_0 h Y_g^{(m)}, \quad \zeta_4^{(0)} = -Y_g^{(m)} \\ Y_g^{(m)} &= i\frac{1+i}{k_0 d}\cot\left((1+i)\frac{H}{d_g}\right). \end{aligned}$$

Now the integral in (8.48) can be calculated. Let

$$\Phi(\xi) = \frac{8}{\pi}\exp(i\xi)\int_0^\infty \frac{\sqrt{t^2 - it}}{(1+2it)^2}\exp(-2\xi t)\,dt, \tag{8.62}$$

with $\mathrm{Re}\sqrt{t^2 - it} > 0$, then

$$\int_0^\infty \frac{\sqrt{\bar{k}_A^2 - \bar{k}^2}}{\bar{k}^2}\exp(-sx)\,ds = \frac{\pi x}{2h}\exp(-ik_A x)\,\Phi(k_A x).$$

Using the table integral [1]

$$\int_0^\infty \sqrt{\bar{k}_A^2 - k_m^2(s)h^2}\exp(-s)ds = i\frac{\pi}{2}\bar{k}_A \exp\left(-ik_A x\right) H_1^{(1)}\left(k_A x\right),$$

where $H_1^{(1)}(k_A x)$ is the Hankel function of the first kind. So, $\mathbf{G}_m$ on the ground surface at large distances from the incident beam axis (see (8.48)) reduces to

$$\mathbf{G}_m = -\frac{1}{\left(k_0 h\zeta_4^{(0)}/\zeta_3^{(0)} - \bar{\tau}_K\right)^2} \times \begin{pmatrix} \frac{k_0 h Y_g^{(m)}}{\zeta_3^{(0)}}\frac{\bar{k}_A}{x}H_1^{(1)} & \operatorname{sign} I\frac{Y}{\Delta_A^{(0)}}\frac{k_0 h Y_g^{(m)}}{\zeta_3^{(0)}}\frac{\bar{k}_A}{x}H_1^{(1)} \\ \operatorname{sign} I\frac{Y}{\Delta_A^{(0)}}k_0^2 h\varepsilon_a\Phi & \left(\frac{Y}{\Delta_A^{(0)}}\right)^2 k_0^2 h\varepsilon_a\Phi \end{pmatrix}, \tag{8.63}$$

where $H_1^{(1)} \equiv H_1^{(1)}\left(k_A x\right)$, and $\Phi \equiv \Phi\left(k_A x\right)$.

The horizontal component of the magnetic field on the ground surface will, in accordance with (8.54), be determined by the sum of integral (8.63) and discrete (8.56) parts. Comparing the corresponding amplitudes in the transmitted Alfvén wave, we conclude that the ground field in this polarization is determined basically by the residue in the pole with the minimal imaginary part. For instance, for the high conductive ground the relation between the amplitudes of the discrete and integral parts is in the order of

$$\left|\frac{G_{mAA}}{G_{0AA}}\right| \approx \left(\frac{k_0 h}{\widetilde{X}}\right)^{3/2} Y^2 \left|\varepsilon_a\right|^{1/2} |\Phi| \ll 1,$$

$$\left|\frac{G_{mAS}}{G_{0AS}}\right| \approx \left(\frac{k_0 h\varepsilon_a}{\widetilde{X}}\right)^{1/2} |\Phi| \ll 1.$$

Let us define the spatial distribution on the large distances $x \gg d$ of the ground magnetic field. To avoid cumbersome expressions, we restrict our consideration to analysis of the high conductive ground:

$$\mathbf{G}_g = \begin{pmatrix} i\bar{k}_A\frac{H_1^{(1)}}{x} - k_0 h\frac{Y^2}{\widetilde{X}}S & -i\bar{k}_A\frac{Y\sin I}{\widetilde{X}}\frac{H_1^{(1)}}{x} - i\frac{Y\sin I}{\widetilde{X}}S \\ -k_0 h\frac{Y}{\sin I}S & -iS \end{pmatrix}, \tag{8.64}$$

where $S = k_{(0)} \exp\left(ik_{(0)}x\right)$. If $k_A x \ll 1$,

$$\frac{H_1(k_A x)}{x} \approx k_A \left(0.5 - i\frac{0.64}{(k_A x)^2}\right).$$

The contribution from a TEM-mode to the ground field becomes significant at distances

$$x_{km} > \left(\frac{h}{k_0}\right)^{1/2} \left(\frac{k_0 h \widetilde{X}}{\varepsilon_a}\right)^{1/4} \approx \begin{cases} 3 \times 10^3 T^{1/4} & \text{day,} \\ 1.5 \times 10^3 T^{1/4} & \text{night.} \end{cases}$$

For instance, for the nighttime $Pi2$ oscillations ($T \sim 50\,\mathrm{s}$) the amplitudes of G_{mAS} and G_{0AS} are comparable at 4000 km.

Expression (8.64) obtained for distances $x \gg d_g$ from a source shows that, far from the source, the wave decays with a distance of approximately x^{-2}. The phase velocity is to be of the order of c_A. At larger distances, where the amplitude distribution is determined by a TEM, the wave amplitude almost does not depend on distance and the phase velocity grows from c_A to the light velocity c.

Of course, in calculating a TEM-mode it is necessary to take into account the curvature of the Earth. Because of low attenuation and large wavelength, the field at the observation point will be the sum of direct and inverse round-the-world echoes.

Non-Monochromatic Waves

Up to now we have studied transformation of monochromatic oscillations. To consider signals finite in time we shall perform the inverse Fourier transformation of (8.64), we assuming a time-dependence in the form of a step-function $\theta(t)$. To simplify computation, assume that $\bar{\tau}_D = k_0 h X \ll 1$ for processes with characteristic times $\tau \gg 10\,\mathrm{s}$ (day) and $\tau \gg 1\,\mathrm{s}$ (night). Then fields on the ground can be presented as

$$\mathbf{G}^{(g)} = \begin{pmatrix} Q - \dfrac{Y^2}{\widetilde{X}} \dfrac{h}{2c} \dfrac{g_1}{t^{3/2}} & -\dfrac{Y \sin I}{\widetilde{X}} Q + \dfrac{Y \sin I}{\widetilde{X}} \dfrac{g_1}{t^{1/2}} \\ -\dfrac{Y}{\sin I} \dfrac{h}{2c} \dfrac{g_1}{t^{3/2}} & \dfrac{g_1}{t^{1/2}} \end{pmatrix}, \tag{8.65}$$

where

$$Q(x,t) = \frac{\theta\,(t_1)}{\pi} \left[\frac{h}{x^2} \frac{t}{t_2} + \frac{h}{c_A^2 t_2 (t + t_2)}\right],$$

$$g_1 = \sqrt{\frac{\varepsilon_a}{\pi h \widetilde{X} c}},\, t_1 = t - \frac{x}{c_A},\, t_2 = \sqrt{t^2 - \frac{x^2}{c_A^2}},$$

$\theta\,(t_1) = 1$ at $t_1 \geq 0$, and $\theta\,(t_1) = 0$ at $t_1 < 0$. In deriving (8.65) we used the fact that phase incursion in a TEM-mode is small ($k_{(0)} x \ll 1$).

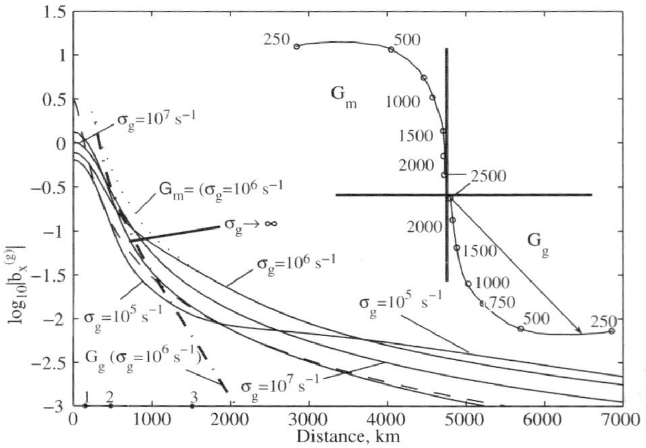

Fig. 8.8. The amplitude of the magnetic field on the ground as a function of distance from the point of entry of the incident Alfvén wave computed with (8.54) (solid lines). Dayside ionosphere. $T = 100\,\mathrm{s}$. The ground is a conductive half-space. $\sigma_g = 10^5\mathrm{s}^{-1}$, $10^6\mathrm{s}^{-1}$, $10^7\mathrm{s}^{-1}$ (thin lines) and perfectly conducting (thick line). The magnetospheric G_m and ground G_g (8.54) are correspondingly shown by the dotted and chain lines. The analytical approximation of the magnetic component (8.29) is the dashed curve. Two curves in the upper right corner show hodographs of the 'ground' and 'magnetospheric' waves versus distance for $T = 100\,\mathrm{s}$ and $\sigma_g = 10^6\,\mathrm{s}^{-1}$. Numbers along the curves mark the distance from the source in kilometers. The radius connect to the origin of the coordinate system and a point on the curves is $\ln A$. Here A is the amplitude of the horizontal magnetic component of the corresponding wave mode and angle between the radius and horizontal axis is a phase of the magnetic component. Values of d_g for $\sigma_g = 10^5, 10^6, 10^7\,\mathrm{s}^{-1}$ are marked by asterisks on the distance axis with numbers $1, 2, 3$, respectively

8.5 Summary

Figure 8.8 shows the amplitude (A) of the ground meridional magnetic component in the log-scale as a function of distance from the axes of the incident Alfvén beam. The magnetic field amplitudes in the beam are given in the form $\exp(-x^2/2l_0^2)$. The spatial distributions for different values of ground conductivity are calculated by (8.54). The integral in (8.54) was calculated with the aid of the Chebyshev–Laguerre quadrature formulae with the weight coefficient numbers 5–30 ([1], [9]). Kramp function $w(V_{m,g})$ was calculated with the method proposed in [8]. For comparison, the curve described by the analytical expression (8.29) with $L_0 = \sqrt{2\pi} l_0$ is shown in the same figure by a dashed line. In the calculations, the half-width of the model source is taken to be equal to $l_0 = 200\,\mathrm{km}$. The oscillation period is $T = 100\,\mathrm{s}$. The calculations are carried out for the dayside ionospheric model in the minimum of the solar activity: $\Sigma_P = 0.12 \times 10^9\,\mathrm{km/s}$, $\Sigma_H = 0.14 \times 10^9\,\mathrm{km/s}$. The ground is assumed to be a half space with specific conductivity $\sigma_g = 10^5 \mathrm{s}^{-1}, 10^6 \mathrm{s}^{-1}, 10^7 \mathrm{s}^{-1}$ (thin lines) and perfectly conducting (thick line).

Two curves in the upper right corner of Fig. 8.8 show hodographs of the 'ground' and 'magnetospheric' waves versus distance. In this example $T = 100\,\mathrm{s}$ and ground conductivity $\sigma_g = 10^6 \mathrm{s}^{-1}$. The hodograph of the ground wave is given in the 4-th quadrant, and the magnetospheric wave in the 2-nd. Numbers along the curves mark the distance from the source in kilometers. The radius connected to the origin of the coordinate system and a point on the curves is $\ln A$. Here A is the amplitude of the horizontal magnetic component of the corresponding wave mode and the angle between the radius and horizontal axis is a phase of the magnetic component.

The hodographs illustrate an interesting feature of both waves. Magnetospheric and ground waves propagate along the ground surface with opposite phases. Hence, the total ground field is determined by the difference of their intensities. In spite of the fact that the intensities of both waves increase sharply with approach to the source, the intensity of the total field is roughly double that of the incident wave with $A = 1$ for the high conductive ground.

Moreover, the intensity of the ground field at distances on the scale of the skin depth is defined mainly by the magnetospheric wave. Values of d_g for low conductive ($\sigma_g = 10^5 \mathrm{s}^{-1}$), moderate $\left(\sigma_g = 10^6 \mathrm{s}^{-1}\right)$ and high conductive $\left(\sigma_g = 10^7 \mathrm{s}^{-1}\right)$ ground are marked by asterisks on the distance axis with numbers $1, 2, 3$, respectively.

It can be seen that the dependence of the field on distance is mainly determined by a ground conductivity. The same figure presents separately the spatial distribution of a 'ground' G_g (8.54) and 'magnetospheric' G_m (8.54) wave for $\sigma_g = 10^6\,\mathrm{s}^{-1}$. In this case the 'ground' wave damps severely from 1000 km, i.e., from distances of the order of skin depth (d_g is marked by asterisks on the distance axis) the field on the ground is formed by the 'magnetospheric' wave. At distances of more than d_g, the least spatial gradients will be found in the curve corresponding to conductivity $\sigma_g = 10^5\,\mathrm{s}^{-1}$. One can say that at distances $x \gg d_g$ the total wave propagates along finitely conductive ground as if it was a perfect conductor.

Figures. 8.9 and 8.10 depict the dependencies of the b_x-component amplitude on distance for dayside (Figure 8.9) and nightside (Fig. 8.10) ionospheric conditions. In this example $\sigma_g = 10^6\,\mathrm{s}^{-1}$. A Gaussian source is chosen, as in Figure 8.8. The values of the monochromatic source periods are indicated near each curve. Comparing Fig. 8.9 and 8.10, it can be seen that the field amplitude reaching the ground surface at night is smaller than daytime. This is a result of the change of the reflection coefficient: the magnetic field is reversed in the transition from day to night roughly from $+1$ to -1. It means that the electric field is almost compensated at ionospheric levels by day and is intensified at night. More precisely, for the chosen model the electric field in the ionosphere is $E \approx 0.1\,\mathrm{mV/m}$ at daytime and $E \approx 1.5\,\mathrm{mV/m}$ at night ($T = 100\,\mathrm{s}$) at the 1 nT amplitude in the incident beam. Going to the ionospheric currents, we obtain that the currents generated by an Alfvén wave

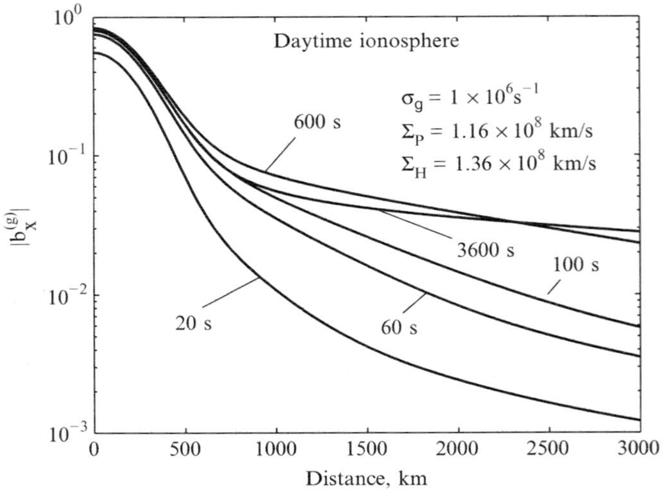

Fig. 8.9. Spatial variations of the ground magnetic component $b_x^{(g)}(x)$ for various periods and for the day time ionospheric conditions. The source as in the Fig. 8.8 is a beam with a Gaussian spatial distribution of the magnetic field in the incident beam. The half-width of the beam $l = 200$ km

at night are about three times smaller than dayside currents. The same correlation must exist between the magnitudes of horizontal magnetic fields as well. Note also the availability of smaller spatial gradients at night in comparison to the dayside model.

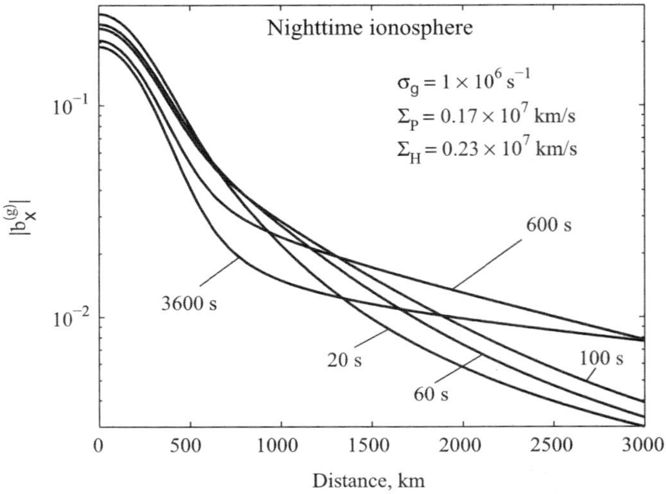

Fig. 8.10. The same as in Fig. 8.9 but for the nightside ionosphere

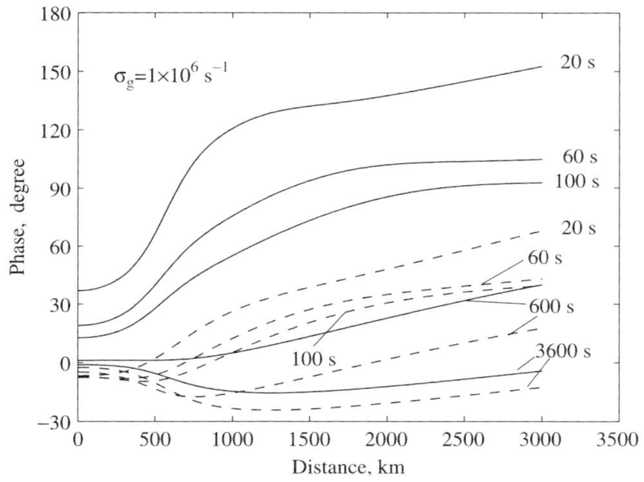

Fig. 8.11. Phase of the horizontal magnetic component $b_x^g(x)$ versus distance from the source for different periods T. Solid lines are for the day ionosphere, dashed lines are for the night ionosphere

The calculations demonstrate also the change in spatial dependencies at large distances. For instance, for oscillations with $T = 20\,\text{s}$ the distance x, where the amplitude of $b_x^{(g)}$-component in the magnetic mode is compared with the amplitude of $b_x^{(g)}$-component in the TEM mode and then remains practically constant, is about 3000 km.

Another unexpected outcome of the calculations is the spatial dependencies of phases. Figure 8.11 illustrates the dependence of phase on distance. Numbers near the curves indicate oscillation periods. Solid lines refer to the dayside ionosphere, dotted lines correspond to the nightside ionosphere. Three significantly different regions exist in the distribution. The first region is close to the source where phase shifts are small. The second region is transitional. Large phase shifts are observed here. The apparent phase velocities turn out to be $\approx$100 km/s for oscillations with $T = 100\,\text{s}$. The behavior of phase velocity in this region is complicated: in going to long periods, regions appear with inverse phase velocities.

In the model considered, the magnetic field on the ground is the sum of the 'ground' and 'magnetospheric' waves. The hodographs for each of these waves presented in Fig. 8.8, demonstrate that the phase of the magnetic field is also the sum of the phases of both waves. Their phases being almost opposite. Therefore situations are possible when at certain distances the amplitudes of these waves are significantly different. Generally, field regions can arise with inverse phase velocities. Figure 8.11 emphasizes that at 'moderate' ground conductivities $\left(\sigma_g = 10^6\,\text{s}^{-1}\right)$ such regions can appear at night for oscillations with $T > 20 - 60\,\text{s}$ and during the day at $T > 600\,\text{s}$. In the

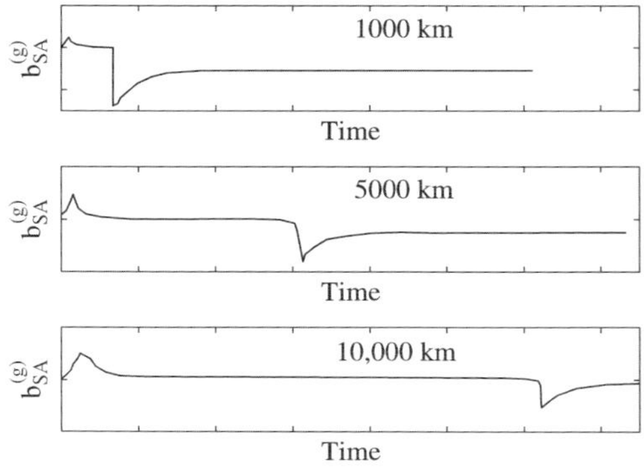

Fig. 8.12. Sketches show the waveforms of the $b_x^{(g)}$-component at 1000, 5000, and 10,000km from a beam with the step-function time dependency $\theta(t)$ in the source

literature on the theory of wave propagation from localized sources, there are numerous references to such effects. For instance, Tamir and Oliner [10] proved an example of long-wave emission from a local source located on a plasma layer that waves can appear with the phase velocity aimed towards the source.

At distances larger than the skin depth d_g, where the $b_x^{(g)}(x)$ is given by (8.64), the wave moves over the surface with Alfvén velocity. At larger distances the TEM-mode magnetic component becomes noticeable and the main. The wave phase velocity then becomes comparable to the speed of light.

The Alfvén wave incident from the magnetosphere onto the ionosphere produces the atmospheric TEM-mode characterized by a finite vertical electric field E_z at the ground. E_z for variations with $T = 100\,\mathrm{s}$ is about $0.3\,\mathrm{mV/m}$ at magnetic amplitude of the incident wave $1\,\mathrm{nT}$. Account taken of sphericity increases E_z by two orders [2]. A peculiarity of the spatial distribution E_z caused by a TEM-mode is the independence of its amplitude of distance.

For non-monochromatic sources the temporal pattern depends significantly on the distance from the source. Figure 8.12 sketches the temporal behavior of the b_x-component at 1000, 5000, $10,000\,\mathrm{km}$ from a beam with the step-function type in the source. At large distances where the TEM-mode becomes noticeable, an inverse pulse appears which could explain the pulse Sc^* [11]. The propagation velocity of this pulse is of the order of light velocity, whereas the basic magnetic signal propagates with Alfvén velocity.

Appendix *Magnetic mode.* $\sigma_g \to \infty$

It follows from (8.38) that

$$\bar{k}_A^2 - \bar{k}^2 = \left(i\bar{k}\coth\bar{k} + \bar{\tau}_K\right)^2 \tag{8.A.1}$$

or

$$\left(i\bar{k}\coth\bar{k}+\bar{\tau}_K\right)^2 = \left(\bar{\tau}-\mathrm{Im}\left(\bar{k}\coth\bar{k}\right)\right)^2+\mathrm{Re}^2\left(\bar{k}\coth\bar{k}\right)$$
$$\geq \mathrm{Re}^2\left(\bar{k}\coth\bar{k}\right) \geq \left(R\cot R\right)^2,$$

where $\left|\bar{k}\right| < R < \pi/2$. The variable $\left|\bar{k}_A^2-\bar{k}^2\right| < R^2+\bar{k}_A^2$ in the circle of radius R. Therefore (8.38) has no roots for $\sqrt{R^2+\bar{k}_A^2} < R\cot R$. Since $\bar{k}_A \ll 1$, the last inequality is true within the circle of radius $R_0 \approx \pi/4$. Hence (8.38) has no roots.

Let $\left|\bar{k}\right| > \pi/4$. Expand $\left(\bar{k}_A^2-\bar{k}^2\right)^{1/2}$into power series of $\left(\bar{k}_A/\bar{k}\right)$:

$$\left(\bar{k}_A^2-\bar{k}^2\right)^{1/2} = \pm i\bar{k}\sqrt{1-\frac{\bar{k}_A^2}{\bar{k}^2}} = \pm i\bar{k}\left(1-\frac{\bar{k}_A^2}{2\bar{k}^2}+\ldots\right)$$

The parameter $\bar{k}_A \propto T^{-1}$ and for $T > 10\,\mathrm{s}$, the value of $\bar{k}_A < 10^{-1}$. We leave the first term in the expansion of the root, then

$$\left(\bar{k}_A^2-\bar{k}^2\right)^{1/2} \approx \pm i\bar{k}.$$

The sign here is chosen according to the following rule: plus is on the physical sheet in the first quadrant and minus is on the nonphysical sheet and vice versa. Therefore (8.38) reduces in the 1-st quadrant on the physical sheet and in the 2-nd quadrant of the nonphysical sheet to

$$\frac{2i\bar{k}}{1-\exp\left(-2\bar{k}\right)} = -\bar{\tau}_K. \tag{8.A.2}$$

In the 2-nd quadrant on the physical sheet and in the 1-st quadrant of the nonphysical one, we have

$$-\frac{2i\bar{k}}{1-\exp\left(-2\bar{k}\right)} = -\bar{\tau}_K. \tag{8.A.3}$$

Equations (8.A.2) and (8.A.3) are tantamount to an equations set for $x = \mathrm{Re}\,\bar{k}$ and $y = \mathrm{Im}\,\bar{k}$:

$$\begin{array}{lll} \text{Phys. sheet } (I) & \Longrightarrow & 2y = \bar{\tau}_K\left(1-\exp\left(-2x\right)\cos 2y\right) \\ \text{Nonphys. sheet } (II) & \Longrightarrow & 2x = -\bar{\tau}_K\exp\left(-2x\right)\sin 2y, \end{array} \tag{8.A.4}$$

and

$$\begin{array}{lll} \text{Phys. sheet } (II) & \Longrightarrow & -2y = \bar{\tau}_K\left(1-\exp\left(2x\right)\cos 2y\right) \\ \text{Nonphys. sheet } (I) & \Longrightarrow & 2x = -\bar{\tau}_K\exp\left(2x\right)\sin 2y. \end{array} \tag{8.A.5}$$

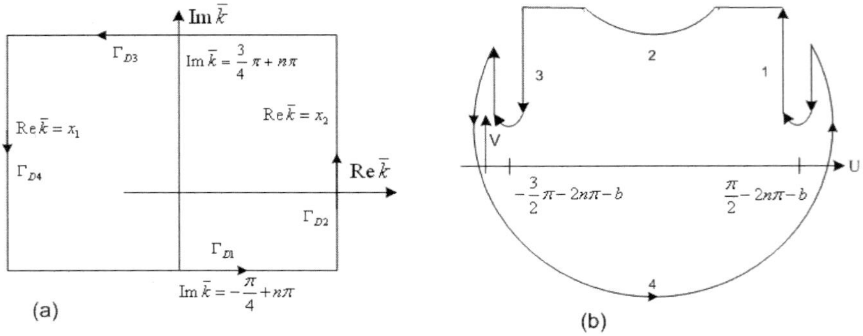

Fig. 8.A.1. (a) A contour Γ_D for $w(\bar{k}) = 2i\bar{k} + b - b\exp(-2\bar{k})$. (b) Hodograph of $w(\bar{k})$ when turning over Γ_D

Let us show that there are no roots on the physical sheet for $\bar{\tau}_K < \pi/2$. The left part of the first equation of (8.A.4) is negative, and is positive for $x < 0$. The second equation of (8.A.4) has roots on the physical sheet in the 1-st quadrant, if they exist, are just for $x > 0$ and $\pi(n + 1/2) < y < \pi(n+1)$. Substituting these values of y into the 1-st equation of (8.A.4), we found that a root can be just for $\bar{\tau}_K > \pi/2$. Considering that (8.A.5) is obtained from (8.A.4) by replacing of $\bar{k}$ by -$\bar{k}$, there are no roots on the whole physical sheet.

Let us prove that roots of (8.A.4) exist in the 3-rd and 4-th quadrants $(\mathrm{Re}\,\bar{k} < 0)$. Consider a function

$$w(\bar{k}) = u + iv = 2i\bar{k} + \bar{\tau}_K - \bar{\tau}_K \exp(-2\bar{k}).$$

According to the argument principle, the number of zeroes of an analytical function without singularities in an area D is equal to an argument increment of this function over 2π obtained when the complete circuit of D is traversed along the contour $\mathit{\Gamma}_D$ containing D. Adopt a contour $\mathit{\Gamma}_D$ shown in Fig. 8.A.1a for the function $w(\bar{k})$. When moving along $\mathit{\Gamma}_{D1}$ and $\mathit{\Gamma}_{D3}$, the values of u and v are

$$u = \bar{\tau}_K + \frac{\pi}{2} - 2n\pi \quad \text{on} \quad \mathit{\Gamma}_{D1},$$
$$u = \bar{\tau}_K - \frac{3\pi}{2} - 2n\pi \quad \text{on} \quad \mathit{\Gamma}_{D3},$$
$$v = 2x + \bar{\tau}_K \exp(-2x) \quad \text{on} \quad \mathit{\Gamma}_{D1} \text{ and } \mathit{\Gamma}_{D3}.$$

Choose large x_1 and x_2, then $w(\bar{k}) \approx -\bar{\tau}_K \exp(-2\bar{k})$ on $\mathit{\Gamma}_{D4}$ and $w(\bar{k}) = 2i\bar{k}$ on $\mathit{\Gamma}_{D2}$. Hodograph of the function $w(\bar{k})$, when turning over $\mathit{\Gamma}_D$, is shown in Fig. 8.A.1b. It follows from the given expressions that the hodograph one time envelopes the origin of the coordinate system, i.e. (8.38) has one root on the nonphysical sheet in the band

$$-\frac{\pi}{4}+n\pi < \operatorname{Im}\bar{k} < \frac{3\pi}{4}+n\pi$$

for $n \neq 0$. When $u = 0$, function $w\left(\bar{k}\right)$ has a root $\bar{k} = 0$ which is not the root of (8.38). Equation (8.A.3) results from (8.A.2) when changing $\bar{k}$ for $-\bar{k}$ Thus, roots of (8.36) do not exist.

Appendix *Magnetic mode. Finite* σ_g

Equation (8.36) at $\sigma_g \neq \infty$ has the form

$$\left(\bar{k}-\frac{i\bar{\tau}_K}{2}\right)\exp 2\bar{k}+\frac{i\bar{\tau}_K}{2}=\frac{i\bar{\tau}_K\bar{k}\bar{d}_g}{\bar{k}\bar{d}_g-\bar{\kappa}_g}. \tag{8.B.1}$$

Equation (8.B.1) is obtained at $\operatorname{Re}\sqrt{\bar{k}_A^2-\bar{k}^2}>0$. We investigate (8.B.1) at $\operatorname{Re}\,\bar{\kappa}_g>0$, $\operatorname{Im}\,\bar{\kappa}_g>0$. Assume that roots satisfy the condition

$$|\bar{k}\bar{d}_g| \gg 1, \tag{8.B.2}$$

then

$$\bar{\kappa}_g=\bar{k}\bar{d}_g\sqrt{1-\frac{2i}{(\bar{k}\bar{d}_g)^2}}\approx \bar{k}\bar{d}_g\left(1-\frac{i}{(\bar{k}\bar{d}_g)^2}\right).$$

Substituting this expression for $\bar{\kappa}_g$ into (8.B.1), we get

$$\exp\xi - A\xi = i\bar{\tau}_K\,\frac{\exp\xi-1}{\xi}, \tag{8.B.3}$$

where $A=\bar{\tau}_K\bar{d}_g^2/2$, $\xi=2\bar{k}$. Equation (8.B.3) is solved by successive approximations with respect to parameter $\bar{\tau}_K$ with fixed A. Then, in the first approximation,

$$\exp\xi_0 = A\xi_0. \tag{8.B.4}$$

We restrict our consideration to the band $0<\operatorname{Im}\bar{k}<\frac{3}{4}\pi$. At $A>e=2.71$ this equation has two real roots, the bigger one can be estimated by the formula

$$\xi_0 \approx \ln\left(A\left(\ln A\left(\ln A\ldots\right)\right)\right). \tag{8.B.5}$$

Complex roots appear at $A<e$. We present (8.B.4) as two real equations for $\operatorname{Re}\xi_0 = x_0$ and $\operatorname{Im}\,\xi_0 = y_0$:

$$\begin{aligned}\exp x_0\ \cos y_0 - Ax_0 &= 0,\\ \exp x_0\ \sin y_0 - Ax_0 &= 0.\end{aligned} \tag{8.B.6}$$

The values of y_0 and x_0 can be found from

$$\exp\left(y_0\cot y_0\right)\frac{\sin y_0}{y_0}=A, \qquad x_0=y_0\cot y_0.$$

As A extends from e to 0, the value of y_0 varies from 0 to π. Correspondingly, x_0 varies from 1 to $-\infty$. The solution of (8.B.3) with an accuracy of up to terms of first-order smallness with respect to $\bar{\tau}_K$, will be of the form:

$$\xi = \xi_0 + \xi_1 = \xi_0 + i\bar{\tau}_K \frac{A\xi_0 - 1}{\xi(\xi_0 - 1)}. \tag{8.B.7}$$

The expression (8.B.7) has been obtained under not very high ground conductivities $|\bar{k}\bar{d}_g| \gg 1$ An analogous investigation of the behavior of roots in all the regions of the four-sheeted Riemannian surface shows that only the root $k_{(1)S}$ is of interest, since the other roots either lie in the region of large $\mathrm{Im}\, k$ and therefore correspond to waves damping, or are remote from the integration contour and are not covered by it.

Appendix *Electric mode*

Electric mode wavenumbers are found from (8.37). Rewrite (8.37) in the form

$$\tan(\kappa_A h) = -i\beta \frac{1 + \dfrac{\kappa_g}{k_0 \varepsilon_g} \dfrac{1-\gamma}{\alpha}}{1 - \gamma + \alpha \dfrac{k_0 \varepsilon_a^2 \kappa_g}{\varepsilon_g \kappa_a^2}} \frac{1}{\kappa_a h}, \tag{8.C.1}$$

where

$$\alpha = \frac{\sin^2 I}{\widetilde{X}}, \qquad \gamma = \frac{Y^2}{\widetilde{X}\, \Delta_S}, \qquad \beta = k_0 h \varepsilon_a \alpha.$$

Expressions for Δ_S is given in (7.102). Note that $\beta \ll 1$. Let us solve (8.C.1) by successive approximations with respect to parameter β. For the first order in β, we get $\tan \kappa_a^{(0)} h \approx 0$, therefore

$$\kappa_{a(n)}^{(0)} \approx \frac{n\pi}{h} \qquad (n = 0, \pm 1, \pm 2, \ldots).$$

Substituting $\kappa_{a(n)}^{(0)}$ into (8.C.1), we find the first approximation

$$\kappa_{a(n)} \approx \frac{n\pi}{h} - i\beta \frac{1 + \dfrac{\kappa_{g(n)}^{(0)}}{k_0 \varepsilon_g} \dfrac{1 - \gamma_{(n)}^{(0)}}{\alpha}}{1 - \gamma_{(n)}^{(0)}} \frac{1}{n\pi}, \qquad n \neq 0, \tag{8.C.2}$$

where $\kappa_{g(n)}^{(0)}$ and $\gamma_{(n)}^{(0)}$ are obtained by substituting $k_{a(n)}^{(0)} = in\pi/h$ into κ_g and γ. The term

$$\alpha \frac{k_0 \varepsilon_a^2 \kappa_g}{\varepsilon_g \kappa_a} \ll 1$$

This was taken into account. Direct examination shows that the correction in (8.C.2) to the zero approximation is of the order of β. We obtain from (8.C.1) the expression for the root $k_{(0)}$ of the fundamental waveguide mode:

$$k_{(0)} = \sqrt{k_0^2 \varepsilon_a - \kappa_{a(0)}^2}, \tag{8.C.3}$$

where

$$\kappa_{a(0)}^2 = \mp i \frac{k_0 \varepsilon_a}{h\sqrt{\varepsilon_g}} \left(1 \pm \frac{\alpha\sqrt{\varepsilon_g}}{1-\gamma}\right).$$

The upper sign refers to the physical and 2-nd sheets and the lower sign to the 3-rd and 4-th sheets (see Table 7.2).

References

1. Abramowitz, M. and I. A. Stegun, (eds), Handbook of Mathematical Functions with Formulas, Graphs, and Mathematical Tables, 9th printing, Dover, New York, 1972.
2. Alperovich, L. S. and E. N. Fedorov, On the propagation of the hydromagnetic wave beams, (in Russian), Izv. Vuzov, *Radiofizika*, **27**, 1238, 1984.
3. Alperovich, L. S. and E. N. Fedorov, The role of the finite conductivity of Earth in the spatial distributions of geomagnetic pulsations, Izv. Ac. Sc. USSR, *Physics of the Solid Earth*, **20** , 858, 1984.
4. Alperovich, L. S. and E. N. Fedorov, The propagation of hydromagnetic waves through the ionospheric plasma and the spatial characteristics of geomagnetic variations, *Geomagn. Aeron.*, **24**, 529, 1984.
5. Bremmer, H., *Terrestrial radio waves*, New York, 1949.
6. Cagniard, L., Basic theory of the magnetotelluric method of geophysical prospecting, *Geophys.*, **18**, 605, 1953.
7. Gugliel′mi, A. V., Hydromagnetic diagnostics and geoelectric prospecting, *Usp. Fiz. Nauk.*, **58**, 605, 1989.
8. Dubovoi, A. P. and A. A. Yaroslavcev, On the calculation of Kramp′s functions, *Study of Structure and Properties of the Space Plasma*, (In Russian), M. IZMIRAN, 1980.
9. Shao, T. S., Chen T. C., and Frank R. M., Tables of zeros and gaussian weight of certain associated Laguerre polynomials and the related generalized polynomials, *Math. Comput.*, **18**, 538, 1964.
10. Tamir, T. and A. A. Oliner, Guided Complex Waves, Parts I and 11, *Proceedings of IEEE (London)*, **110**, No. 2, February, 310, 1963.
11. Kikuchi, T. and T. Araki, Horizontal transmission of the polar electric field to the equator, *J. Atmosph. Terr. Physics*, **41**, 927, 1979.
12. Nikolaenko, A. P. and M. Hayakawa, Resonances in the Earth-Ionosphere Cavity, Kluwer Academic Publishers, Dordrecht/Boston/London, 2002.
13. Tichonof, A.N., On the definition of electric characteristics of deep layers of the Earth crust, (In Russian), *Doklady Ak.Nauk USSR*, **73**, 295, 1950.

9

Inhomogeneous Ionosphere

This chapter deals with horizontally inhomogeneous ionosphere. First, we will show that the quasi-DC (direct current) approximation is appropriate for the description of ionospheric currents caused by FMS and Alfvén waves. Next, we will discuss how this approximation can be applied to an ionosphere containing irregularities of different scales. Then we will show several examples of global distributions of magnetic variations produced either by the field-aligned currents of an incident Alfvén wave or by the electric field of an FMS-wave.

We have shown that FMS-wave is almost insensitive to the ionospheric conductivity. The transition region between the dayside and nightside ionosphere is one example of a strong horizontal gradient of electron concentration, it is therefore also an example of a strong horizontal anomaly of the conductivity. We have also shown that on the contrary, the dependency of transmission on ionospheric conductivity becomes an important factor for Alfvén waves. The behavior of the ground magnetic field at sunrise and sunset is then discussed and its dependence on the initial MHD-mode is examined.

9.1 Quasi-Stationary Approximation

Basic Equations

The horizontal gradients of integral Pedersen and Hall conductivities become large at sunrise and at sunset, in the auroral zones and in the equatorial region. These gradients can considerably influence the passage and reflection of MHD-waves. The exact solution of diffraction problems on such inhomogeneities is an intricate mathematical problem and probably cannot be achieved for the general case. However, strong plasma anisotropy (infinite longitudinal conductivity) enables us to use an approximate field description and to obtain a comparatively simple analytical or numerical solution of the problem.

The ULF-wave skin depth is much more than the thickness of the ionosphere. So, the approximation of thin ionosphere is applicable. Chapter 7 presents a description for fields in this approximation. The wave electric field $\mathbf{E}$ is a sum of the Alfvén wave electric field $\mathbf{E}_A$ and the FMS-electric field electric field $\mathbf{E}_S$:

$$\mathbf{E} = \mathbf{E}_A + \mathbf{E}_S. \tag{9.1}$$

The field $\mathbf{E}_A$ of the Alfvén beam is 2D curl-free, and of $\mathbf{E}_S$ is the curl field. The influence of the vortex part on the curl-free part is negligible in the horizontally homogeneous ionosphere and can be ignored under the condition (7.135) which is written as

$$X_K k_0 L_A \ll 1,$$

where L_A is the horizontal scale of the Alfvén beam, and X_K is defined in (7.107).

Both Hall conductivity and the inhomogeneities cause an FMS-wave with the vortex field. As for the horizontally homogeneous ionosphere the influence of this field on the electric field of the Alfvén wave can be ignored for small-scale inhomogeneities. This scale $L_\perp$ must satisfy the same condition except that, L_A has to be replaced by $L_\perp$

$$X_K k_0 L_\perp \ll 1.$$

It means that the phase incursion at the scale $L_\perp$ is small. The horizontal scale L_F of the FMS-wave field in the ionosphere must be larger than $L_\perp$.

Let us estimate L_F. Let horizontal wavenumber $k \gg k_A$, then (8.36) reduces to

$$2ik + k_0 X_K = 0.$$

Then the scale of the FMS-wave $L_F = 1/k$ is

$$L_F = \frac{cT}{\pi X_K}. \tag{9.2}$$

In the equations for the Alfvén electric field (potential part), FMS electric field (vortex part) can be omitted under the condition

$$\frac{X_K L}{Tc} \ll 1, \quad X_K = \frac{4\pi}{c}\left(\Sigma_P + \frac{\Sigma_H^2}{\Sigma_P}\right). \tag{9.3}$$

The ratio of the correction $E_A^{(1)}$ caused by the FMS-wave to the initial Alfvén electric field $E_A^{(0)}$ is given likewise (7.139) by

$$\frac{E_A^{(1)}}{E_A^{(0)}} \sim \frac{4\,\pi\,L_\perp}{T\,c^2}\left(\Sigma_P + \frac{\Sigma_H^2}{\widetilde{\Sigma}_P}\right), \tag{9.4}$$

is small.

Now, let the external magnetic field $\mathbf{B}_0$ be orthogonal to the ionosphere. Use the Cartesian coordinate system (x, y, z) with z oriented along the magnetic field with the origin at the ionosphere current layer. The curl-free electric field is

$$\mathbf{E}_\perp = -\boldsymbol{\nabla}_\perp \Phi_0, \tag{9.5}$$

where $\boldsymbol{\nabla}_\perp$ denotes projection of a gradient on the plane orthogonal to $\mathbf{B}_0$, $\Phi_0(x, y) = \Phi(x, y, z = 0)$ is the potential at the ionospheric level.

The continuity equation for the electric current is

$$\boldsymbol{\nabla} \cdot \mathbf{j} = 0. \tag{9.6}$$

Considering the following two conditions:

- current cannot leak into the atmosphere and therefore its longitudinal component on the lower ionosphere boundary is nil;
- field line in the ionosphere is the equipotential line.

Integrate (9.6) along a field line that traverses the ionosphere, we obtain

$$j_\parallel = \boldsymbol{\nabla}_\perp \cdot (\boldsymbol{\Sigma} \cdot \boldsymbol{\nabla}_\perp \Phi_0), \tag{9.7}$$

where $\boldsymbol{\Sigma}$ is the integral conductivity tensor of the ionosphere, operators $\boldsymbol{\nabla}_\perp \cdot$ and $\boldsymbol{\nabla}_\perp$ are the divergence and gradient in the $\{x, y\}$ plane. The field-aligned current can be found from the Ampere's law

$$j_\parallel = \frac{c}{4\pi} (\boldsymbol{\nabla} \times \mathbf{b})_z .$$

Substituting $\mathbf{b} = \boldsymbol{\nabla} \times \mathbf{E}/(ik_0)$ into the above and expressing the electric field in terms of the potential (9.5), we find

$$j_\parallel = \frac{c}{4\pi i k_0} \boldsymbol{\nabla}_\perp \cdot \frac{\partial \mathbf{E}_\perp}{\partial z} = -\frac{c}{4\pi i k_0} \boldsymbol{\nabla}_\perp^2 \frac{\partial \Phi}{\partial z}, \tag{9.8}$$

where $\boldsymbol{\nabla}_\perp^2 = \dfrac{\partial^2}{\partial x^2} + \dfrac{\partial^2}{\partial y^2}$ is a transverse Laplacian. Substituting the latter equation into (9.7), we obtain

$$\boldsymbol{\nabla}_\perp \cdot \left[\left(\frac{4\pi}{c} \boldsymbol{\Sigma} + \frac{1}{ik_0} \frac{\partial}{\partial z} \right] \boldsymbol{\nabla}_\perp \Phi \right) \Bigg|_{z=0} = 0.$$

If the Alfvén field is a sum of the incident and reflected waves, then $\Phi(x, y, z)$ in the magnetosphere is given by

$$\Phi(x, y, z) = \Phi^{(i)}(x, y) \exp(-ik_A z) + \Phi^{(r)}(x, y) \exp(ik_A z), \tag{9.9}$$

where $k_A = \omega / c_A$. Substitution of (9.9) into (9.7) with

$$\Phi_0(x, y) \equiv \Phi(x, y, z = 0) = \Phi^{(i)}(x, y) + \Phi^{(r)}(x, y),$$

yields

$$\nabla_{\perp} \cdot \left(\widetilde{\mathbf{X}}_1(x,y) \nabla_{\perp} \Phi_0(x,y) \right) = 2\sqrt{\varepsilon_m} \nabla_{\perp}^2 \Phi^{(i)}(x,y), \tag{9.10}$$

where

$$\widetilde{\mathbf{X}}_1 (x,y) = \frac{4\pi}{c} \mathbf{\Sigma} + \sqrt{\varepsilon_m} \mathbf{1}.$$

1 is a unit diagonal matrix. The integration of (9.10) results in

$$-\widetilde{\mathbf{X}}_1 (x,y) \nabla_{\perp} \Phi_0 (x,y) = \frac{4\pi}{c} \nabla_{\perp} \times (\Psi (x,y)\, \hat{\mathbf{z}}) - 2\sqrt{\varepsilon_m} \nabla_{\perp} \Phi^{(i)} (x,y)\,. \tag{9.11}$$

where Ψ is a function defined by the boundary conditions.

Equations (9.7) and (9.11) enable us to find the ionospheric current and from it the magnetic field under the ionosphere ([4], [5]). These equations have been generalized for a more realistic model of the magnetosphere (see e.g. [9]). Equation (9.11) may also be used to investigate relations between the ground and magnetospheric magnetic fields in the presence of stochastic ionospheric irregularities (see Chapter 10).

Explicit Solutions

Contact of two Half-Planes

In a number of important cases (9.11) is integrable. For example, such solutions can be found if the ionosphere consists of strips with piecewise constant tensor integral conductivities $\mathbf{\Sigma}_n$. In the n-th strip the potential is of the form

$$\Phi_0 (x,y) = \frac{2\sqrt{\varepsilon_m}}{\widetilde{X}_{1n}} \Phi^{(i)} (x,y) + \widetilde{\Phi}_n (x,y)\,. \tag{9.12}$$

Here

$$\widetilde{X}_{1n} (x,y) = \frac{4\pi}{c} \Sigma_{Pn} + \sqrt{\varepsilon_m}\,.$$

Substituting (9.12) into (9.11), we obtain

$$-\widetilde{\mathbf{X}}_{1n} \nabla_{\perp} \Phi_n = \nabla_{\perp} \times \widetilde{\Psi}_n \hat{\mathbf{z}}, \tag{9.13}$$

where

$$\widetilde{\Psi}_n (x,y) = \frac{4\pi}{c} \Psi_n + 2\sqrt{\varepsilon_m} \frac{Y_n}{\widetilde{X}_{1n}} \Phi^{(i)}.$$

Equation (9.13) is to be supplemented by the boundary conditions between adjacent regions. The first condition follows from the continuity of the tangential component of the electric field. The second follows from the continuity equation for the electric current (9.6). The normal component of

$$-\widetilde{\mathbf{X}}_{1n} (x,y) \nabla_{\perp} \Phi_n (x,y)$$

is continuous. This gives

$$\left(\widetilde{\Phi}_{n+1} - \widetilde{\Phi}_n\right)\Big|_{\Gamma_{n,n+1}} = 2\sqrt{\varepsilon_m}\left(\widetilde{X}_{1,n}^{-1} - \widetilde{X}_{1,n+1}^{-1}\right)\Phi^{(i)}\Big|_{\Gamma_{n,n+1}},$$
$$\left(\widetilde{\Psi}_{n+1} - \widetilde{\Psi}_n\right)\Big|_{\Gamma_{n,n+1}} = 2\sqrt{\varepsilon_m}\left(Y_{n+1}\widetilde{X}_{1,n+1}^{-1} - Y_k\widetilde{X}_{1,n}^{-1}\right)\Phi^{(i)}\Big|_{\Gamma_{n,n+1}}. \quad (9.14)$$

Field Line Resonance and Terminator

Let us assume that we have two half-planes with different conductivities (day-side and nightside ionospheres)

$$\begin{array}{lll} \mathbf{\Sigma} = \mathbf{\Sigma}_+ = \text{const} & \text{if} & y > 0 \\ \mathbf{\Sigma} = \mathbf{\Sigma}_- = \text{const} & \text{if} & y < 0 \end{array}$$

Let us also assume that potential $\Phi^{(i)}$ describes resonant Alfvén oscillations (the first term of the series in (8.15)). Then horizontal fields under the ionosphere are

$$\begin{aligned} b_x^{\pm}\big|_{z=-0} &= A\left(\frac{g_{1\pm}\cdot\delta_i}{x+i\delta_i} + \frac{g_{2\pm}\cdot\delta_i}{x+i\left(|y|+\delta_i\right)}\right), \\ b_y^{\pm}\big|_{z=-0} &= \pm A\frac{ig_{2\pm}\cdot\delta_i}{x+i\left(|y|+\delta_i\right)} \\ g_{1\pm} &= -\frac{Y_\pm}{\widetilde{X}_\pm} \\ g_{2\pm} &= -i\left[\left(\frac{1}{\widetilde{X}_+} - \frac{1}{\widetilde{X}_-}\right) \pm \left(\frac{Y_+}{\widetilde{X}_+} - \frac{Y_-}{\widetilde{X}_-}\right)\frac{1}{\widetilde{X}_\pm \mp iY_\pm}\right] \\ &\quad\times\left(\frac{1}{\widetilde{X}_+ + iY_\pm} + \frac{1}{\widetilde{X}_- - iY_-}\right)^{-1}. \end{aligned} \quad (9.15)$$

where '+' and '−' signs correspond to $y > 0$ and $y < 0$, δ_i is the half-width of the resonance region in the ionosphere, A-amplitude factor.

If the ionosphere is homogeneous, $g_{2\pm} = 0$, and the magnetic vector makes a $\pi/2$ turn. Ionospheric inhomogeneities distort the angle from $\pi/2$. The ground magnetic field becomes elliptically polarized, even though the initial field was linearly polarized. A numerical example is shown in Fig. 9.1 and the chosen parameters are given in Table 9.1

The ellipses are shown for $x = 0$ and $y/\delta_i = -10, -8, \ldots, 8, 10$. The magnetic vector rotation sense is opposite on the two sides of $y/\delta_i = 0$. Deviation of the main axis decreases from $\approx 15°$ for $y/\delta_i \approx 1$ to $\approx 5°$ for $y/\delta_i \approx 10$ to the right of the discontinuity. To the left of $y/\delta_i = 0$ the main axis is $\approx -15°$ off for $y/\delta_i \approx -1$ and is $\approx -5°$ off for $y/\delta_i \approx -10$. Far from the discontinuity, the polarization becomes linear.

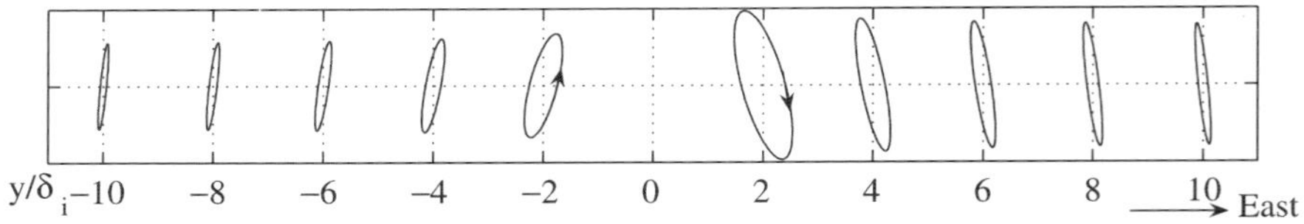

Fig. 9.1. The horizontal magnetic polarization ellipses below the FLR-shell versus meridional coordinate y. The ionosphere consists of two half-planes. Σ_P and Σ_H for the nightside and dayside ionospheres are given in Table 9.1. The distance is in the half-width δ_i of the shell. Arrows show sense of rotation of the horizontal magnetic component for $y/\delta_i = -2$ (nightside ionosphere) and $+2$ (dayside ionosphere)

Distributed Inhomogeneities

Let the conductivity depend only on x and let $\mathbf{\Sigma} = \mathbf{\Sigma}(x)$ be a continuous function of x. Consider an incident wave with the electric field directed along $\nabla_\perp \mathbf{\Sigma}(x)$ and described by the potential $\Phi^{(i)} = \Phi^{(i)}(x)$. Equation (9.11) enables us to find an analytical solution for the currents on the ionosphere and for the magnetic field below the ionosphere. Let

$$\mathbf{E}^{(i)} = -\nabla_\perp \Phi^{(i)}(x)$$

be the electric field in the incident wave. Then, from (9.11), for components of the total ionospheric electric field $\mathbf{E}$ and surface current $\mathbf{I} = (I_x, I_y)$, we obtain

$$\begin{aligned}
E_x(x) &= \frac{2\sqrt{\varepsilon_m}}{\widetilde{X}(x)} E^{(i)}(x), \qquad E_y = 0, \\
I_P(x) &= I_x(x) = \Sigma_P(x) \cdot E_x(x) = \frac{2\sqrt{\varepsilon_m}}{\widetilde{X}(x)} \Sigma_P(x) E^{(i)}(x), \\
I_H(x) &= I_x(x) = -\Sigma_H(x) \cdot E_x(x) = -\frac{2\sqrt{\varepsilon_m}}{\widetilde{X}(x)} \Sigma_H(x) E^{(i)}(x), \\
\widetilde{X}(x) &= \frac{4\pi}{c} \Sigma_P(x) + \sqrt{\varepsilon_m}.
\end{aligned} \tag{9.16}$$

It is obvious that the maximum of $I_x(x)$ and, thus, of $b_y(x)$ is shifted from $E^{(i)}(x)$ maximum ([4], [5], [10]).

Table 9.1. Parameters of dayside and nightside ionospheres utilized for the calculation of the series of ground horizontal magnetic variation ellipses shown in Fig. 9.1 near a terminator. Subindex '+' refers to dayside ionosphere modeled by a half-plane and '-' to the nightside

Σ_{P+}	1.5×10^8 km/s	Σ_{P-}	1.5×10^7 km/s
Σ_{H+}	1.75×10^8 km/s	Σ_{H-}	1.75×10^7 km/s
g_{1+}	1.1	g_{1-}	0.79
g_{2+}	$0.83 + i1.36$	g_{2-}	$0.54 + i0.88$

Electric and magnetic components in the reflected FMS-wave can be written as (see Chapter 7)

$$b_x = b_x^{(r)} = R_{SA} b_y^{(i)} \exp\left(-\sqrt{k^2 - k_A^2}\, z\right),$$

$$E_y = E_y^{(r)} = \frac{ik_0}{\sqrt{k^2 - k_A^2}} b_x.$$

where $b_y^{(i)} = -\sqrt{\varepsilon_m} E^{(i)}$. To find the ground magnetic field, we represent the current by its spatial Fourier decomposition

$$I_x = \frac{1}{\sqrt{2\pi}} \int\limits_{-\infty}^{+\infty} I_x(k) \exp(ikx)\, dk, \qquad (9.17)$$

where $I_x(k)$ is the Fourier component of $I_x(x)$. Let the ratio of the spectral electric and magnetic components on the ground be

$$\left.\frac{E_y}{b_x}\right|_{z=-h} = Z_g^{(m)}$$

and

$$\left.\frac{E_y^-}{b_x^-}\right|_{z=-0} = Z_a^{(m)} \qquad (9.18)$$

is the impedance under the ionosphere. With the impedance $Z_a^{(m)}$, the electric and magnetic fields on the ground surface can be obtained from those under the ionosphere, as it was done in Chapter 7. On the ionosphere, the horizontal electric field is continuous and discontinuity of the horizontal magnetic field is proportional to the total electric current, i.e.,

$$E_y^- = E_y^+, \qquad b_x^+ - b_x^- = \frac{4\pi}{c} I_y, \qquad (9.19)$$

where the plus and minus superscripts denote the components above and below the ionosphere, respectively. Then from (9.19)

$$\frac{ik_0}{\sqrt{k^2 - k_A^2}} R_{SA} b_y^{(i)} = E_y^-, \qquad R_{SA} b_y^{(i)} - b_x^- = \frac{4\pi}{c} I_y,$$

$$E_y^- = Z_a^{(m)} b_x^-. \qquad (9.20)$$

Hence

$$R_{SA} b_y^{(i)} = \frac{4\pi}{c} I_y \left(1 - \frac{ik_0}{\sqrt{k^2 - k_A^2}} \frac{1}{Z_a^{(m)}}\right)^{-1}, \qquad (9.21)$$

and the final expressions for the total horizontal magnetic and electric components below the ionosphere are

$$
\begin{aligned}
b^{-}(k) &= \frac{4\pi}{c} I_y(k) \frac{ik_0}{\sqrt{k^2 - k_A^2 Z_a^{(m)}} - ik_0}, \\
E^{-}(k) &= Z_a^{(m)} b^{-}.
\end{aligned}
\tag{9.22}
$$

The next step is to make the transition from (9.22) to expressions for ground fields and to the inverse-Fourier transform of the obtained field.

9.2 Numerical Modeling

Let the ionosphere have a complicated spatial distribution of its tensor conductivity including inclination of $\mathbf{B}_0$. This leads to the growth of ionospheric conductivity tensor components on the geomagnetic equator. The distribution also includes a significant conductivity difference between the dayside and nightside ionospheres.

The main idea of the numerical simulation is to use quasi-static field-aligned currents (Alfvén wave) and electric fields (FMS-wave) as the sources of ground magnetic variations.

In the examples considered above, the Alfvén wave was incident on the ionosphere. This allowed us to uniquely determine the longitudinal currents flowing into and out of the ionosphere. We could also determine the field-aligned currents generated by horizontal inhomogeneities of the ionospheric conductivity. The longitudinal currents arising thereby are proportional to $\sqrt{\varepsilon_m}$ and connected with the transverse polarizing current by

$$
\frac{\varepsilon_m}{4\pi} \frac{\partial \mathbf{E}}{\partial t}.
$$

To find the currents spread out over the ionosphere requires us to solve the problem of MHD-wave propagation in the magnetosphere-ionosphere system. The spread-out currents on one of the ionospheres are determined not only by the properties of this ionosphere, but also by the conductivity of the conjugate ionosphere. A solution of this problem is only possible for very simplified models. An example of such a model is the box model with constant Alfvén velocity inside the box and with variable integral conductivity on its walls. However, consideration of horizontal changes of integral conductivity in this model obviously exceeds its 'accuracy' and can hardly help us to understand the situation in the real magnetosphere.

The situation is simplified for quasi-static fields, the potential along a field line is constant and determined by (9.10). In this case the problem with two conjugate ionospheres can be reduced to one ionosphere conductivity of which is the sum of their integral conductivities. This approach is only true for frequencies

$$
\omega \ll \omega_1,
$$

where ω_1 is the frequency of the fundamental FLR-harmonic.

The above condition is too rigid a restriction for the pulsation range. This is why in constructing models of interaction between MHD-waves and the ionosphere, the following approaches are more preferable. In one approach, already used in this book, we set the incident MHD-wave above the ionosphere and use it to determine the ground electromagnetic field and field line currents. In another approach we proceed from a given system of longitudinal currents and use (9.7). A third approach is possible, justified only for perturbations that change quickly enough in the transverse direction. More precisely, the transverse scale of perturbations $k_\perp^{-1}$ satisfies

$$L_\perp \gg k_\perp^{-1},$$

where $L_\perp$ is the scale of change of magnetosphere and ionosphere parameters in the transverse direction (see Chapter 10).

The equation for the potential (9.7) is ([9], [11])

$$\mathbf{\Sigma}\left(-\mathbf{\nabla}_\perp \Phi_0 + \mathbf{E}_0\right) = -\mathbf{\nabla}_\perp F + \mathbf{\nabla} \times \mathbf{\Psi}. \tag{9.23}$$

Here F is a scalar determining the potential part of the horizontal current and linked with the field-aligned currents by

$$\mathbf{\nabla}_\perp^2 F = j_\parallel .$$

$\mathbf{\Psi}$ provides the vortex part of the horizontal current. $\mathbf{E_0}$ denotes external sources of the electric field not connected with field-aligned currents, for instance, $\mathbf{E_0}$ can be transported from the magnetosphere to the ionosphere by an incident FMS-wave or generated by neutral wind due to dynamo action. In the latter case

$$\mathbf{E_0} = -\frac{1}{c}\left[\mathbf{v}_n \times \mathbf{B}_0\right].$$

When condition (9.3) holds it is possible to neglect electric fields and corresponding currents of the secondary Alfvén waves caused by the FMS-wave (see (9.4)). Let $L_\perp \sim 1000$ km and $T \sim 100$ s, $\Sigma_H \sim \Sigma_P = 0.7 \times 10^8$ km / s for the dayside ionosphere. In the night, the conductivity is smaller at least by a factor of 10. Then, (9.4) gives

$$\frac{E_A^{(1)}}{E_A^{(0)}} \sim 0.2 \qquad \text{day}, \tag{9.24}$$

$$\frac{E_A^{(1)}}{E_A^{(0)}} \sim 0.02 \qquad \text{night}. \tag{9.25}$$

The resulting estimations (9.24) and (9.25) may be considered as a foundation for the investigation of geomagnetic variations using the mathematical formalism suggested in the theory of long-period geomagnetic variations [5].

The above can also be used to study quasi-stationary currents and electric fields ([9], [11]). These estimations allow us to find the equivalent current systems associated with Alfvén and FMS-waves separately.

The distribution of field-aligned current $j_{\|}(\theta', \lambda')$ associated with Alfvén waves is connected with the horizontal ionospheric current $\mathbf{I}_\tau$ through the two-dimensional divergence operator

$$-\boldsymbol{\nabla}_\perp \cdot \mathbf{I}_\tau = \boldsymbol{\nabla}_\perp \cdot (\hat{\boldsymbol{\Sigma}}\, \boldsymbol{\nabla}_\perp \Phi_0) = j_{\|}, \tag{9.26}$$

where Φ_0 is a scalar electrostatic potential of the polarization electric fields $\boldsymbol{\nabla}_\perp \Phi_0$; θ' and λ' are latitude and longitude, respectively. Solving (9.26) one can find $\Phi_0(\theta', \lambda')$ and currents $-\hat{\boldsymbol{\Sigma}}\, \boldsymbol{\nabla}_\perp \Phi_0$ induced by the Alfvén wave.

In the case of FMS-waves we put $j_{\|} = 0$ in the right-hand side of (9.26) and take into consideration the external electric field $\mathbf{E_0}$ as the main source of ionospheric electric fields and currents:

$$\boldsymbol{\nabla}_\perp \cdot (\hat{\boldsymbol{\Sigma}}\, (-\boldsymbol{\nabla}_\perp \Phi_0 + \mathbf{E_0})) = 0. \tag{9.27}$$

To solve (9.26)–(9.27) the following boundary conditions for electrostatic potential on the poles can be used:

$$\Phi_0 = 0. \tag{9.28}$$

In order to describe the geomagnetic variations produced by Alfvén and FMS-waves we use their representation in terms of equivalent current systems. These are two-dimensional solenoidal currents in the ionosphere, with a magnetic effect equal to the effect observed on the ground. Such fictitious currents are defined by

$$\mathbf{j}_{\rm eq} = \boldsymbol{\nabla}_\perp \Psi \times \mathbf{n} \tag{9.29}$$

where $\mathbf{j}_{\rm eq}$ is the equivalent current density; $\mathbf{n}$ is a unit vector of the normal to the ionospheric surface (positive upward); Ψ is called the current function and the current is parallel to the lines $\Psi = \text{const}$.

If there are no field-aligned currents, the equivalent current is considered to be the real ionospheric current. In a three-dimensional case, the equivalent and real currents are different because the magnetic effect on the ground is produced by field-aligned currents as well. To calculate the equivalent currents in this case we can use the following algorithm.

Alfvén Wave

Let us assume that the solenoidal ionospheric current system is generated by a localized field-aligned current $\mathbf{j}_{\|}$ which flows into a homogeneous high-latitude ionosphere with nonzero Pedersen and Hall conductivities. It is easy to show that Ψ_{sol} of this current system depends on the horizontal coordinate R as $\ln R$ and that the current density and magnetic field are proportional to R^{-1}.

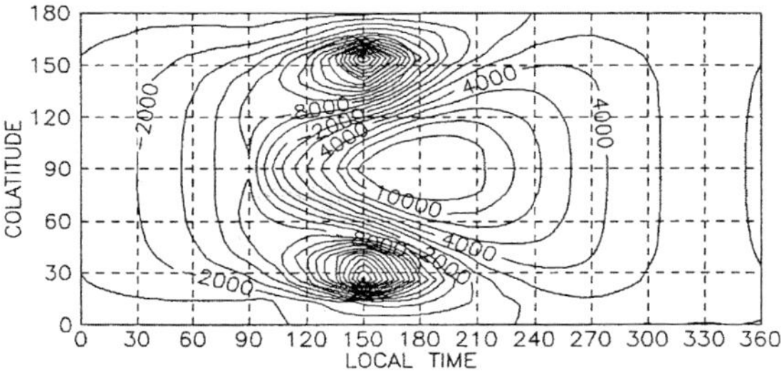

Fig. 9.2. The ionospheric equivalent current induced by a field-aligned current flowing into the Northern hemisphere at 09.00 LT and outflowing at the same time from the Southern hemisphere. Equinox. The geomagnetic field is dipole-like. Σ_P and Σ_H depend on latitude and local time as $\Sigma_P(\theta', t) = \Sigma_{P0}\sqrt{\cos Z}$ (day), $\Sigma_P = \Sigma_{P0}/30$ (night); $\Sigma_H(\theta', t) = \sqrt{\cos Z}$ (day), $\Sigma_H = \Sigma_{H0}/30$ (night); $\Sigma_{P0} = 3\,\mathrm{Ohm}^{-1}$, $\Sigma_{H0} = 5\,\mathrm{Ohm}^{-1}$. The sunrise and sunset terminators coincide with longitudes 90° and 270° respectively, or with the local time meridian 06.00 LT and 18.00 LT

If the ionospheric conductivity distribution is a continuous function of the coordinates, the spatial distribution of solenoidal currents Ψ_{sol} and magnetic field as well are quite complicated, even for the localized field-aligned currents and a simple spatial dependence of Σ_p and Σ_H. Let us assume that there is a system of two field-aligned currents with one of them flowing out from the Southern Hemisphere, and the other flowing into the Northern Hemisphere, each of them generates a solenoidal current system in the ionosphere. Figure 9.2 shows the solenoidal current system Ψ_{sol} for equinox conditions with field-aligned currents as specified above, and localized at latitudes $67°N$ and $67°S$, on local time meridian 150°. The geomagnetic field is dipole-like.

Elements of the ionospheric conductivity tensor depend on the Sun's zenith angle. Even in such a simple case, when conductivities are symmetrical with respect to the noon meridian, there is no a symmetrical spatial distribution of the solenoidal current system Ψ_{sol}.

At the sunrise hours (∼90°) the solenoidal current system is transformed more significantly because its source is close to the terminator. The equivalent current turns near the sunrise terminator and aligns with the low and equatorial latitudes along the terminator. There is an additional current vortex on the sunlit part of the ionosphere associated with the mutual penetration of equivalent currents in both hemispheres due to the interaction between the current systems in them.

The magnetic field in the equatorial region is directed along a parallel and the terminator does not influence the rotation of the magnetic vector under the ionosphere. The latter statement is valid, of course, for arbitrary odd disturbances during the equinox. The odd disturbances in terms of field-aligned currents mean that a pair of currents of the opposite direction in the conjugate points in the Northern and Southern Hemispheres is considered.

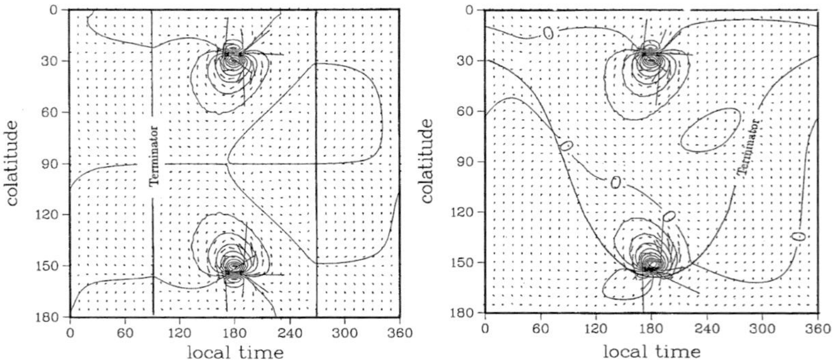

Fig. 9.3. Spatial distribution of the equivalent current intensity and its vectors induced by the incident Alfvén wave with the 'bell-like' coordinate dependence of its magnetic field intensity $\propto \exp(-x^2/2\,L^2)$. x is the southward coordinate; $L \sim$ 100 km. The azimuthal extension of the source was chosen to be 20°. The beam of Alfvén waves is incident on the high-latitude ionosphere near the noon meridian. Equinox (a) and Summer in the Northern hemisphere (b)

Contrary to the above case, the even disturbances are associated with field-aligned currents of the same direction in the conjugate hemispheres.

Let us assume, for example, that there is a geomagnetic pulsations maximum at high latitudes (~67°). Let the latitude dependence of these pulsations intensity be described by the 'bell'-like function $\propto \exp(-x^2/2\,L^2)$, where x is the southward coordinate, and L is the characteristic half-width of the 'bell'. Using such a function, it is possible to approximate an amplitude distribution of many types of oscillations, for example $Pi2$. This distribution of the magnetic field in an Alfvén wave corresponds to a pair of field-aligned currents of opposite directions extended along a parallel, and separated by a space scale L in latitude. The respective system of ionospheric equivalent currents Ψ_{sol} is presented in left-hand panel in Fig. 9.3 for equinox. In this case, the source (field-aligned current) is localized on the noon meridian in both hemispheres.

When the disturbances are small-scale along a parallel, a spatial distribution of the solenoidal current system Ψ_{sol} is dipole-like. Such a current system makes the equivalent current and magnetic field dependent on the distance, like R^{-2} at long distances from its ionospheric source, which is a projection of the Alfvén magnetospheric source on the ionospheric surface in the case under consideration. In the middle and high latitudes, the current's stream line (line of equal current) aligns along the terminator, and two current vortices arise in the middle and low latitudes in both hemispheres. One of these vortices captures almost all the sunlit part of the ionosphere and spreads to the nightside. The other vortex is localized in the middle latitudes and has a focus at colatitude ≈75° at local time 15 LT (225°). At the equator, as well as in the case of a localized point source, there are no terminator effects. Terminator influence becomes more or less significant beginning at colatitude ≈60°.

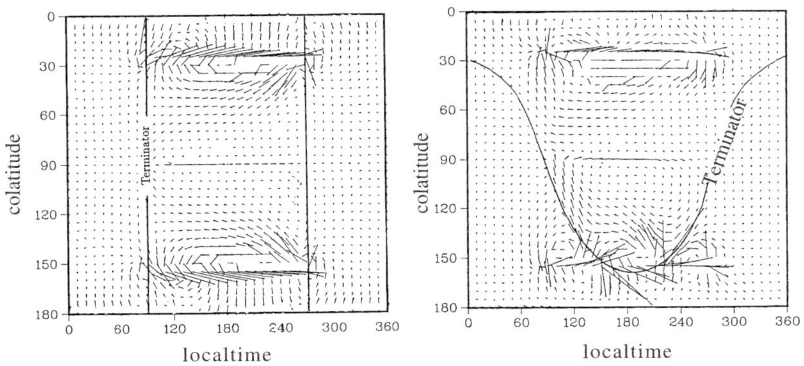

Fig. 9.4. (a) The same as in Fig. 9.3, but for source of 180° azimuthal width. Equinox. (b) The same as on left frame, but for the summer in the Northern Hemisphere

The spreading pattern changes abruptly, going from the equinox to summer or winter. Figure 9.3 (right frame) shows the current distribution for the 'bell'-like magnetic component of the magnetospheric source of the Alfvén wave for summer in the Northern Hemisphere. The bold solid line approximately represents the terminator position. One can see a strong transformation of the current in the Southern Hemisphere due to the vicinity of the terminator to the equivalent current source. Due to the dipole character of the source, and as its sequence due to the fast decreasing of current when moving away from the source, the terminator does not influence the ionosphere current distribution or the magnetic field under the ionosphere. In the Northern Hemisphere, winter conditions with the same source, symmetrical with respect to the equator, we obtain the same equivalent currents distribution Ψ_{sol} as in the Southern Hemisphere (Fig. 9.3).

When the pulsation source is large-scale along a parallel (this is valid for Alfvén waves with small azimuthal wavenumbers) and its length is comparable with the distance to the equator, the spatial spreading of such field-aligned currents changes abruptly. The intensity of the spreading currents and their magnetic field on the ground grow in the equatorial region (see Figure 9.4a for the equinox, and Fig. 9.4b for the summer in the Northern Hemisphere). An additional maximum in the latitudinal pulsations intensity distribution appears in the equatorial region. A strong current flows along the terminator and closes upon the source through the equator due to the weak space decrements, contrary to the case of high wavenumbers in which there is a strong interference at adjacent source edges. In Fig. 9.4b the source is in the Southern Hemisphere. An anomaly in the magnetic vector rotation is clearly seen at the equator. A maximum deviation of about 1 hour takes place before the arrival of the terminator. It is clear that for other seasons (summer in the Southern Hemisphere), the magnetic vector will turn symmetrically about the equator.

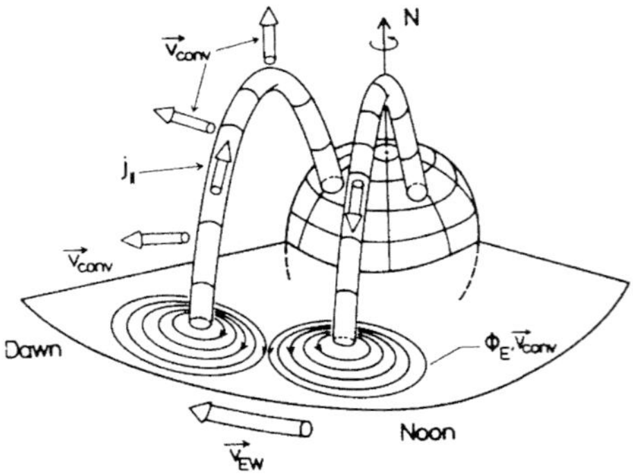

Fig. 9.5. Twin-vortex ionospheric equivalent current system generated by the reconnection of the interplanetary magnetic field with the geomagnetic magnetic field. Two field-aligned currents flow into each hemisphere. After Glassmeier *et al.* [7]

Thus, it is possible to argue, that the seasonal anomaly of the rotation angle of the magnetic field, main polarization axis may be explained within the framework of the model considered above.

Near the equator where the current system of the Southern Hemisphere is closed, additional regions appear, where the terminator effect significantly decreases. In our simulations, these regions are at colatitude ≈85°.

Twin-Vortex Current

Glassmeier [8] extracted a special type of high-latitude long-period pulsations. Their magnetic effect on the ground may be expressed in terms of an equivalent twin-vortex current system that moves along the Earth's surface at 2–5 km/s. One vortex rotates clockwise and the other counter-clockwise (see Fig. 9.5). Pulsations such as these were said to be the result of the reconnection of geomagnetic and interplanetary magnetic fields on the dayside of the magnetosphere. Next, these ULF MHD perturbations are carried by an Alfvén wave to the conjugate hemisphere. The pair of field-aligned currents brought by the Alfvén wave form a closed circuit through the ionosphere. To summarize, the equivalent twin-vortex current system that had been created in the dayside magnetosphere reappears in the ionosphere.

During the equinox (Figure 9.5), part of the west vortex closes on the middle and high latitudes. Due to the low conductivity during nighttime, the low-latitude part of the current turns southward to the terminator and flows along the equator. Thereafter, the current closes on the equatorial ionosphere. Currents in the Northern and Southern Hemispheres have the same direction in the equatorial region. Therefore, the total equatorial current is doubled in the equatorial region.

Seasonal changes do not significantly influence the general distribution of the equivalent current system. Much as it does during equinox, the current

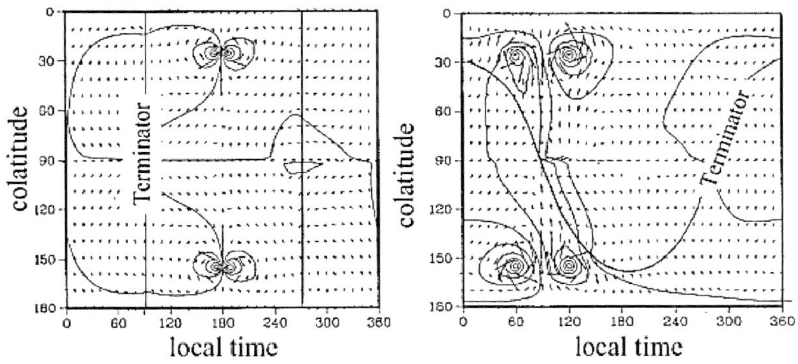

Fig. 9.6. Twin-vortex ionospheric equivalent current system arising due to the source at the noon meridian. Equinox (left panel). In right panel are shown corresponding system for the source shifted to the sunrise hours. Summer in the Northern hemisphere

spatial distribution depends on the terminator location. However during non-equinox conditions, due to conductivity differences between the hemispheres, currents flow between them. This is not true during an equinox when as it is well know, the hemisphere's conductivities are equal. In the sunrise/sunset hours, an appreciable azimuthal component of the magnetic field, as well as an anomaly of the main polarization axis, should appear. If the source field-aligned current system is shifted from the noon meridian, let say, to the morning hours (Fig. 9.6, summer in the Northern Hemisphere), the equator disturbance intensity can reach 25 % of its value near the source.

FMS-Wave

Let us consider currents in the inhomogeneous ionosphere associated with the FMS-wave incident on the equator. Longitudinal electric current in the FMS-wave is nil, therefore the electric and magnetic components in such wave can be expressed in terms of the potential ψ (see Chapter 7). If we adopt an $\{x, y, z\}$-coordinate system where the local orientation of the axis are southward, east and upward along the Earth's radius, then the horizontal electric $E_y^{(i)}$ and magnetic $b_x^{(i)}$ components can be written in the form:

$$b_x^{(i)} = \psi^{(i)} \left(\frac{\omega^2}{c_A^2} - k_x^2 - k_y^2 \right) \sin \theta'$$

and the electric component as

$$E_y^{(i)} \approx \psi^{(i)} \frac{\omega}{c} \sqrt{\frac{\omega^2}{c_A^2} - k_x^2 - k_y^2} \cos \theta'.$$

Here $\psi^{(i)}$ is the potential of the incident wave, k_x and k_y are horizontal latitudinal and azimuthal wavenumbers, respectively and θ' is the colatitude. Since the maximum horizontal scale on the ground is the equator length, then for the typical values of Alfvén velocity $c_A \sim 10^3$ km, frequency $\omega = 2\pi/T \sim 6\times10^{-2}$ and horizontal wavenumber $k \sim 10^{-4}\,\mathrm{km}^{-1}$ the term ω^2/c_A^2 in the expressions for b_x and E_y is negligibly small.

Let us assume that the azimuthal dependence is absent in the wave, e.g., the azimuthal wavenumber is $k_y = 0$. Then, the electric fields and currents in the ionosphere are proportional to the frequency for the fixed intensity of the incident magnetic component. The horizontal magnetic component has a magnitude discontinuity that depends on the ionospheric current generated by the wave electric component. When going to lower ω, the value of the electric field will decrease as long as the magnetic component intensity stays the same. This leads to a decrease in the ionospheric current. The influence of the ionosphere on the FMS-wave will also decrease. The FMS-wave electric field induces both ionospheric and ground currents.

The ground current is concentrated mainly within the skin depth

$$d_g = \frac{c}{2\pi}\sqrt{\frac{T}{\sigma_g}}$$

The total electric surface current $I_S^{(g)}$ driven by the horizontal electric field $E_S^{(g)}$ of the FMS-wave in the Earth is

$$I_S^{(g)} \sim \sigma_g d_g E_S^{(g)} = \frac{c}{2\pi}\sqrt{T\sigma_g}E_S^{(g)}.$$

The height integrated ionospheric current is

$$I_S^{(ion)} \sim \Sigma_P E_S^{(ion)}$$

Hence

$$\frac{I_S^{(ion)}}{I_S^{(g)}} \sim \frac{\Sigma_P}{\sigma_g d_g}\frac{E_S^{(ion)}}{E_S^{(g)}} = \frac{\Sigma_P}{c}\frac{2\pi}{\sqrt{T\sigma_g}}\frac{E_S^{(ion)}}{E_S^{(g)}}. \tag{9.30}$$

For FMS-waves with frequency ω and horizontal wavenumber k (7.16), (7.34) and (7.113) give

$$\frac{E_S^{(ion)}}{E_S^{(g)}} = \frac{\zeta_3(0)}{\zeta_3(h)} = \cosh kh - i\frac{k_0}{k}Y_g^{(m)}\sinh kh. \tag{9.31}$$

For disturbances with horizontal scales > 100 km the inequality $kd_g \gg 1$ holds almost everywhere (horizontal scale is big in comparison with the skin depth). So the admittance $Y_g^{(m)} = \sqrt{\varepsilon_g} = (1+i)(k_0 d_g)^{-1}$. Let also $kh \ll 1$, then (9.31) becomes

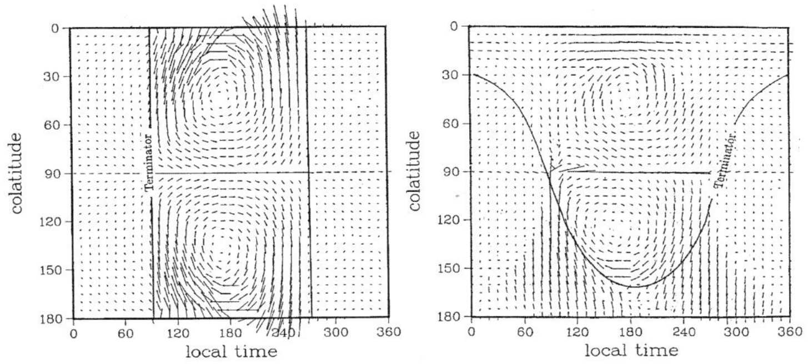

Fig. 9.7. Left panel. Ionospheric equivalent current system excited by the FMS-wave with the $E_0 \propto \sin\theta'$. Azimuthal wavenumber is assumed to be 0. Equinox. Right panel. Same as left panel, but for the summer in the Northern hemisphere

$$\frac{E_S^{(ion)}}{E_S^{(g)}} \approx 1 + \frac{h}{d_g} - i\frac{h}{d_g}. \qquad (9.32)$$

Therefore the ratio

$$\left|\frac{I^{(ion)}}{I^{(g)}}\right| = 4 \times 10^{-2} - 3 \times 10^{-2}$$

for the dayside ionosphere with $\Sigma_P = 0.9 \times 10^8$ km/s, oscillations with period $T = 60$ s under variation of the ground conductivity from $\sigma_g = 10^6\,\mathrm{s}^{-1}$ to $10^7\,\mathrm{s}^{-1}$. It follows from (9.30) and (9.32) that with T increasing the role of the ionospheric current diminishes.

The left panel of Fig. 9.7 shows the ionospheric current distribution induced by FMS-waves. Two current, oppositely directed vortices appear on the sunlit part of the ionosphere in the Northern and Southern Hemispheres, and close to the poles and the equator. These currents and their magnetic field increase at the equator. The terminator effect takes place everywhere in the ionosphere, and is especially strong at the geomagnetic equator.

The seasonal difference between the hemispheres leads to a significant enhancement of the equator oscillations intensity near the terminator (Fig. 9.7, right panel). The magnetic vector makes a $\pi/2$ turn when going from nightside SW - NE to dayside SE - NW.A transition from summer to winter in the Northern Hemisphere leads to the symmetrical rotation of the current density vectors with respect to the equator, and to significant seasonal variations of the polarization ellipse main axis. This effect should not appear under equinox conditions.

However, owing to the finite conductivity of the Earth, the magnetic effect of the FMS-wave reflected from the Earth's surface will be much more significant than that associated with the incident wave. Therefore, the total magnetic field of the FMS-wave (incident + reflected) will not represent some equatorial features near the terminator, contrary to the case when only the incident FMS-wave is under consideration.

9.3 Experimental Verification of the MHD-Wave Polarization

Experimental data based on the observations of pulsations at low latitudes, especially in the region of the equatorial electrojet and near the terminator, provide a unique possibility to classify oscillations not only according to morphological criteria, but also primarily, according to physical features and mechanisms of their excitation. The advanced ideas and numerical results show essential differences between the mechanism of ionospheric transformation of Alfvén polarized MHD-waves and that of FMS-polarized waves.

The isotropic homogeneous ionosphere shields the Earth from the magnetosphere almost completely for low-frequency Alfvén waves. A deviation from isotropy, i.e. when Hall conductivity exists or when the ionosphere is inhomogeneous, makes the ionosphere becomes transparent and allows the magnetic field to be seen below the ionosphere from the ground. Since the Pedersen and Hall conductivities have approximately the same order of magnitude, an increase will take place in the transmission coefficient of the Alfvén wave in the terminator zone which separates the high-conductive sunlit ionosphere from the low-conductive night ionosphere. The initial azimuthal magnetic component of the incident wave makes a $\pi/2$ turn under the ionosphere due to Hall currents.

In an homogeneous ionosphere moving from the dayside to nightside, the angle does not change by more than a few degrees. On the other hand, at the terminator itself, the horizontal magnetic vector may rotate by more than 10 degrees.

Another anomaly occurs in the region of equatorial electrojets. This region is extremely sensitive to whether the Alfvén wave mode is even or odd. The equator is a special area, even in the event, that waves originate in remote magnetospheric regions and penetrate from high to middle and low latitudes. This is primarily due to the large Cowling conductivity, and also owing to the change of the Hall conductivity sign as a result of the proximity of the ionospheres of the Northern and Southern Hemispheres. The current systems generated by the symmetric and antisymmetric modes of the Alfvén wave display an essential difference on the equator.

When field-aligned currents have the same direction in conjugate hemispheres (the symmetric mode), a ground magnetic variations anomaly appears on the equator, about a quarter of the variation below the source field-aligned currents. A well-defined seasonal effect of the terminator also takes place, as it follows from Figures 9.4–9.5. Long-period pulsations due to the reconnection of the geomagnetic field and the IMF [8] may be considered as an example of such oscillations. The seasonal effect must be larger on the sunrise for $Pc3, 4$ oscillations whose maximum occurrence frequency is shifted to the morning hours.

The Alfvén wave is a source of field-aligned currents which generate Hall and Pedersen currents systems. In the more general case of an arbitrary spatial dependence of conductivities, the total ionospheric current system caused by the magnetospheric Alfvén wave, may be divided into curl-free and solenoidal parts. The former are reflected into the magnetosphere. And therefore an ground observer will only see the solenoidal part.

Transition from dayside to nightside does not influence the solenoidal current as long as the incident wave intensity stays the same. In other words, setting the intensity of the incident wave magnetic component is the same as setting the longitudinal current flowing into the ionosphere. And since the wave is incident on low conductive ionosphere and $j = \Sigma E$, a very large electric field must be accordingly induced.

The FMS-wave has no field-aligned current. The electric field is proportional to ω. When going to lower ω, the value of the electric field will decrease as long as the magnetic component intensity stays the same. This leads to a decrease in the ionospheric current. The influence of the ionosphere on the FMS-wave will also decrease. For $\omega \to 0$ the ionosphere becomes transparent for FMS-waves. Therefore, on the ground, the initial plus induced in the Earth magnetic component will be measured.

Therefore, to better understand what type of waves are observed on the ground, it is necessary to define the frequency response of the polarization vector as it depends on ionospheric conductivity and on local time. The main effect must take place near the terminator (e.g. [12], [13], [14]).

The location of the inflowing field-aligned currents associated with Alfvén waves depends on the location of the generator in the magnetosphere. Then the magnetic field on the ground may be seen as a field induced by ionospheric currents. Spatial and time distortions of the current and of the ground magnetic field, induced due to the anomalies of ionospheric conductivity, do not have to be frequency dependent.

The daily variation of the orientation angle may be considered to be a feature of FMS-waves that may be used to differentiate them apart from Alfvén waves.

Figure 9.8 demonstrates the dependence of the orientation angle of the horizontal magnetic vector on local time LT for three frequency ranges: 0–50, 50–150, 150–500 s ([1], [3]). The sunrise effect takes place within all ranges. The typical angle of magnetic vector rotation near the terminator is about 45°. Solid lines in the middle and bottom panels demonstrate the orientation angle behavior at the equator taken from Fig. 9.3(a) and Fig. 9.6(a), respectively. It is assumed that the pulsations in the period range $50 < T < 150$ s originate due to some bounded Alfvén beams incident on the high-latitude ionosphere (Fig. 9.3(a)), whereas the long period pulsations with $150 < T < 500$ s are caused by a twin-vortex current system (Fig. 9.6(a)).

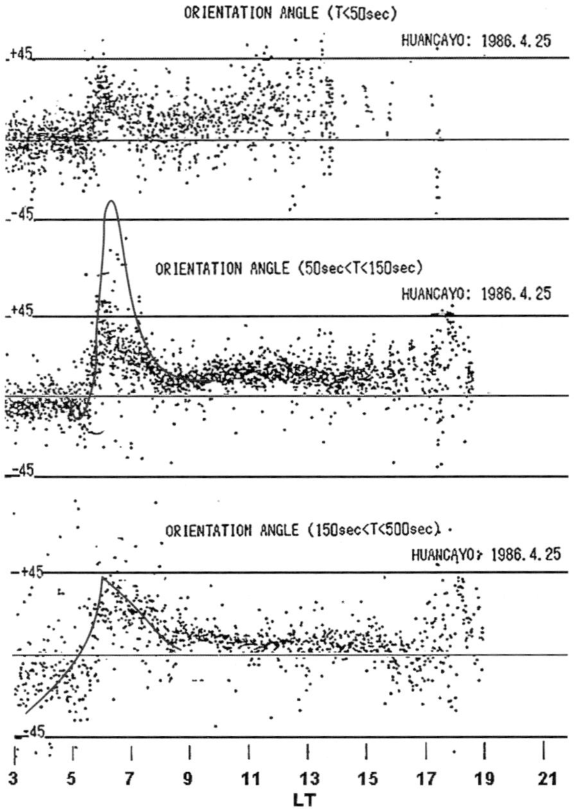

Fig. 9.8. Daily variation of the orientation angle (in degrees) of the pulsation horizontal magnetic field for pulsations within three different ranges $T < 50\,\mathrm{s}$; $50 < T < 150\,\mathrm{s}$ and $150 < T < 500\,\mathrm{s}$. The equatorial observatory Huancayo, 04/25/1986

For the same date (04/25/86) the time dependence of the logarithm of the H-component power was studied within the frequency range 0–0.06 Hz. The largest intensity found was recorded for long-period $T \sim 150-500\,\mathrm{s}$ oscillations with clearly expressed diurnal variations of the H-component values with a maximum at noon, and with an abrupt fading before sunrise. Maximal intensity of the oscillations with periods $T \sim 50-150\,\mathrm{s}$ is closer to $T \sim 150\,\mathrm{s}$. Pulsations with $T < 50\,\mathrm{s}$ fade abruptly. Despite the significant scattering behavior of the pulsation intensity with $T < 50\,\mathrm{s}$ and $T > 150\,\mathrm{s}$ a clear ($\approx 45°$) anomaly was found in the orientation angle. The orientation angle anomaly corresponds mainly to 50 s oscillations and to oscillations with periods 130–140 s.

To summarize one can conclude that oscillations within the range 50–500 s are generated by Alfvén waves. Shorter $T \approx 40\,\mathrm{s}$ oscillations are generated by FMS-waves. A more detailed analysis of the experimental data indicates that

$T \approx 40$ s oscillations are not sensitive to sunrise variations of the ionospheric conductivity. Therefore, we suggest that these oscillations are in fact FMS-waves.

The intensity of $T \leq 30$ s oscillations increases in the morning and almost disappears during the day. These events may indicate that either the magnetospheric source of these pulsations is close to the morning hours or that the terminator is an active ionospheric generator of oscillations within this range.

To further clarify the incident wave type, it seems reasonable to carry out simultaneous observations at least at two observatories which are far enough from each other along a parallel. One of them should be located at a low resistance cross-section and the other at a high resistance cross-section.

If the low latitude pulsations are caused by poloidal oscillations within the plasmasphere, the first observatory may be chosen as the standard, because anomalies of ionospheric conductivity will not influence the polarization features of pulsations at its location.

If the initial wave is an FMS-wave, the orientation angle must not change as a function of local time within a wide pulsation frequency range. When the pulsations orientation angle is sensitive to the presence of the terminator in a wide frequency range, they may treated as Alfvén waves. Waves may be classified as FMS-waves on the other hand, if their orientation angle is dependent on frequency and is insensitive to the presence of the terminator. These waves may be measured at observatories located on high-resistance cross-sections. An example of spectral-temporal analysis of pulsations at the equatorial observatory in Huancayo was demonstrated to confirm the proposed experimental technique. A weak dependence was found of the orientation angle anomaly on the frequency near the terminator. The latter is evidence for the dominant contribution of Alfvén waves to the low latitude and equatorial oscillations.

References

1. Alperovich, L., B. Fidel, and O. Saka, A determination of the hydromagnetic waves polarization from their perturbations on the terminator, *Ann. Geophys.*, **14**, 647, 1996.
2. Alperovich, L., B. Fidel, and O. Saka, Determination of Magnetospheric wave polarization using ground-based sunrise geomagnetic observations, *J. Geomagn. Geoelectr.*, **48**, 79, 1996.
3. Alperovich, L., B. Fidel, and O. Saka, Ground-based diagnostics of the magnetospheric MHD-waves polarization using sunrise geomagnetic observations, *J. Geomag. Geoelectr.*, **48**, 79, Japan, 1995.
4. Glassmeier, K. H., Reflection of MHD-waves in the Pc4–5 period range at ionospheres with non-uniform conductivity distributions, *Geophys. Res. Lett.*, **10**, 678, 1983.
5. Glassmeier, K. H., On the influence of ionospheres with nonuniform conductivity distribution on hydromagnetic waves, *Geophys. J.*, **54**, 125, 1984.

6. Glassmeir, K. H., Ground-based observations of field-aligned currents in the auroral zone: Methods and results, *Ann. Geophys.*, **5**, 115,1987.
7. Glassmeier, K. H., M. Honisch, and J. Untiedt, Ground-based and satellite observations of travelling magnetospheric convection twin-vortices, *J. Geophys. Res.,* **94**, 2520, 1989.
8. Glassmeir, K. H., Traveling magnetospheric convection twin-vortices: Observations and theory, *Ann. Geophys.,* **10**, 547 1992.
9. Gurevich, A. V., A. L. Krylov, and E. E. Tsedilina, Electric fields in the Earth's magnetosphere and ionosphere, *Space Sci. Rev.*, **19**, 59, 1976.
10. Itonaga, M. and T. Kitamura, Effect of non-uniform ionospheric conductivity distributions on $Pc3-5$ magnetic pulsations – Alfvén wave incidence, *J. Geomag. Geoelectr.*, **40**, 1413, 1988.
11. Kamide, Y., A. D. Richmond, and S. Matsushita, Estimation of ionospheric electric fields, ionospheric currents and field-aligned currents from ground magnetic records, *J. Geophys. Res.*, **86**, 801, 1981.
12. Saka, O., T.-J. Ijima, and T. Kitamura, Ionospheric control of polarization of low-latitude geomagnetic micropulsations, *J. Atmos. Terr. Phys.*, **42**, 517, 1980.
13. Saka, O., M. Itonaga, and T. Kitamura., Ionospheric control of polarization of low-latitude geomagnetic micropulsations at sunrise, *J. Atmos. Terr. Phys.,* **44**, 703, 1982.
14. Saka, O. and L. Alperovich, Sunrise effect on dayside pulsations at the dip equator, *J. Geophys. Res.*, **98**, 13779, 1993.

10

Effective Conductivity of a Cloudy Ionosphere

10.1 Introduction

Up to this point, we have treated the ionosphere as a thin anisotropic layer. As indicated in Chapter 1, to obtain the integral conductivity expression, we use the three-fluid hydrodynamics model for electron, ion and neutral gases and derive expressions for their specific conductivities. Due to the high conductivity along the external magnetic field $\mathbf{B}_0$, the longitudinal electric field vanishes and the transversal electric component is necessarily constant. This allows us to simplify the detailed treatment and to use the height-integrated conductivity $\boldsymbol{\Sigma}(x, y)$ instead of the specific conductivity $\boldsymbol{\sigma}(x, y, z)$.

This implies that distributed, isolated or random inhomogeneities can be considered to be perturbations of $\boldsymbol{\Sigma}$. Hence, small electron concentration perturbations and integral conductivity perturbations both produce small perturbations of local Pedersen and Hall currents. These currents flow through an inhomogeneous ionosphere and as a consequence produce magnetic perturbations above and below the ionosphere ([3], [4]).

Using $\boldsymbol{\Sigma}(x, y)$ instead of $\boldsymbol{\sigma}(x, y, z)$, we in fact replace a real physical object by a thin sheet current. This invariably leads to some significant plasma features to be lost. The lack of these features can, in general, have a drastic impact on the current and as consequence on the value of $\boldsymbol{\Sigma}$. Specific conductivity over the whole ionosphere is defined by both ions and electrons. It can be presented as a sum of tensor electron $\boldsymbol{\sigma}_e(x, y, z)$ and ion $\boldsymbol{\sigma}_i(x, y, z)$ conductivities. Both of these conductivities have Pedersen $\boldsymbol{\sigma}_{P(e,i)}(x, y, z)$ and Hall $\boldsymbol{\sigma}_{H(e,i)}(x, y, z)$ components, four tensor components in whole. The relative contribution of these four components to the total $\boldsymbol{\sigma}_P(x, y, z)$ and $\boldsymbol{\sigma}_H(x, y, z)$ is defined primarily by electron β_e and ion β_i magnetization parameters (see Fig. 2.5).

Over the whole ionosphere from 80 km, $\beta_e \gg 1$, but β_i can be larger or smaller than 1. As a result, Hall conductivity is governed mainly by electrons while σ_P is determined by either electrons or ions. In the low ionosphere where $\beta_i \ll 1$, the electric conductivity is therefore determined only by electrons, i.e. both Pedersen $\sigma_P(x, y, z)$ and Hall $\sigma_H(x, y, z)$ conductivities are electron conductivities, respectively, $\sigma_{Pe}(x, y, z)$ and $\sigma_{He}(x, y, z)$, with $\sigma_{He}(x, y, z) \gg \sigma_{Pe}(x, y, z)$. Hence, an electric field applied to such plasma produces a large Hall current $\mathbf{j}_H$ and a very small Pedersen current $\mathbf{j}_P$.

Since our concern is the integral magnetic field on the ground, caused by a large-scale MHD-wave that can simultaneously 'cover' a vast number of small random inhomogeneities, the magnetic effect caused by these irregular currents is equivalent to the magnetic effect of an average current flowing over the ionosphere with an effective conductivity σ^{eff}. Therefore the problem reduces to the following question: how can we define σ^{eff} if we have detailed information either about every inhomogeneity or about the correlation properties of the random inhomogeneous distribution of the specific conductivities?

At first glance it would seem that σ^{eff} is the sum of the mean conductivity and its variability. But, this is not necessarily so. The reason for this is that the specific conductivity is a tensor. And therefore the current is a product of the electric field vector and this tensor. The resulting current is rotated with respect to the applied electric field. In the degenerate case, in which σ_P is much smaller than the σ_H and the plasma is homogeneous, only Hall current is generated by the applied electric field. A small inhomogeneity creates a polarized electric field and a Pedersen current collinear with the applied electric field. These are, of course, extensively simplifying assumptions. In reality, it turns out that for magnetized electron plasma the total j_P is defined by $\delta N_e \cdot \beta_{e.}$ [17].

This effect occurs because the two tensor elements of the electron conductivity depend differently on strong magnetic field ($\beta_e \gg 1$). In the case of a medium containing only one sort of charge carriers, let us say electrons, $\langle \sigma_{Pe} \rangle \propto \sigma_0 / B_0^2$ and $\langle \sigma_{He} \rangle \propto \sigma_0 / B_0$. The polarization fields produce additional Pedersen and Hall currents. Their contributions to the effective Pedersen conductivity are not equal, since $\delta\sigma_P \propto \varepsilon / B_0^2$ while $\delta\sigma_H \propto \varepsilon / B_0$ where

$$\varepsilon = \frac{\left\langle \left(N_e(\mathbf{r}) - \langle N_e(\mathbf{r}) \rangle\right)^2 \right\rangle^{1/2}}{\langle N_e(\mathbf{r}) \rangle} \equiv \frac{\left\langle \delta N_e(\mathbf{r})^2 \right\rangle^{1/2}}{\langle N_e(\mathbf{r}) \rangle} = \left(\frac{\langle \delta\sigma^2 \rangle}{\langle \sigma \rangle^2} \right)^{1/2} \tag{10.1}$$

is a value which characterizes inhomogeneities.

At height $h \approx 100$ km $\beta_e \approx 100$ (see Fig. 2.5). This means that a 10% perturbation of the electron density N_e can produce a Pedersen conductivity anomaly $\propto \varepsilon \cdot \beta_e = 10$ times greater than the unperturbed conductivity. Obviously, the connection between the initial $\mathbf{j}_H$ and the induced $\mathbf{j}_P$ and with it the effect of the anomalous σ^{eff}, disappears if instead of a finitely thick ionosphere we use the 'thin sheet' approximation, and instead of σ we use Σ.

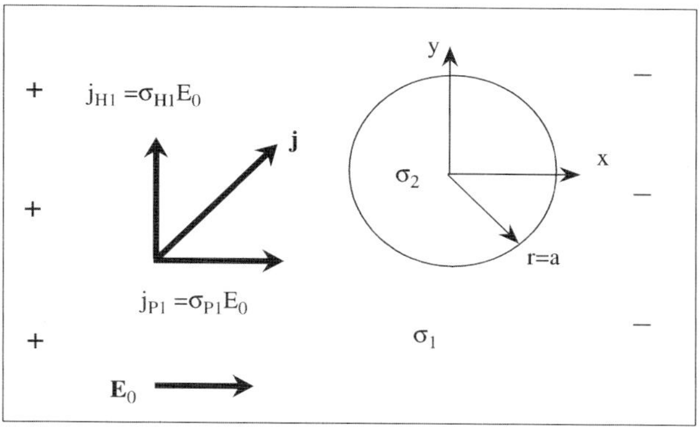

Fig. 10.1. Isolated inhomogeneity

Alfvén and Felthammer demonstrated in one specific example that changing the local specific conductivity in a plasma slab by 20% compared with the background specific conductivity, can change the effective conductivity in a strong magnetic field by an order of magnitude [1].

Local Inhomogeneity Example

The unusual behavior of σ^{eff} in a strong $\mathbf{B}_0$ can be demonstrated using a simple example. An isolated cylindrical inhomogeneity of tensor conductivity $\boldsymbol{\sigma}_2(z)$ is placed in a medium of $\boldsymbol{\sigma}_1(z)$ conductivity. For the sake of simplicity, we suppose that the external $\mathbf{B}_0$ is applied along the cylindrical axis z and that the electric field $\mathbf{E}_0$ is perpendicular to the cylinder. We also assume that currents do not penetrate from one height level to another. This would be the situation for two conjugate ionospheres with identical properties. Another example is a conductive layer bounded by nonconductive media. Then the considered example is identical to a circle inhomogeneity with a tensor conductivity $\boldsymbol{\sigma}_2$ placed on a thin conductive sheet with conductivity $\boldsymbol{\sigma}_1$ (see Fig. 10.1). A magnetic field $\mathbf{B}_0$ is perpendicular to the sheet plane. Let the applied electric field be $\mathbf{E}_0$. This electric field produces the dc electric current $\mathbf{j}_1 = \mathbf{j}_{P1} + \mathbf{j}_{H1}$ where $\mathbf{j}_{P1}$ is the Pedersen current and $\mathbf{j}_{H1}$ is the Hall current. Let axis x the Cartesian coordinate system $\{x, y\}$ with its origin in the center of the circle and x-axis along the current $\mathbf{j}_1 = (j_1, 0)$. We also use the cylindrical coordinate system $\{\rho, \varphi\}$. The inhomogeneity causes an anomalous current $\mathbf{j}(r, \varphi) = (j_r, j_\varphi)$.

The boundary conditions are: continuity of the radial current component j_r and of the tangential electric field component E_φ at $r = a_0$. Then j_r and j_φ outside the inhomogeneity are

$$\begin{pmatrix} j_r \\ j_\varphi \end{pmatrix} = j_1 \left[\mathbf{1} + \frac{a_0^2}{r^2} \begin{pmatrix} c & -b \\ b & c \end{pmatrix} \right] \begin{pmatrix} \cos\varphi \\ \sin\varphi \end{pmatrix}, \qquad (10.2)$$

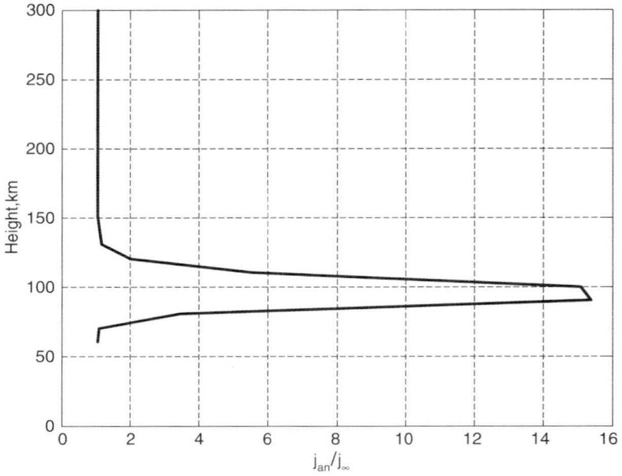

Fig. 10.2. Height dependency of the ratio of anomalous j_{an} current to the background undisturbed current j_∞

where I_1 is the current far from the inhomogeneity,

$$c = \left[\rho_1^2 - \rho_2^2 - (\delta_1 - \delta_2)^2\right] g^{-1}, \qquad b = 2\rho_1(\delta_1 - \delta_2)g^{-1},$$
$$j_1 = E_0\sqrt{\sigma_{P1}^2 + \sigma_{H1}^2}, \quad \delta_{1,2} = \sigma_{H1,2}h_{1,2}^{-1}, \qquad \rho_{1,2} = \sigma_{P1,2}h_{1,2}^{-1},$$
$$g = (\rho_1 + \rho_2)^2 + (\delta_1 - \delta_2)^2, \quad h_{1,2} = \sigma_{H1,2}^2 + \sigma_{P1,2}^2.$$

Here $\sigma_{P,H}$ is the Pedersen or Hall specific conductivities outside the inhomogeneity (index '1') or of the inhomogeneity itself (index '2'). The x-component of the current is

$$j_x = j_{P1} + \frac{a_0^2}{r^2}\left[j_{P1}(\rho_1^2 - \rho_2^2) + 2j_{H1}\rho_1(\delta_1 - \delta_2)\right] g^{-1}, \tag{10.3}$$

where $j_{P1} = \sigma_{P1}E_0$ is the Pedersen current far from the inhomogeneity. The second term in (10.3) is a component defined by the inhomogeneity, it consists of two parts. The first one, proportional to $(\rho_1 - \rho_2)$, is the anomalous Pedersen current and the second one is the term defined by the Hall conductivity anomaly.

Figure 10.2 shows the ratio of anomalous current $j_{an} = j_x - j_{P1}$ to the background undisturbed current j_{P1} as a function of ionospheric height. The local electron concentration perturbation is 10%. A conductivity change of 10% results in a 10–20% anomalous current on the bottom of the D-layer (h = 60–70 km). At altitudes 70–100 km, the anomalous Pedersen current is 5–10 times greater than the background current due to increasing β_e and a small β_i. Here σ_P is defined by electrons alone, $\sigma_P \approx \sigma_{Pe}$. One can see that a small conductivity perturbation significantly affects the Pedersen current.

The basic idea is adequately illustrated by the present elementary example and shows how, with increasing β_e, the effective Pedersen conductivity σ_P^{eff} can grow beyond proportions for small local inhomogeneities.

Since the discovery of the strong influence of small perturbations on the effective transport properties of magnetized media, many physicists have been intrigued by this effect (e.g., [11], [12], [13], [19]). The key questions was: how is the integral current caused by an external electric field applied to randomly inhomogeneous medium?

This chapter is organized as follows: in the first section we briefly review existing theories of effective conductivity both for regular media with scalar local conductivity and for magnetized media. Next we present an expression for the effective conductivity of partially ionized plasma. Then, we derive expressions for σ^{eff} for small perturbations of local conductivity in a strong magnetic field. Theoretical outcomes are compared with results of laboratory experiments on silicic inhomogeneous films placed into a strong magnetic field.

10.2 Existing Theories

Assume that the ionosphere at a certain altitude is replaced by an anisotropic sheet with a constant conductivity $\boldsymbol{\sigma}_1$ containing rarefied inhomogeneities of conductivity $\boldsymbol{\sigma}_2$. Let us consider a region of the ionospheric sheet of volume V. Suppose also that the inhomogeneities occupy a volume V_k, then $p = \sum V_k/V$ is the inhomogeneity concentration.

An effective conductivity σ^{eff} defines the relation between the volume average current density $\langle \mathbf{j} \rangle$ and the electric field $\langle \mathbf{E} \rangle$:

$$\langle \mathbf{j}(\mathbf{r}) \rangle = \sigma^{eff} \langle \mathbf{E}(\mathbf{r}) \rangle . \tag{10.4}$$

σ^{eff} is actually measured in experiments and appears in averaged Maxwell's equations. $\mathbf{j}(\mathbf{r})$ and $\mathbf{E}(\mathbf{r})$ in (10.4) are the local current and electric field. It would appear reasonable that the spatial distribution of local electron concentration inhomogeneities and hence, of conductivity $\sigma(\mathbf{r})$ are both random.

σ^{eff} does not in general coincide with the mean conductivity $\langle \sigma \rangle$ and can greatly differ from it. A wide range of theoretical approaches has been applied to this problem. One such approach concerns the case of sparse inhomogeneities (i.e. $p \ll 1$). In this case the mutual impact of inhomogeneities may be neglected and we may assume that only $\langle \mathbf{E} \rangle$ influences the inhomogeneities and therefore

$$\langle \mathbf{j}(\mathbf{r}) \rangle = \sigma_1 \langle \mathbf{E} \rangle + p\,(\sigma_2 - \sigma_1) \langle \mathbf{E}_2 \rangle ,$$

where $\langle \mathbf{E} \rangle = (1-p) \langle \mathbf{E}_1 \rangle + p \langle \mathbf{E}_2 \rangle$, and

$$\langle \mathbf{E}_1 \rangle = \frac{1}{V - V_2} \int\limits_{V-V_2} \mathbf{E}(\mathbf{r}) dv, \qquad \langle \mathbf{E}_2 \rangle = \frac{1}{V_2} \int\limits_{V_2} \mathbf{E}(\mathbf{r}) dv.$$

V_1 is the total volume of the system and V_2 is volume occupied by inhomogeneities. $\mathbf{E_1}$ and $\mathbf{E}_2$ are the fields outside and inside of an inhomogeneity, respectively. The averaging here is performed over a volume much larger than the scale size of an inhomogeneity. For rare spherical inhomogeneities (the volume concentration $p \ll 1$) of the conductivity σ_2 embedded into the medium of the conductivity σ_1, the effective conductivity σ^{eff} is given by Landau and Lifshitz [21]

$$\sigma^{eff} = \sigma_1 + p\frac{3\left(\sigma_2 - \sigma_1\right)\sigma_1}{\sigma_2 + 2\sigma_1}. \tag{10.5}$$

This relation is obtained in assumption of arbitrary σ_1 and σ_2 including the situation 'metal – dielectric'.

The next generalization phase treats dense inhomogeneities ('gas' approximation) and considers a media in which pieces with conductivity σ_1 are homogeneously mixed with pieces with conductivity σ_2. Brugemann [10] has devised a method of so-called self-consistent field. This method assumes that the inhomogeneities create an average electric field $\langle \mathbf{E} \rangle$ and that each of the inhomogeneities is embedded into an 'effective medium' of conductivity σ^{eff}. The average field created by the inhomogeneities polarizes each of them. The total electric field induced by the inhomogeneities has to be equal to $\langle \mathbf{E} \rangle$. The effective conductivity is found from this equality. For example, for the binary mixture of spherical inhomogeneities of conductivity σ_1 and arbitrary conductivity σ_2 and concentrations p_1 and $p_2 = 1 - p_1$, the equation for σ^{eff} becomes

$$\sigma^{eff} = \frac{f + \left(f^2 + 8\sigma_1\sigma_2\right)^{1/2}}{4}. \tag{10.6}$$

where

$$f = (3p_2 - 1)\,\sigma_2 + (3p_1 - 1)\,\sigma_1$$

If we now assume that $p \ll 1$ and σ_1 and σ_2 are close and substitute in (10.6), it easily follows, as it should, (10.6) tends to (10.5). If one of the components is an insulator (e.g., $\sigma_2 = 0$), σ^{eff} approaches zero. There exists a critical concentration p_c ('percolation threshold') below which $\sigma^{eff} = 0$. In the case of spherical inhomogeneities $p_c = 1/3$. The value of percolation threshold is a characteristic of a wide variety of heterogenous systems. The behavior of effective conductivity close to the percolation threshold is described by

$$\sigma^{eff} = \sigma_1 \left(p - p_c\right)^t,$$

where t is a critical index which was defined by modeling the percolation problem on different lattices ([18], [22], [24]).

The outlined calculation methods for σ^{eff} are based on the assumption that the mixed phases are separated by well-shaped edges. On numerous occasions, however, and in particular in space plasma and other systems with irregular spatial distribution, the conductivity σ is a continuous function of the coordinate $\mathbf{r}$. Herring [17] has given a theoretical treatment of the influence of random inhomogeneities on the effective conductivity of such systems.

A numerical approach has been proposed in which all fluctuated values are expanded as a Fourier series in spatial wavenumbers **k**:

$$\sigma(\mathbf{r}) = \langle\sigma\rangle + \sum_{\mathbf{k}\neq 0} \sigma_{\mathbf{k}} \exp(i\mathbf{kr}). \qquad (10.7)$$

Here $\sigma_{\mathbf{k}}$ is an amplitude of the harmonics of spatial wavenumber **k**, and $\langle\sigma\rangle$ is the mean conductivity. It has been found that in the lowest approximation in fluctuated conductivities, the effective conductivity is

$$\sigma^{eff} = \langle\sigma\rangle \left(1 - \frac{\sum_{\mathbf{k}} \sigma_{\mathbf{k}}\sigma_{-\mathbf{k}}}{\langle\sigma\rangle^2}\right) \qquad (10.8)$$

and can be rewritten in the more descriptive form

$$\sigma^{eff} = \langle\sigma\rangle \left(1 - \varepsilon^2\right). \qquad (10.9)$$

In order to calculate σ^{eff}, the spatial distribution of $\sigma(\mathbf{r})$ needs to be known. Obviously, in the case of random inhomogeneities, we would first like to theoretically determine the spectrum of inhomogeneities and only then to calculate the resulting effective conductivity. Unfortunately in the case of laboratory or space plasmas, we generally do not have to hand the actual spatial correlation properties of various plasma instabilities. But, independent of the actual form of $\sigma(\mathbf{r})$, the magnitude of σ^{eff} for small perturbations according to (10.9) is always less than the mean conductivity $\langle\sigma\rangle$.

The situation changes drastically in magnetized media. Here, the Pedersen component of the tensor σ_{ij}^{eff} becomes larger than the mean Pedersen conductivity $\langle\sigma_P\rangle$, even for small local conductivity perturbations.

Dreizin and Dychne [11] shown that the Pedersen effective conductivity σ_P^{eff} of the magnetized disorder medium is given by

$$\delta\sigma_P^{eff}(\beta_e) = A \left(\frac{\varepsilon}{\beta_e}\right)^{\mu} \sigma_0. \qquad (10.10)$$

Here σ_0 is the longitudinal conductivity along $\mathbf{B}_0$, A is a constant independent of β_e. For a 3D system the exponent is $\mu = 4/3$.

Kvyatkovsky [19] also indicated that in 2D system a specific size effect can appear – a dependence of the conductivity on the scale-size L_z along $\mathbf{B}_0$. Moreover, in very strong magnetic fields σ_P^{eff} again becomes inversely proportional to $|\mathbf{B}_0|$:

$$\delta\sigma_P^{eff}(\beta_e) = \left(\frac{\varepsilon}{\beta_e}\right) \cdot \left(\frac{l}{L_z}\right)^{\frac{1}{2}} \sigma_0 \qquad (10.11)$$

where l is the size of inhomogeneities. The method of ([11], [19]) was developed in ([5], [6]) to calculate $\sigma_\perp^{eff}$ for an electron-ion ionospheric plasma with random inhomogeneities.

10.3 Inhomogeneous Plasma

Stochastic Inhomogeneities

Let us consider a stationary current in a medium in which the conductivity tensor $\boldsymbol{\sigma}(\mathbf{r})$ is a random function of the coordinates. We will restrict our consideration to analysis of small-scale disturbances with length-scale much smaller than the size of the system. Then Ohm's law connecting local current density $\mathbf{j}(\mathbf{r})$ and local electric field $\mathbf{E}(\mathbf{r})$ can be written as

$$\mathbf{j}(\mathbf{r}) \;=\; \boldsymbol{\sigma}(r)\mathbf{E}(\mathbf{r}). \tag{10.12}$$

In addition, $\mathbf{j}(\mathbf{r})$ and $\mathbf{E}(\mathbf{r})$ satisfy the equations

$$\boldsymbol{\nabla}\cdot\mathbf{j}(\mathbf{r}) \;=\; 0, \quad \boldsymbol{\nabla}\times\mathbf{E}(\mathbf{r}) = 0, \quad \mathbf{E} = -\boldsymbol{\nabla}\varphi, \tag{10.13}$$

where φ is a potential.

Much more interesting, however, is not the relation between specific local conductivity and local electric field (10.12) but rather the relation (10.4) between average current $\langle j_i\rangle$ and average electric field $\langle E_k\rangle$:

$$\langle j_i\rangle = \sigma_{ik}^{eff}\langle E_k\rangle. \tag{10.14}$$

Here σ_{ik}^{eff} is the effective conductivity of a spatially inhomogeneous anisotropic system. We define the fluctuation $\delta y(\mathbf{r})$ and the mean-free value $\langle y\rangle$ of $y(\mathbf{r})$ as such that

$$y(\mathbf{r}) = \langle y\rangle + \delta y(\mathbf{r}), \tag{10.15}$$

where $y(\mathbf{r})$ is any variable. For example, the potential φ can be written as

$$\varphi = \langle\varphi\rangle + \delta\varphi \;=\; -\langle E_k\rangle\, x_k + \delta\varphi, \tag{10.16}$$

then

$$\frac{\partial\varphi}{\partial x_k} = -\langle E_k\rangle + \frac{\partial\delta\varphi}{\partial x_k}. \tag{10.17}$$

Assuming that

$$\sigma_{ik} = \langle\sigma_{ik}\rangle + \delta\sigma_{ik} \tag{10.18}$$

we have for j_i:

$$\begin{aligned} j_i &= -\sigma_{ik}\frac{\partial\varphi}{\partial x_k} = -\left(\langle\sigma_{ik}\rangle + \delta\sigma_{ik}\right)\left(-\langle E_k\rangle + \frac{\partial\delta\varphi}{\partial x_k}\right) \\ &= \langle\sigma_{ik}\rangle\langle E_k\rangle - \langle\sigma_{ik}\rangle\frac{\partial\delta\varphi}{\partial x_k} + \delta\sigma_{ik}\langle E_k\rangle - \delta\sigma_{ik}\frac{\partial\delta\varphi}{\partial x_k}. \end{aligned}$$

Then an expression for the mean current $\langle j_i\rangle$ can be written as

$$\langle j_i\rangle = \langle\sigma_{ik}\rangle\langle E_k\rangle - \langle\sigma_{ik}\rangle\left\langle\frac{\partial\delta\varphi}{\partial x_k}\right\rangle + \langle\delta\sigma_{ik}\rangle\langle E_k\rangle - \left\langle\delta\sigma_{ik}\frac{\partial\delta\varphi}{\partial x_k}\right\rangle. \tag{10.19}$$

Taking into account that the spatially averaged mean values of fluctuated variables are vanish, that is

$$\langle \delta\sigma_{ik} \rangle = 0, \text{ and } \left\langle \frac{\partial \delta\varphi}{\partial x_k} \right\rangle = 0$$

then the equation mean current (10.19) becomes

$$\langle j_i \rangle = \langle \sigma_{ik} \rangle \langle E_k \rangle - \left\langle \delta\sigma_{ik} \frac{\partial \delta\varphi}{\partial x_k} \right\rangle . \tag{10.20}$$

The physical meaning of two terms is the following. The first term is a part of the total current defined by the mean electric field and mean conductivity. The second term is the contribution into the total current due to the inhomogeneities. Therefore, the problem reduces to finding $\delta\varphi(\mathbf{r})$ with known coordinate dependency $\sigma_{ik}(\mathbf{r})$ or, equivalently, $\delta\sigma_{ik}(\mathbf{r})$. The unknown variable is fluctuations of the electric potential $\delta\varphi$ which can be found from the condition that j_i is divergence free current. From $\nabla \cdot \mathbf{j}(\mathbf{r}) = 0$ follows that

$$\frac{\partial}{\partial x_i}\left(\sigma_{ik}\frac{\partial \varphi}{\partial x_k}\right) = \frac{\partial}{\partial x_i}(\delta\sigma_{ik})\frac{\partial \varphi}{\partial x_k} + \sigma_{ik}\frac{\partial^2 \varphi}{\partial x_i \partial x_k} = 0 \tag{10.21}$$

and with (10.17) the last equation becomes

$$\frac{\partial(\delta\sigma_{ik})}{\partial x_i} \cdot \frac{\partial \delta\varphi}{\partial x_k} + \langle \sigma_{ik} \rangle \frac{\partial^2 \delta\varphi}{\partial x_i \partial x_k} + \delta\sigma_{ik}\frac{\partial^2 \delta\varphi}{\partial x_i \partial x_k} - \frac{\partial(\delta\sigma_{ik})}{\partial x_i}\langle E_k \rangle = 0. \tag{10.22}$$

Therefore, the problem reduces to finding $\delta\varphi(\mathbf{r})$ with known coordinate dependency $\sigma_{ik}(\mathbf{r})$ or, equivalently, $\delta\sigma_{ik}(\mathbf{r})$.

In the Appendix it is shown a formal solution procedure reducing to an expansion of the fluctuating variables into a Fourier series. Substitution of the series into (10.22) yields a much more tractable equation for an individual spatial Fourier harmonic and after considerable manipulation of Ohm's law for the average current (10. A.5):

$$\langle j_i \rangle = \langle \sigma_{ik} \rangle \langle E_k \rangle + \sum_{\mathbf{q}} \delta\sigma_{il}(-\mathbf{q}) \frac{B_l(\mathbf{q})}{\langle \sigma_{im} \rangle q_i q_m} \langle E_l \rangle q_l \tag{10.23}$$

which is a basis for the calculation of effective conductivity $\sigma_{ik}^{\mathit{eff}}$. Here $B_i(\mathbf{q})$ is defined by (10. A.3). By definition, $\sigma_{xx}^{\mathit{eff}} = \sigma_P^{\mathit{eff}}$ is the transverse effective conductivity for $\mathbf{E} \parallel \hat{\mathbf{x}}$ and $\mathbf{B}_0 \parallel \hat{\mathbf{z}}$. Equation (10. A.5), can be rewritten as

$$\langle j_k \rangle = \sum_{i=x,y} \left\{ \langle \sigma_{ki} \rangle + \sum_{\mathbf{q}} \frac{\delta\sigma_{kl}(\mathbf{q})\ q_l\ B_i(\mathbf{q})}{\langle \sigma_{mn} \rangle q_m q_n} \right\} \langle E_i \rangle ; \tag{10.24}$$

Here, $k = x, y$ for the corresponding current components.

From (10.24) follows

$$\sigma_{xx}^{eff} = \langle \sigma_{xx} \rangle + \delta \sigma_{xx} \tag{10.25}$$

where

$$\delta \sigma_{ij} = \sum_{\mathbf{q}} \frac{\delta \sigma_{il}(-\mathbf{q}) q_l B_j(\mathbf{q})}{\langle \sigma_{mn} \rangle q_m q_n}, \; (i, j \equiv x, y). \tag{10.26}$$

Equation (10.26) was analyzed in details in series of papers (see, e.g., [11], [19]) for strong magnetic fields $\beta_e \gg 1$.

Let σ_0 is the longitudinal conductivity. Then taking into account that $\langle \sigma_{xxe} \rangle = \sigma_0/\beta_e^2$ and $\langle \sigma_{xye} \rangle = \sigma_0/\beta_e$ one can suppose that the fluctuations $\delta\sigma_{xxe}(q)$ and $\delta\sigma_{xye}(q)$ in (10.26) are proportionally respectively to β_e^{-2} and β_e^{-1}. It follows that the fluctuations of the electron Hall conductivity will be essentially stronger than the fluctuations of the electron Pedersen conductivity. Formally, the anisotropy of a partially ionized gas under the strong magnetic field leads to that the electron terms, including the Hall's part, prevail in (10.25) and (10.26).

The influence of $\mathbf{B}_0$ on σ_{Pi} is extremely weak when $\beta_i \ll 1$. The average ion Pedersen conductivity $\langle \sigma_{xxi} \rangle$ in this case is independent of $\mathbf{B}_0$. Hall ion conductivity $\langle \sigma_{xyi} \rangle$ is proportional to the parameter of β_i. This is $\langle \sigma_{xyi} \rangle = \langle \sigma_{xxi} \rangle \beta_i \ll \langle \sigma_{xxi} \rangle$. In so doing, the fluctuations $\delta\sigma_{Pi}$ will be proportional to the fluctuations of the relative electron concentrations $\delta\sigma_{xxi}(q) \propto \varepsilon$ and the perturbations of Hall component of the ion conductivity $\delta\sigma_{xyi}$ will be lesser than Pedersen conductivity because of weak ion magnetization, this is $\delta\sigma_{xyi}(q) \propto \varepsilon\beta_i$. This means that $\delta\sigma_{xxi}$ gives the main contribution to the fluctuated ion part. It independent of the magnetic field and gives the correction to σ_P^{eff} proportional to $\varepsilon^2 \ll 1$, that is considerably lesser than $\langle \sigma_{xxi} \rangle$.

The crossed terms with the products of the correlation functions of the fluctuations of ion and electron conductivities appear also during the calculations of $\delta\sigma_{xx}^{eff}$. It follows from estimations that their contribution is essentially smaller than the "pure" electron contribution.

Thus, the main contribution to the expression for $\delta\sigma_{ik}^{eff}$ in (10.25) is defined by electrons. Let us write out the first term in the series for $\delta\sigma_{xx}^{eff}$ substituting the first approximation of $B_k(\mathbf{q})$ in (10.25) from (10.26) : $B_k(\mathbf{q}) = -\delta\sigma_{ik}(\mathbf{q})q_i$. Instead of $\delta\sigma_{il}$ we consider only electron Hall component $\delta\sigma_{xye}$. An estimate of the first order correction $\delta\sigma_{xx}^{(1)eff}$ gives [11]:

$$\delta\sigma_{xx}^{(1)eff} \simeq \frac{\sigma_0 \varepsilon^2}{\beta_e}.$$

As a result, we have:

$$\langle \sigma_{xxe} \rangle + \delta\sigma_{xx}^{(1)eff} + \cdots = \langle \sigma_{xxe} \rangle (1 + \beta_e \varepsilon^2 + \cdots).$$

That means that even for weak inhomogeneities $\varepsilon^2 \ll 1$ in the strong magnetic field ($\beta_e \gg 1$) the product $\beta_e \varepsilon^2$ may be greater than 1. The last circumstance forces us to take into account all serial terms. The summing was performed by Dreizin and Dychne [11] for electron plasma. An analogous expression for the effective conductivity σ_{xx}^{eff} of a two-component plasma containing electrons and ions is ([5], [8]):

$$\sigma_{xx}^{eff} = \langle\sigma_{xx}\rangle + \delta\sigma_{xx}^{eff}, \tag{10.27}$$

where

$$\langle\sigma_{xx}\rangle = \langle\sigma_{xxe}\rangle + \langle\sigma_{xxi}\rangle, \quad \delta\sigma_{xx}^{eff} = \sigma_0\left(\frac{\varepsilon}{\beta_e}\right)^{4/3},$$

and

$$\sigma_0 = Ne^2/m\nu_{en}, \quad \langle\sigma_{xxe}\rangle = \sigma_0/\beta_e^2, \quad \langle\sigma_{xxi}\rangle = \sigma_0\frac{m\nu_{en}}{M\nu_{in}},$$

m and M denote electron and ion mass.

Taking into consideration that (10.27) is obtained without using perturbation theory, it is possible to consider the case $\langle\sigma_{xx}^e\rangle \ll \delta\sigma_{xx}^{eff}$. Then from (10.27) we have

$$\sigma_{xx}^{eff} = \sigma_0\left(\frac{\varepsilon}{\beta_e}\right)^{4/3} + \langle\sigma_{xxi}\rangle. \tag{10.28}$$

The relative value of the part of effective conductivity caused by inhomogeneities for a 3-dimensional model of a medium with stochastic inhomogeneities is

$$\frac{\delta\sigma_{xx}^{eff}}{\langle\sigma_{xx}\rangle} = \frac{(\beta_e\varepsilon^2)^{2/3}}{1+\langle\sigma_{xxi}\rangle/\langle\sigma_{xxe}\rangle}. \tag{10.29}$$

We notice that for $\varepsilon \sim \beta_e^{-1/2}(\langle\sigma_{xxi}\rangle/\langle\sigma_{xxe}\rangle)^{3/4}$ the contribution of the fluctuating part $\delta\sigma_{xx}^{eff}$ becomes comparable with the background conductivity $\langle\sigma_{xx}\rangle$.

Expression (10.26) is obtained under very general assumptions about the magnetic field. Therefore, it is valid for both strong and weak fields.

Weak Field

For weak fields when $\beta_e \ll 1$, the component $\langle\sigma_{xx}\rangle \gg \langle\sigma_{xy}\rangle$. The main contribution to (10.26) is provided by $\delta\sigma_{xx}(q)$. From (10.26) we obtain equation that

$$\delta\sigma_{xx} = \sum_{\mathbf{q}}\frac{\delta\sigma_{xx}(-\mathbf{q})q_xB_x(\mathbf{q})}{\langle\sigma_{mn}\rangle\, q_mq_n} \tag{10.30}$$

In the lowest-order approximation to the fluctuations $\delta\sigma_{xx}(q)$, it follows from (10. A.3) that

$$B_x(\mathbf{q}) \approx -\delta\sigma_{xx}(\mathbf{q})q_x. \tag{10.31}$$

And the addition to Pedersen conductivity can be written as

$$\delta\sigma_{xx} \approx -\sum_{\mathbf{q}} \frac{\delta\sigma_{xx}(-\mathbf{q})\delta\sigma_{xx}(\mathbf{q})q_x^2}{\langle\sigma_{mn}\rangle\, q_m q_n}.$$

We obtain, in view of $\langle\sigma_{xx}\rangle \gg \langle\sigma_{xy}\rangle$, that

$$\delta\sigma_{xx} \approx -\frac{\langle\delta\sigma_{xx}^2(\mathbf{r})\rangle}{\langle\sigma_{xx}\rangle} \tag{10.32}$$

Substituting (10.32) into (10.25), we get an expression for the effective conductivity in the form of (10.9) in which $\langle\sigma\rangle$ should be replaced by $\langle\sigma_{xx}\rangle = \sigma_0/(1+\beta_e^2)$. It is clear that the effective conductivity is less than the average conductivity in the case of a weakly magnetized plasma ($\beta_e < 1$).

Strong Field

In strong magnetic fields ($\beta_e \gg 1$) but ($\beta_i \ll 1$) from (1.93, 1.94) follows

$$\sigma_{xx} \approx \frac{\sigma_0}{\beta_e^2}, \qquad \sigma_{xy} \approx \frac{\sigma_0}{\beta_e}. \tag{10.33}$$

The main term in (10.26) is defined by the perturbed Hall component $\delta\sigma_{xy}$. Therefore, the correction $\delta\sigma_{xx}$ is

$$\delta\sigma_{xx} \approx \sum_{\mathbf{q}} \frac{\delta\sigma_{xy}(-\mathbf{q})q_y B_x(\mathbf{q})}{\langle\sigma_{mn}\rangle\, q_m q_n} \tag{10.34}$$

and

$$B_x(\mathbf{q}) \approx -\delta\sigma_{yx}(\mathbf{q})q_y.$$

Noting the asymmetry of $\delta\sigma_{ij}$ for $i \neq j$, that is $\delta\sigma_{xy} = -\delta\sigma_{yx}$, we obtain

$$\delta\sigma_{xx} \approx +\sum_{\mathbf{q}} \frac{|\delta\sigma_{xy}(\mathbf{q})|^2\, q_y^2}{\langle\sigma_{mn}\rangle\, q_m q_n}.$$

Hence, it follows that the contribution from inhomogeneities is positive and in the case of $\beta_e \gg 1$, the effective conductivity always exceeds $\langle\sigma_{xx}\rangle$.

Effective 2D Medium

We have found relations appropriate for distributed inhomogeneities. We now return to $\boldsymbol{\sigma}^{eff}$, and use the method of 'effective medium' ([10], [20]) for a brief treatment of a thin conductive layer model containing a binary mixture of circle inhomogeneities with conductivities $\boldsymbol{\sigma}_1$ and $\boldsymbol{\sigma}_2$. The second term on the right-hand side of (10.2) are r- and φ-components of a dipole moment of an isolated circle inhomogeneity placed into crossed electric **E** and magnetic $\mathbf{B}_0$ fields. The total dipole moment of the ensemble of such inhomogeneities is easily derived. Let us suppose that there are N_1 inhomogeneities of radius a and conductivity $\boldsymbol{\sigma}_1$, and N_2 inhomogeneities of radius b and conductivity $\boldsymbol{\sigma}_2$. Assuming that the whole region is occupied by non-intersecting inhomogeneities and denoting $x_1 = N_1 a^2$, $x_2 = N_2 b^2$, we see that x_1 and x_2 must satisfy

$$x_1 + x_2 = 1.$$

The total polarization caused by the inhomogeneities should vanish. After some algebraic manipulations with expressions in brackets of (10.2), we can write a system of equations to find σ_P^{eff} and σ_H^{eff}:

$$\sum_{i=1,2} x_j \frac{\sigma_{Pi}^2 - \left(\sigma_P^{eff}\right)^2 + \left(\sigma_H^{eff} - \sigma_{Hi}\right)^2}{\left(\sigma_P^{eff} + \sigma_{Pi}\right)^2 + \left(\sigma_H^{eff} - \sigma_{Hi}\right)^2} = 0\,, \tag{10.35}$$

$$\sum_{i=1,2} x_j \frac{\sigma_{Hi} - \sigma_H^{eff}}{\left(\sigma_P^{eff} + \sigma_{Pi}\right)^2 + \left(\sigma_H^{eff} - \sigma_{Hi}\right)^2} = 0. \tag{10.36}$$

Here $j = 2, 1$ respectively for $i = 1, 2$.

The 3D case of isotropic media ($\sigma_{H1} = \sigma_{H2} = 0$) was studied by Landauer [20] for spherical inhomogeneities.

Figure 10.3 shows the dependency of the ratio of effective Pedersen conductivity σ_P^{eff} to average Pedersen conductivity $\langle\sigma_P\rangle$ on the magnetization parameter β_e. The conductivities $\sigma_{P1}\,(\sigma_{H1})$ and $\sigma_{P2}\,(\sigma_{H2})$ refer to the local Pedersen (Hall) conductivities of the inhomogeneities of the 1-st- and 2-nd kind, respectively. One of the curves with $x_1 = 0.5$ relates to the case when the areas of the two components are equal. The curve with $x_1 = 0.1$ shows $\sigma_P^{eff}(\beta_e)\,/\,\langle\sigma_P(\beta_e)\rangle$ in which 10% of the whole area is occupied by a highly conductive component with σ_1, whereas the rest of the mixture is $\sigma_2 = 0.9\sigma_1$. The effective conductivity is 5 times larger than the average conductivity, even when the medium is weakly perturbed. The ratio $\sigma_P^{eff}(\beta_e)/ < \sigma_P(\beta_e) >$ can be broadly estimated from

$$\frac{\sigma_P^{eff}(\beta_e)}{\langle\sigma_P(\beta_e)\rangle} \sim \beta_e \varepsilon, \tag{10.37}$$

where ε is defined in (10.1).

Using the same approach used to derive (10.3) for an isolated inhomogeneity, we can construct ‘an effective bounded medium’ in a strong magnetic field. By virtue of the fact that in this case the dependency on magnetic field drops out, for σ^{eff} we can use (10.35) putting $\sigma_{Hi} = \sigma_H^{eff} = 0$. Equation (10.35)

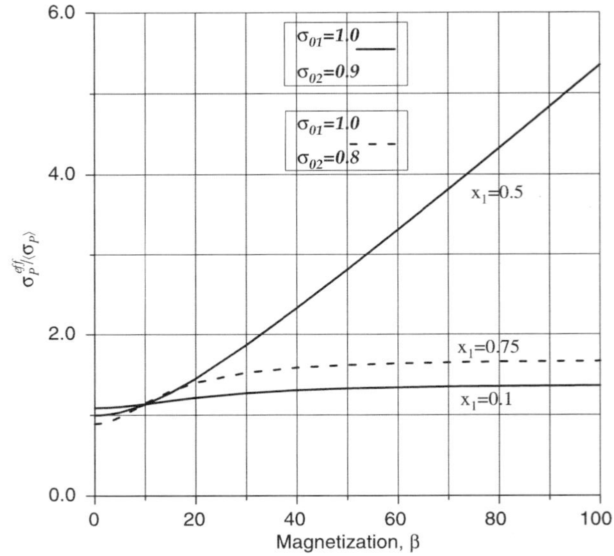

Fig. 10.3. The ratio of effective Pedersen conductivity σ_P^{eff} to spatial average Pedersen conductivity $\langle\sigma_P\rangle$ as a function of magnetization parameter $\beta = \beta_e = \omega_{ce}/\nu_e$. σ_1 and σ_2 refer to the local conductivities of inhomogeneities of the 1-st and 2-nd kind respectively. The curve with $x_1 = 0.5$ relates to the case when areas of two phases are equal. The curve with $x_1 = 0.1$ shows $\sigma_P^{eff}(\beta_e)/\langle\sigma_P(\beta_e)\rangle$ in which 10% of the whole area is occupied by highly conductive component of σ_{P1}, whereas the rest of the mixture is $\sigma_{P2} = 0.9\sigma_{P1}$

becomes

$$\sigma_P^{eff} = (\sigma_{P2} - \sigma_{P1})\left(x_1 - \frac{1}{2}\right) + \frac{1}{2}\sqrt{(\sigma_{P1} + \sigma_{P2})^2 - 4x_1x_2(\sigma_{P1} - \sigma_{P2})^2}. \tag{10.38}$$

Hence, for example, for a mixture with equal portions of two components $x_1 = x_2 = 0.5$, we have $\sigma_P^{eff} = \sqrt{\sigma_{P1}\sigma_{P2}}$. If $\sigma_{P2} = 0.9\sigma_{P1}$, as in Fig. 10.3, then

$$\frac{\sigma_P^{eff}}{\langle\sigma_P\rangle} = 2\frac{\sqrt{\sigma_{P1}\sigma_{P2}}}{\sigma_{P1} + \sigma_{P2}} = 0.6$$

independently of the intensity of the applied magnetic field as distinguished from an open system in which $\sigma_P^{eff}/\langle\sigma_P\rangle \sim 2–5$ for $\beta_e \sim 30–100$.

Experimental Laboratory Simulation

The type of behavior predicted in the preceding theoretical considerations has been confirmed in experiments with semiconductor films in strong magnetic

Appendix A

Let us expand fluctuating variables in the Fourier series in spatial harmonics with wavenumbers $\mathbf{q}$:

$$\begin{pmatrix} \delta\sigma_{ik} \\ \partial(\delta\sigma_{ik})/\partial x_i \\ \partial^2\delta\varphi/\partial x_i \partial x_k \\ \partial(\delta\sigma_{ik})/\partial x_i \\ \partial\delta\varphi/\partial x_k \end{pmatrix} = \sum_{\mathbf{q}} \begin{pmatrix} \delta\sigma_{ik}(\mathbf{q}) \\ iq_i\delta\sigma_{ik}(\mathbf{q}) \\ -q_i q_k \delta\varphi(\mathbf{q}) \\ iq_i\delta\sigma_{ik}(\mathbf{q}) \\ iq_k\delta\varphi(\mathbf{q}) \end{pmatrix} \exp i(\mathbf{qr}). \qquad (10.\ \text{A}.1)$$

Substituting these relations into (10.22), we get

$$\begin{aligned} &\sum_{\mathbf{q}} iq_i\delta\sigma_{ik}(\mathbf{q}) \exp i(\mathbf{qr}) \langle e_k \rangle + \langle \sigma_{ik} \rangle \sum_{\mathbf{q}} q_i q_k \delta\varphi(\mathbf{q}) \exp i(\mathbf{qr}) \\ &+ \sum_{\mathbf{q}} \sum_{\mathbf{p}} q_i p_k \delta\sigma_{ik}(q) \delta\varphi(\mathbf{p}) \exp i(\mathbf{q}+\mathbf{p})\mathbf{r} \\ &+ \delta\sigma_{ik} \sum_{\mathbf{q}} q_i q_k \delta\varphi(\mathbf{q}) \exp i(\mathbf{qr}) = 0. \end{aligned}$$

It can be rewritten as

$$\begin{aligned} &iq_i\delta\sigma_{ik} \langle e_k \rangle + q_i q_k \delta\varphi(\mathbf{q}) \langle \sigma_{ik} \rangle + \sum_{\mathbf{p}} (q_i - p_i) p_k \delta\sigma_{ik}(\mathbf{q}-\mathbf{p}) \delta\varphi(\mathbf{p}) \\ &+ q_i q_k \delta\sigma_{ik}(\mathbf{q}-\mathbf{p}) \delta\varphi(\mathbf{q}) = 0. \end{aligned}$$

From this we deduce

$$\delta\varphi(\mathbf{q}) = i \frac{B_k(\mathbf{q})}{\langle \sigma_{im} \rangle q_i q_m} \langle e_k \rangle, \qquad (10.\ \text{A}.2)$$

where $B_k(\mathbf{q})$ satisfies the equation

$$B_k(\mathbf{q}) = -\delta\sigma_{ik}(\mathbf{q}) q_i - \sum_{\mathbf{q}' \neq 0} \frac{q_i \delta\sigma_{il}(\mathbf{q}-\mathbf{q}') q_l'}{\langle \sigma_{mn} \rangle q_m' q_n'} B_k\left(\mathbf{q}'\right) \qquad (10.\ \text{A}.3)$$

If we make substitution for $\delta\sigma_{ik}$ and $\partial\delta\varphi/\partial\delta x_k$ from (10. A.1), then the relationship (10.20) between the total current and perturbations of conductivity and potential becomes

$$\begin{aligned} \langle j_i \rangle &= \langle \sigma_{ik} \rangle \langle e_k \rangle - \left\langle \sum_{\mathbf{q}} \delta\sigma_{ik}(\mathbf{q}) \exp i(\mathbf{qr}) \sum_{\mathbf{p}} ip_k \delta\varphi(\mathbf{p}) \exp i\mathbf{pr} \right\rangle \\ &= \langle \sigma_{ik} \rangle \langle e_k \rangle - i \sum_{\mathbf{q}} \delta\sigma_{ik}(-\mathbf{q}) \delta\varphi(\mathbf{q}) q_k \end{aligned}$$

Because, by definition,

$$\langle f(\mathbf{r}) \rangle = \frac{1}{V} \int f(\mathbf{r})\, d\mathbf{r}$$

then

$$\begin{aligned} \langle \exp i(\mathbf{q}+\mathbf{p})\mathbf{r} \rangle &= 1, \quad \text{if} \quad \mathbf{q}+\mathbf{p}=0, \\ \langle \exp i(\mathbf{q}+\mathbf{p})\mathbf{r} \rangle &= 0, \quad \text{if} \quad \mathbf{q}+\mathbf{p}\neq 0\,. \end{aligned} \tag{10. A.4}$$

Substituting $\delta\varphi(\mathbf{q})$ from (10. A.2) we have

$$\langle j_i \rangle = \langle \sigma_{ik} \rangle \langle e_k \rangle + \sum_{\mathbf{q}} \delta\sigma_{il}\,(-\mathbf{q})\,\frac{B_l(\mathbf{q})}{\langle \sigma_{im} \rangle\, q_i q_m}\,\langle e_l \rangle\, q_l \tag{10. A.5}$$

which is the basis for calculation of the effective conductivity σ_{ik}^{eff}.

References

1. Alfvén, H. and C. G. Falthammer, *Cosmical electrodynamics*, 2nd edn., Oxford, 1963.
2. Alperovich L. S. and E. N. Fedorov, Generation of low-frequency electromagnetic oscillations by the field of a powerful radio wave in the ionosphere, *Radiophysics and Quantum Electronics*, **24**, 190, 1981.
3. Alperovich, L. S., N. I. Gershenson, and A. L. Krylov, The fluctuations of the quasistationary electric and magnetic fields as a result of the stochastic irregularities of the ionospheric conductivity, *Geom. Aeron.*, **26**, 928–932, 1986.
4. Alperovich, L., N. I. Gershenson, and A. L. Krylov, Connection between spatial and temporal spectrums of the ionospheric wave disturbances, *Geom. Aeron.*, **26**, 406–410, 1986.
5. Alperovich, L., and I. Chaikovsky, On effective conductivity of the ionosphere with random irregularities, *Ann. Geophys.*, **13**, 339, 1995.
6. Chaikovsky, I., and L. Alperovich, A method for determining microinhomogeneities in semiconductors, *J. Appl. Phys.*, American Inst. of Phys., **83**, 5277, 1998.
7. Alperovich, L., S. Grachev, Yu. Gurvich, L. Litvak-Gorskaya, A. Melnikov, and I. Chaikovsky, An optical method for simulating of non-uniform systems, *JETP Lett.*, **65**, 224, 1997.
8. Alperovich, L. I. Chaikovsky, On the effective conductivity of the magnetized bounded partially ionised plasma with random irregularities, *Plasma Phys. Control. Fusion*, **41**, 1071–1090, 1999.
9. Balagurov, B. Ya., Theory of the galvanomagnetic properties of two-dimensional two-component systems, *Sov. Phys. JETP*, **108**, 2202, 1995.
10. Brugemann, D.A.Q., *Ann. Phys, Lpz.*, **24**, 636, 1935.
11. Dreizin, Yu. A., and A. M. Dichne, Anomalous conductivity of inhomogeneous media in strong magnetic field, *Sov. Phys. JETP*, **36**, 127–136, 1973.
12. Dychne, A. M., *Sov. Phys. JETP*, **59**, 641, 1970.
13. Dychne, A. M. and I. M. Rusin, *Phys. Rev.*, **B**, **50**, 2369, 1994.
14. Galperin, Yu. M., and B. D. Laichtman, Effect of microinhomogeneities on kinetic phenomena in high-mobility semiconductors, *Sov. Phys. Sol. State,* **13**, 1760, 1972.
15. Gershman, B. N., *Dynamics of the Ionospheric Plasma* (in Russian), Nauka, Moscow, 1974.

16. Gurevich, A. V., *Nonlinear Plasma Phenomena in the Ionosphere,* Springer-Verlag, New York-Heidelberg-Berlin, 1978.
17. Herring, C., Effect of random inhomogeneities on electrical and galvanomagnetic measurements, *J. Appl. Phys.*, **31**, 107, 1961.
18. Kirkpatric, S., Classical transport in disordered media: Scaling and effective-medium theories, *Phys. Rev. Lett.,* **27**, 1722, 1971.
19. Kvyatkovskiy, O. E., Effective conductivity of an inhomogeneous medium in a strong magnetic field, *Sov. Phys. JETP*, **58**, 120, 1983.
20. Landauer, R., The electrical resistance of binary metallic mixtures, *J. Appl. Phys.*, **23**, 779, 1952.
21. Landau, L. D., and E. M. Lifshitz, *Electrodynamics of continuous media,* Pergamon Press, Oxford, 1960.
22. Last, B. J., and Thouless D. J., Percolation theory and electrical conductivity, *Phys. Rev. Lett.*, **27**, 1719, 1971.
23. Shalimov, S., C., Haldoupis and K. Schlegel, Large polarization electric fields associated with midlatitude sporadic *E*, *J. Geophys. Res.*, **103**, 11617, 1998.
24. Watson, B. P. and P. I. Leath, Conductivity in the two-dimensional-site percolation problem, *Phys. Rev. B.*, **9**, 4893, 1974.

11

ULF-Sounding of Magnetosphere and Earth

11.1 Introduction

The investigation of the pulsation spatial structure was based on the solution of direct problems of MHD-wave propagation. In this chapter, we shall show how the developed concepts can be adapted to the inverse problems, to determine the magnetospheric cold plasma distribution and to study the Earth's crust conductivity by using either ground or satellite ULF-observations.

We will consider a method of ground-based magnetospheric plasma monitoring based on the FLR-theory (see Chapter 6). In this method the distribution of Alfvén velocity and, thus, plasma density, along a field line is obtained from the FLR-frequencies. To this end, the inverse problem is reduced to a spectral problem. The FLR-frequencies are found from the geomagnetic pulsations.

The Earth's crust conductivity is determined by methods known as magnetotelluric sounding (MTS)([10], [32]). The method is based on simultaneous measuring of electric and magnetic fields in the ULF-range. Numerous papers and monographs have been devoted to this topic. However, little attention has been devoted to the influence of magnetospheric peculiarities of MHD-wave propagation and excitation. We shall study the role of resonance magnetic shells in interpreting data of ground-based MTS as well as the possibility of an above ionosphere MTS from low-orbit satellites.

11.2 Inverse Problem of FLR

Obayshi [26] and Dungey [14] were the first to suggest that the plasma parameters can be reconstructed from the FLR-spectrum of a field-line. While in general this is a very difficult task, under ideal MHD, the spectrum of the magnetosphere oscillations splits into two parts (see Chapter 6): a discrete spectrum of global magnetospheric oscillations and a continuous FLR-spectrum.

The spectra of toroidal and poloidal oscillations can be found from the solution of 1D-eigenvalue problems (6.47) and (6.50).

1) The toroidal mode

$$\frac{\partial}{\partial s}\frac{h_2}{h_1}\frac{\partial E_1}{\partial s} + \frac{\omega^2}{c_A^2}\frac{h_2}{h_1}E_1 = 0, \qquad \left.\frac{\partial E_1}{\partial s}\right|_{s=0;S} = 0, \tag{11.1}$$

2) The poloidal mode

$$\frac{\partial}{\partial s}\frac{h_1}{h_2}\frac{\partial E_2}{\partial s} + \frac{\omega^2}{c_A^2}\frac{h_1}{h_2}E_2 = 0, \quad \left.\frac{\partial E_2}{\partial s}\right|_{s=0;S} = 0, \tag{11.2}$$

where h_1 and h_2 are Lamé coefficients of the coordinate system (6.27) related to the geomagnetic field, the field-aligned coordinate s is a distance along the field-line, S is the length of the field line between two conjugate ionospheres.

Let L be a McIllwain parameter, λ is a geomagnetic longitude of the field line footpoint on the ground, then the eigenvalues of boundary problems (11.1) and (11.2) form two discrete sets continuously dependent on L and λ:

$$\omega_n^T(L,\lambda) \quad \text{and} \quad \omega_n^P(L,\lambda).$$

Each field line is characterized by two sets of resonance frequencies completely determined by the field-aligned plasma density distribution and field line geometry. Field-aligned distribution of plasma density can be unambiguously determined from the spectra of these boundary problems (11.1) and (11.2). The problem of the hydromagnetic diagnostics of the magnetosphere is reduced to the experimental determination of FLR-frequencies and solution of the inverse problems (11.1)–(11.2) ([7], [9], [18], [20]).

The usefulness of the inverse problem hydromagnetic diagnostics is limited by the instability of the solution, i.e., its sensitivity to small disturbances of the initial data. Such disturbances always exist because it is impossible to get a complete FLR-spectrum from measured geomagnetic pulsations. Not only because of measurement inaccuracy but also because of the variability of the magnetosphere itself as plasma density changes during observation time. To make the inverse problem stable, it is necessary to select a class of functions C_ρ for the sought plasma density distributions $\rho(s)$. Then the distribution found from the inverse problem will have a bounded confidence band $\{\rho(s)\}$. The bandwidth will be narrower for a narrower class of functions C_ρ and for more accurate and full initial data. Such a technique can be used if the class of functions C_ρ can be determined from the information about the plasma density distribution obtained with other experimental methods.

Summarizing, the plan for the solution of the problem of hydromagnetic diagnostics is as follows:

1. Find FLR-frequencies from measured ULF-fields.
2. Select class of functions C_ρ for the plasma density distribution from independent experimental data.

3. Find a distribution $\rho(s)$ within the class C_ρ with minimal difference between theoretical and experimental FLR-frequencies using a fitting procedure.

ULF-Magnetospheric Diagnostics

Below, we describe a technique to determine experimentally the necessary parameters of the ULF-structure. It follows from (8.26) that upon transmission through the ionosphere, the width δ of the resonance peak, as observed on the ground, is increased compared to its value above the ionosphere, δ_i, namely $\delta = \delta_i + h$, where h is the height of the ionospheric E-layer. We rewrite (8.26) for the amplitude and phase characteristics of the magnetic H-component at the ground in the form

$$H(x, f) = \frac{b_R(f)}{1 - i\zeta}, \tag{11.3}$$

where $\zeta = (x - x_r)/\delta$ denotes the normalized distance from the FLR-point $x_R(f)$. $b_R(f)$ is the amplitude of the pulsations at the FLR-point, and x is the coordinate of a magnetic shell, as measured along the geomagnetic meridian.

A principal problem of the experimental determination of FLR-parameters is that in most events the input into the ULF-spectral content from the resonant magnetospheric response and the one from the 'source' are comparable. So in most cases, a spectral peak does not necessarily correspond to a local resonant frequency, and the width of a spectral peak cannot be directly used to determine the Q-factor of the FLR. This ambiguity can be resolved with the help of the experimental methods listed below and in particular the gradient method.

Gradient Method

Using precise measurements of the gradients of spectral amplitude and phase along a small baseline, one can exclude the influence of the source spectrum and reveal the relatively weak resonant effects. The following simple relationships, stemming from the properties of function (11.3), describe the specific features of the ratio $G(f)$ between amplitude spectra and the difference of phase spectra $\Delta\psi(f)$ of magnetic H-components, recorded at points x_1 and x_2 ($\Delta x = x_1 - x_2 > 0$):

$$G(f) = \frac{|H(x_1, f)|}{|H(x_2, f)|} = \left(\frac{1 + \zeta_2^2}{1 + \zeta_1^2}\right)^{1/2}, \tag{11.4}$$

$$\Delta\psi(f) = \arctan\left(\frac{\zeta_2 - \zeta_1}{1 + \zeta_1\zeta_2}\right). \tag{11.5}$$

The typical features of functions (11.4), (11.5), schematically shown in Fig. 11.1, are

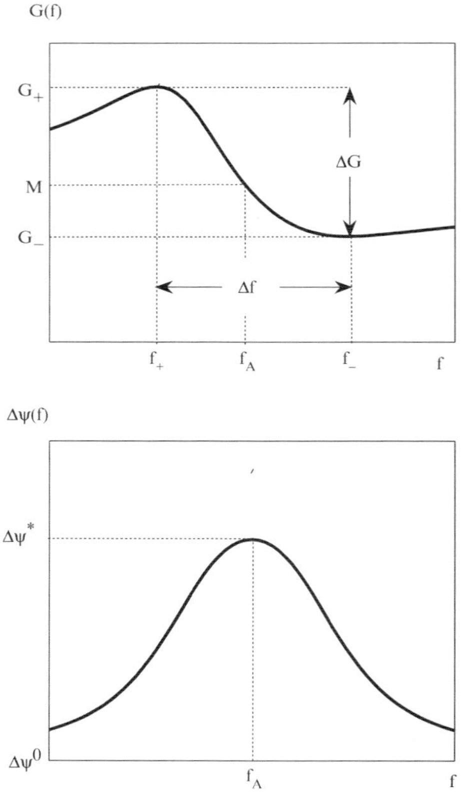

Fig. 11.1. Schematic plots of the amplitude spectral ratio $G(f)$ and phase spectral difference $\Delta\psi(f)$ between two near-by stations, as predicted by the resonance theory

A. $G(f_A) = M = 1$ for frequency $f_c = f_A(x_c)$, where the point

$$x_c = (x_1 + x_2)/2$$

is located at the midpoint between the stations;

B. $G(f)$ reaches extreme values G_+ and G_- at frequencies f_+ and f_- which correspond to the points

$$x_r(f_\pm) = x_\pm = x_c \pm [\delta^2 + (\Delta x/2)^2]^{1/2};$$

C. $G_+G_- = 1$ and $G_+ - G_- = \Delta x/\delta$;

D. $\Delta\psi(f)$ reaches an extreme value $\Delta\psi^* = 2\arctan(\Delta x/2\delta)$ at frequency $f_A = f_r(x_r)$.

The properties (A) and (D) of functions $G(f)$ and $\Delta\psi(f)$ enable us to estimate the FLR-frequency between the stations. We can also find the width of the resonant peak by applying the extreme values of the phase difference (D) to get

$$\delta = \frac{\Delta x}{2} \cot\left(\frac{\Delta\psi^*}{2}\right).$$

However, the developed method can be applied only if both observation points are located in the same geoelectrical conditions. The lateral geoelectrical inhomogeneity, especially when the condition of a strong skin effect is violated, may substantially distort a resonant structure of pulsations. The disturbing influence of geoelectrical cross-section would be mainly seen in the behavior of the magnetic component oriented across the structure's extension. Above a rock with higher conductivity, an pulsation magnetic H-component increases by ΔH. An additional phase shift $\Delta\psi^{(0)} > 0$, as compared with the incident field, appears that can reach values of $\Delta\psi^{(0)} \sim (180/\pi)\Delta H/H$.

A theoretical modification of the gradient method of MHD-diagnostics for the case of a crust geoelectrical inhomogeneity requires numerical calculations of a complicated self-consistent problem. To eliminate the geological influence, it is possible to use the following simple phenomenological method. Suppose that the influence of a geoelectrical inhomogeneity can be expressed as some coefficient M, which the ratio G is multiplied by; and an additional phase shift $\Delta\psi^{(0)}$, which is added to the phase difference $\Delta\psi = \psi^{(1)} - \psi^{(2)}$. Then the experimentally measured functions G' and $\Delta\psi'$ will be presented in the form: $G' = MG$ and $\Delta\psi' = \Delta\psi + \Delta\psi^{(0)}$. Hereafter we neglect a weak dependence of the unknown coefficients $M(f)$ and $\Delta\psi^{(0)}(f)$ on the frequency in a bounded frequency band near the resonant frequency, as compared with $G(f)$ and $\Delta\psi(f)$ (11.4), (11.5). The coefficient M can be found experimentally from the following set of simple relationships resulting from the properties (C) of function $G(f)$:

$$G'_+ = MG_+, \quad G'_- = MG_-, \quad G_-G_+ = 1. \tag{11.6}$$

The correction coefficient for amplitude ratio can be estimated as

$$M = \sqrt{G'_-G'_+}.$$

As a rough estimate of $\Delta\psi^{(0)}$, the constant level of phase shift away from resonant phase excursion can be taken. So, the amplitude ratio (11.4) and phase difference (11.5) will be shifted to the new levels M and $\Delta\psi^{(0)}$ (see Fig.11.1).

A meridional gradient of the Alfvén frequency can be estimated from the data of gradient measurements as

$$\frac{\partial f}{\partial x} \simeq \frac{f_+ - f_-}{x_+ - x_-} = \frac{2\Delta f}{\left[\delta^2 + (\Delta x/2)^2\right]^{1/2}}. \tag{11.7}$$

The gradient method can even be used to restore a smooth $f_r\,(L)$ profile in a limited interval of latitudes. The properties (A, B, C) of G provide a way to estimate the width δ of FLR, and three resonant frequencies f_A, f_+, f_- of the field-lines crossing the meridian between two stations ($x = x_c$) and at some points to the north and to the south ($x = x_-$, $x = x_+$).

Inverse Problem of MHD-Sounding

We choose a class of functions C_ρ for the field-aligned plasma density distribution in the form

$$\rho = \rho_0(L)\left(\frac{LR_E}{r}\right)^q,$$

where $\rho_0(L)$ is the plasma density at the top of the field line with the McIllwain L parameter, r is the distance along the field line from the Earth's center to a point on the field line L, LR_E is the distance from the Earth's center to the top of this field-line. Assume $n = \rho/m_p$, where m_p is the proton mass. For a multicomponent plasma

$$n = \sum_\alpha n_\alpha \frac{m_\alpha}{m_p}, \tag{11.8}$$

where n_α and m_α are the concentration and mass of α-ions. Let the distribution of the plasma concentration in the magnetospheric equatorial plane be a power function of L:

$$n_0 = aL^{-s}, \quad \rho_0(L) = m_p n_0. \tag{11.9}$$

Consider only the inner magnetosphere where the dipole model of the external magnetic field is valid and rather a narrow L range with a negligible variation of the parameters a and s.

Below, two variants of the hydromagnetic diagnostics are treated. In the first method q, and n_0 are obtained from frequencies of two or more FLR-harmonics at one magnetic shell. In the second one, the parameters a and s are found from the FLR-harmonic measured at two or more L-shells (11.9).

FLR-resonance frequencies for a dipole magnetic field and high integral ionospheric conductivity are found from the boundary problem (6.79), (6.80). New dependent variables are chosen. Let

$$e_1 = y_1, \quad e_2 = i\frac{L^2 R_E^2}{c}\,\Omega\, y_2, \quad \Omega(n_0, L) = \frac{B_{eq}}{2\sqrt{\pi m_p n_0}\, L^4 R_E},$$

where $\Omega(n_0, L)$ is frequency scale (6.93), B_{eq} is the magnetic field on the Earth's equator. The boundary problem (6.79), (6.80) for the field-aligned plasma density distribution (11.8) becomes

$$\begin{aligned} \frac{dy_1}{dw} &= -\bar{\omega}(1 - w^2)^{6-q} y_2, \\ \frac{dy_2}{dw} &= \bar{\omega} y_1, \end{aligned} \tag{11.10}$$

with $y_2|_{w=\pm w_0} = 0$. Here $\bar{\omega} = \omega/\Omega$ is a normalized frequency and $w_0 = (1-1/L)^{1/2}$. Integrating the set (11.10) with Cauchy's initial data

$$y_1(-w_0, \bar{\omega}) = 1, \qquad y_2(-w_0, \bar{\omega}) = 0,$$

we get the function $\Delta(\bar{\omega}, q, w_0) = y_2(w_0, \bar{\omega})$. Equating Δ to zero, we get

$$\Delta(\bar{\omega}, q, w_0) = 0. \tag{11.11}$$

for the FLR-frequencies. The roots of (11.11) determine the dependence of FLR-frequencies $\omega_r^{(k)} = \Omega(n_0, w_0)\,\bar{\omega}_k(q, w_0)$ on q, the parameter that controls the field-aligned plasma density distribution. Note that the ratio of frequencies of FLR-higher harmonics to the FLR-fundamental frequency

$$l_k(q, w_0) = \frac{\omega_r^{(k)}}{\omega_r^{(1)}} = \frac{\bar{\omega}_k(q, w_0)}{\bar{\omega}_1(q, w_0)}$$

depends only on q and w_0.

Single-Field Line MHD-Diagnostics

Suppose that the FLR-fundamental $f_r^{(1)}$ and second $f_r^{(2)}$ harmonic frequencies are known for a central point $\Phi^* = (\Phi_1 + \Phi_2)/2$ located between two close points at geomagnetic latitudes $\Phi_1 > \Phi_2$, ($\Phi_1 - \Phi_2 \approx 2°$) along a geomagnetic meridian where pulsations are measured. It is convenient to use as initial parameters the ratio of the first two harmonics $f_r^{(2)}/f_r^{(1)}$ and one of the resonance frequencies, e.g. $f_r^{(1)}$. For q we have the equation

$$W(q, w_0) = l_2(q, w_0) - \frac{f_r^{(2)}}{f_r^{(1)}} = 0. \tag{11.12}$$

It can be solved graphically or numerically. The initial approximation $q^{(1)}$ is then improved with the iterative procedure

$$q^{(n+1)} = q^{(n)} - W(q^{(n)}, w_0)\left(\frac{\partial W(q^{(n)}, w_0)}{\partial q}\right)^{-1} \underset{n\to\infty}{\rightarrow} q(w_0).$$

Hence, we may obtain $\Omega = \omega_r^{(1)}/\bar{\omega}_1$ as a function of the measured fundamental frequency and calculated normalized frequency on which Ω depends. From (6.103) the frequency $\Omega = 106/n_0^{1/2}L^4$ and for the concentration n_0 we have

$$n_0 = \left(\frac{106}{\Omega L^4}\right)^2. \tag{11.13}$$

Multi-Field-Lines MHD-Diagnostics

In this case, a number of FLR-frequencies, for example fundamental harmonics $\omega_j^{(1)} = 2\pi f_r^{(1)}(L_j^*)$ at several points along a geomagnetic meridian are taken as input parameters. Let the distribution of plasma density have a form (11.8), (11.9). In this model the plasma density distribution is determined by three parameters q, a and s. We find the dependence of the normalized frequency of the FLR-fundamental harmonic $\bar{\omega}_j^{(1)}(q)$ on q for each shell $L = L_j^*$ from the boundary problem (11.10). Equating calculated and measured FLR-frequencies, we obtain

$$\Omega_j(q) = \frac{\omega_j^{(1)}}{\bar{\omega}_j^{(1)}(q)}, \qquad j = 1, \ldots, N, \tag{11.14}$$

where $N \geqslant 2$ is a number of measured FLR-frequencies. Plasma concentration at the top of the field line L_j is expressed in terms of Ω_j from (11.13) as

$$n_{0j}(q) = \left(\frac{106}{\Omega_j L_j^4}\right)^2. \tag{11.15}$$

Let the plasma distribution in the equatorial plane n_0 obey a power law (11.9):

$$n_{0j} = aL_j^{-s}, \quad j = 1, \ldots, N.$$

The logarithm of this set is an overdetermined linear system

$$x_1 + (\ln L_j)x_2 = \ln n_{0j}, \quad j = 1, 2, \ldots, N,$$

where $x_1 = \ln a, \quad x_2 = -s$. The variables x_1 and x_2 can be computed by a least-squares fit:

$$F(x_1^*, x_2^*) = \min_{x_1,\ x_2} F_1(x_1,\ x_2), \tag{11.16}$$

where

$$F(x_1,\ x_2) = \sum_{j=1}^{N} |x_1 + (\ln L_j)x_2 - \ln n_{0j}|^2 .$$

Equating to zero the $x_1,\ x_2$-derivatives of $F(x_1,\ x_2)$ we get the equation set

$$\mathbf{Ax} = \mathrm{U}(q), \tag{11.17}$$

where

$$\mathbf{A} = \begin{pmatrix} N & \sum_{j=1}^{N} \ln L_j \\ \sum_{j=1}^{N} \ln L_j & \sum_{j=1}^{N} \ln^2 L_j \end{pmatrix}, \quad \mathbf{U} = \begin{pmatrix} \sum_{j=1}^{N} \ln n_{0j} \\ \sum_{j=1}^{N} \ln n_{0j} \ln L_j \end{pmatrix},$$

$$\mathbf{x} = \begin{pmatrix} \ln a \\ -s \end{pmatrix}. \tag{11.18}$$

We reconstruct the plasma density using the data from Table 11.1 with the following 2 methods of hydromagnetic diagnostics.
A solution of (11.18) determines the dependencies $x_1(q), x_2(q)$, $a(q)$ and $s(q)$. The parameter q in the field-aligned plasma density distribution can be determined at $N \geqslant 3$ minimizing the sum of the squares of the deviations of the calculated and measured FLR-frequencies.

$$\Psi_q(q) = \sum_{j=1}^{N} \left(\Omega_j(q)\bar{\omega}_j^{(1)}(q) - \omega_j^{(1)} \right)^2 .$$

However, this method is not effective for finding q because of the weak dependence of Ψ_q on q and resulting in severe mistakes in q. Therefore a different method should be used to find q, e.g., the ratio of two FLR-harmonic frequencies as described above.

Numerical Example

As an example, we estimate the parameters of the magnetospheric plasma using the data of a meridional chain of 5 stations in England [16]. Four pairs of adjacent stations allow us to use the gradient method to find the resonant frequencies of four field-lines crossing the meridian in the centers of arcs between station pairs. These frequencies are from the spectra of $Pc3,4$ pulsations measured on April 22, 1976. The geomagnetic latitudes of corresponding field line footpoints are given in Table 11.1 [7].

1 method. Table 11.1 gives $f_r^{(1)} = 6.9$ mHz, $f_r^{(2)} = 17.3\,\text{mHz}$, $\Phi^* = 61.25°$, $w_0 = \sin\Phi^* = 0.8767$. The numerical solution of the equation $l_2(q, w_0) = 17.3/6.9 = 2.51$ at $w_0 = 0.8767$ gives $q \approx 1.15$. A root can also be obtained graphically, as shown in Fig. 11.2. At $q = 1.15$ and $w_0 = 0.8767$ from (11.11), we have $\bar{\omega}_1 = 2.15$ and for $\Omega = \omega_r^{(1)}/\bar{\omega}_1 = 2\pi f_r^{(1)}/\bar{\omega}_1 = 2.02 \times 10^{-2}$ s^{-1}. Substituting these results into (11.13), we obtain $n_0 \approx 227\ \text{cm}^{-3}$.

2 method. Assume that $q = 1.15$ obtained from the first method is constant within the latitude range 55°–60°. We get the normalized FLR-frequencies $\bar{\omega}_j^{(1)}(q = 1.15) = (2.15, 2.18, 2.20, 2.24)\,\text{s}^{-1}$ corresponding to the geomagnetic latitudes $\Phi_j^* = 61.25°,\ 59.45°,\ 58.00°,\ 55.85°$ from the boundary problem (11.10). Then we compute $\Omega_j = \omega_j^{(1)}/\bar{\omega}_j^{(1)} = (2.02 \times 10^{-2},\ 2.85\times 10^{-2}, 3.68 \times 10^{-2}, 5.29 \times 10^{-2})$ from (11.14) and from (11.15) get the plasma

Table 11.1. FLR-frequencies extracted from the observations at the meridional chain of 5 stations in England [16]

Φ_j^*	61.25	59.45	58.00	55.85
L_j^*	4.32	3.87	3.56	3.17
$f_r^{(1)}$ mHz	6.9	9.9	12.9	18.9
$f_r^{(2)}$ mHz	17.3	–	–	–

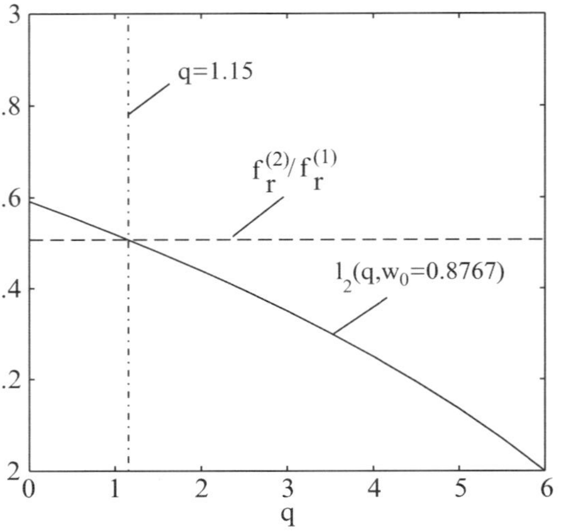

Fig. 11.2. Graphic calculation of (11.12)

concentration at the top of the field line $n_{0j} = \left(106/\Omega_j L_j^4\right)^2 = 226.9, 273.7, 321.2, 390.3\,\mathrm{cm}^{-3}$. The parameters a and s of the power approximation (11.9) of the equatorial plasma density distribution are found from (11.18) as $a = 3 \times 10^3\,\mathrm{cm}^{-3}$, $s = 1.767$.

The dependence of n_0 on the geomagnetic latitude is shown in Fig. 11.3. The solid circles in Fig. 11.3 are calculated according to (11.15). The approximation (11.9) is shown with the line.

The obtained values of plasma concentration n_0, and the parameters q and s are in good agreement with those obtained from the satellite and whistler measurements under weak and moderate geomagnetic activity at $L = 3$–4 (e.g., [11], [27], [31]). The analysis of the accuracy of the estimates of plasma density with hydromagnetic diagnostics [15] shows that the accuracy of this method is close to that of whistler measurements [28]. Besides, the MHD-diagnostics combined with whistler measurements allow us to estimate the concentration of heavy ions in the magnetosphere.

11.3 Ground-Based Magnetotelluric Sounding

Impedance

The horizontal electric and magnetic field components on the ground are linked by a linear operator $\mathbf{Z}$ which is a functional from the distribution of the ground conductivity $\sigma_g(\mathbf{r})$. For a layered medium providing the basis for the magnetotelluric model, operator $\mathbf{Z}$ can be written in the form of an integro-differential transformation. The kernel $\mathbf{G}$ of the operator by virtue of

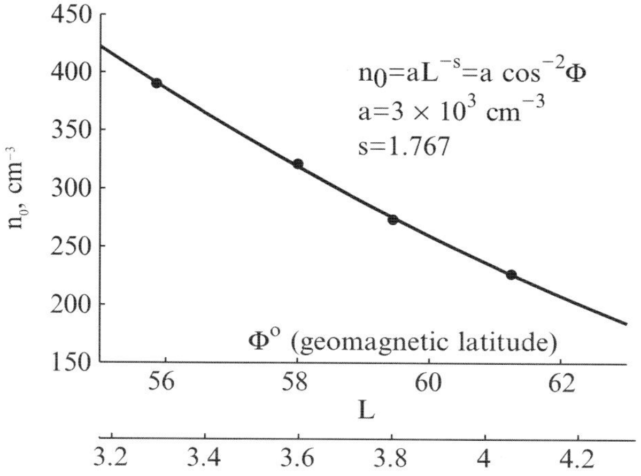

Fig. 11.3. Equatorial distribution of the plasma concentration n_0 recalculated to the geomagnetic latitudes by (11.15). The circles are calculated according to (11.15). The approximation (11.9) is shown by the line

horizontal homogeneity depends only on the difference between the arguments, i.e.

$$\mathbf{E}_\tau(\mathbf{r})|_{r=R_E} = \int \mathbf{G}\left(\mathbf{r}-\mathbf{r}'\right)\left[\mathbf{n}\times\mathbf{b}_\tau\left(\mathbf{r}'\right)\right] ds', \tag{11.19}$$

where $\mathbf{n}$ is a unit vector normal to the ground and integrating is over the ground surface. Kernel $\mathbf{G}$ is determined from the solution of the electrodynamic problem for a half-space with the predetermined conductivity.

Applying the spatial x, y-Fourier transform to (11.19), we obtain the relation between spectral components of the electric $\mathbf{E}_\tau$ and magnetic $\mathbf{b}_\tau$ fields on the ground

$$\mathbf{E}_\tau\left(\omega;\ k_x, k_y\right) = \mathbf{Z}\left(\omega;\ k_x, k_y\right)\mathbf{b}_\tau\left(\omega;\ k_x, k_y\right), \tag{11.20}$$

where $\mathbf{Z}\left(\omega;\ k_x, k_y\right)$ is the spectral impedance matrix. It is convenient to present the ground and atmospheric fields as a superposition of two modes – electric (index e) and magnetic (h) (see Chapter 7). From (7.30) we get

$$\mathbf{Z}\left(\omega;\ k_x, k_y\right) = \frac{1}{k^2}\begin{pmatrix} k_x k_y\left(Z_e - Z_h\right) & -k_x^2 Z_e - k_y^2 Z_h \\ k_y^2 Z_e + k_x^2 Z_h & k_x k_y\left(Z_h - Z_e\right)\end{pmatrix}, \tag{11.21}$$

where $Z_h\left(\omega;\ k_x, k_y\right)$ and $Z_e\left(\omega;\ k_x, k_y\right)$ are spectral impedances of the magnetic and electric types. Z_h and Z_e coincide if k vanishes

$$\lim_{k\to 0}\mathbf{Z}_h\left(\omega;\ k_x, k_y\right) = \lim_{k\to 0}\mathbf{Z}_e\left(\omega;\ k_x, k_y\right) = \mathbf{Z}_0 + 0\left(k^2\right),$$

where $k^2 = k_x^2 + k_y^2$. In this case (11.20) yield

$$\mathbf{E}_\tau = \mathbf{Z}_0\left(\omega\right)\mathbf{b}_\tau \tag{11.22}$$

with

$$\mathbf{Z}_0 = \begin{pmatrix} 0 & -Z_0 \\ Z_0 & 0 \end{pmatrix}.$$

This equation is strictly valid for a vertically incident plane wave. But its application proves to be valid in a wide range of incidence angles for highly conductive geoelectrical cross-sections. These angles or the corresponding horizontal wavenumbers can be estimated using a simple model of homogeneous half-space with conductivity σ_g. The spectral impedances reduce to

$$Z_h\left(\omega;\ k_x, k_y\right) = Z_0 \left(1 + i\frac{k^2 d_g^2}{2}\right)^{-1/2},$$
$$Z_e\left(\omega;\ k_x, k_y\right) = Z_0 \left(1 + i\frac{k^2 d_g^2}{2}\right)^{1/2}, \qquad (11.23)$$

where

$$Z_0 = \omega(4\pi\sigma_g)^{-1/2} \exp\left(-i\frac{\pi}{4}\right)$$

and d_g is the skin depth. The spectral impedances coincide with the impedance Z_0, if the horizontal spatial scale of the field is much larger than the skin depth (i.e. $kd_g \ll 1$) [34]. This approximation is widely used for deep ULF-electromagnetic sounding of the Earth ([10], [32]).

With linear dependence of field components on coordinates, the integral operator in (11.20) reduces to a matrix multiplication operator and condition (11.22) proves to be suitable for small-scale field perturbations as well. Numerical estimations showed that integral operator (11.19) is indeed reduced to (11.22) for field perturbations on a wide range of horizontal scales $\infty-$ 100–200 km and time scales 10–10^3 s ([12], [34]).

For low-conductivity cross-sections the applicability conditions of the model of a vertically incident wave can be violated. The most critical, in that sense, may be polar regions. Near the resonance magnetic shell the pulsation phase and amplitude are sharply changed at distances comparable to the height of the ionosphereic conductive layer ($\approx$100 km). In this case, this model also fails.

Sounding Near an FLR-Shell

The amplitude of the azimuthal component of the magnetic field $b_y^{(g)}$ of pulsations near the FLR-shell is given on the ground by

$$\frac{b_y^{(g)}}{b_0} = \frac{\delta_i + h}{x - x_r\left(f\right) + i\left(\delta_i + h\right)} + C_1 k_y^2 \log k_y \left(x - x_r\left(f\right) + i\left(\delta_i + h\right)\right) + \cdots, \qquad (11.24)$$

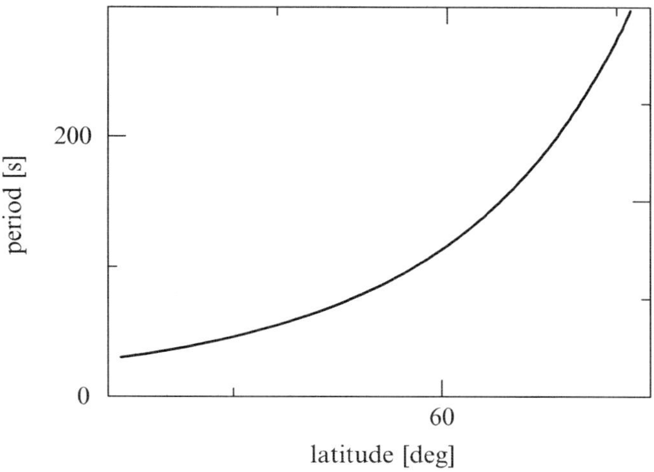

Fig. 11.4. Latitudinal dependency of the FLR-period

where δ_i is the half-width of the FLR-region in the ionosphere, h is the E-layer height, x is the distance along the meridian, $x_r(f)$ is the location of the FLR. Coefficient C_1 is defined by the curvature of the FLR-line. It vanishes in the 'box' model with a straight field (see Chapter 6). Such a model provides reasonable accuracy with experimental data at $x_r \leq 500\,\text{km}$.

Let a half-width of an FLR-area on the ionospheric level be $\delta_i = 20\,\text{km}$. Then a characteristic spatial scale of the resonance structure on the ground is about 100–200 km. It is evident that there are geoelectrical cross-sections for which the effective skin depth d_g in the $Pc3, 4$ range and a horizontal field scale are comparable. Therefore phase and amplitude features of the initial waves noticeably influence the results of the magnetotelluric soundings of such geological profiles.

Let Φ, Λ be the geomagnetic coordinates of an observation point. Assume that the observed field at some frequency f results from FLR-oscillations of the corresponding FLR-shell located at distance

$$x_r(f) = 6370 \cdot (\Phi_r(f) - \Phi) \cdot \frac{180}{\pi}$$

from the observer. Here $\Phi_r(f)$ is the latitude of the FLR-shell with resonance frequency $f_r = f$ (see Fig.11.4). Latitude-dependence of FLR-frequency is shown in Fig.11.4. The ground model is a four-layered geoelectrical cross-section with $\rho_1 = 30\,\text{Ohm} \cdot \text{m}$, $h_1 = 3\,\text{km}$; $\rho_2 = 3 \times 10^3\,\text{Ohm} \cdot \text{m}$, $h_2 = 50\,\text{km}$; $\rho_3 = 3 \times 10^2\,\text{Ohm} \cdot \text{m}$, $h_3 = 50\,\text{km}$ and lower half-space with $\rho_4 = 0$. Ionospheric conductivities are $\Sigma_P = 13\,\text{Ohm}^{-1}$, $\Sigma_H = 14^{-1}\,\text{Ohm}^{-1}$.

The height of the 'thin' ionosphere is $h = 105\,\text{km}$. The calculations are carried out for wavenumber $k_y = 0$. Period $T = 100\,\text{s}$ is the FLR for the geomagnetic latitude $\Phi = 60°$. The dip is $I = 74°$. There were computed

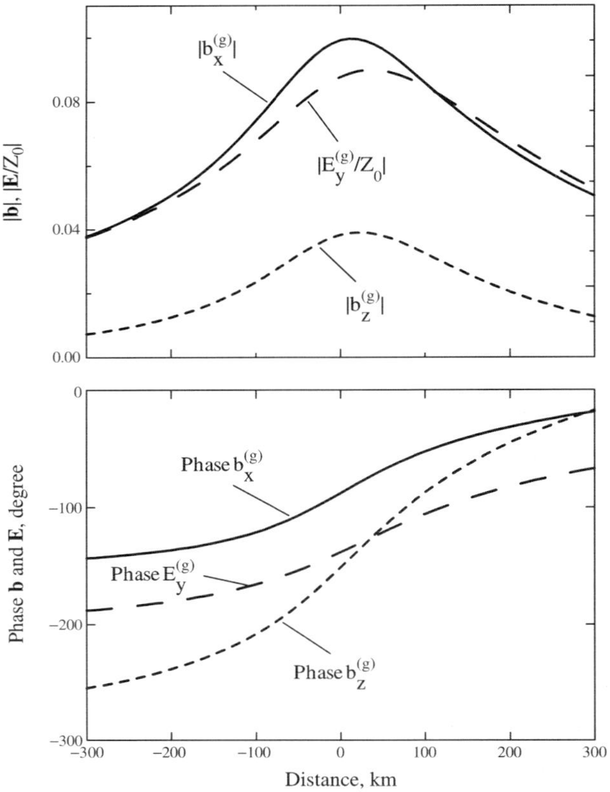

Fig. 11.5. The computed meridional distribution of amplitudes (upper panel) and phases (bottom panel) of electric ($E_y^{(g)}/Z_0$–long dashed line) and magnetic components ($b_x^{(g)}$– solid line, $b_z^{(g)}$– short dashed line). Geomagnetic latitude is $\Phi = 60°$, inclination is $I = 74°$. Distance $= 0$ is the base of the FLR-line. A four-layered geoelectrical cross-section with $\rho_1 = 30\,\text{Ohm}\cdot\text{m}$, $h_1 = 3$ km; $\rho_2 = 3\times10^3\,\text{Ohm}\cdot\text{m}$, $h_2 = 50\,\text{km}$; $\rho_3 = 3\times10^2\,\text{Ohm}\cdot\text{m}$, $h_3 = 50$ km and the perfect conductive lower half-space. $\Sigma_P = 13\,\text{Ohm}^{-1}$, $\Sigma_H = 14\,\text{Ohm}^{-1}$. The electric and magnetic components are normalized with the incident Alfvén wave

coordinate dependencies of the electric $E_y^{(g)}(x)$ and magnetic $b_x^{(g)}(x)$ components and their ratio as functions of distance from the FLR-shell

$$Z_g^*(x) = \frac{E_y^{(g)}(x)}{b_x^{(g)}(x)}\,. \tag{11.25}$$

Figure 11.5 shows the dependencies of amplitudes and phases on distance from the base of the FLR-shell for field components $b_x^{(g)}$, $E_y^{(g)}/Z_0$ and $b_z^{(g)}$.

The computations revealed a number of important features of a ground-based field:

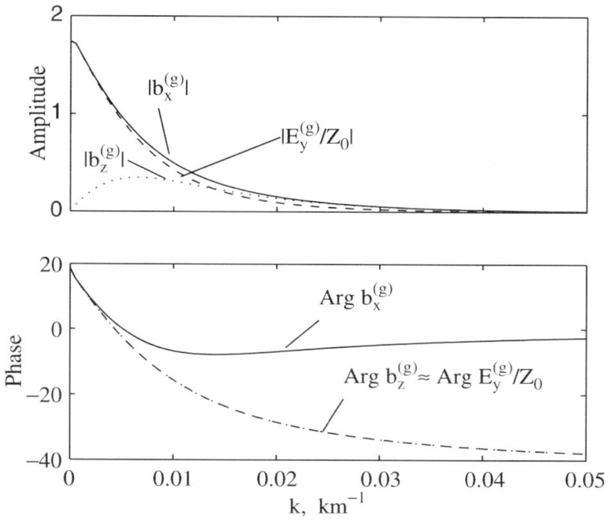

Fig. 11.6. Dependence of the amplitude (upper panel) and phase (bottom panel) of the ground magnetic $b_x^{(g)}$ and electric $E_y^{(g)}$ components on the horizontal wavenumber k. $E_y^{(g)}$ are normalized with the impedance Z_0. The model is the same as in Fig. 11.5

- Both electric and magnetic component maxima are shifted out of the field line base. In the example given this shift is ≈40 km .
- These maxima are shifted away from each other (≈20 km).
- Amplitude dependencies are asymmetrical with respect to positions of $b_x^{(g)}$, $E_y^{(g)}$ maxima.
- Phase discontinuity of $b_x^{(g)}$ and $E_y^{(g)}$ through the resonance region is ≈180°, while for $b_z^{(g)}$ it is ≈270° Taking into account the logarithmic term in formula (8.15) reduces the magnitude of the phase discontinuity to ≈100°.

These features account for the special characteristics of the coefficient $T_{SA}(k)$ for transformation of an Alfvén wave into a magnetic mode of a ground field. Figure 11.6 illustrates the $T_{SA}(k)$-dependence. The labeling of the curves is the same as in Fig. 11.5. The spatial spectrum of FLR is nonzero only in the positive semi-axis and is concentrated in the k_x range from 0 to $\sim 10^{-1}\,\mathrm{km}^{-1}$. As a result, the spectrum at the ground contains harmonics from 0 to $10^{-2}\,\mathrm{km}^{-1}$. In this region, the phase of $T_{SA}(k)$ exhibits strong variations attributable primarily to the fact that the horizontal magnetic field scale is comparable to the depth of the skin depth in the ground. It is convenient to assume that the phase change $\Phi(k)$ in the indicated range of wavenumbers is

$$\Phi(k) \approx (x + x_0)\,k + \Phi_0,$$

where

$$x_0 \approx \frac{\Delta\Phi}{\Delta k}.$$

The amplitude maximum is displaced northward by a value on the order of x_0. The direction of the displacement is determined by the sign of $\partial\Phi/\partial k$. k_x-dependencies of the phases of transfer functions for electric and magnetic fields show an appreciable difference. This also leads to a relative displacement in the spatial distributions.

Changes of the geoelectrical parameters of the model will first affect the $E_y^{(g)}(x)$ and $b_x^{(g)}(x)$ dependence. On the other hand, the $b_x^{(g)}(x)$ maximum position depends slightly on the $\rho(z)$ variations. For example, for a cross-section that includes a layer of high resistance ($\rho_1 = 10^4\,\mathrm{Ohm\cdot m}$, $h_1 = 200\,\mathrm{km}$; $\rho_2 = 0$, $h_2 = \infty$) and resonance period $T = 100\,\mathrm{s}$, the displacement of the electric component maximum is $\approx 100\,\mathrm{km}$ and $\approx 30\,\mathrm{km}$ for the magnetic component.

Z_g^* tends to Z_0 for highly conductive ground if the horizontal scale is larger than the ground skin depth (see (11.23)). Let us introduce the so-called 'apparent resistivity' ρ_{ap} (or ρ_{ap}^*) associated with Z_0 (or Z_g^*) by

$$\rho_{ap} = 2T\left|Z_0\right|^2, \qquad \rho_{ap}^* = 2T\left|Z_g^*\right|^2. \tag{11.26}$$

ρ_{ap} is frequency-dependent. For a frequency ω a layered geoelectrical cross-section has the same surface impedance as the homogeneous half-space of the resistivity $\rho_{ap}(\omega)$. The applicability of the model of the plane normally incident wave can be checked by comparing curves $\rho_{ap}(T)$ and $\rho_T^*(T)$. Upper panel of Fig.11.7 presents $\rho_{ap}(T)$ (solid line) and $\rho_{ap}^*(T)$ at observation latitudes of $\Phi = 60°$ (long dashed) and $\Phi = 55°$ (short dashed) with resonance periods of 100 and 46 s, respectively. The low panel of Fig. 11.7 shows phase curves of Z_0 and Z_g^*. It can be seen that taking into account the field's peculiarities near the resonance shells results in the appearance of additional extrema for both high-latitude and middle-latitude observations. Deviation of $\rho_{ap}(T)$ from $\rho_{ap}^*(T)$ reaches 30% for the high-latitude curve and is caused by the displacement of $E_y^{(g)}(x)$ and $b_x^{(g)}(x)$ with respect to each other. The appearance of such peculiarities on curves $\rho_{ap}(T)$ should always be expected on the FLR-frequencies of the observation point. Anomalies in $\rho_{ap}(T)$ increase significantly in transition to high-resistance cross-sections without a sedimentary cover.

*Pc*3,4 pulsation fields on the ground within a 500 km zone along the FLR-shell cannot be approximated with linear functions. Here the interpretation of MTS-observations within the framework of traditional conceptions may result in erroneous conclusions. Special caution is needed in interpreting sounding data on high-resistance cross-sections without a sedimentary cover. To obtain a reliable result it is necessary to find the spatial structure of the pulsation fields (see Sect. 11.2).

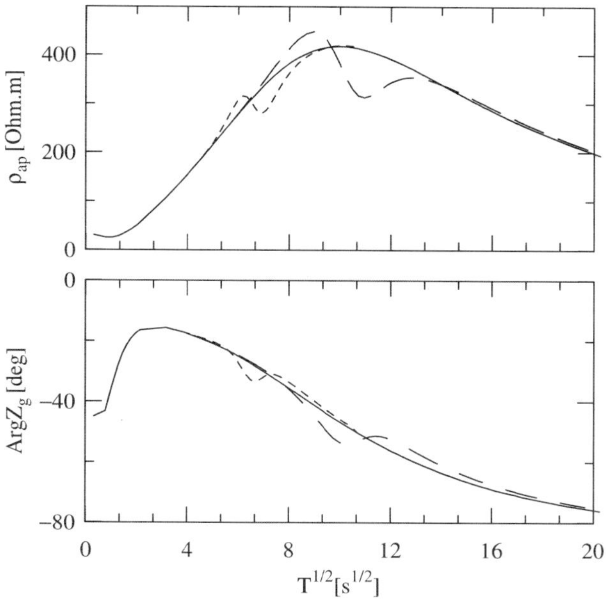

Fig. 11.7. Plots of the impedance and apparent resistivity near resonance shell as functions of $T^{1/2}$. Solid curve refers to the apparent resistivity $\rho_{ap}(T)$ for the normally incident initial wave. The long dashed and short dashed lines are the apparent resistivities $\rho^*_{ap}(T)$ calculated for latitudes $\Phi = 60°$ (long dashed) and $\Phi = 55°$ (short dashed) where the resonance periods are 100 s and 46 s, respectively

11.4 The Satellite Electromagnetic Sounding of Earth

Sounding on a Spatial Harmonic

There are two main difficulties inherent in the method of ground magnetotelluric sounding. The first is the existence of horizontal subsurface inhomogeneities. Such inhomogeneities cannot only complicate, but often make it impossible to interpret the observed data. The diffraction field of the waves scattered on the inhomogeneities, and the reflected wave from the ground layers superimpose on the initial wave field. If there are inhomogeneities, the approach employed in the different geophysical methods is moving away from the irregularities and using height observations. The other complicating factor is the necessity of prior assumption about the spatial structure of the initial wave. For example, an assumption of a plane wave which is incident normally on the ground. Moving the observations above the ionosphere, solves the second problem due to the peculiar ionospheric reflection of an Alfvén wave.

The method of satellite electromagnetic sounding of the Earth may be based on the investigation of a horizontal homogeneous model of the media

[5]. It may be summarized in a more physical language as follows. If an Alfvén or an FMS wave is incident on the ionosphere both Alfvén and FMS waves are reflected. An Alfvén wave has a longitudinal electric current. This wave reflects into the same wave with the reflection coefficient (7.126)

$$R_{AA}^{(0)} = \frac{X - \sqrt{\varepsilon_m}\sin I}{X + \sqrt{\varepsilon_m}\sin I}, \qquad X = \frac{4\pi}{c}\Sigma_P. \tag{11.27}$$

The total electric field of the Alfvén wave creates the Pedersen and Hall current systems. The Pedersen system provides the reflection of the Alfvén wave into another Alfvén wave. The Hall current creates a reflected FMS-wave and a transmitting wave which penetrates into the atmosphere and into the Earth. The electric current of the wave, penetrated into the Earth, generates the atmospheric and magnetospheric fields. As a result, the electric and magnetic field in the reflected FMS-wave depend on the Earth's conductivity.

Let us estimate the influence of the Earth's conductivity on the longitudinal component of the magnetic field $b_\parallel$. Let, as before, z-axes be vertically upwards so that x- and y- axes are horizontal. The medium is assumed to be horizontally stratified, which means that the conductivity is function only the vertical coordinate z. And let any field component F be given by

$$F = F(z)\exp(ik_x x + ik_y y)$$

where k_x and k_y are horizontal wavenumbers in meridional and longitudinal directions, respectively. Then the horizonal components **E** and **b** in (7.16) may be written

$$E_x = -k_0 k_y \Psi, \quad E_y = k_0 k_x \Psi, \quad b_x = ik_x\frac{d\Psi}{dz}, \quad b_y = ik_y\frac{d\Psi}{dz},$$

and (7.80) become

$$b_\parallel = -\sin I\,(k - ik_x\cot I)k\Psi,$$

where $k = \sqrt{k_x^2 + k_y^2}$. The substitution of (7.148) in the last equation gives

$$b_\parallel = -\frac{Y}{2}\exp(iI\,\mathrm{sign}\,k_x)[1 - R_g\exp(-2kh)]E_A\,,$$

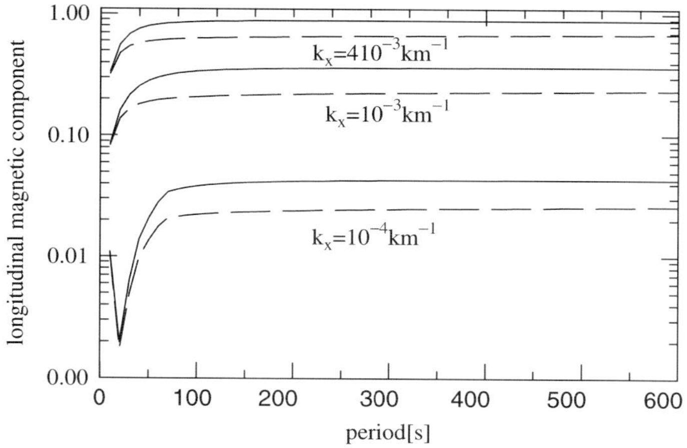

Fig. 11.8. Amplitude of the above-ionosphere field-aligned magnetic component $b_{\|}$ versus oscillation period. The curves correspond to three horizontal wavenumbers and refer to two geoelectric models. The first one is the one-layered one (solid line) and the second one is the two-layered model (dashed line). In the 1-st model the resistivity of the 1-st layer is $\rho = 10^3$ Ohm $\cdot$ m and thickness is $h_1 = 100$ km. In the two-layered model the resistivity of the upper layer is $\rho_1 = 10^1$ Ohm $\cdot$ m, and the lower layer is $\rho_2 = 10^3$ Ohm $\cdot$ m. Respectively, thicknesses are $h_1 = 50\,\text{km}$, and $h_2 = 50\,\text{km}$. The basement is the perfect conductor in the both models

where

$$R_g = \frac{1 - \dfrac{ik}{k_0} Z_g^{(m)}}{1 + \dfrac{ik}{k_0} Z_g^{(m)}} .$$

The angle I is the dip of the magnetic field, $Z_g^{(m)}$ is a surface impedance of the ground, E_A is the electric field amplitude of the Alfvén wave in the ionosphere (see (7.136)).

Thus it is possible to get information about the geoelectrical cross-section by using the properties of waves reflected from the ionosphere. This statement is correct only if the scale of the initial perturbation will be approximately equal to or exceed the thickness of the atmosphere h. An attenuation of the wave is proportional to $\exp(-kh)$. Therefore for a large $\mathbf{k}$ the wave loses information about the surface impedance of the ground.

The dependencies of the longitudinal component $b_{\|}$ on the period for wavenumbers $k_x = 10^{-4}$, 10^{-3} and $4 \times 10^{-3}\,\text{km}^{-1}$, are shown in Fig. 11.8 by solid lines for a one-layer model and for a two-layer model (dashed line). The basis is a perfect conductor in both cases. It can be seen that moving from one model to another, $b_{\|}$ is approximately double for small $\mathbf{k}$.

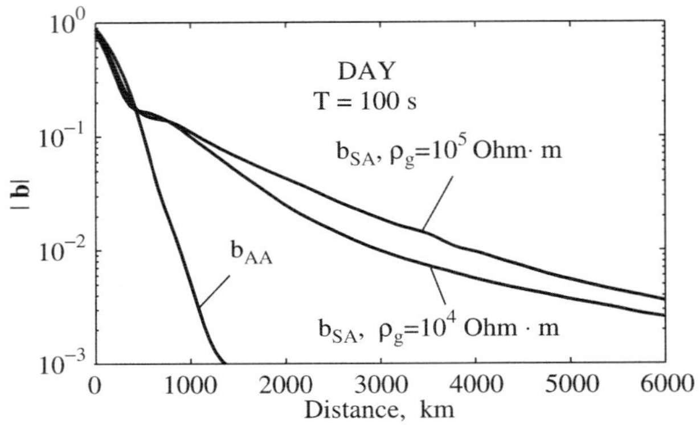

Fig. 11.9. The horizontal magnetic component $b_{SA}(x)$ in the FMS-wave above the ionosphere. The initial incident wave is an Alfvén beam with the horizontal magnetic component $b^{(i)} = \exp\left(-x^2/2L^2\right)$. Oscillations period is $T = 100\,\mathrm{s}$, $L = 100\,\mathrm{km}$. The ground is a homogeneous half-space of the resistivity ρ_g. The ionospheric conductivities were chosen for the dayside conditions and the dip $I = \pi/2$. The distance is shown from the axis beam

MHD-Wave Beams

Let us abandon the spectral presentation of fields in favor of a coordinate presentation for sources of two types: a localized wave bell-beam in the high-latitude ionosphere and a resonance magnetic shell. The relations for the ground ($\mathbf{b}^{(g)}, \mathbf{E}^{(g)}$) and magnetospheric ($\mathbf{b}^{(m)}, \mathbf{E}^{(m)}$) fields are defined by the inverse Fourier transformation as was discussed in Chapter 8. Assume that an Alfvén monochromatic bell-beam $b_A^{(i)}(x)$ is incident on the polar ionosphere with a vertical magnetic field. Coordinate dependence of the intensity in the beam is

$$b_A^{(i)}(x) \propto \exp\left(-\frac{x^2}{\delta_i^2}\right),$$

where δ_i a half-width of the beam. It is firstly useful to assume that the Earth is a homogeneous half-space with finite specific resistivity ρ_g. Let the values of the integral Hall and Pedersen conductivities correspond to dayside conditions, $\Sigma_P = 1.2 \times 10^8\,\mathrm{km/s}$, $\Sigma_H = 1.4 \times 10^8\,\mathrm{km/s}$. Fig. 11.9 shows the spatial dependencies of horizontal magnetic component $b_{SA}(x)$ of the reflected FMS-wave and the total $b_{AA}(x)$ of the Alfvén wave. The curves are given for two values of the resistivity $\rho_g = 10^5\,\mathrm{Ohm \cdot m}$ and $\rho_g = 10^4\,\mathrm{Ohm \cdot m}$. It follows from the figure that just the reflected FMS-wave is sensitive to the geoelectric resistivity. The horizontal magnetic field of the reflected FMS-wave roughly doubles when ground resistivity changes by an order of magnitude.

The longitudinal magnetic component $b_{\|}$ is a distinctive feature of an FMS-wave. Figure 11.10 demonstrates the spatial distribution above the ionosphere of the longitudinal magnetic component for the ground resistivities shown

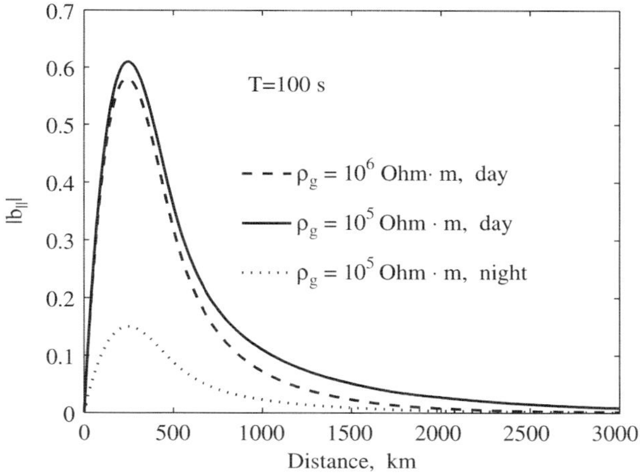

Fig. 11.10. The estimated spatial distribution of the vertical component in the reflected FMS-wave for two values of the ground resistivity. The ionospheric models respect the dayside and midnight conditions. The incident wave is the Alfvén beam of $T = 100\,\mathrm{s}$

in the figure. Maximum of the $b_{\|}$ is shifted with respect to the beam axis. The far regions, beginning at distances of $\sim$1000 km, are the most sensitive to variations of ground resistivity (compare the solid and dashed lines). An order of magnitude change in the ground resistivity, changes the $b_{\|}$ some 1.5–2 times.

Suppose now that the field above the ionosphere is due to the FLR. In the calculations the resonance period is 100 s, the geomagnetic dip is $I = 60°$. The height integrated Pedersen and Hall conductivities are $\Sigma_P = 1.2 \times 10^8$ km/s, $\Sigma_H = 1.4 \times 10^8$ km/s. This corresponds to dayside ionospheric conditions. The curves are computed for two geoelectric cross-sections (see Table 11.2). For simplicity the wave is independent on longitude, $k_y = 0$.

Figure 11.11 shows the magnetic components $b_x(x), b_y(x), b_z(x)$ (left frame) above the ionosphere in the vicinity of the FLR-line for two different geoelectric models (see Table 11.2). b_y represents a sum of the incident

Table 11.2. Two models of the geoelectrical cross-sections

1 Model		2 Model	
Resistivity, Ohm · m	Thickness, km	Resistivity, Ohm · m	Thickness, km
$\rho_1 = 30$	$h_1 = 3$	$\rho_1 = 3 \times 10^3$	$h_1 = 3$
$\rho_2 = 3 \times 10^3$	$h_2 = 50$	$\rho_2 = 3 \times 10^4$	$h_2 = 50$
$\rho_3 = 3 \times 10^2$	$h_3 = 50$	$\rho_3 = 3 \times 10^3$	$h_3 = 50$

$\rho_{\mathrm{basement}} = 3 \times 10^{-2}$

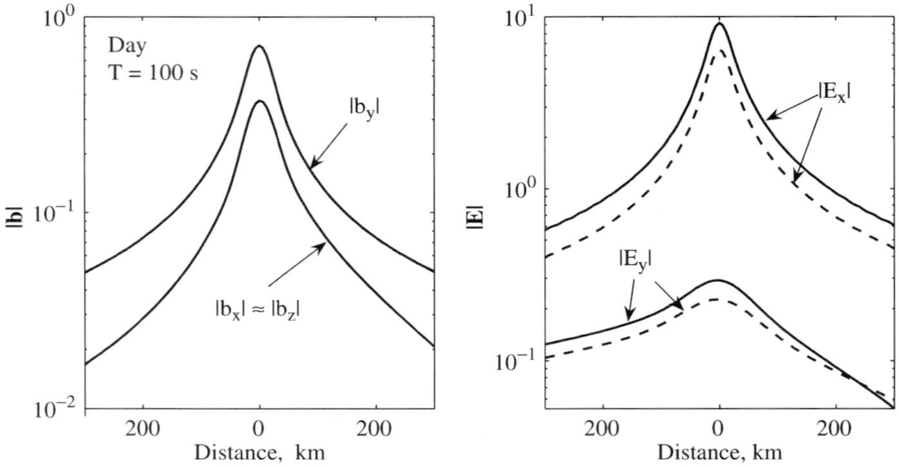

Fig. 11.11. The spatial distribution of the above-ionosphere magnetic (left panel) nearby a resonance region for an incident likewise FLR-wave beam. b_y is a sum of the incident and reflected Alfvén wave magnetic components. $b_x \approx b_z$ is the magnetic field transformed into the FMS-wave. The right panel shows the same dependencies for the electric components. E_x corresponds to the total horizontal electric field in the Alfvén polarization, and E_y is the total electric field in the FMS-polarization. The curves are computed for two geoelectric cross-sections (see Table 11.2). The curves on the left panel represent both models. Solid lines on the right panel refer to the 1-st model and the dashed to the 2-nd. The resonance period is 100 s. The geomagnetic dip is $I = 60°$. Dayside ionosphere: $\Sigma_P = 1.2 \times 10^8$ km/s, $\Sigma_P = 1.4 \times 10^8$ km/s

and reflected Alfvén wave magnetic components. $|b_x| \approx |b_z|$ is the magnetic field of the FMS-wave caused by the initial Alfvén wave. Their distributions coincide virtually at any distances from the base of the resonance shell $x = 0$. The right panel shows the same dependencies for the electric wave components. E_x corresponds to the total horizontal electric field in the Alfvén polarization, and E_y is the total electric field in the FMS-polarization.

Note that the magnetic field is almost insensitive to the geoelectric properties. So the curves on the left panel represent both models. Solid lines on the right panel refer to the 1-st model and the dashed to the 2-nd. It can be seen that E_x has the same half-width as the initial wave. The influence of the geoelectric properties on the intensity of the electric component E_x is clearly visible.

Such dependencies enable us to construct the above ionospheric spatial curves of apparent resistivity $\rho_{ap}(x)$. Similar to the relations (11.22) and (11.26) on the ground, one can define an apparent resistivity $\rho_{ap}(x)$ above the ionosphere determined as

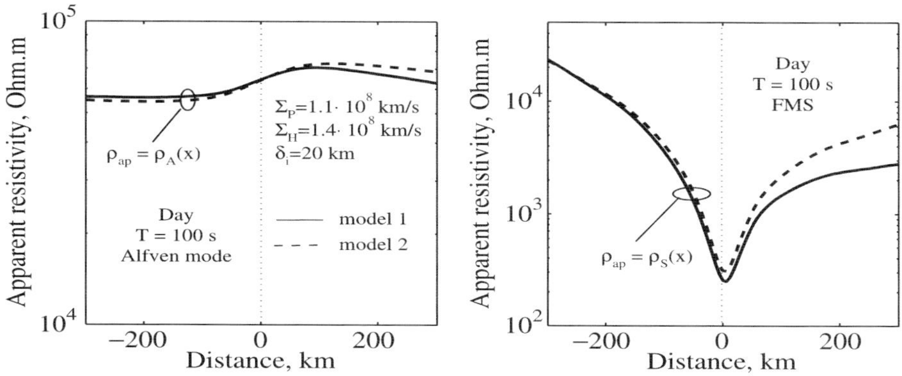

Fig. 11.12. The curves of apparent resistivity versus a distance x from the resonance beam axis for the Alfvén ρ_A and FMS ρ_S modes. Resistivities and thicknesses for two geoelectrical models are given in Table 11.2. δ_i is the half-width of the FLR. Solid and dashed lines are $\rho_A(x)$ and $\rho_S(x)$ for the 1-st and 2-nd geoelectrical cross-sections, respectively

$$\rho_{ap}\,(x) = 0.2T \left(\frac{E_y(x)}{b_x(x)}\right)^2 . \tag{11.28}$$

The apparent resistivities are different for two MHD-wave modes. Let $\rho_{ap} \equiv \rho_A\,(x)$ for the Alfvén mode and $\rho_{ap} \equiv \rho_S\,(x)$ for the FMS mode. The dependencies of $\rho_A\,(x)$ are shown in the left frame of Fig. 11.12 and $\rho_S\,(x)$ in the right. The solid line respects to the 1-st and the dashed line to the 2-nd geoelectrical cross-section (Table 11.2). $\rho_A\,(x)$ depends weakly on coordinate x. It follows from (11.27) that

$$\rho_A\,(x) \approx \frac{c}{4\pi\Sigma_P} .$$

The weak dependency of ρ_A on the horizontal coordinate and the ground resistivity can be explained as follows. The longitudinal downgoing and upgoing currents of the Alfvén wave cannot penetrate into the highly resistive atmosphere. They spread over the ionosphere and are enclosed in it. So, $\rho_A(x)$ is defined by the ionospheric conductivity and magnetospheric wave conductivity. The influence of Hall currents on the Alfvén wave is small (see Chapter 7).

The apparent resistivity $\rho_S(x)$ contrary to $\rho_A(x)$ depends strongly on both x and on geoelectrical properties. In the vicinity of the FLR-shell there is a noticeable deep minimum, as shown by the right frame of Fig. 11.12. The north ($x < 0$) and south ($x > 0$) sides of the curve $\rho_S\,(x)$ are asymmetrical, and the south side is more sensitive to changes of the ground electrical properties. In the vicinity of the resonance shell, the horizontal scale is of the order of δ_i. So, the small-scale harmonics $k^{-1} \approx \delta_i$ contribute mostly to the spatial structure at the minimum of $\rho_S(x)$. If δ_i lesser than the thickness of the atmosphere h then the atmospheric field decreases so rapidly that there is no contribution of the ground resistivity into ρ_S.

In the reflected FMS-wave the horizontal wavenumber on these scales exceeds the Alfvén wavenumber $k_A = k_0\sqrt{\varepsilon_m} = \omega/c_A$. For a reflected FMS-wave the impedance Z is

$$Z = \frac{k_0}{\sqrt{k^2 - k_A^2}}.$$

At the minimum of $\rho_S(x)$ the value of $k \gg k_A$. Then impedance

$$Z \approx \frac{k_0}{k},$$

and

$$\rho_S = T\,|Z|^2 \approx \frac{8\pi}{Tc^2}\delta_i^2\,.$$

It is evident that it is possible to define the scale size of the half-width of the FLR-region δ_i by measuring $\rho_S(x)$.

The asymmetry of $\rho_{ap}(x)$-curves is connected only to the wave-phase velocity in the FLR-shell. At large distances from the axis, the curves $\rho_S(x)$ flatten out at an asymptotic value. By measuring $\rho_S(x)(\omega)$ it is possible to reconstruct the geoelectrical profile $\rho_g(z)$.

References

1. Alperovich, L. S. and E. N. Fedorov, Effect of the ionosphere on the propagation of beams of MHD waves, *Radiophysics and Quantum Electronics*, **27**, 864, 1984.
2. Alperovich, L. S. and E. N. Fedorov, The role of the finite conductivity of the ground in the spatial distributions of geomagnetic pulsations, Izv. Ac. Sc. USSR, *Physics of the Solid Earth*, **20**, 858, 1984.
3. Alperovich, L. S. and E. N. Fedorov, The propagation of hydromagnetic waves through the ionospheric plasma and spatial characteristics of the geomagnetic variations, *Geomagn. Aeron.*, **24**, 529, 1984.
4. Alperovich, L. S. and E. N. Fedorov, On hydromagnetic wave beams propagation through the ionosphere, *Ann. Geoph.*, **10**, 647, 1992.
5. Alperovich, L. S. and E. N. Fedorov, The satellite electromagnetic sounding of the Earth, *Adv. Space Res.*, **13**, (11)23–(11)34, 1993.
6. Alperovich, L. S., E. N. Fedorov, and T. B. Osmakova, Characteristics of telluric field near resonance magnetic shell, *Izvestia Ac. Sci. USSR, Physics of the Solid Earth*, **7**, 23, 1991.
7. Baransky, L. N., Yu. E., Borovkov, M. B. Gochberg, S. M. Krylov, and V. A. Troitskaya, High resolution method of direct measurement of the magnetic field-lines eigenfrequencies, *Planet. Space Sci.*, **33**, 1369,1985.
8. Baransky, L. N., S. P. Belokris, Y. E. Borovkov, M. B. Gokhberg, E. N. Fedorov, and C. A. Green, Restoration of the meridional structure of geomagnetic pulsation fields from gradient measurements, *Planet. Space Sci.*, **37**, 859, 1989.

9. Baransky, L. N., S. P. Belokris, U. E. Borovkov, and C. A. Green, Two simple methods for the determination of the resonance frequencies of magnetic field-lines, *Planet. Space Sci.*, **38,** 1573–1576, 1990.
10. Cagniard, L., Basic theory of the magnetotelluric method of geophysical prospecting, *Geophysics*, **18**, 605–635, 1953.
11. Chappel, C. R. and K. K. Harris, A study of the influence of magnetic activity on the location of the plasmapause as measured by OGO-5, *J. Geophys Res.*, **75**, 50, 1970.
12. Dmitriev, V. I. and M. N. Berdichevsky, The fundamental model of magnetotelluric soundings, *Proc. IEEE*, **67**, 1033, 1979.
13. Dmitriev, V. I. and M. N. Berdichevsky, Fundamental model of magnetotelluric soundings, In *Geomagnetic Investigations, M., Radio i sviaz*, No. 27, 5, 1982.
14. Dungey, J. W., The structure of the exosphere, or, adventures in velocity, in *Geophysics, The Earth's Environment* (*Proceedings of the 1962 Les Houches Summer School*), 503, C. DeWitt, J. Hieblot, and A. Lebeau, eds, Gordon and Breach, NY, 1963.
15. Fedorov, E. N., B. N. Belenkaya, M. B. Gokhberg, S. P. Belokris, L. N. Baransky, and C. A. Green, Magnetospheric plasma density diagnosis from gradient measurements of geomagnetic pulsations, *Planet. Space Sci.*, **38**, 269–272, 1990.
16. Green, C., Meridional characteristis of *Pc*-4 micropulsation event in the plasmasphere, *Planet. Space Sci.*, **26**, 955, 1978.
17. Green, A. W., E. W. Worthington, L. N. Baransky, E. N. Fedorov, N. A. Kurneva, V. A. Pilipenko, D. N. Shvetzov, A. A. Bektemirov, and G. V. Philipov, Alfvén field line resonances at low latitudes (L = 1.5), *J. Geophys. Res.*, 98(A9), 15693, 1993.
18. Guglielmi, A. V., Diagnostics of the magnetosphere and interplanetary medium by means of pulsations, *Space Sci. Rev.*, **16**, 331, 1974.
19. Gugliel'mi, A. V., Hydromagnetic diagnostics and geoelectric prospecting, *Usp. Fiz. Nauk.*, **158**, 605, 1989.
20. Gugliel'mi, A. V., Diagnostics of the plasma in the magnetosphere by means of measurement of the spectrum of Alfvėn oscillations, *Planet. Space Sci.*, **37**, 1011, 1989.
21. Hughes, W. J. and D. J. Southwood, The screening of micropulsation signals by the atmosphere and ionosphere, *J. Geophys Res.*, **81**, 3234, 1976.
22. Kaufman, A. A., *Electromagnetic Soundings*, P. Hoekstra (Editor), Elsevier Science & Technology Books, Ser.: Methods in Geochemistry and Geophysics, 2001.
23. Lifshicz, A. E. and E. N. Fedorov, Hydromagnetic oscillations of the magnetosphere–ionosphere resonator. *Doclady of the USSR Academy of Sciences*, **287**, No.1, 90, 1986.
24. Marchenko, V. A., *Sturm-Liouville operators and application*, Naukova dumka, Kiev, 1977.
25. Neubauer, F. M. and K. H. Glassmeier, Use of an array of satellites as a wave telescope. *J. Geophys. Res.*, **95**, 19115, 1990.
26. Obayshi, T., Geomagnetic pulsations and the Earth's outer atmosphere, *Ann. Geophy.*, **14**, 464, 1958.
27. Park, C. G. and D. L. Carpenter, Electron density in the plasmasphere: whistler data on solar cycle, annual and diurnal variations, *J. Geophys Res.*, **83**, 3137, 1978.
28. Park, C. G., *Technical report, N 3451-1*: Stanford University, Stanford, California, 1972.

29. Pilipenko, V. A. and E. N. Fedorov, Magnetotelluric sounding of the crust and hydromagnetic monitoring of the magnetosphere with the use of ULF waves, in *Solar Wind Sources of Magnetospheric Ultra-Low-Frequency Waves*, Ed. M. J. Engebretson, K. Takahashi and M. Scholer, Geophysical Monograph, **81**, AGU, 283, 1994.
30. Pilipenko, V., K. Yumoto, E. Fedorov, and N. Yagova, Hydromagnetic spectroscopy of the magnetosphere with *Pc*3 geomagnetic pulsations at 210 meridian, *Ann. Geophys.*, **17**, 53, 1999.
31. Takahashi, K., R. E. Denton, R. R. Anderson, and W. J. Hughes, Frequencies of standing Alfvén wave harmonics and their implication for plasma mass distribution along geomagnetic field-lines: Statistical analysis of CRRES data, *J. Geoph. Res.*, **109 (A8)**, A08202, 2004.
32. Tichonov, A. N., On the definition of electric characteristics of deep layers of the Earth crust, (In Russian), *Doklady Ak.Nauk USSR*, **73**, 295–297, 1950.
33. Walker, A. D. M. and R. A. Greenwald, Pulsation structure in the ionosphere derived from auroral radar data, *J. Geomag. Geoelec.*, **32** , Suppl. II. 111, 1980.
34. Wait, J. R., Telluric currents and the Earth's magnetic field, *Geophysics*, **19**, 1954.

12

MHD-Wave Exposure on the Ionosphere

Another point of interest is the influence of an MHD-wave on ionospheric plasma. It is apparent that the electric field of an MHD-wave causes the movement of the background electron component and of ionospheric plasma inhomogeneities. The field-aligned currents in an Alfvén wave are directly related to the electron transfer along the external field. The FMS-wave magnetic pressure results in perturbations in plasma density. All these effects can be detected by ground-based and satellite methods using radio waves to investigate the ionosphere. Electron movement and density change cause displacement in the frequency and a turn of the polarization plane of the radio signal radiated either from the ground or a satellite. Both these effects are detected using Doppler and Faraday sounding.

12.1 The Doppler Effect Provoked by an MHD-Wave

In Doppler sounding the transmitted pulse filled by a wave with a predominant frequency, reflects from the ionosphere and returns to the ground. In the simplest situation of vertical sounding, the transmitter and receiver are located in the same place. Here the frequencies of the upgoing and downgoing waves are compared and subtracted from each other. The small difference between them versus time is called a 'dopplergram'. Variation in the dopplergram are defined either by displacement of the reflected layer and by variations of the refraction index by the way of propagation of the radio wave to the reflection point and back.

The coordinated observations of short-period fluctuations in the frequency of stable UF-transmitters and ULF-variations ([6], [7], [8]) have demonstrated that almost any magnetic perturbations give a response in the radio wave Doppler oscillations. Rishbet and Garriott [16] and Jacobs and Watanabe [9] suggested a theory to explain radio frequency oscillations by reflection point displacement as a result of electron drift in the electric field of the Alfvén wave.

Let V^* be the velocity of the reflecting layer, f_r be the frequency on which radio sounding is performed, $f(t)$ be the frequency of the received signal reflected from the ionosphere. Then

$$\Delta f_r = f\,(t) - f_r \tag{12.1}$$

is the observed frequency deviation. If the radio wave is incident onto the ionosphere normally, then the relation between these two values is determined as

$$\Delta f_r = -2 f_r \frac{V^*}{c}. \tag{12.2}$$

This theory was used to interpret observations in most of the subsequent works. Watermann [20] showed, however, that it is impossible to explain all experimental results in the framework of this concept. A revision of the earlier conceptions was carried out in ([13], [14], [17], [18]). Numerical calculations of the Alfvén wave transformation and of the corresponding perturbations in ionospheric plasma demonstrated the need for a thorough modification of the views expounded in ([9], [16]).

Figure 12.1 demonstrates an example of variations in H- and D-components of $Pi2$ oscillations (panels (a) and (b) and of simultaneous Doppler displacements in the $\Delta f(t)$ signal on frequency $f_r = 4.94\,\text{MHz}$, (panel (c)). The measurements were performed at a middle-latitude observatory (49°40′N, 36°50′E) [2].

The task of detecting MHD-wave effects in dopplergrams requires the utilization of fine enough methods of signal extraction at a high noise level. The basic idea of [2] consisted in $\Delta f\,(t)$ filtration by an optimal filter constructed by $Pi2$ pulsations. The filtered Doppler displacements (panel (d)) and the magnetic oscillations are depicted in Fig. 12.1 (H-component in the (e) panel and D-component in the (f) panel). The filtered magnetograms frames have been extracted using the Doppler trains (d) as a basis for the digital adaptation filter for the total initial horizontal magnetic components. The amazing similarity can be seen of the simultaneous $Pi2$ pulsations on the ground and Doppler displacements of radio frequency.

The Doppler Velocity

The total change of the phase in the wave transmitted from the ground surface to the reflection point and back is [5]

$$\Phi = \frac{4\pi f_r}{c} \int_0^{z_0} \mu\, dz - \frac{\pi}{2}, \tag{12.3}$$

where $\mu = \operatorname{Re} n$ is the real part of the refractive index n; the radio wave reflection point z_0 is located at altitude z at which $\mu = 0$. Displacement of the

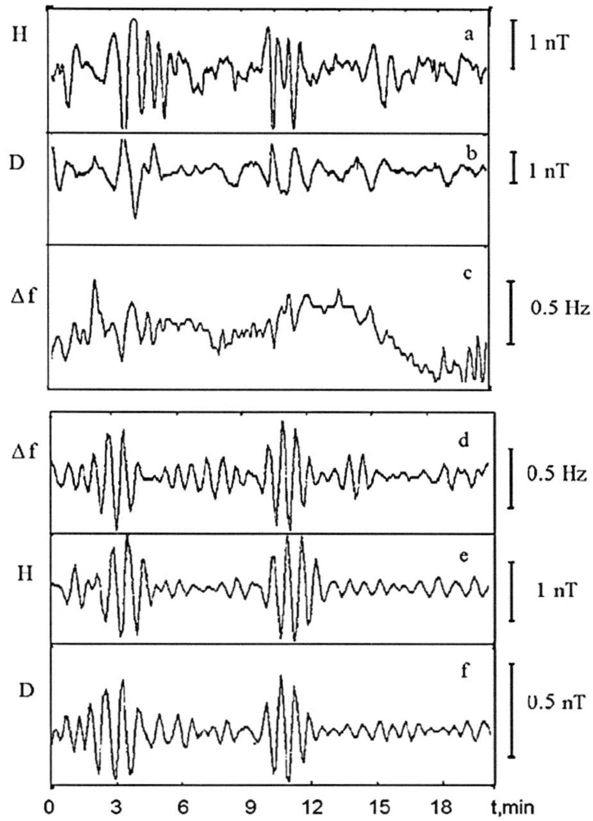

Fig. 12.1. The analyzed series of $Pi2$ and concurrent ionospheric disturbances: frames (a, b)–magnetograms of H and D components; (c)–variations of instantaneous probing radio frequency from the transmitter at $f_0 = 4.940\,\text{MHz}$; (d, e, f)–filtered Doppler frequency shift, and H and D pulsation components. After [2]

instantaneous radio signal frequency due to a phase change in the received signal is

$$\Delta f_r\,(t) = -\frac{1}{2\pi}\frac{d\Phi}{dt} = -2\frac{f_r}{c}\frac{d}{dt}\int\limits_0^{z_0} \mu\; dz\,. \tag{12.4}$$

Substituting (12.4) into (12.2), find that the Doppler velocity V^* is equal to the derivative with respect to time from the radio signal phase path

$$V^*\,(t) = -\frac{c}{2}\frac{\Delta f}{f_r} = \frac{d}{dt}\int\limits_0^{z_0} \mu\, dz\,. \tag{12.5}$$

The refractive index n of a medium containing free electrons with the external magnetic field $\mathbf{B_0}$ is determined by the Appleton–Hartree formula [20]

$$n_{\pm}^2 = 1 - \frac{X}{1 + iZ - \dfrac{Y_T^2/2}{1 - X + iZ} \pm \left(\dfrac{Y_T^2/4}{1 - X + iZ} + Y_L^2 \right)^{1/2}}, \tag{12.6}$$

where

$$Y_L = \frac{eB_L}{mc}\frac{1}{\omega} \quad \text{and} \quad Y_T = \frac{eB_T}{mc}\frac{1}{\omega},$$

and B_L, B_T are projections of the vector $\mathbf{B}$ on the longitudinal and transverse to the radio-wave propagation direction;

$$X = \frac{\omega_{pe}^2}{\omega_r^2}, \qquad Z = \frac{\nu_e}{\omega_r}, \qquad \omega_r = 2\pi f_r,$$

$\omega_{pe}^2 = 3.18 \times 10^9 N_e$ is square of the electron plasma frequency; ν_e is the collision frequency of an electron with all the other particles.

The magnetic field

$$\mathbf{B(t)} = \mathbf{B_0} + \mathbf{b(t)},$$

where $\mathbf{B_0}$ is the background geomagnetic field and $\mathbf{b(t)}$ is the pulsations magnetic field. Let us rewrite (12.5) in the form

$$V^* = \frac{d}{dt} \int_0^{z_0} \mu \left(\mathbf{B}_L(z,t), \mathbf{B}_T(z,t), N_e(z,t) \right) dz. \tag{12.7}$$

As the refractive index turns into zero at the reflection point, the last equation can be written

$$\begin{aligned} V^* &= \int_0^{z_0} \left(\frac{\partial \mu}{\partial \mathbf{B_L}} \frac{\partial \mathbf{B_L}}{\partial t} + \frac{\partial \mu}{\partial \mathbf{B_T}} \frac{\partial \mathbf{B_T}}{\partial t} + \frac{\partial \mu}{\partial N_e} \frac{\partial N_e}{\partial t} \right) dz + \mu\left(z_0, t\right) \frac{dz_0}{dt} \\ &= \int_0^{z_0} \left(\frac{\partial \mu}{\partial \mathbf{B_L}} \frac{\partial \mathbf{B_L}}{\partial t} + \frac{\partial \mu}{\partial \mathbf{B}_T} \frac{\partial \mathbf{B}_T}{\partial t} + \frac{\partial \mu}{\partial N_e} \frac{\partial N_e}{\partial t} \right) dz. \end{aligned} \tag{12.8}$$

The electron concentration change in time is found from the continuity equation

$$\frac{\partial N_e}{\partial t} = -\boldsymbol{\nabla} \cdot (N_e \mathbf{v}_e) + Q - L, \tag{12.9}$$

where $\mathbf{v}_e$ is the electron drift velocity in the MHD-wave field, Q and L are the production and loss rates, respectively. Combining (12.8) and (12.9),

Poole *et al.* [14] expressed V^* as the sum of four addends

$$V^* = V_1 + V_2 + V_3 + V_4, \tag{12.10}$$

where

$$V_1 = \int\limits_0^{z_0} \left(\frac{\partial \mu}{\partial \mathbf{B_L}} \frac{\partial \mathbf{B_L}}{\partial t} + \frac{\partial \mu}{d\mathbf{B_T}} \frac{\partial \mathbf{B_T}}{\partial t} \right) dz, \tag{12.11}$$

$$V_2 = -\int\limits_0^{z_0} \frac{\partial \mu}{\partial N_e} \left(\mathbf{v}_e \cdot \boldsymbol{\nabla}\right) N_e \, dz, \tag{12.12}$$

$$V_3 = -\int\limits_0^{z_0} \frac{\partial \mu}{\partial N_e} N_e \, \boldsymbol{\nabla} \cdot \mathbf{v}_e \, dz, \tag{12.13}$$

$$V_4 = \int\limits_0^{z_0} \frac{\partial \mu}{\partial N_e} \left(Q - L\right) dz. \tag{12.14}$$

V_1 Mechanism

For a vertically propagating radio signal

$$\frac{\partial}{\partial t} B_L = \frac{\partial}{\partial t} \left(B_{0z} + b_z\right) = \frac{\partial b_z}{\partial t}.$$

The time derivative of the transverse component B_T can easily be found in the linear approximation with respect to $|\mathbf{b}|/\mathbf{B_0}$:

$$\begin{aligned} B_T &= \sqrt{\left(B_{0T} + b_x\right)^2 + b_y^2} \\ &= B_{0T} \sqrt{1 + 2\frac{b_x}{B_{0T}} + \left(\frac{b_x}{B_{0T}}\right)^2 + \left(\frac{b_y}{B_{0T}}\right)^2} \approx B_{0T} + b_x. \end{aligned}$$

Therefore

$$\frac{\partial B_T}{\partial t} = \frac{\partial b_x}{\partial t}.$$

Substitution this relation into (12.11) for harmonic perturbations $\propto \exp(-i\omega t)$ leads to

$$V_1 = -i\omega \int\limits_0^{z_0} \left(\frac{d\mu}{dB_L} b_z + \frac{d\mu}{dB_T} b_x \right) dz. \tag{12.15}$$

Equation (12.15) describes the contribution to the Doppler velocity V^* associated with the dependence of the refractive index on magnetic field variations.

$\mathbf{V_2}$ Mechanism

Addends V_2 and V_3 are entirely determined by the electron velocity equal in the first approximation to the drift velocity

$$\mathbf{v}_e = c\frac{\mathbf{E} \times \mathbf{B_0}}{B_0^2} . \tag{12.16}$$

In the general case, (12.16) must be supplemented by terms that take into account, first, collisions in the ionosphere and, second, electron velocity along the magnetic field, which is not equal to zero in the Alfvén waves. The directional electron velocity $\mathbf{v}_e$ is

$$\mathbf{v}_e = -\frac{1}{N_e e}\left(\mathbf{j}_{\parallel e} + \mathbf{j}_{\perp}\right) , \tag{12.17}$$

$$\mathbf{j}_{\parallel e} = \sigma_{\parallel e}\mathbf{E}_{\parallel}, \quad \mathbf{j}_{\perp} = \sigma_{Pe}\mathbf{E}_{\perp} - \sigma_{He}\frac{\mathbf{E}_{\perp} \times \mathbf{B}_0}{B_0} , \tag{12.18}$$

where $\sigma_{\parallel e}$, σ_{Pe} and σ_{He} are longitudinal, Pedersen and Hall electron conductivities, respectively. $\mathbf{E}_{\parallel}$ and $\mathbf{E}_{\perp}$ are components of the wave electric field, parallel and transverse to $\mathbf{B}_0$. For the electron conductivities, we have from (1.93), (1.94), and (1.98)

$$\sigma_{\parallel e} = \sigma_0 = \frac{N_e e^2}{m_e \nu_e}, \qquad \sigma_{Pe} = \frac{\sigma_0}{1 + {\beta_e}^2} , \qquad \sigma_{He} = \beta_e \sigma_{Pe} , \tag{12.19}$$

where β_e is the electron magnetization parameter, ν_e is the electron collision frequency.

The mobility along field-lines exceeds transverse mobility by 5–6 orders. Hence $\mathbf{E}_{\parallel} = 0$. This gives

$$E_z = -E_x \cot I. \tag{12.20}$$

Using transformation (7.1) and (12.20) from (12.18) we obtain

$$j_{xe} = \sigma_{Pe}E_x + \sigma_{He}E_y \sin I + j_{\parallel e} \cos I, \tag{12.21}$$

$$j_{ye} = -\frac{\sigma_{He}}{\sin I}E_x + \sigma_{Pe}E_y, \tag{12.22}$$

$$j_{ze} = -\sigma_{Pe}E_x \cot I - \sigma_{He}E_y \cos I + j_{\parallel e} \sin I, \tag{12.23}$$

The main contribution to the longitudinal current $j_{\parallel}$ comes from electrons. Indeed, total $j_{\parallel}$ is a sum of electron $j_{\parallel e}$ and ion $j_{\parallel i}$:

$$j_{\parallel} = j_{\parallel e} + j_{\parallel i} \qquad \text{and} \qquad j_{\parallel i} = \frac{\sigma_{\parallel i}}{\sigma_{\perp e}} j_{\parallel e},$$

where $\sigma_{\parallel i}$, $j_{\parallel i}$ are longitudinal ion conductivity and current. Hence

$$j_{\|e} = j_\| \frac{1}{1 + \sigma_{\|i}/\sigma_{\|e}}.$$

But

$$\frac{\sigma_{\|i}}{\sigma_{\|e}} \sim \frac{m_e}{m_i}\frac{\nu_e}{\nu_i} \sim \sqrt{\frac{m_e}{m_i}} \ll 1,$$

where m_i being the average ion mass. Therefore

$$j_\| \approx j_{\|e}.$$

This allows us to express $j_\|$ in terms of the magnetic pulsations components.

It follows from the Ampere's law that the vertical current density component j_z is related with the magnetic wave components as

$$j_z = \frac{c}{4\pi}\left(\frac{\partial b_y}{\partial x} - \frac{\partial b_x}{\partial y}\right). \tag{12.24}$$

From an equation analogous to (12.23), but written for the total current, we find

$$j_\| \sin I = j_z + \sigma_P E_x \cot I + \sigma_H E_y \cos I.$$

Taking into account that $j_{\|e} \approx j_\|$, we rewrite (12.21)–(12.23)

$$\begin{aligned}
j_{xe} &= j_z \cot I + \left(\sigma_{Pe} + \sigma_P \cot^2 I\right) E_x + \left(\sigma_{He} + \sigma_H \cot^2 I\right) E_y, \\
j_{ye} &= -\frac{\sigma_{He}}{\sin I} E_x + \sigma_{Pe} E_y, \\
j_{ze} &= j_z + \sigma_{Pi} E_x \cot I + \sigma_{Hi} E_y \cos I.
\end{aligned} \tag{12.25}$$

For the electron velocity, we obtain from (12.25)

$$\begin{aligned}
v_{xe} &= -\frac{j_z \cot I}{N_e e} - \frac{1}{N_e e}\left[\sigma_{Pe} + (\sigma_{Pe} + \sigma_{Pi}] \cot^2 I\right) E_x \\
&\quad + \frac{1}{N_e e}\left[\sigma_{He} + (\sigma_{He} + \sigma_{Hi}) \cot^2 I\right] E_y \sin I \\
v_{ye} &= \frac{1}{N_e e}\frac{\sigma_{He}}{\sin I} E_x - \frac{1}{N_e e}\sigma_{Pe} E_y, \\
v_{ze} &= -\frac{j_z}{N_e e} - \frac{1}{N_e e}\sigma_{Pi} E_x \cot I - \frac{1}{N_e e}\sigma_{Hi} E_y \cos I.
\end{aligned} \tag{12.26}$$

Above the E-layer, the electron and ion Pedersen conductivities are small. Neglecting collisions, we get

$$\sigma_{He} = -\sigma_{Hi} = -c\frac{N_e e}{B_0}.$$

Then (12.26) becomes

$$v_{xe} = \frac{c}{B_0} E_y \sin I - \frac{j_z}{N_e e} \cot I,$$

$$v_{ye} = -\frac{c}{B_0 \sin I} E_x,$$

$$v_{ze} = \frac{c}{B_0} E_y \cos I - \frac{j_z}{N_e e}, \tag{12.27}$$

or in the vector form

$$\mathbf{v}_e = c \frac{\mathbf{E} \times \mathbf{B}_0}{B_0^2} - \frac{j_\parallel}{N_e e} \frac{\mathbf{B}_0}{B_0}. \tag{12.28}$$

The expression for V_2 in view of only electron drift can be written

$$V_2 = -c \int_0^{z_0} \frac{\partial \mu}{\partial N_e} \frac{\partial N_e}{\partial z} \frac{\cos I}{B_0} E_y \, dz. \tag{12.29}$$

Equation (12.29) describes the so-called 'motor' part of dynamo action [16], in which V^* equals to the vertical component of electron velocity under the action of the east-west component of the electric field. This mechanism can be identified by a reflection level motion. Indeed,

$$\frac{\partial \mu}{\partial z} = \frac{\partial \mu}{\partial N_e} \frac{\partial N_e}{\partial z} + \frac{\partial \mu}{\partial B_L} \frac{\partial B_L}{\partial z} + \frac{\partial \mu}{\partial B_T} \frac{\partial B_T}{\partial z}. \tag{12.30}$$

B_L and B_T change slowly with altitude z. Simple numerical estimations show that the terms containing $\partial B_L/\partial z$ and $\partial B_T/\partial z$ in (12.30) can be neglected. Therefore,

$$\frac{\partial \mu}{\partial z} \approx \frac{\partial \mu}{\partial N_e} \frac{\partial N_e}{\partial z}. \tag{12.31}$$

E_y changes slowly with altitude, therefore it follows from (12.29) that

$$V_2 = -c \frac{\cos I}{B_0} E_y \int_0^{z_0} d\mu = c \frac{\cos I}{B_0} E_y. \tag{12.32}$$

Thus V_2 is indeed equal to the vertical component of electron velocity and by virtue of being independent of altitude, coincides with the velocity of reflection level motion.

V_3 Mechanism

Transform now the expression for V_3. We neglect changes in B_0. Then

$$\begin{aligned}\nabla \cdot \mathbf{v}_e &= \frac{c}{B_0^2}\nabla \cdot (\mathbf{E} \times \mathbf{B}_0) = \frac{c}{B_0^2}\mathbf{B}_0 \cdot (\nabla \times \mathbf{E}) \\ &= i\omega\frac{(\mathbf{B}_0 \cdot \mathbf{b})}{B_0^2} = i\omega\frac{b_\parallel}{B_0} = -i\omega\nabla \cdot \xi_\perp ,\end{aligned}$$

where $b_\parallel = (\mathbf{b} \cdot \mathbf{B}_0)$ is a pulsation field component parallel to $\mathbf{B}_0$. Hence

$$V_3 = -i\omega \int_0^{z_0} \frac{\partial \mu}{\partial N_e} N_e \frac{b_\parallel}{B_0}\, dz \tag{12.33}$$

is proportional to the degree of plasma compression

$$-\frac{b_\parallel}{B_0} = \nabla \cdot \xi_\perp .$$

V_4 Mechanism

Velocity V_4 is associated with the change in rates of ion production and loss when the plasma volume moves by the action of the MHD-wave field. Poole and Sutcliffe [13] showed that the oscillatory part of Q and L caused by an MHD-wave is small and, in the first approximation, it can be set as

$$V_4 = 0.$$

Quantitative Treatment

MHD-waves propagating in the meridional plane contain the following non-zero components (see 7.66):

Alfvén wave	$b_y,\ E_x,\ E_y$
FMS	$b_x,\ b_z,\ E_y$

The Alfvén wave electric field lies in the magnetic meridian plane and does not cause vertical electric drift. If we ignore the longitudinal current transferred by the Alfvén wave, then only the FMS-wave causes vertical electron drift and plasma compression.

We now find the relation between Doppler velocity and pulsations on the ground. Let $b_y^{(i)}$ be the amplitude of magnetic field perturbations in the incident Alfvén wave, $b_y = b_y^{(i)} \exp(-i\omega t + ikx - ik_A z)$. Equations (7.95), (7.96) for the reflected FMS-wave in the Cartesian coordinates $\{x, y, z\}$ (see (7.2)) then gives

$$b_\parallel(z) = -R'_{SA} b_y^{(i)} \exp(iI) \exp(i\kappa_S z), \tag{12.34}$$

$$E_y(z) = -\frac{k_0}{\kappa_S} R'_{SA} b_y^{(i)} \exp(i\kappa_S z). \tag{12.35}$$

For the coefficient transformation R'_{SA} of Alfvén wave into an FMS-wave, (7.99) gives

$$R'_{SA} = \frac{\kappa_S}{k_0}\frac{2Y}{\Delta}\mathrm{sign}I\,.$$

The transmission coefficient T'_{SA} connects the magnetic fields on the ground $b_x^{(g)}$ and in the ionosphere $b_y^{(i)}$:

$$b_x^{(g)} = T'_{SA} b_y^{(i)}. \tag{12.36}$$

where (7.105) for T'_{SA} gives

$$T'_{SA} = -\frac{Y_g^m}{\zeta_3}\frac{2Y}{\Delta}\mathrm{sign}I\,.$$

Now we can find electric field $E_y(z)$ in the ionosphere from the ground magnetic field dividing (12.35) by (12.36), so that

$$\frac{E_y(z)}{b_x^{(g)}} = -\frac{k_0}{\kappa_S}\frac{R'_{SA}}{T'_{SA}}\exp(i\kappa_S z) \tag{12.37}$$

$$\frac{R'_{SA}}{T'_{SA}} = -\frac{\kappa_S}{k_0}\left(Z_g^{(m)}\cosh(kh) - i\frac{k_0}{k}\sinh(kh)\right], \tag{12.38}$$

where

$$h = \int \frac{\sigma_H}{\Sigma_H}\, z\, dz.$$

$Z_g^{(m)}$ is the magnetic impedance on the ground surface.

Let σ_g = const and let a horizontal scale k^{-1} less than 10^3 km. Then impedance $Z_g^{(m)} = \kappa_g/k_0 \approx ik/k_0$ (see (7.28)) and $\kappa_S \approx ik$. Equation (12.38) becomes

$$\frac{R'_{SA}}{T'_{SA}} = -\sinh(kh) - \frac{kd_g}{\sqrt{k^2 d_g^2 - 2i}}\cosh(kh). \tag{12.39}$$

Hence the ratio of the electric field in the reflected Alfvén wave to the ground magnetic field is independent of the integral ionospheric conductivity. This follows at once from the continuity of the tangential components of the electric field in the thin ionospheric model.

Using (12.29), (12.33), (12.34) and (12.37), we find the relations between Doppler velocity components and magnetic variation on the ground

$$\frac{V_2}{b_x^{(g)}} = -i\frac{\omega}{k}\frac{\cos I}{B_0}\frac{R'_{SA}}{T'_{SA}}\int_0^{z_0}\frac{\partial\mu}{\partial N_e}\frac{\partial N_e}{\partial z}\exp(-kz)\,dz\,, \tag{12.40}$$

$$\frac{V_3}{b_x^{(g)}} = i\omega\frac{\exp(iI)}{B_0}\frac{R'_{SA}}{T'_{SA}}\int_0^{z_0}\frac{\partial\mu}{\partial N_e}N_e\exp(-kz)\,dz\,. \tag{12.41}$$

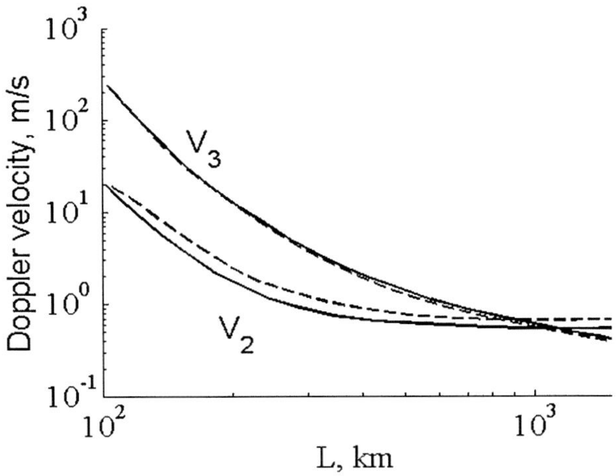

Fig. 12.2. Comparison between numerical (broken line) [14] and analytical calculations (solid line). L is the north-south scale length, and V_2 and V_3 are given by (12.40) and (12.41)

Analytical estimates by (12.40)–(12.41) and the numerical results by Poole *et al.*[14] are shown in Fig. 12.2, respectively, by the solid and the dashed lines. The curves as for V_2 as for V_3 are close to each other. Both theory and numerical experiment predict that, for almost all the interesting scales $L_\perp$ velocity,

$$V_2(L) < V_3(L).$$

It also follows from (12.40) and (12.41) that the effects of both convection and compressibility are proportional to the MHD-wave frequency.

Sutcliffe and Poole [17] calculated the height profiles of the pulsation electric **E** and magnetic **b** fields of an MHD-wave propagating in a real ionosphere and the electron and ion velocities caused by the wave. It was shown that the Jacobs and Watanabe theory [9] has to be modified. Specifically, it is necessary to take into account, besides simple electron drift in crossed electric and magnetic fields, both the longitudinal current transported by the Alfvén wave and the plasma compression effect in the reflected FMS-wave.

When a localized Alfvén wave beam is incident on the ionosphere, a contribution to the radio-wave frequency displacement will, as before, come from plasma compression in the reflected FMS-wave and from the total longitudinal current of the incident and reflected Alfvén waves. Far from the beam axis, the Alfvén wave intensity falls rapidly. The contribution from plasma compression by the FMS-wave generated by spread-out Hall currents becomes predominant.

In contrast to localized beams, the longitudinal current amplitude in FLR-oscillations decreases more slowly than the amplitude of spread-out Hall currents generated by them. Consequently, the contribution to the Doppler frequency displacement of the FLR-oscillations from plasma compression and from longitudinal currents will be of the same order at any distances from the resonance shell.

12.2 TEC Modulation by an MHD-Wave

At the above ionosphere radio wave sounding, the radio wave transmitted from the magnetosphere penetrates through the ionosphere and reaches a ground receiver. The geomagnetic field results in the ionosphere becoming an anisotropic medium. It means that there are in general two refractive indices associated with two different wave polarizations. Suppose that a linear radio wave is transmitted from the magnetosphere. The linear wave may be resolved into two circular components with an opposite rotation sense. Since the refraction indices (n) of these waves are different then their phase velocities are different as well. Thus polarization plane of the downgoing wave turns below the ionosphere with respect to the initial polarization plane. The value of the angle (the Faraday rotation angle) is defined by the total electron content within the column between the magnetospheric emitter and the ground receiver.

Short-period fluctuations of N_T were discovered in experiments to determine the Total Electron Content N_T (TEC) of the ionosphere with the help of a radio beacon on a geostationary satellite ([7], [11], [4]). The appearance of these fluctuations was associated with the intensification of $Pc4$ pulsations. The so-called differential Doppler method with very high sensitivity was used to determine N_T along the radio path. Comparison of simultaneous TEC records and of the H (or D) pulsation-field component demonstrates a clear connection [11]. The mechanism of plasma compression in the wave field stimulated by the presence of a longitudinal MHD magnetic component $b_{\|}$ proved the most effective [13]. The differential Doppler method enables integral electron concentration along a radio beam between emitter (S) and receiver (R) to be determined from observations of satellite radio beacon signals:

$$N_T = \int_R^S N_e \, dz \, . \tag{12.42}$$

Determination of N_T from Faraday Rotation

The method based on measuring the Faraday rotation angle provides information on the weighted-mean value of TEC. The rotation angle of the polarization plane of the VHF radio wave on the way between the geostationary satellite and the ground depends on the number of free electrons along the ray path. A linearly polarized wave in an anisotropic plasma can be presented as sum of the right-hand and left-hand polarized waves with different propagation velocities. Phase difference $\Delta\Phi$ between these two modes changes at the propagation path at

$$\Delta\Phi = \frac{\omega}{c}\int_R^S \Delta\mu ds, \tag{12.43}$$

where the integral is taken along the ray path; $\Delta\mu$ is the real part difference of the refractive index of two modes. Let the frequency $\omega \gg \omega_{pi}$ and $\omega \gg \omega_{ci}$, then in the case of quasi-longitudinal propagation (see, e.g., [20]), from (12.6) we get

$$\Delta\mu = \frac{\omega_{pe}^2}{\omega^2}\frac{\omega_{ce}}{\omega}\cos\theta.$$

where θ is the angle between the ray path and $\mathbf{B}_0$. Let α be the angle between the ray path and the vertical, then $ds = dz/\cos\alpha$. Now (12.43) can be written in the form

$$\Delta\Phi = \frac{1}{c\omega^2}\int_0^{h_s} \omega_{pe}^2\omega_{ce}\frac{\cos\theta}{\cos\alpha}ds, \tag{12.44}$$

where the integration is performed up to the satellite height (h_s).

The Faraday rotation angle Ω_F is

$$\Omega_F = \frac{1}{2}\Delta\Phi.$$

Substitution of $\Delta\Phi$ from (12.44) into the last equation yields ([19], [21])

$$\Omega_F = \frac{1}{2\omega^2 c}\int_0^{h_s} \omega_{pe}^2\omega_{ce}\frac{\cos\theta}{\cos\alpha}dz = k\int_0^{h_s} N_e M dz, \tag{12.45}$$

where

$$k = \frac{e^3}{2\pi c^2 m_e^2 f^2} = \frac{2.36\times10^4}{f^2}, \qquad M = B_0\frac{\cos\theta}{\cos\alpha}.$$

Equation (12.45) is written in the Gaussian system. The numerical value of the coefficient k in SI is the same: $k = 2.36\times10^4/f^2\,\mathrm{m}^{-1}$.

The main contribution to (12.42) and (12.45) comes from the altitude region of up to 2×10^3 km [19]. Therefore, it can be assumed that M is constant up to some limiting height h_F, and zero above this height. Then (12.45) is rewritten as

$$\Omega_F = kMN_F, \tag{12.46}$$

where $N_F = \int_0^{h_F} N_e\, dz$. With $h_F = 2000$ km it was found that the value of M varies by only ±3% [19]. Therefore (12.46) can be used, and changes in Ω_F correspond to changes in $N_F \approx N_T$.

Estimations of N_T Modulation

An MHD-wave coming from the magnetosphere to the ground produces TEC-variations N_T the time derivative of which is

$$\frac{\partial N_T}{\partial t} = \int_S^R \frac{\partial N_e}{\partial t}\, dz, \tag{12.47}$$

where $\partial N_e/\partial t$ is found from the continuity equation (12.9). For the times considered fluctuations in the rates of ion production Q and loss L prove to be insignificant, (12.9) becomes

$$\frac{\partial N_e}{\partial t} = -N_e \nabla \cdot \mathbf{v}_e - (\mathbf{v}_e \cdot \nabla) N_e. \tag{12.48}$$

The second term in (12.48) does not contribute to the time variations N_T if $N_e = 0$ in the boundary points [13]. It is easy to confirm by integrating (12.48) over z for $N_e = 0$ at the end points of the integration. The main effect comes from the first term in the right-hand part of (12.48) describing the plasma compression in the wave field (for details, see [13]).

In the F-layer, which gives the dominant contribution into N_T, we can assume the velocity to be given by the $\mathbf{E} \times \mathbf{B}$ drift, so the electron velocity

$$\mathbf{v}_e = \frac{c}{B_0^2} \mathbf{E} \times \mathbf{B_0}.$$

In order to simplify consideration, let us assume that $\mathbf{B_0}$ is homogeneous and directed vertically upward. Then

$$\nabla \cdot \mathbf{v}_e = \frac{c}{B_0} (\nabla \times \mathbf{E})_z\,. \tag{12.49}$$

The right hand of the last equation can be rewritten using Faraday's law. Then, (12.48) becomes

$$\frac{1}{N_e}\frac{\partial N_e}{\partial t} = \frac{1}{B_0}\frac{\partial b_{\|}}{\partial t}, \tag{12.50}$$

where $b_{\|} = (\mathbf{b} \cdot \mathbf{B}_0)/B_0$.

To simplify calculations, we make the inessential assumption that the magnetospheric radio beacon is located in the zenith above the receiver R. Then

$$\frac{\delta N_T}{N_T} = \int N_e \frac{b_{\|}}{B_0} dz \Big/ \int N_e dz\,. \tag{12.51}$$

Integrating in (12.51) is performed in fact from the E-layer ($z = 0$) up to 2×10^3 km. Let us make use of the results of Chapter 7.

Alfvén Wave

Consider the case of an incident Alfvén wave with transverse wavenumber $\mathbf{k} = (k_x > 0, k_y = 0)$ [12]. The ratio of the ground magnetic signal $b_x^{(g)}$ to the incident wave amplitude $b_y^{(i)}$ is the transmission coefficient:

$$T'_{SA} = \frac{b_x^{(g)}}{b_y^{(i)}}. \tag{12.52}$$

The field-aligned FMS-mode amplitude $b_{\|}^{(r)}$ above the ionosphere is associated with $b_y^{(i)}$ as

$$b_{\|}^{(r)}(z) = -b_y^{(i)} R'_{SA} \exp(iI - kz), \tag{12.53}$$

where R'_{SA} is the reflection coefficient. The FMS-mode near the ionosphere has a magnetostatic character and for $k_y = 0$ its dispersion equation is reduced to

$$k_x^2 + k_z^2 \approx 0.$$

Then from $\boldsymbol{\nabla} \cdot \mathbf{b} = 0$ we obtain

$$b_{\|}^{(r)} = -(b_x^{(r)} \cos I + i b_z^{(r)} \sin I) = -b_x^{(r)} \exp iI.$$

This explains the appearance of an additional phase factor $\exp(iI)$ in (12.53). From (12.52) and (12.53) it follows that

$$\frac{b_{\|}^{(r)}}{b_x^{(g)}} = -\frac{R'_{SA}}{T'_{SA}} \exp(iI - kz). \tag{12.54}$$

In the simple case of the high conductive ground, $kd_g \ll 1$ and the terms associated with the skin effect in (7.124) and (7.125) can be omitted. Then the surface impedance $Z_g^{(m)} = 0$. For the MHD-wave with the horizontal scale $L_\perp \ll 1/k_A$ it can be set $(k_A^2 - k^2)^{1/2} \approx ik$. Then (12.38) becomes

$$\frac{R'_{SA}}{T'_{SA}} \approx -\sinh(kh). \tag{12.55}$$

For qualitative estimations set them so that the electron concentration decreases rapidly with height as

$$N(z) \propto \exp\left(-\frac{z}{H}\right),$$

where H is the magnetospheric scale height. Then, from (12.51)–(12.55) we have

$$\frac{\delta N_T}{N_T} = \frac{b_x^{(g)}}{B_0} \sinh(kh) \exp(iI) \frac{1}{1 + kH}. \tag{12.56}$$

Typically,

$$k \approx \frac{2\pi}{300}\,\mathrm{km}^{-1}, \qquad H \approx 10^3\,\mathrm{km}, \qquad B_0 \approx 2\times 10^4\,\mathrm{nT}.$$

It follows from (12.56) that the recorded fluctuations

$$\frac{\delta N_T}{N_T} \approx 5\times 10^{-5} - 3\times 10^{-4}$$

can be caused by geomagnetic pulsations with amplitudes on the ground of

$$b_x^{(g)} \approx 5 - 30\,\mathrm{nT}.$$

Such pulsation amplitudes are not frequent, but quite realistic at middle latitudes.

The formulae presented above are of an evaluating character. For more precise quantitative conclusions, one can conduct numerical calculations according to the following scheme. Firstly, we calculate the vertical profile of electron concentration, the integral conductivities and the altitude of the thin ionosphere for a given set of geophysical parameters in accordance with the IRI-model [3]. Using the values obtained, we calculate the vertical distribution of the MHD-wave field with given transverse scales. And lastly, we find the TEC-perturbation N_T caused by the MHD-wave by integrating (12.48) numerically. Figure 12.3 demonstrates the results of the computations for the dayside and nightsight ionospheres.

The presented amplitude (Figures 12.3a, c) and phase (Figures 12.3b, d) relations between the $b_x^{(g)}$-component of the ground magnetic variation and TEC N_T perturbation correspond to the middle-latitude ionosphere. It was assumed that the pulsation period $T = 40\,\mathrm{s}$, the azimuthal wavenumber $k_y = 2\pi/5000\,\mathrm{km}^{-1}$. The value $\delta N_T/N_T$ presented in the plots is normalized on the value of $b_x^{(g)} = 1\,\mathrm{nT}$.

The sharp rise at large k_x (Fig. 12.3a) is connected with the exponential decreasing of the ground wave magnetic field at the horizontal scales comparable to the ionosphere altitude. At horizontal scales $3\times 10^2\,\mathrm{km}$ typical for middle-latitude $Pc3,4$ pulsations, the relative fluctuations

$$\frac{\delta N_T}{N_T} \approx 10^{-4}\, b_x^{(g)}\,(\mathrm{nT})\,. \qquad (12.57)$$

Quasi-periodic fluctuations of TEC with amplitudes of $\delta N_T/N_T \approx 5\times 10^{-5} - 3\times 10^{-4}$ were discovered experimentally ([7], [11]). One can believe that geomagnetic pulsations of the $Pc3,4$ type with amplitudes from fractions to several nT at middle latitudes can indeed cause the observed periodic TEC variations.

The dependence $\delta N_T/N_T$ on the wavenumber in the nighttime ionosphere (Figures 12.3c,d) is qualitatively of similar nature. It follows from Fig. 12.3c

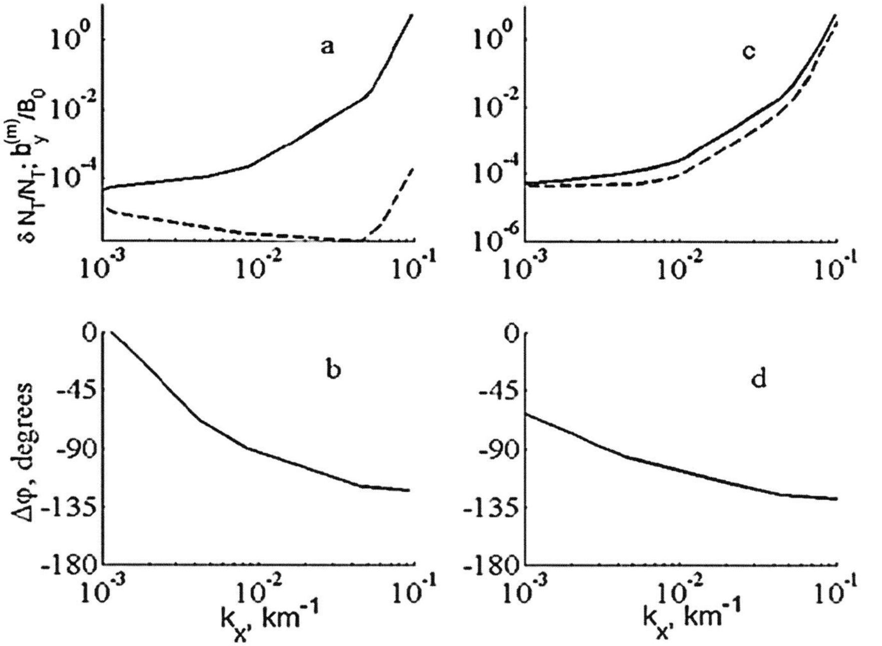

Fig. 12.3. $N_T(k_x)$ modulation dependence on the transverse wavenumber of the MHD-wave, reflected from the dayside ionosphere: (a) amplitude relation for $\delta N_T/N_T$ (the solid line) and $b_y^{(m)}/B_0$ (the dashed line) for the ground signal $b_x^{(g)} = 1\,\mathrm{nT}$; ($b$) the phase difference between the N_T variations and $b_x^{(g)}$; panels (c) and (d) are the same as (a) and (b) but for the nighttime ionosphere

that at $k_x \approx 2\pi/\,(300)\,\mathrm{km}^{-1}$ the modulation rate is $\delta N_T/N_T \approx 2 \times 10^{-6} b_x^{(g)}$. For the nighttime pulsations $Pi2$ at the typical amplitudes of $b_x^{(g)} \approx 10\,\mathrm{nT}$, the magnitude of the expected effect is $\delta N_T/N_T \approx 2 \times 10^{-5}$. Thus, according to the numerical model, the modulation effects will be more clearly manifested for daytime than for nighttime pulsations [4].

FMS-Wave

There is an additional possibility of a more effective coupling between an MHD-wave and TEC-perturbations. FMS-wave incident on the ionosphere produces, contrary to the transversal Alfvén wave, compression and rarefaction of the charged components. The expected modulation effects can be estimated as

$$\frac{\delta N_T}{N_T} \approx \frac{b_{\|}}{B_0}$$

The observed perturbations of electron concentration $\delta N_T/N_T \approx 2 \times 10^{-4}$ can be caused by FMS-waves in the magnetosphere with amplitudes of

$b_{\|} \approx 0.1-1\,\mathrm{nT}$. Thus, the reflected non-propagating FMS-mode resulting due to the incident Alfvén wave, causes quasi-periodic modulation of electron concentration. Theoretical estimates and numerical calculations have shown that pulsations in the $Pc3,4$ range can result in a modulation of the relative TEC of up to 0.01%.

References

1. Alperovich, L. S., and E. N. Fedorov, Effect of the ionosphere on the propagation of beams of MHD waves, *Radiophysics and Quantum Electronics*, **27**, 864, 1984.
2. Alperovich, L. S., E. N. Fedorov, A. V. Volgin, V. A. Pilipenko, and S. N. Pokhil'ko, Doppler sounding as a tool for the study of the MHD wave structure in the magnetosphere, *J. Atmos. Terr. Phys*, **53**, 581-586, 1991.
3. Bilitza D., Solar-terrestrial models and application software, *Planet. Space Sci.*, **40**, 541, 1992.
4. Bondarenko, N. M., Bud'ko, N. I., Gul'elmi, A. V., and Sinelnikov, V. M., On the connection of radio signal fluctuations from a coherent transmitter on an ATS-6 satellite with geomagnetic pulsations. [In Russian]. *Wave processes in near-Earth plasma*, 61, Nauka: Moscow, 1977.
5. Budden, K. G., *Radio Waves in the Ionosphere*, Cambridge University Press, 1966.
6. Chan, K. L., P. Kanellakos, and O. G. Villard, Correlation of short-period fluctuations of the Earth's magnetic field and instantaneous frequency measurements,*J.* Geophys. Res., **67**, 2066, 1962.
7. Davies, K. and G. K. Hartman, Short-period fluctuations in total columnar electron content, *J. Geophys. Res.*, **81**, 3431, 1976.
8. Duffus, H. J., and G. M. Boyad, The association between ULF geomagnetic fluctuations and Doppler ionospheric observations, *J. Atmos. Terr. Phys*, **30**, 481, 1968.
9. Jacobs, J. A., and T. Watanabe, Doppler frequency changes in radio waves propagating through a moving ionosphere, *Radio Sci.*, **1**, 257, 1966.
10. Liu, J. Y., A note on the phase relationship between ULF geomagnetic pulsations and HF-Doppler oscillations owing to the compressional mechanism, *J. Geomagn. Geoelectr.*, **43**, 777, 1991.
11. Okuzawa, T. and K. Davies, Pulsations in total columnar electron content, *J. Geophys. Res.* **86**, 1355, 1981.
12. Pilipenko, V. A. and E. N. Fedorov, Modulation of the total electron content in the ionosphere by geomagnetic pulsations, *Geomagnetism and Aeronomy.*, **34**, 516, 1995.
13. Poole, A. W. V. and P. R. Sutcliffe, Mechanisms for observed total electron content pulsations at mid latitudes, *J. Atmos. Terr. Phys.*, **49**, 231, 1987.
14. Poole, A. W. V., P. R. Sutcliffe, and A. D. M. Walker, The relationship between ULF geomagnetic pulsations and ionospheric Doppler oscillations: Derivation of a model, *J. Geophys. Res.*, **93**, 14656, 1988.
15. Ratcliffe, J. A., *Magneto-ionic Theory*, Cambridge University Press, 1956.
16. Rishbeth, H. and O. K. Garriott, Relationship between simultaneous geomagnetic and ionospheric oscillations, *Radio Sci. J. Res. NBS*, **68D**, 339, 1964.

17. Sutcliffe, P. R. and A. W. V. Poole, Ionospheric Doppler and electron velocities in the presence of ULF waves, *J. Geophys. Res.*, **94**, 505, 1989.
18. Sutcliffe, P. R. and A. W. V. Poole, The relationship between ULF geomagnetic pulsations and ionospheric Doppler oscillations: Model predictions, *Planet. Space Sci.*, **38**, 1581, 1990.
19. Titheridge, J. E. , Determination of ionospheric electron content from the Faraday rotation of geostationary satellite signals, *Planet. Space Sci.*, **20**, 353, 1972.
20. Watermann, J., Observations of correlated ULF fluctuations in the geomagnetic field and in the phase path of ionospheric HF soundings, *J. Geophys.*, **61**, 39, 1987.
21. Yeh, K. C., and V. H. Gonzalese, Note on the geometry of the Earth magnetic field useful to Faraday effect experiments, *J. Geophys. Res.*, **65**, 3209, 1960.

Additional articles

22. Baddeley, L. J., T. K. Yeoman, and D. M. Wright, HF doppler sounder measurements of the ionospheric signatures of small scale ULF waves, *Ann. Geoph.*, **23 (5)**, 1807, 2005.
23. Kim, K. H. and K. Takahashi, Statistical analysis of compressional *Pc*3,4 pulsations observed by AMPTE CCE at L = 2–3 in the dayside magnetosphere, *J. Geoph. Res.*, **104 (A3)**, 4539, 1999.
24. Marshall, R. A., and Menk F. W., Observations of *Pc*3,4 and *Pi*2 geomagnetic pulsations in the low-latitude ionosphere, *Ann. Geoph.*, **17 (11)**, 1397, 1999.
25. Menk, F. W., Yeoman T. K., Wright D. M., Lester M., and Honary F., High-latitude observations of pulse-driven ULF pulsations in the ionosphere and on the ground, *Ann. Geoph.*, **21 (2)**, 559, 2003.
26. Motoba, T., Kikuchi T., Shibata T. F., and Yumoto K., HF Doppler oscillations in the low-latitude ionosphere coherent with equatorial long-period geomagnetic field oscillations, *J. Geoph. Res.*, **109 (A6)**, 1, 2004
27. Ponomarenko, P. V., F. W. Menk, C. L. Waters, and M. D. Sciffer, *Pc*3,4 ULF waves observed by the SuperDARN TIGER radar, *Ann. Geoph.*, **23 (4)**, 1271, 2005.
28. Provan, G. and T. K Yeoman., A comparison of field line resonances observed at the Goose Bay and Wick radars, *Ann. Geoph.*, **15 (2)**, 231-235, 1997.
29. Su, S. Y., K. Y. Chen, J. M. Wu , H. C. Yeh , and C. K. Chao, ROCSAT observation of the field line resonance effect in a plasma pulsation at topside ionosphere, *J. Geoph. Res.*, **110 (A1)**, A01303, 2005.
30. Wright, D. M. and Yeoman T. K., High-latitude HF Doppler observations of ULF waves: 2. Waves with small spatial scale sizes, *Ann. Geoph.*, **17 (7)**, 868, 1999.
31. Zanandrea, A., J. M. Da Costa, S. L. G. Dutra, N. B. Trivedi, T. Kitamura, K. Yumoto, H. Tachihara, M. Shinohara, and O. Saotome, *Pc*3,4 geomagnetic pulsations at very low latitude in Brazil, *Plan. and Space Sci.*, **52 (13)**, 1209, 2004.
32. Ziesolleck, C. W. S., Feng Q., and McDiarmid D. R., *Pc*5 ULF-waves observed simultaneously by GOES 7 and the CANOPUS magnetometer array, *J. Geoph. Res.*, **101 (A3)**, 5021, 1996.
33. Ziesolleck, C. W. S., D. R. McDiarmid, and Q. Feng, A comparison between *Pc*3,4 pulsations observed by GOES 7 and the CANOPUS magnetometer array, *J. Geoph. Res.*, **102 (A3)**, 4893–4909, 1997.

13

MHD-Wave Generation by HF-Heating

13.1 Introduction

Since the wavelength of the ULF-signal in the ground is an of the order to hundreds kilometers, therefore to produce a signal in the ULF-range, one needs a tremendous size of radiated antenna. For example, in order to produce the large-scale signal at 1 Hz and amplitude of about 0.05 nT we need a ground loop with a radius of about 100 km with a current of 1 kA ([8], [13]). An important restriction of this method of generation, besides the enormous size of the radiating system and the large currents, is associated with the conducting ground, which may reduce the indicated value to 10^{-4} nT.

The main point of the man-made ULF-generation is the attempt to lift up a radiation system above the ground, transferring it to either the ionospheric or magnetospheric heights.

In this chapter we examine an intensity of conceivable artificial magnetic disturbances which could be produced by a powerful HF-wave. One can choose the frequency of a strong HF-wave so that its pump frequency ω will be close to one of the resonance ionospheric frequencies. The powerful HF-wave is incident onto the ionosphere results in a rise of the electron temperature and in fluctuations of the ionized component ([11], [12]). In doing so the electron temperature T_e can increase in 20–40 times. It leads to a change in the collision frequency of electron with neutrals. This HF-wave is modulated in amplitude. The modulation frequency Ω can be varied within very wide limits, depending on the range of the geomagnetic variations of interest. As a result, geomagnetic variations occur with the frequency determined by the modulation frequency of the strong HF-wave. The conductivity changes due to an increase of the electron collision frequency ν_e with T_e increase.

There is also another possibility of the variation of the conductivities: increase in T_e with a consequent displacement of the ionization-recombination balance and changing of the electron concentration N_e. An anomalous region in the background ionospheric current results in a magnetic disturbance.

Choosing different regimes of the amplitude modulations of the HF-pumping wave one can generate an ULF-magnetic signal of the desire geomagnetic pulsation type. The first results on the producing and extraction of an artificial ULF-signal were obtained in ([4], [10], [15], [22], [25]).

13.2 Ionospheric Heating

The basic regularities in the behavior of the non-stationary chemical kinetics of the E-layer, numerical and analytical estimations of the intensity of geomagnetic variations for various powers of the emitting system and various regimes of the pump wave frequency, amplitude modulation, etc., are briefly presented here. The following procedure is used: first, the distortions of the ionospheric parameters at the axes of the nonlinear wave beam radiated from the ground are calculated; then the perturbation of the integral conductivity within the heated ionospheric spot is found; and, finally, the intensity of the produced variation on the ground is estimated.

To estimate the disturbances of conductivity and, hence, the values of anomalous currents, we use Ohm's law (1.82), (1.83). We concentrate on the most conductive ionospheric region, i.e. E-layer. In Ohm's law, we neglect the neutral-ion collisions in comparison with the modulation frequency ω. This assumption is valid for the processes with scale less than 1 hour, virtually irrespective of altitude. We also assume that the electron-neutral collision frequency, ν_{en}, to be equal to the sum of electron collision frequencies, ν_{ek}, with k^{th} neutral components of the ionospheric plasma, each ν_{ek} being dependent on the electron temperature, T_e.

A modulated HF-wave comes to the ionosphere and when it reaches heights where electrons are magnetized, it splits into two components, the ordinary and extraordinary waves. In the case of the longitudinal propagation along the geomagnetic field, the ordinary wave has right-hand circular polarization and rotates clockwise together with the electrons. The wave energy stored in the wave is being converted into heating, mainly in the electron component of the ionospheric plasma.

The rate of electron temperature change must satisfy the equation of energy conservation for the E-layer [14]

$$\frac{dT_e}{dt} = -\frac{2}{3} e\mathbf{v}_e \cdot \mathbf{E} - \delta_{en}\nu_{en} \left(T_e - T_n\right), \tag{13.1}$$

where δ_{en} is a portion of the energy transferred through the collision of an electron with neutral particles, $\mathbf{v}_e$ is the directed velocity obtained by an electron under joint action of the electric and magnetic fields. For elastic collisions, $\delta_{en} \approx 2m_e/m_n$. Let the right-hand and the left-hand circularly polarized waves are incident normally onto the ionosphere with a vertical $\mathbf{B}_0$. Then a solution of (13.1) for the electron temperature T_e with initial $T_e = T_{e0}$

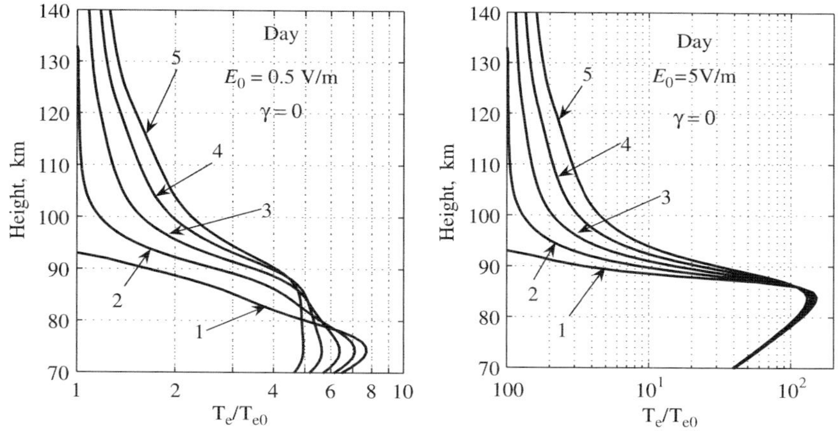

Fig. 13.1. Plots of the height dependency of the electron temperature $T_e(z)$ caused by the right-polarized HF-wave in the dayside ionosphere. The intensities of the wave electric field are 0.5 V/m (left) and 5 V/m (right) at the bottom ionosphere. The HF pump wave frequencies ω are shown at lines and given in units of 10^7 rad/s

is given by

$$T_e = T_{e0} + \frac{e^2}{3m\delta_e} \frac{E^2}{(\omega \pm \omega_{ce})^2 + \nu_{en}^2 (T_e)}. \tag{13.2}$$

For the definition of a part of the total Pedersen and Hall conductivities associated with heating of the ionosphere by a high-powered HF-wave far from the reflection point be defined at the 65 km which is chosen as the bottom of the ionosphere. In the non-linear geometrical optics approximation, the HF-wave intensity is described by [14]

$$\frac{dE}{dz} + \frac{\omega}{c} \kappa (E) E = 0. \tag{13.3}$$

Here $\kappa = \operatorname{Im} n$, the refractive index n is defined by the Appleton–Hartree formula (see Chapter 12). The calculation procedure utilized in calculations of Σ_P and Σ_H is the following. An intensity of the electric field $E(z)$ of a non-linear wave (the right-hand and left-hand) within the ionosphere is computed at every step of the integration of (13.3). Then, using (13.2), electron temperatures is found at the corresponding heights.

Figure 13.1 shows the calculated height dependency of ratio T_e/T_{e0} of the perturbed electron temperature T_e to the background T_{e0}. Both frames show this ratio for the dayside ionosphere. The left frame is for the applied electric field $E = 0.5$ V/m and the right frame is for $E = 5$ V/m at the bottom of the ionosphere $z = 65$ km. E (V/m) at the height z (km) and an effective radiation power W_{eff} (kW) of a ground equivalent transmitting dipole are related as $E = 0.3\sqrt{W}/z$. So the chosen E correspond, respectively, to

170 and 1700 MW the effective power in a source. The electron concentration N_e has been assumed to be independent of T_e. Parameter γ indicates a depth of the low-frequency amplitude modulation of the pump frequency ω. In the shown examples, $\gamma = 0$ means that it is considered to be a monochromatic non-modulated wave. The pump frequencies measured in MHZ are shown at lined in the Fig. 13.1.

At heights over 85–90 km the jump of $\triangle T_e = T_e - T_{e0}$ decreases with frequency. This is caused by decrease of both the amplitude of the electric field E, and the efficiency of energy transfer from the pump-wave to electrons with ω increasing. Besides, E-amplitude decreases due to decreasing radio-wave damping in the D-layer with frequency. Starting with ≈85 km we have $\omega \gg \nu_e$ and ν_e can be omitted in the denominator at the right-side of (13.2). Hence,

$$\triangle T_e = \frac{e^2}{3m\delta_e} \frac{E^2}{(\omega \pm \omega_{ce})^2} \, .$$

Thus the jump of the electron temperature decreases in inverse proportion to $(\omega \pm \omega_{ce})^2$. The electron temperature is almost independent of wave frequency at heights $\lesssim 85$ km for $E_0 = 5$ V/m. At these altitudes the electron collision frequency increases under heating by the wave of such amplitude up to $\nu_e \gtrsim 10^8\,\mathrm{s}^{-1} \gg \omega$. Then in (13.2) in the denominator at right-side the term $(\omega - \omega_{ce})^2$ can be omitted. Hence the frequency of the pump-way does not influence the electron temperature.

Equation (2.1) shows that increasing of the electron temperature T_e in the HF wave pump field leads to the electron-neutral collision frequencies, ν_{en}, rise and to change of the conductivities σ_P (1.86) and σ_H (1.87). We obtain a new height profile of $\sigma_P(z)$ and $\sigma_H(z)$. The ionospheric conductivity is perturbed up to the reflection point of the wave pump.

For the wave of small amplitude, the reflection point is located where the frequency of the incident wave ω is equal to the plasma frequency ω_{pe}. The height of penetration of the non-linear wave depends not only on the sounding frequency ω but also on the wave amplitude because the wave changes the medium it propagates over.

Integrating σ_P and σ_H over height, we find Σ_P and Σ_H of the heated ionosphere. Let perturbations of the background Σ_{P0} and Σ_{H0} before HF wave pump heating be $\delta\Sigma_{P0}$ and $\delta\Sigma_{H0}$, that is $\delta\Sigma_P = \Sigma_{P1} - \Sigma_{P0}$ and $\delta\Sigma_H = \Sigma_{H1} - \Sigma_{H0}$.

The relative variations of the Pedersen and Hall conductivities

$$\frac{\delta\Sigma_{P,H}}{\Sigma_{P0,H0}} = \frac{\Sigma_{(P,H)1} - \Sigma_{(P,H)0}}{\Sigma_{P0,H0}}$$

are shown in percents in Fig. 13.2 as functions of the frequency f for the applied electric field $E = 0.5$ V/m at the bottom of the ionosphere $z = 65$ km. The dayside (left panel) and nightside (right panel) summer ionospheric models (IRI-2001 [29]) were taken. The calculations were performed to both the right-hand and the left-hand polarized waves.

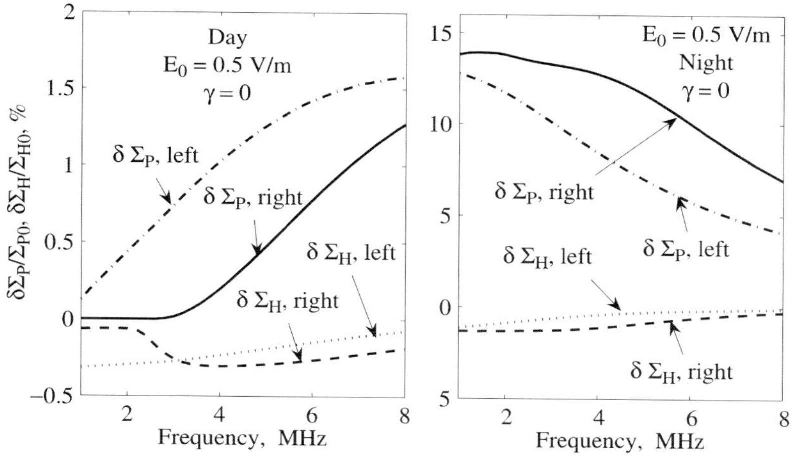

Fig. 13.2. The dependencies of the relative deviations of Σ_P and Σ_H against the frequency of the pump wave for the dayside ionospheric model (left frame) and for the night ionosphere (right frame). The perturbations are caused by an increase in T_e

Figure 13.2 shows that $\delta\Sigma_H$ is negative in both models as for the right (dashed line) as for the left polarized (dotted line) pump waves. $\delta\Sigma_H$ are of the order of tenths of percent in the daytime and less than 1% in the nighttime. $\delta\Sigma_P$ are greater and $\sim$1−2% in the daytime and $\sim$10−15% in the nighttime. Σ_P grows with ω in the daytime and decreases in the nighttime. By day, the counterclockwise polarized mode causes stronger perturbations of Σ_P than the clockwise polarized mode. In the afternoon, the left-handed wave leads to greater perturbation of Σ_P than the right-handed wave. a wave. At night, on the contrary, the right-handed wave causes greater perturbations of Σ_P. Switch from night to day decreases the heating effect by exponent of magnitude because of strong attenuation in the lower ionosphere with a fairly small contribution to the total conductivity.

These and others similar calculations demonstrated that we should not expect the maximum absolute deviation in the existing current systems to exceed 1% in the daytime and 15% in the nighttime when the effective power of an equivalent dipole is $\approx 10^3$ MW. The low daytime $\delta\Sigma_P$ and $\delta\Sigma_H$ would be expected with simple arguments. The portion of energy lost at elastic collisions is approximately equal to the ratio of masses of the colliding particles. Therefore the main heated component is the electron gas.

Since mainly the electrons are the heating, the deviations of the conductivities are due to β_e, then (see (1.93), 1.94))

$$\delta\sigma_P = \delta\sigma_{Pe}, \quad \delta\sigma_H = \delta\sigma_{He}, \tag{13.4}$$

$$\sigma_{Pe} = c\frac{Ne}{B_0}\frac{\beta_e}{1+\beta_e^2}, \quad \sigma_{He} = c\frac{Ne}{B_0}\frac{\beta_e^2}{1+\beta_e^2}, \tag{13.5}$$

where

$$\beta_e \approx \frac{\omega_{ce}}{\nu_e(T_{e0})}\sqrt{\frac{T_e}{T_{e0}}},$$

Here, the collision frequencies are given by

$$\nu_{en} \sim \nu_{eno}\sqrt{\frac{T_e}{T_{e0}}}.$$

instead of more accurate (2.1). σ_{Pe} is small for both large and small β_e and maximal when $\beta_e = 1$.

Let $\beta_e(h_0) = 1$ at height h_0. Since $\nu_e \propto \exp(-z/l_0)$ with the scale-height l_0 of the order of a few kilometers, then the main contribution into σ_{Pe} comes from the interval 5−10 km around h_0. With increasing of T_e the collision frequency goes up and β_e goes down. Therefore, h_0 increases with heating. Since the electron concentration of the D- and E-layers increases with height then $\sigma_{Pe}(z)$ also increases as it can be seen from numerical calculations. For example, the unperturbed height $h_0 \approx 70$km as for dayside as for nightiside ionospheric models. HF heating with initial amplitude of the electric field $E_0 = 0.5$ V/m leads to increasing h_0 to 78 km in daytime and to 87 km at night. Function $\beta_e^2/(1+\beta_e^2)$ decreases monotonically with decreasing of β_e. Therefore, the Hall conductivity goes down with increasing of ν_e and, consequently, with decreasing of β_e (see (13.5)).

Absorbtion of the pump wave in the dayside D-layer reduces essentially the effect of the HF heating of E-layer. Figure 13.2 makes it clear that the maximal disturbances in the ionosphere are found at local nighttime, in the absence of the D. The maximum value of $\delta\Sigma_P/\Sigma_{P0}$, about 14%, occurs in the right polarized wave at frequency $f = 1{-}2$ MHz. So that the anomaly conductivities depend weakly on T_e. It is, therefore, clear that excitation of ULF-oscillations with this mechanism is low.

13.3 Kinetics of the E-Layer in a Strong HF-Wave

Let us consider the changes in the electron concentration within E-layer caused by changes in the effective recombination rate. To clarify the role of non-stationarity during heating of the ionospheric gas, one can consider the chain of reactions of the chemical kinetics of the E-layer (see Table 13.1) ([6], [17]).

The notation N^* is used in Table 13.1 for the excited nitrogen $= N(^2D)$; p and p_1, are dimensionless parameters characterizing the fraction of the generation of $N\left(^4S\right) = N$ as a result of recombination of NO^+ and collisional dissociation of $N_2 + e$. The reaction rates γ_k, α_k of Table 13.1 and all other parameters needed to calculate the equations of chemical kinetics of the perturbed E-layer, are given in Table 13.2 ([6], [17]).

Table 13.1. A general scheme of chemical reactions in the E-layer

$N_2^+ + O_2 \xrightarrow{\gamma_1} O_2^+ + N_2$	$N_2^+ + O \xrightarrow{\gamma_2} NO^+ + N$
$N_2^+ + O \xrightarrow{\gamma_2^*} NO^+ + N^*,$	$O^+ + N_2 \xrightarrow{\gamma_3} NO^+ + N$
$O^+ + O_2 \xrightarrow{\gamma_4} O_2^+ + O$	$O_2^+ + N \xrightarrow{\gamma_5} NO^+ + O$
$O_2^+ + N^* \xrightarrow{\gamma_5} NO^+ + O$	$O_2^+ + N_2 \xrightarrow{\gamma_6} NO^+ + NO$
$O_2^+ + NO \xrightarrow{\gamma_7} NO^+ + O_2$	$N^* + O \xrightarrow{\gamma_8} N + O$
$N + O_2 \xrightarrow{\gamma_{10}} O + NO$	$NO \xrightarrow{\gamma_9} N + O$
$NO + N \xrightarrow{\gamma_{12}} N_2 + O$	$N^* \xrightarrow{\gamma_{11}} N$
$O_2^+ + e \xrightarrow{\alpha_{D1}} O_2$	$NO^+ + e \xrightarrow{p\alpha_{D2}} N + O$
$NO^+ + e \xrightarrow{(1-p)\alpha_{D2}} N^* + O$	$N_2^+ + e \xrightarrow{p_1\alpha_{D3}} 2N$
$N_2^+ + e \xrightarrow{(1-p_1)\alpha_{D3}} 2N^*$	$N^* + e \xrightarrow{\alpha_{D4}} N + e$

Since the maximum dynamo effect can occur in the E-layer at 120–150 km, to estimate the significance of the various terms in Table 13.1, we assign the following values of the concentrations of the main components:

$$[N_2] \approx 5.8 \times 10^{11} \text{cm}^{-3}, \qquad [O_2] \approx 1.2 \times 10^{11} \text{cm}^{-3},$$
$$[O] \approx 7.6 \times 10^{10} \text{cm}^{-3}, \qquad [N_e] \approx 10^5 \text{cm}^{-3}$$

We use the notation [X] for the concentration of a component X. A detailed discussion on a redistribution of the chemical kinetics in the field of the strong HF-wave and some close questions connected with transport of various ionospheric components, particularly the problem of penetration of the initiated electric field from the point of generation, etc., are beyond the scope of this book. For detailed description of these questions we refer the reader to [2]. Here we resume that, in a first approximation, one can ignore ion-molecular reactions in the E-layer. Then chemical kinetics equations corresponding to Table 13.1 are simplified, and then they can be written in the

Table 13.2. The rates of the chemical reactions in the E-layer (cm^3s^{-1}) ([6], [17])

$\gamma_1 = 10^{-10}$	$\gamma_2 = \gamma_2^* = 2.5 \times 10^{-10}$
$\gamma_3 = 5 \times 10^{-13}$	$\gamma_4 = 2 \times 10^{-11}$
$\gamma_5 = 2 \times 10^{-10}$	$\gamma_6 = 10^{-16}$
$\gamma_7 = 7 \times 10^{-10}$	$\gamma_8 = 10^{-12}$
$\gamma_9 = 10^{-5}$	$\gamma_{10} = 1.2 \times 10^{-11} \exp(-3523/T_0)$
$\gamma_{11} = 1.8 \times 10^{-10}$	$\gamma_{12} = 2.2 \times 10^{-11}$
$\alpha_{d1} = 2.2 \times 10^{-7} \times 300/T_e^{0.7}$	$\alpha_{d2} = 5 \times 10^{-7} \times (300/T_e)$
$\alpha_{d3} = 3.1 \times 10^{-6} \times (T_e)^{-0.39}$	$\alpha_{d4} = 4 \times 10^{-12} \times (T_e)^{0.8}$

form

$$\frac{\partial\left[\mathrm{O}_2^+\right]}{\partial t} = I_1 - A\left[\mathrm{O}_2^+\right] - \alpha_{d1}(t)\left[\mathrm{O}_2^+\right]\left[\mathrm{N}_e\right], \tag{13.6}$$

$$\frac{\partial\left[\mathrm{NO}^+\right]}{\partial t} = I_2 + A\left[\mathrm{O}_2^+\right] - \alpha_{d2}(t)\left[\mathrm{NO}^+\right]\left[\mathrm{N}_e\right], \tag{13.7}$$

where

$$\begin{aligned} I_1 &= J_{\mathrm{O}_2} + \gamma_1\left[\mathrm{N}_2^+\right]\left[\mathrm{O}_2\right] + \gamma_4\left[\mathrm{O}^+\right]\left[\mathrm{O}_2\right], \\ I_2 &= J_{\mathrm{NO}} + \gamma_3\left[\mathrm{O}^+\right]\left[\mathrm{N}_2\right] + \gamma_2\left[\mathrm{N_2}^+\right]\left[\mathrm{O}\right], \\ A &= \gamma_6\left[\mathrm{N}_2\right] + \gamma_7\left[\mathrm{NO}\right] \end{aligned}$$

and $[\mathrm{N}_e] = \left[\mathrm{O}_2^+\right] + [\mathrm{NO}^+]$, since $\left[\mathrm{N}_2^+\right]$ and $[\mathrm{O}^+] \ll \left[\mathrm{O}_2^+\right]$ and $[\mathrm{NO}^+]$ everywhere in the E-layer.

The electron temperature, T_e, is assumed to be periodically dependent on time, with a period $\tau_T = \tau/2 = \pi/\Omega$. Then the recombination rates can be represented in the form of Fourier series:

$$\alpha_{D1} = a_0 + \sum_n a_n \cos 2n\Omega t + b_n \sin 2n\Omega t, \tag{13.8}$$

$$\alpha_{D2} = c_0 + \sum_n c_n \cos 2n\Omega t + d_n \sin 2n\Omega t. \tag{13.9}$$

We will also seek the steady-state solution of (13.6)–(13.7) in the form of Fourier series:

$$\left[O_2^+\right] = \alpha_0 + \sum_n \alpha_n \cos 2n\Omega t + \beta_n \sin 2n\Omega t, \tag{13.10}$$

$$\left[NO^+\right] = \gamma_0 + \sum_n \gamma_n \cos 2n\Omega t + \delta_n \sin 2n\Omega t. \tag{13.11}$$

Consider two limiting cases:

1. the period of variation of $T_e(t)$ (τ_T) is much more than the chemical equilibrium time (τ_{chem}), i.e.,

$$\tau_T \gg \tau_{chem};$$

2. opposite case

$$\tau_T \ll \tau_{chem}.$$

$[\mathrm{N}_e]$ in the 1-st case is determined from

$$\alpha_{D1}^3\left[\mathrm{N}_e\right]^3 + A\left[\mathrm{N}_e\right]^2 - \alpha_{D1}\left(\frac{I_1}{\alpha_{D1}} + \frac{I_2}{\alpha_{D2}}\right)\left[\mathrm{N}_e\right] - A\frac{I_1 + I_2}{\alpha_{D2}} = 0. \tag{13.12}$$

Let us introduce in the 2-nd case the small parameter $\varepsilon = \tau_T/\tau_{chem} \ll 1$. Then, using Bogolyubov's method of averaging [5], for $\left[\mathrm{O}_2^+\right]$ and $[\mathrm{N}_e]$ we obtain

$$\left[O_2^+\right] = \alpha_0^{(0)} \left[1 + \frac{\alpha_0^{(0)} + \gamma_0^{(0)}}{2\Omega} \left(b_1 \cos 2\Omega t - a_1 \sin 2\Omega t\right) + \cdots \right], \qquad (13.13)$$

$$\begin{aligned}[N_e] = \left(\alpha_0^{(0)} + \gamma_0^{(0)}\right) \Big\{1 + \frac{1}{2\Omega} \Big[\left(\alpha_0^{(0)} b_1 + \gamma_0^{(0)} d_1\right) \cos 2\Omega t \\ - \left(\alpha_0^{(0)} a_1 + \gamma_0^{(0)} c_1\right) \sin 2\Omega t\Big] + \cdots \Big\}, \qquad (13.14)\end{aligned}$$

where $\alpha_0^{(0)}$ and $\gamma_0^{(0)}$ are found from the averaged equations (13.6), (13.7):

$$\begin{aligned} I_1 - A\alpha_0^{(0)} - a_0 \alpha_0^{(0)} \left(\alpha_0^{(0)} + \gamma_0^{(0)}\right) = 0, \\ I_2 + A\alpha_0^{(0)} - c_0 \alpha_0^{(0)} \left(\alpha_0^{(0)} + \gamma_0^{(0)}\right) = 0. \qquad (13.15)\end{aligned}$$

Suppose now that the pump wave amplitude is modulated by the harmonic oscillations $\sin \Omega t$. Then T_e fluctuations are proportional to $\sin^2 \Omega t$. The total electron temperature T_e is given by $T_e = T_0 + T_1 \sin^2 \Omega t$. We find $\alpha_0^{(0)}$ and $\gamma_0^{(0)}$ from (13.15), substitute them into (13.13) and (13.14) and obtain the time dependencies of $\left[O_2^+\right]$ and $[N_e]$. Figure 13.3 (curves 7 and 8 for periods $\tau_T = 10$ s and $\tau_T = 100$ s, respectively) shows the dependence of the depth of modulation $p_m = \delta N_e / [N_e(\max) + N_e(\min)]$ of N_e on T_1/T_0.

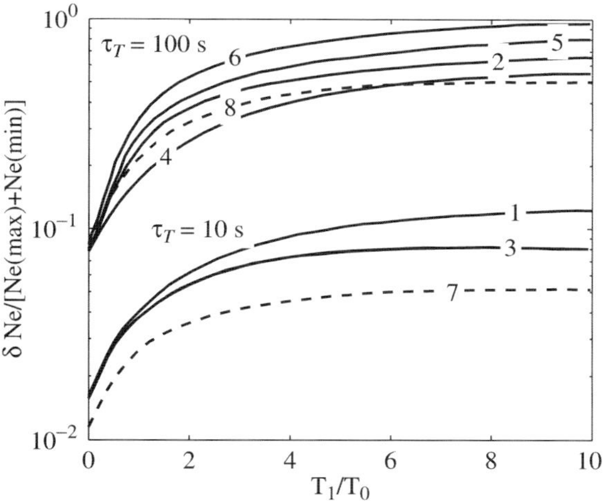

Fig. 13.3. Relative amplitude of oscillations of the electron concentration $p_m = \delta N_e / [N_e(\max) + N_e(\min)] = [N_e(\max) - N_e(\min)]/[N_e(\max) + N_e(\min)]$ as a function of the temperature modulation T_1/T_0 for $\tau_T = 10$ and $\tau_T = 100$ s. Curves 1,2 correspond to 10/100s minimum solar activity. Curves 3,4 are for maximum activity. Curves 6 and 5 represent the dependency $p_m\,(T_1/T_0)$ for the long time modulation $(\tau_T/\tau_{chem} \gg 1)$ for the minimum and maximum solar activity, respectively. The dashed lines labeled 7 and 8 respect to the fast modulation $(\tau_T/\tau_{chem} \ll 1)$ are the integrand of (13.13)–(13.14) for $\tau_T = 10$ and 100s correspondingly

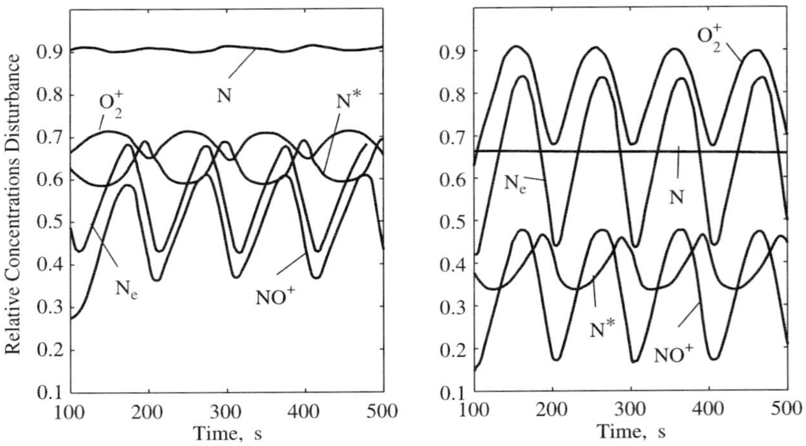

Fig. 13.4. The ratio of an ion concentration to the corresponding background value as a function of time for minimum (left) and maximum solar activity (right) for $\tau_T = 100\,\mathrm{s}$

The results of the numerical solution of (13.6)–(13.7) for different conditions of solar activity and periods of modulation $\tau_T = 10$ and $100\,\mathrm{s}$ are given in Fig. 13.3 (curves 1 and 2 correspond to maximum solar activity and periods of 10 s and 100 s; curves 3 and 4 correspond to minimum solar activity and the same periods of 10 and 100 s). Two families of curves, corresponding to heating with 10 and 100 s modulation, are distinguished in this figure.

The time dependencies of various components of the ionospheric plasma for the minimum and maximum solar activity in the case of 100 s modulation are shown in Fig. 13.4. Here the normalized concentrations of the various components of the ionospheric plasma are laid out along the ordinate axis. Different depths of modulation for different components are seen from the curves. The effect of nonlinearity is manifested most clearly in N^* and O_2^+. The curves in Fig. 13.4 are constructed for $T_1 = 2000°$ K. Analogous curves for 10 s temperature modulation with an amplitude $T_1 = 1000°$ K in a period of minimum solar activity are presented in Fig. 13.5 (left panel). The transitional mode for N_e with a characteristic time of the order of 100 s is clearly seen in this figure. The time dependencies of the concentrations of the various components for the solar activity maximum are given in the right panel of Fig. 13.5.

Summarizing the results of the calculations of chemical kinetics for the E-layer, we conclude that

- the deepest modulation of concentration appears in the quasi-stationary case;
- 100 s temperature oscillations yield a depth of modulation of N_e differing by not more than two to three times from the quasi-stationary case;

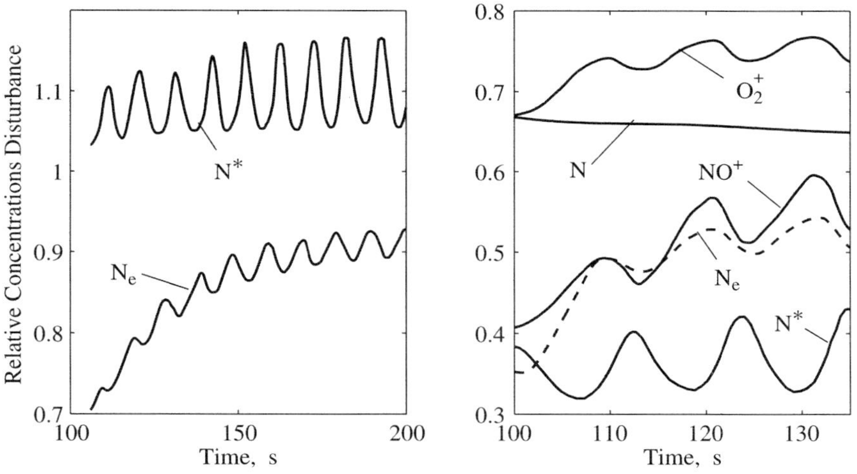

Fig. 13.5. Time variations of normalized ion components for minimum (left) and maximum solar activity (right) for $\tau_T = 10\,\mathrm{s}$

- 10 s temperature oscillations yield depths of modulation of N_e, an order of magnitude smaller than in the case of 100 s oscillations.

13.4 Ionospheric Conductivity

We assume that a powerful HF-wave travels through a medium with a new established value of N_e determined by the temperature dependence of the effective recombination rate. Let

$$N_e = N_{e0}\sqrt{1 + \gamma\frac{T_e - T_{e0}}{T_{e0}}}, \tag{13.16}$$

with the parameter γ is introduced to take into account the dependence of the N_e variations on the modulation period of the HF-wave. For the estimates we put $\gamma = 0.5$ and $\gamma = 1$, which correspond roughly to a period at 100 s and to the quasi-stationary case.

Figure 13.6 shows the relative disturbances of $\delta\Sigma_P$ (solid and chain lines) and $\delta\Sigma_H$ (dashed and dotted lines) determined from a numerical solution of (13.3) with N_e dependent on T_e in accordance with (13.16). Here, as well as in Fig. 13.2, the applied electric field is $E = 5\,\mathrm{V/m}$. Comparing Figures 13.2 and 13.6, we see that the displacement of the ionization balance in the pump wave leads to a growth of the conductivity to 10–20% in the daytime and to 50% in the nighttime.

Of course, the self-action of a strong wave, especially in the night ionosphere, leads to distortion of its modulation. We will not consider these

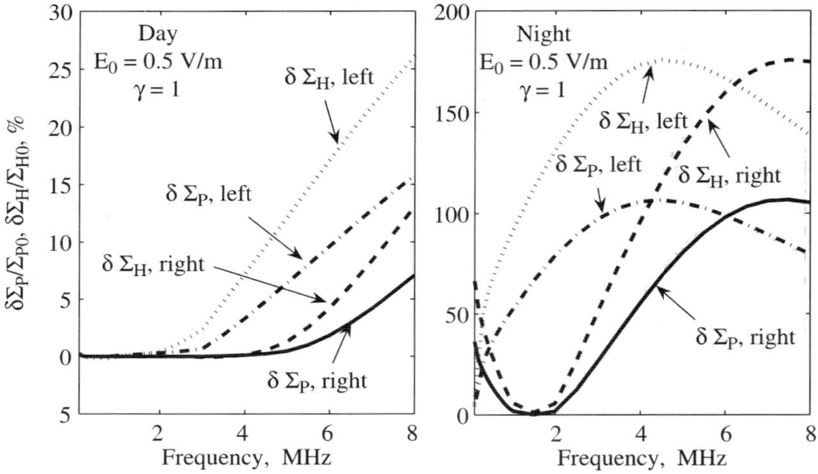

Fig. 13.6. The frequency dependencies of the relative disturbances of $\delta\Sigma_P$ (solid and chain lines) and $\delta\Sigma_H$ (dashed and dotted lines) caused by a displacement of the chemical equilibrium of the E-layer in the strong HF wave. The electric field of the pump wave at the bottom ionosphere $z = 65\,\mathrm{km}$ is $5\,\mathrm{V/m}$

effects here and restrict ourselves to estimates of the conductivity disturbances obtained above.

Geomagnetic Variations

To estimate the magnitude of the artificial magnetic variations, we return to the simple model of the circle inhomogeneity with Σ_{H1}, Σ_{P1} of radius a considered in Chapter 10. The inhomogeneity is placed into a homogeneous background current in the ionosphere with Σ_{H0}, Σ_{P0} . A magnetic perturbation is caused by the perturbed ionospheric current (10.2). The radial b_r and azimuthal b_φ magnetic components below such an inhomogeneity, are given by [1]:

$$\delta b_r = \frac{2a}{\Delta}\{[2\rho_0(\rho_0+\rho_1)-\Delta]\sin\varphi - 2\rho_0(\delta_0-\delta_1)\cos\varphi\} \times \left[\frac{1}{r}I(1,1,-1) - I(1,0,0)\right],$$

$$\delta b_\varphi = -\frac{2a}{\Delta}\{[2\rho_0(\rho_0+\rho_1)-\Delta]\cos\varphi + 2\rho_0(\delta_0-\delta_1)\sin\varphi\} \times I(1,1,-1), \quad (13.17)$$

where

$$\rho_{0,1} = \frac{\Sigma_{P_{0,1}}}{\Sigma^2_{P_{0,1}} + \Sigma^2_{H_{0,1}}}, \qquad \delta_{0,1} = \frac{\Sigma_{H_{0,1}}}{\Sigma^2_{P_{0,1}} + \Sigma^2_{H_{0,1}}},$$

$$I(\mu,\nu,p) = \int\limits_0^\infty \lambda^p \cdot J_\mu(a\lambda) \cdot J_\nu(r\lambda) \cdot \exp(-\lambda h) \cdot d\lambda,$$

$$\delta b_r = \frac{b_r - b_{0r}}{b_0}, \qquad \delta b_\varphi = \frac{b_\varphi - b_{0\varphi}}{b_0},$$

$$\Delta = (\rho_0 + \rho_1)^2 + (\delta_0 - \delta_1)^2$$

and h is the thickness of the atmosphere.

Let us estimate the anomalous magnetic field below the center of the inhomogeneity ($r = 0, \varphi = 0$, $\delta b_r = \delta b_x, \delta b_\varphi = \delta b_y$). Let also assume that the disturbances $\delta\Sigma_P$ and $\delta\Sigma_H$ of Σ_{P0} and Σ_{H0} are small. Then (13.17) can be simplified and reduced to

$$\begin{aligned} \delta b_\varphi &= \frac{q_P}{4}\left(\frac{a}{z}\right)^2, \\ \delta b_r &= -\frac{\delta_0 q_H}{4\rho_0}\left(\frac{a}{z}\right)^2, \end{aligned} \tag{13.18}$$

where

$$\begin{aligned} q_P &= \frac{\delta\Sigma_P}{\Sigma_{P0}} - 2\frac{\Sigma_{P0}\delta\Sigma_P + \Sigma_{H0}\delta\Sigma_H}{\Sigma_{P0}^2 + \Sigma_{H0}^2}, \\ q_H &= \frac{\delta\Sigma_H}{\Sigma_{H0}} - 2\frac{\Sigma_{P0}\delta\Sigma_P + \Sigma_{H0}\delta\Sigma_H}{\Sigma_{P0}^2 + \Sigma_{H0}^2}. \end{aligned} \tag{13.19}$$

Since

$$\left(\frac{a}{z}\right)^2 \approx \frac{4}{\Gamma},$$

where Γ is the antenna gain,

$$\delta b_\varphi = \frac{q_P}{\Gamma}, \qquad \delta b_r = -\frac{\Sigma_{H0}}{\Sigma_{P0}}\frac{q_H}{\Gamma}. \tag{13.20}$$

Let us consider the variation of the conductivity near the reflection point. By virtue of the foregoing, the disturbances of the conductivities due to high T_e are proportional to the electron concentration $Ne = N_e(T_e)$, or

$$\delta\Sigma_P = \Sigma_{P0\Delta}\frac{\delta N_e}{N_{e0}} l_h \quad \text{and} \quad \delta\Sigma_H = \Sigma_{H0\Delta}\frac{\delta N_e}{N_{e0}} l_h,$$

where l_h is the thickness of the heated layer near the reflection point while $\Sigma_{P0\Delta}$ and $\Sigma_{H0\Delta}$ are Σ_P and Σ_H of an undisturbed 1 km layer, respectively. Then

$$q_P \sim q_H \sim \frac{\delta N_e}{N_{e0}},$$

since $\delta N_e = \delta N_e(T_e)$ or $\delta N_e = \delta N_e(E)$, where E is the electric field of the wave at the height z. So long as the amplitude E_0 is not too high, we can

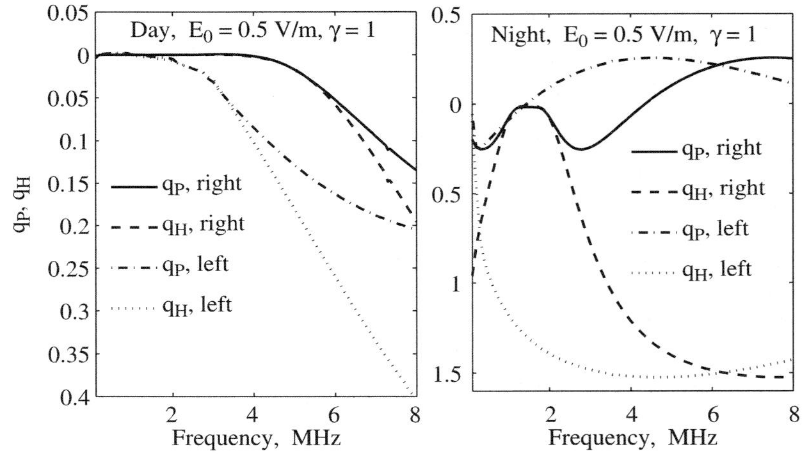

Fig. 13.7. q_P and q_H (see (13.19)) for the dayside ionosphere (left frame) and for the nightside ionosphere (right frame) as functions of the pump frequency

assume that

$$E \propto \sqrt{\Gamma W_0},$$

where W_0 is the power radiated by the antenna in kW. It is seen from Fig. 13.3 that δN_e is roughly proportional to T_e: $\delta N_e/N_{e0} \approx T_e \approx E^2 \approx \Gamma W_0$. Therefore, $q_P \sim \Gamma$ and $q_H \sim \Gamma$. Substituting the latter relation into (13.20), we find that the magnitude of the magnetic field disturbances does not depend in this case on the antenna gain. However, for a strong field, nonlinear effects result in the fact that a strong dependence of the antenna gain develops.

Figure 13.7 presents the values of q_P and q_H in percents for the dayside ionosphere (left frame) and for the nightside ionosphere (right frame) as functions of the pump frequency. In the calculations 0.5 V/m.

The maximal magnetic disturbances δb in daytime are $\sim 10^{-3}$ and $\sim 10^{-1}$ in nighttime for $E_0 = 0.5\,\mathrm{V/m}$. The same calculations for $E_0 = 5\,\mathrm{V/m}$ show that $\delta b \sim 10^{-4}$ in daytime and $\sim 10^{-3}$ in nighttime. The reduction of the magnetic effect when passing from relatively low electric field to the stronger field is a result of strong attenuation of the nonlinear pump wave in the low ionosphere. Going from this region, the wave comes to the E-layer with low intensity and cannot heat this region significantly. On the other hand, the strong effect in night ionosphere in comparison with the dayside ionosphere is explained by the absence of a D-layer in the night ionosphere where the wave decays appreciably.

At middle latitudes, the magnitude of the background S_q variation is of the order of 50 nT in daytime and 1 nT at night. Then the magnitudes of the expected variations generated during the irradiation of the ionosphere with a strong HF wave $\sim 5 \times (10^{-2} - 10^{-3})$nT in daytime and $\sim 10^{-1}$ nT in nighttime for $W_0 \approx 10^2$ MW. The amplitudes of the man-made magnetic variations at modulation period ≈10 s can reach the magnitude of the natural 10 s pulsations. These values are correct only under the center of the generated inhomogeneity. Near the inhomogeneity, the magnetic field decreases approximately as r^{-1} (r is the distance to the center of the inhomogeneity) and as r^{-2} far from it. An increase in the antenna gain at a given transmitter power leads to a decrease in the magnitude of the variations generated, as follows from the magnetic field estimates obtained.

These currents produce also an MHD-wave going to the magnetosphere to the conjugate ionosphere. Estimates of the emitted Alfvén wave caused by such oscillated ionospheric current will be given in the next chapter.

References

1. Abramov, L. A. and L. S. Alperovich, The estimation of influence of local inhomogeneities on the ionospheric current system, *Cosmical Researches*, (in Russian), n. 1, **8**, 74–78, 1970.
2. Alperovich, L. S. and E. N. Fedorov, Generation of low-frequency electromagnetic oscillations by the field of a powerful HF wave in the ionosphere, *Radiophysics and Quantum Electronics*, **24**, 3, 190, DOI: 10.1007/BF01035368, Springer, New York, 1981.
3. Belinda, J., A. C. Fraser-Smith, O. G. Villard, and J. R. Storey, Controlled artificial generation of geomagnetic pulsations, *Nature*, 258, 311, doi:10.1038/258311b0, 1975.
4. Bilichenko, S. B., Yu. A. Dreizin, A. V. Zotov, A. N. Kozlov, B. P. Murashov, P. P. Pushkarev, Yu. P. Sizov, R. I. Turbin, and V. M. Chmyrev, Observation of Artificial Geomagnetic disturbances at distances of 500–700 km from the source in the 'Khibiny' Geophysical experiment, *Geom. and Aeron.*, **19**, 514, 1979.
5. Bogolyubov, N. N. and Yu. A. Mitropol'skii, Asymptotic Methods in the Theory of Nonlinear Oscillations [in Russian], Nauka, Moscow, 1974.
6. Chakrabarty, P., D. K. Chakrabarty, and L. Bjorn, *J. Atmos. Terr. Phys.*, **40**, 81, 1978.
7. Ermakova, E. N., D. S. Kotik, L. A. Sobchakov, *et al.*, Experimental Studies of Propagation of Artificial Electromagnetic Signals in the Range 0.6–4.2 Hz, *Radiophysics and Quantum Electronics*, **48**, 700, DOI: 10. 1007/ s11141-005-0114-6, Springer, New York, 2005.
8. Frazer-Smith, A. C. and D. M. Bubenik, *J. Geophys. Res.,* **79**, 1038, 1974.
9. Gershman, B. N., *Dynamics of the Ionospheric Plasma* (in Russian), Nauka, Moscow, 1974.
10. Getmantsev, G. G., A. V. Gulelmi, B. I. Klain, *et al.*, Excitation of magnetic pulsations under the action on the ionosphere of radiation of a powerful short-wave transmitter, *Radiophysics and Quantum Electronics*, **20**, 703, DOI: 10.1007/BF01040635, Springer, New York, 1981.
11. Ginzburg, V. L., *The Propagation of Electromagnetic Waves in Plasmas*, Pergamon, Oxford, 1970.
12. Ginzburg, V. L. and V. A. Gurevich, Nonlinear phenomena in a plasma located in an alternating electromagnetic field, *Sov. Phys. Ups.* (English translation), **3**, 115 and 175, 1960.

13. Greifinger, C., Feasibility of ground-based generation of artificial micropulsations, *J. Geophys. Res.*, **77**, 6761, 1972.
14. Gurevich, A.V., *Nonlinear Plasma Phenomena in the Ionosphere,* Springer-Verlag, New York-Heidelberg-Berlin, 1978.
15. Kotik, D. S., S. V. Polyakov, V. O. Rapoport, and V. V. Tamoikin, In: *Influence of Powerful Radio Emission on the Ionosphere* [In Russian], Apatity, 1979.
16. Lyatskaya, A. M., V. B. Lyatskii, and Yu. P. Mal'tsev, Geomagnetic Pulsations and Modulation of Charged. Particle Fluxes,*Geomagn. Aeron.*, **16**, 566, 1976.
17. McEwan, M. J. and L. F. Phillips, *Chemistry of the Atmosphere*, Edward Arnold, London, 1975.
18. Meltz, G., L. H. Hollway, Jr., and N. M. Tomlianovlch, *Radio Sci.*, **9**, 1049, 1974.
19. Pashin, A. B., A. L. Kotikov, T. Yeoman, *et al.*, Electric field variations in the magnetosphere originated from ionospheric heated volume, *Physics of Auroral Phenomena*, Proc. XXVII Annual Seminar, Apatity, pp. 73, 2004.
20. Ratcliffe, J. A., *Magneto-ionic Theory*, Cambridge University Press, 1956.
21. Schmeltekopf, A. L., Fehsenfeld, F. C., Gilman, G. I., Ferguson, E. E.,Reaction of atomic oxygen ions with vibrationally excited nitrogen molecules, *Planet. Space Sci.*, **1**, 401, 1967.
22. Stubbe, P. and H. Kopka, Modulation of polar electrojet by powerful HF waves, *J. Geophys. Res.*, **82**, 2319, 1977.
23. Stubbe, P., H. Kopka, H. Lauche, *et al.*, Ionospheric modification experiments in northern Scandinavia, *J. Atmos. Terr. Physics*, **44**, No. 12, 1025, 1982.
24. Stubbe, P., H. Kopka, M. T. Rietveld, R. L. Dowden, ELF and VLF wave generation by modulated HF heating of the current carrying lower ionosphere, *J. Atm. Terr. Physics,* **44**, 1123 and 1133, 1982.
25. Willis, J. W. and J. R. Davis, Radio frequency heating effect on electron density in the lower E region, *J. Geophys. Res.,* **78**, 5710, 1973.

Additional Articles on the HF Ionospheric Heating

26. Andreev, A. D., N. F. Blagoveshchenskaya, V. A. Kornienko, Ionospheric wave processes during HF heating experiments, *Adv. Space Res.*, **15**, (12)45-(12)48, 1995.
27. Barr, R., P. Stubbe and M.T. Rietveld, ELF wave generation in the ionosphere using pulse modulated HF heating: initial tests of a technique for increasing ELF wave generation efficiency, *Ann. Geophys.*, **17**, 759, 1999.
28. Belenov, A. F., L. M. Erukhimov, P. V. Ponomarenko, and Y. M. Yampolski, Interaction between artificial ionospheric turbulence and geomagnetic pulsations *J. Atmos. Terr. Phys.*, **59**, 2367, 1997.
29. Bilitza, D., Solar-terrestrial models and application software, *Planet. Space Sci.*, **40**, 541, 1992.
30. Blagoveshchenskaya, N. F., V. A. Kornienko, A. V. Petlenko, A. Brekke, and M. T. Rietveld, Geophysical phenomena during an ionospheric modification experiment at Tromse, *Ann. Geophys.*, **16**, 1212, 1998.
31. Djuth, F. T., R. J. Jost, and S. T. Noble, *et al.*, Observations of E region irregularities generated at auroral latitudes by a high power radio wave, *J. Geophys. Res.*, **90**, 12293, 1985.
32. Inhester, B., Thermal modulation of the plasma density in ionospheric heating experiments, *J. Atmos. Terr. Phys.*, **44**, 1049, 1982.

33. James, H. G., R. L. Dowden, M. T. Rietveld, P. Stubbe, and H. Kopka, Simultaneous observations of ELF waves from an artificially modulated auroral electrodes in space and on the ground, *J. Geophys. Res.*, **89**, 1655, 1984.
34. James, H.G., U.S. Inan, and M.T. Rietveld, Observations on the DE-1 space craft of ELF/VLF waves generated by an ionospheric heater, *J. Geophys. Res.*, **95**, 12187, 1990.
35. Lyatsky, W. B., E. G. Belova, and A. B. Pashin, Artificial magnetic pulsation generation by powerful ground-based transmitter *J. Atmos. Terr. Phys.*, **58**, 407, 1996.
36. Rietveld, M. T., P. Stubbe, and H. Kopka, On the frequency dependence of ELF/VLF waves produced by modulated ionospheric heating, *Radio Sci.*, **24**, 270, 1989.
37. Stubbe, P. and H. Kopka, Modification of the polar electrodes by powerful HF waves. *J. Geophys. Res.*, **82**, 2319, 1977.
38. Stubbe, P., Review of ionospheric modification experiments at Tromsø, *J. Atmos. Terr. Phys.*, **58**, 349, 1996.
39. Trakhtengerts, V. Y., P. P. Belyaev, S. V. Polyakov, A. G. Demekhov, and T. Bösinger, Excitation of Alfvén waves and vortices in the ionospheric Alfvén resonator by modulated powerful radio waves, *J. Atmos. Terr. Phys.*, **62**, 267, 2000.
40. Yeoman, T. K., Lester M., Milling D. K., Orr, and D. Polarization, propagation and MHD wave modes of *Pi*2 pulsation: SABRE/SAMNET results, *Planet. Space Sci.*, **39**, 983–998, 1991.
41. Yeoman, T. K., D. M. Wright, T. R. Robinson, J. A. Davies, and M. Rietveld, High spatial and temporal resolution observations of an pulse-driven field-lines resonance in radar backscatter artificially generated with the Tromse heater, *Ann. Geophys.*, **15**, 634, 1997.

14

Active Cloud Releases and MHD-Emission

14.1 Introduction

In this chapter we describe MHD-perturbations generated through the release into the ionosphere of plasma producing products. Careful consideration needs to be given to the dynamics of the release cloud and to various geophysical effects associated with this. Detailed discussions of these problems may be found in ([1], [2], [3], [4], [5], [6], [10], [11], [13]). Our purpose in this chapter is to show, based on a simplified dynamic model, how electromagnetic pulses can be generated by a man-made ionospheric plasma release.

Let us consider an injection of plasma producing compounds at heights of 150–200 km. Initially the emission speed of the gas cloud is about 2–3 km/s. Due to the high cloud density, conditions typical for an ionospheric dynamo area hold and therefore intensive horizontal dynamo currents can be generated.

First, the emitted release products (initially, the cloud is composed of neutral particles with speed $\mathbf{v}_n$) induce an electrical current

$$\mathbf{j} = \sigma_P \left[\frac{\mathbf{v}_n}{c} \times \mathbf{B_0}\right] + \frac{\sigma_H}{B_0}\left[\mathbf{B_0} \times \left[\frac{\mathbf{v}_n}{c} \times \mathbf{B}_0\right]\right].$$

Since the release plasma is more conductive than the ionospheric plasma, the radial component of the current cannot be compensated and its radial component is closed by the longitudinal currents $\mathbf{j}_{\|}$ out-flowing into the low ionosphere and magnetosphere.

A conductivity gradient caused by a plasma cloud placed into the ionospheric electric field and current, is the second mechanism that produces longitudinal current. The redistribution of the current systems creates $\mathbf{j}_{\|}$ and a polarized field $\mathbf{E}$. As such currents are non-stationary, they produce MHD-waves and in particular Alfvén waves. The Alfvén wave can traverse large-distances along the field-lines with no significant distortion or loss. The current pulse propagates with a velocity close to the Alfvén velocity along a force tube that

Table 14.1. The main parameters of the 'Trigger' experiment

place: Kiruna ($67.9° N$, $21.1° E$)	Height: 164 km
Total mass: 12 kg, 6% of Cs^+	Degree of ionization 1.6×10^{-3} at τ
Number of molecules: 1.7×10^{25}	Conductivity for $N_e = 3.1 \times 10^{12}\,\mathrm{m}^{-3}$
Initial phase of the emission: $\tau \approx 0.6\,\mathrm{s}$	$\sigma_P = 5 \times 10^{-3}\,\mathrm{S/m}$
Cloud radius $R_0 = 1.2\,\mathrm{km}$ at τ	$\sigma_H = 5.7 \times 10^{-3}\,\mathrm{S/m}$
Number density N versus radius r :	Integral ionospheric conductivity:
$N(r) = N(r=0)\exp\left(-r^2/R_0^2\right)$	$\Sigma_P = 18\,\mathrm{S}$, $\Sigma_H = 20\,\mathrm{S}$

pierces the cloud. Generation and propagation of a current pulse $j_{\|}$ and the pulse's roles in electron precipitations are described in [9].

The parameters of the 'Trigger' experiment (11.02.1977) (see Table 14.1 after [7]) can be used in order to estimate the ULF-frequency range electromagnetic effects caused by a plasma release. In this experiment, a cesium cloud (Cs^+) was released at a height of 164 km.

14.2 MHD-Pulse Initiation

The efficiency of MHD-wave excitation at various stages of scattering into the ionosphere depends on the amount of mobile neutral component. It is also dependant on how the gas components were emitted. The most convenient way to simplify this rather complicated problem is to suppose an instant emission. Further, it is convenient to consider the movement of neutral cloud components such as expansion of gas into vacuum. As the gas slows down it rakes up ionospheric plasma and almost stops when the energy of the displaced background plasma is equal to the release energy.

Neutral Components Dynamics

The injected neutral component mobilizes the electron and ion gases which have arisen at the initial stage (in the container itself), and at later stages under the influence of solar UV radiation. Due to the weak ionization 1 s after emission, the feedback action of ionized components on a neutral motion can be neglected.

Let us suppose that the expanding cloud of neutral gas is a sphere of radius $R_n(t)$. Let the radial component v_n of the neutral velocity in a cloud be linearly dependent on the running radius r, i.e.

$$v_n(t,r) = v_{n0}(t)\frac{r}{R_n(t)}, \qquad v_{n0}(t) = \frac{dR_n(t)}{dt}. \tag{14.1}$$

Let the concentration distribution $N_n(r,t)$ on the time-scale of the process be:

$$N_n(r,t) = N_0(t) f\left(\frac{r}{R_n(t)}\right), \tag{14.2}$$

where $f(\xi)$ is normalized by the condition

$$4\pi \int_0^1 f(\xi)\, \xi^2 d\xi = 1.$$

and $f(\xi) = 0$ for $\xi >\geq 1$.

We neglect the change of concentration of neutral gas due to processes of ionization and recombination. The total number of injected particles be $N_T = \text{const}$. This allows us to find a relation between $N_0(t)$ and $R_n(t)$. The total number of particles is

$$N_T = \int_0^{R_n(t)} N_n(r,t)\, 4\pi r^2 dr = 4\pi N_0(t)\, R_n^3(t) \int_0^1 f(\xi)\, \xi^2 d\xi.$$

The last equation and the normalization requirement yield

$$N_0(t) = \frac{N_T}{R_n^3(t)}. \tag{14.3}$$

Equation (14.3) and dependencies $R_n(t)$ allow us to study the time dependencies of electron and ion magnetization parameters, $\beta_e = \omega_{ce}/\nu_{em}$ and $\beta_i = \omega_{ci}/\nu_{im}$. Initially the neutral density in the cloud is so strong that the electron-neutral collision frequencies exceed ω_{ce} and ω_{ci} considerably, and so $\beta_e \ll 1$ and $\beta_i \ll 1$. As the cloud continues to expand, the neutral density drops and electrons become completely frozen that is $\beta_e \gg 1$. At this time the ions move together with the neutral gas. With the further expansion, both electrons and ions become frozen into the magnetic field. For emission at 150–210 km electrons are frozen after $\tau_e \sim 10\,\text{ms}$ and ions are frozen after $\tau_i \sim$ 0.5–1 s.

The basic conclusion is that virtually instantly after emission electrons become magnetized and ions remain non-magnetized until ~0.5 s. The hydromagnetic pulse, radiated due to the dynamo field, has a characteristic duration of 0.5 s. The time change of β_e and β_i defines three stages in ionized component dynamics: non-magnetized electrons and ions, magnetized electrons and non-magnetized ions and electrons and ions frozen into the magnetic field.

Ionized Cloud Component Dynamics

The feedback action of charged components on neutral-gas motion can be ignored at times $\tau \leqslant 10^3$ s. Let (14.1) be valid for the neutral gas speed, v_n.

For emissions time-scale of ~1 s it is possible to neglect the electron/ion inertia in the equations of motion for electrons/ions in crossed electric and magnetic fields ((1.44)–(1.45) and restrict our consideration to the quasi stationary approximation. Then for the electron and ion drift velocities we have

$$\mathbf{v}_{e0} = \mathbf{v}_n - \frac{\boldsymbol{\sigma}_e}{eN_e}\mathbf{E}'_\mathbf{0},$$

$$\mathbf{v}_{i0} = \mathbf{v}_n + \frac{\boldsymbol{\sigma}_i}{eN_i}\mathbf{E}'_\mathbf{0}. \tag{14.4}$$

Here $\boldsymbol{\sigma}_e$ and $\boldsymbol{\sigma}_i$ are tensors of the electron and ion conductivities (see electron and ion parts of (1.86)–(1.87) for $\omega = 0$). The sum of the external field $\mathbf{E_0}$ and dynamo-field is

$$\mathbf{E}'_\mathbf{0} = \mathbf{E}_0 + \frac{1}{c}\mathbf{v}_n \times \mathbf{B}_0. \tag{14.5}$$

Perturbations of the electron and ion concentrations are connected with the corresponding fluxes through the equations of a continuity

$$\frac{\partial N_e}{\partial t} + \boldsymbol{\nabla} \cdot \mathbf{J}_e = q, \tag{14.6}$$

$$\frac{\partial N_i}{\partial t} + \boldsymbol{\nabla} \cdot \mathbf{J}_i = q, \tag{14.7}$$

where

$$\mathbf{J}_e = N_e\mathbf{v}_e, \quad \mathbf{J}_i = N_i\mathbf{v}_i$$

are the electron and ion fluxes; q is defined by the processes of the ion production, recombination etc.

In an inhomogeneous plasma, the internal concentration gradients, amongst others, create particle flows that in turn produce electric and magnetic fields. Moreover, an inhomogeneity placed in an external field of any kind (i.e. temperature, electric, etc.) produces polarization electric fields. Upon emission, electron-ion gas pressure is significantly lower than magnetic pressure and all velocities are substantially lower than the Alfvén velocity. In such conditions, motion of the ionized gas is unaffected by the magnetic perturbations.

The electric field $\mathbf{E}_p$ caused by the plasma inhomogeneity is given by

$$\boldsymbol{\nabla} \cdot \mathbf{E}_p = -4\pi e\left(N_e - N_i\right). \tag{14.8}$$

The electron and ion fluxes are

$$\mathbf{J}_e = N_e\mathbf{v}_{e0} - \frac{\boldsymbol{\sigma}_e\mathbf{E}_p}{e}, \tag{14.9}$$

$$\mathbf{J}_i = N_i\mathbf{v}_{i0} + \frac{\boldsymbol{\sigma}_i\mathbf{E}_p}{e}. \tag{14.10}$$

$\mathbf{B}_0$ is so strong that the charged particles cannot move in the transverse direction and transport across $\mathbf{B}_0$ is small. This is why the fluxes proportional to $\boldsymbol{\nabla}_\perp N_{e,i}$ and $\boldsymbol{\nabla}_\perp T_{e,i}$ in (14.9) and (14.10) are not included.

Let us simplify the mathematics, using the quasi-neutrality condition

$$N = N_e \approx N_i .$$

The quasi-neutrality condition only holds for $|\nabla N| \ll N/r_D$. In this case, it is possible to replace (14.6) and (14.7) with

$$\frac{\partial N}{\partial t} + \nabla \cdot \mathbf{J} = 0. \tag{14.11}$$

This equation is equivalent to (14.6) and (14.7) under the condition of

$$\nabla \cdot \mathbf{J}_e = \nabla \cdot \mathbf{J}_i = \nabla \cdot \mathbf{J}. \tag{14.12}$$

Substitution of (14.9) and (14.10) into (14.12) for the electric field $\mathbf{E}_p$, gives

$$\nabla \cdot (\boldsymbol{\sigma} \mathbf{E}_p) = -\nabla \cdot (\boldsymbol{\sigma} \mathbf{E}'_0) , \tag{14.13}$$

where $\boldsymbol{\sigma} = \boldsymbol{\sigma}_e + \boldsymbol{\sigma}_i$.

The MHD-signal travels in the ionosphere at Alfvén velocity $c_A \gtrsim 100$ km/s. It therefore takes the signal much less than the process scale-time to cover the newly created ~1 km inhomogeneity. Hence there is almost no phase delay across the MHD-signal and the quasi-stationary approximation can be used (see Section 9.1). In this approximation, neglecting any induced magnetic fields, it is possible to assume that $\mathbf{E}_p$ is curl-free and therefore $\mathbf{E}_p = -\nabla \varphi_p$. Then (14.13) becomes

$$\nabla \cdot \left(\boldsymbol{\sigma} \nabla \varphi_p\right) = -\nabla \cdot (\boldsymbol{\sigma} \mathbf{E}'_0) . \tag{14.14}$$

The potential φ_p obtained from the solution of (14.14) allows us to find $N_e - N_i$ from

$$N_e - N_i = -\frac{1}{4\pi e} \nabla^2 \varphi_p .$$

Since, $\nabla \cdot \mathbf{J}_e = \nabla \cdot \mathbf{J}_i$, $\nabla \cdot \mathbf{J}$ in (14.11) can be written as

$$\nabla \cdot \mathbf{J} = \frac{1}{2} \nabla \cdot (\mathbf{J}_e + \mathbf{J}_i) .$$

Substituting this relation into (14.11), we obtain

$$\frac{\partial N}{\partial t} + \nabla \cdot \left[N \left(\mathbf{v}_n + \frac{\boldsymbol{\sigma}_i - \boldsymbol{\sigma}_e}{2Ne} \mathbf{E}'_0 \right) \right] = 0. \tag{14.15}$$

One can divide into three stages of electron/ion dynamics. At the first stage when the neutral concentration is large, the collision frequency of electrons with neutrals is higher than the corresponding cyclotron frequency, the same is true for ions, albeit for a different collision frequency ($\nu_{en} \gg \omega_{ce}$, $\nu_{in} \gg \omega_{ci}$). And therefore the Hall conductivity is small. The charged particles move together with the neutral component and the flux caused by the dynamo-field is small in comparison with $\mathbf{v}_n$. From (14.15) we obtain

$$\frac{\partial N}{\partial t} + \boldsymbol{\nabla} \cdot (N\mathbf{v}_n) = 0.$$

From the above, it is also clear that the charged particles are completely dragged, i.e. without slipping, by the neutral flow. For their part, closed currents due to Pedersen conductivity, while not small, do not change the charged particles concentration N.

At the second stage when $\nu_{em} \ll \omega_{ce}$ and $\nu_{in} \gg \omega_{ci}$, electrons are magnetized but ions move together with neutrals. The charges separation produces an electric field, which in turns excites an electric current. In this case, the Hall conductivity is defined by electrons. The dynamo field and the Hall conductivity produce a radial current, causing perturbations of the charged particles concentration N.

At the third stage electrons and ions are both magnetized, and neutral component continues to extend. The Pedersen conductivity is small, the electron and ion Hall conductivities are equal in magnitude and opposite in sign. As a result, the expression in the parentheses of (14.15) vanishes: $\partial N/\partial t = 0$, i.e. the ionized component concentration is almost constant.

Let us consider (14.15) for an axially symmetric emission, neglecting external electric field effects $\mathbf{E}_0$. In the cylindrical coordinate system (ρ, φ, z), the radial and azimuthal fluxes of charged components are

$$J_\rho = NU, \quad U = \alpha v_n, \tag{14.16}$$

$$J_\varphi = \frac{\sigma_{eP} - \sigma_{iP}}{2eN} \frac{v_n}{c} B_0, \tag{14.17}$$

where

$$\alpha = 1 - \frac{\sigma_{eH} - \sigma_{iH}}{2eN} \frac{B_0}{c}.$$

The Pedersen flux forms a ring and $\partial J_\varphi/\partial\varphi = 0$, therefore the Pedersen flux does not contribute to the perturbation of N. Then (14.15) is

$$\frac{\partial N}{\partial t} + \frac{1}{\rho}\frac{\partial}{\partial \rho}(\rho N U) = 0. \tag{14.18}$$

with the initial condition

$$N|_{t=0} = \begin{cases} N_0, & \rho < R_n(0), \\ 0, & \rho > R_n(0). \end{cases}$$

Let

$$R_i(t), \quad v_{i0}(t) = \frac{dR_i(t)}{dt}$$

be the radius of the region occupied by the ionized component and the expansion speed of this region's boundary, respectively. Then from the second equation in (14.16) we get

$$\frac{dR_i(t)}{dt} = \alpha(t)v_n(R_i(t), t)$$

Substituting (14.1) in the last equation obtain

$$\frac{dR_i(t)}{dt} = v_{n0}\left(t\right)\frac{R_i\left(t\right)}{R_n\left(t\right)}\cdot\alpha\left(t\right). \tag{14.19}$$

Integrating (14.19) we find

$$R_i\left(t\right) = R_i(0)\exp\left(\int_0^t \frac{v_{n0}(t)}{R_n\left(t\right)}\cdot\alpha\left(t\right)dt\right), \tag{14.20}$$

where $R_i\left(0\right) = R_n\left(0\right)$. Note that for $\alpha\left(t\right) = 1$, $R_i\left(t\right) = R_n\left(t\right)$. Since the total number of charged particles is constant, the time dependency of $N\left(t\right)$ can be found from the relation

$$N\left(t\right)R_i^2\left(t\right) = N_0R_0^2. \tag{14.21}$$

Equation (14.18) is obtained on the assumption that the cloud of the injected gas is a cylinder extended along $\mathbf{B}_0$. For more realistic models of a release, for example, spherically symmetric emission, can be found that an equation is similar to (14.18) but for the concentration is integrated over the cloud thickness. Resultant formulas enable us to define space-time distribution of the charged particles and to estimate changes of the integral conductivities in the cloud.

Induced MHD-Wave

The electric fields excited by the plasma cloud release generate electric currents which in turn produce hydromagnetic emission from the region occupied by the injected plasma [14]. The current density $\mathbf{j}_d$ caused by the dynamo-field $\mathbf{E}_d$ is

$$\mathbf{j}_d = \boldsymbol{\sigma}\mathbf{E}_d, \quad \mathbf{E}_d = \frac{1}{c}\left[\mathbf{v}_n \times \mathbf{B}_0\right]. \tag{14.22}$$

The electromagnetic field is found by plugging the external currents into Maxwell's equations. The concepts developed in Chapter 7 allow us to estimate the intensity of the MHD-wave emitted from the man-made plasma release. One can use the small-scale approximation, also discussed in Chapter 7, to uncouple the general system of MHD-equations to yield (7.136) and (7.146) for the Alfvén and FMS-waves, respectively.

To make the problem tractable, we consider an axially symmetrical perturbation in the polar region where the magnetic inclination is $I = \pi/2$. In cylindrical coordinates with the origin at the center of the cloud, we have

$$-\frac{\partial b_\varphi}{\partial z} = \frac{1}{c_A}\frac{\partial E_\rho}{\partial t} + \frac{4\pi}{c}\sigma_P E_\rho + \frac{4\pi}{c^2}\sigma_H v_n B_0,$$
$$\frac{\partial E_\rho}{\partial z} = -\frac{1}{c}\frac{\partial b_\varphi}{\partial t} \tag{14.23}$$

with the radiation condition at $z \to +\infty$ and the impedance condition at the ionosphere

$$\left.\frac{E_\rho}{b_\varphi}\right|_{z=-h} = -\frac{1}{X}. \tag{14.24}$$

The initial conditions for the electric and magnetic fields are

$$E_\rho(z, t=0) = 0, \qquad b_\varphi(z, t=0) = 0. \tag{14.25}$$

Integrating (14.23) over the cloud thickness, we obtain a homogeneous equation for $z \lessgtr 0$, and conditions on the discontinuities of b_φ and E_ρ for $z = 0$:

$$\{b_\varphi\}_0 + X_c E_\rho|_0 = -q,$$
$$\{E_\rho\}_0 = 0, \tag{14.26}$$

where $\{Q\} = Q|_{z=+0} - Q|_{z=-0}$ denotes discontinuity of a quantity Q,

$$q = -Y_c\frac{v_n}{c}B_0, \quad X_c = \frac{4\pi}{c}\Sigma_P^c, \quad Y_c = \frac{4\pi}{c}\Sigma_H^c, \quad \Sigma_{P,H}^c = \int \sigma_{P,H}\, dz.$$

Using equations (14.23), boundary conditions and initial data (14.24), (14.25), one can find the hydromagnetic pulse from the region occupied by the injected plasma.

Consider an example from which it is possible to get an estimate of the field-aligned current in the Alfvén pulse. Let the background plasma above and below the cloud be characterized by the dielectric permeability $\varepsilon_m = (c/c_A)^2$. Then the wave equation for $b_\varphi\,(\rho, z, t)$ is

$$\frac{\partial^2 b_\varphi}{\partial z^2} - \frac{1}{c_A^2}\frac{\partial^2 b_\varphi}{\partial t^2} = 0. \tag{14.27}$$

The solution of (14.27) is

$$b_\varphi = \begin{cases} u_+\left(\rho, t - \dfrac{z}{c_A}\right), & z > 0, \\ u_-\left(\rho, t + \dfrac{z}{c_A}\right), & z < 0. \end{cases} \tag{14.28}$$

Then the electric field E_ρ is

$$E_\rho = \begin{cases} \dfrac{1}{\sqrt{\varepsilon_m}} u_+ \left(\rho, t - \dfrac{z}{c_A}\right), & z > 0, \\ -\dfrac{1}{\sqrt{\varepsilon_m}} u_- \left(\rho, t + \dfrac{z}{c_A}\right), & z < 0. \end{cases} \tag{14.29}$$

Substitution (14.28) and (14.29) into (14.26) yields

$$u_+ (\rho, t) = -\frac{\sqrt{\varepsilon_m}}{X_c + 2\sqrt{\varepsilon_m}} q (\rho, t) ,$$
$$u_- (\rho, t) = -u_+ (\rho, t) .$$

The magnetic and electric fields for $z > 0$ are given by

$$b_\varphi = -\frac{\sqrt{\varepsilon_m}}{2\sqrt{\varepsilon_m} + X_c (\rho, t - z/c_a)} q (\rho, t - z/c_a) ,$$
$$E_\rho = b_\varphi / \sqrt{\varepsilon_m}, \tag{14.30}$$

and the longitudinal current density by

$$\mathbf{j}_\parallel = \frac{c}{4\pi} (\nabla \times \mathbf{B})_\parallel = \frac{c}{4\pi} \frac{1}{\rho} \frac{\partial}{\partial \rho} (\rho b_\varphi) . \tag{14.31}$$

Let us now consider the influence of the conductive ionospheric layer on the generation of the Alfvén pulse; here, as before, $c_A = \text{const}$. According to (14.28)–(14.29, the magnetic and electric fields are

$$b_\varphi = \begin{cases} u_1 \left(\rho, t - \dfrac{z}{c_A}\right), & z > 0, \\ u_2 \left(\rho, t - \dfrac{z + 2h}{c_A}\right) + u_3 \left(\rho, t + \dfrac{z}{c_A}\right), & z < 0, \end{cases}$$

$$E_\rho = \begin{cases} \dfrac{1}{\sqrt{\varepsilon_m}} u_1 \left(\rho, t - \dfrac{z}{c_A}\right), & z > 0, \\ \dfrac{1}{\sqrt{\varepsilon_m}} u_2 \left(\rho, t - \dfrac{z + 2h}{c_A}\right) - u_3 \left(\rho, t + \dfrac{z}{c_A}\right), & z < 0. \end{cases} \tag{14.32}$$

Then, from (14.24) and (14.26), it follows that

$$u_2 \left(\rho, t - \frac{h}{c_A}\right) = R_{AA} u_3 \left(\rho, t - \frac{h}{c_A}\right) ,$$

$$u_2 \left(\rho, t - \frac{2h}{c_A}\right) - u_3 (\rho, t) = u_1 (\rho, t) ,$$

$$\left(1 + \frac{X_c}{\sqrt{\varepsilon_m}}\right) u_1 (\rho, t) - u_2 \left(\rho, t - \frac{2h}{c_A}\right) - u_3 (\rho, t) = -q (\rho, t) , \tag{14.33}$$

where R_{AA} is the reflection coefficient of the Alfvén wave into itself (see (11.27)). From (14.33) we find

$$u_1(\rho,t) = \frac{2R_{AA}}{2+X_c/\sqrt{\varepsilon_m}}u_3\left(\rho, t-\frac{2h}{c_A}\right) - \frac{q(\rho,t)}{2+X_c/\sqrt{\varepsilon_m}},$$

$$u_3(\rho,t) = \frac{X_c/\sqrt{\varepsilon_m}}{2+X_c/\sqrt{\varepsilon_m}}R_{AA}u_3\left(\rho, t-\frac{2h}{c_A}\right) + \frac{q(\rho,t)}{2+X_c/\sqrt{\varepsilon_m}}. \quad (14.34)$$

This allows us to study the Alfvén wave emission from the cloud. Until $t < 2h/c_A$, the downward propagating Alfvén wave does not yet 'know' that below it, at distance h, there is a conductive layer. At $t = 2h/c_A$ the wave has reached the ionosphere and has returned. Between $2h/c_A < t < 4h/c_A$, the emission is a superposition of the direct wave and the wave reflected from the ionosphere; between $4h/c_A < t < 6h/c_A$, the waves twice reflected from the ionosphere and once from the cloud are added, etc.

Let us suggest that the MHD-pulse is induced by a plasma release with parameters like 'Trigger' (see Table 14.1). At ≈1 s after the emission 500 m from it, the dynamo-field $\mathbf{E}_d$ (14.22) amplitude reaches 50–100 mV/m. The height variations of the estimated electric field between 150 and 300 km did not exceed ≈20 mV/m. Thus the numerical simulations demonstrate that E_d is weakly dependant on the release altitude. For example, peak values of the electric field are ≈25 mV/m at an altitude of 150 km, and ≈40 mV/m at 300 km.

From (14.32) and (14.34) one can find the electric and magnetic fields in the Alfvén wave radiated from the cloud. The field is reduced from ~100 nT at $t \sim 0.1$ s to 10 nT at $t \sim 1$ s, this confirms the validity of the qualitative dynamics pattern of the expanding cloud which was discussed above.

The longitudinal currents can be estimated from (14.31). It follows from the calculations that the current in the radiated pulse 1 s after release is 10 − −20 μA / m^2. Currents of such magnitude, obviously lead to the development of various plasma instabilities, to the appearance of anomalous resistance along the field-lines, to the excitation of appreciable longitudinal electric fields and to the acceleration and precipitation of particles.

Up to now, we have not considered the finite longitudinal resistance. By substituting collision frequencies and electron concentrations typical for ≈150 km into (1.127), we find that a finite longitudinal resistivity must be included in the consideration if the transversal scale is less than ≈1 km. Including the longitudinal resistivity reduces by several orders of magnitude the electric and magnetic fields and the longitudinal current. To summarize, at the initial stage, up until ~10 ms, the equations we have used break down. And at ~1 s it is possible to have an order of magnitude estimation. The amplitude values given in the estimation above are 2–3 times larger than the expected values. Neglecting the longitudinal resistivity is only possible after ~10 s. And it is only after such a time, that the theoretical concepts developed in the preceding chapters may be correctly applied.

References

1. Baker, K. D. and J. C. Ulwick, Measurements of electron density structure in striated barium clouds, *Geophys. Res. Lett.*, **5**, 723, 1978.
2. Blagoveshchenskaya, N. F., T. D Borisova., O. V. Kolosov, V. A. Kornienko, Ionospheric effects caused by barium releases, *Adv. Space Res.*, **15** (12), 111, 1995.
3. Blagoveshchenskaya, N. F., T. D. Borisova, and V. A. Kornienko, Mediumscale ionospheric wave processes during the CREES barium releases, *Adv. Space Res.*, **21**, 749, 1998.
4. Fitzgerald T. J., P. E. Argo, and R. C. Carlos, Effects of artificially modified ionosphere on HF propagation: Negative ion carbon release experiment 2 and CRRES Coqui experiments, *Radio Sci.*, **32**, 579, 1997.
5. Gaidukov, V. Iu., S. A. Namazov, M. A. Nikitin, and Yu. A. Romanovskii, The 'Equatorial Trigger' experiment – Stimulated development of plasma instabilities and irregularities in the equatorial ionosphere, *Kosmicheskie Issledovaniia (ISSN 0023-4206)*, **31**, no. 1, 63, 1993.
6. Grebnev, I., M. Deminov, Y. A. Romanovsky, and A. Tcema, Trigger processes in the artificially modified auroral ionosphere, *Adv. Space Res.*, **21**, 761, 1998.
7. Holmgren G., R. Boström, M. C. Kelley, P. M. Kintner, R. Lundin, U. V. Fuhleson, E. A. Bering, and W. R. Sheldon, Trigger, an active release experiment that stimulated auroral particle precipitation and wave emissions, *J. Geophys. Res.*, **85**, 5043, 1980.
8. Fuhleson, U. V., E. A. Bering, W. R. Sheldon, Trigger, an active release experiment that stimulated auroral particle precipitation and wave emissions, *J. Geophys. Res.*, **85**, 5043, 1980.
9. Kelly, M. C., U. V. Fahleson, G. Holmgren et al., Generation and propagation of an electromagnetic pulse in the Trigger experiment and its possible role in electron acceleration, *J. Geophys. Res.*, **85**, A10, 5055, 1980.
10. Kelley M. C., K. D. Baker, and J. C. Ulwick, Late time barium cloud striations and their possible relationship to equatorial spread-F, *J. Geophys. Res.*, **94**, 1979, 1989.
11. Kinter, P. M., M. C. Kelly, G. Holmgren, and R. Bostrom The observations and production of ion acoustic waves during the trigger experiment, *J. Geophys. Res.*, **85**, A10, 5071, 1980.
12. Namazov, S. A., Y. A. Romanovsky, and V. B Ivanov, Simultaneous wave effects in the ionosphere and in an artificial ion cloud after its formation, *Adv. Space Res.*, **21**, 765, 1998.
13. Narcisi, R. S., E. P. Szcuszczewicz, Plasma composition and structure characterization of an ionospheric Barium cloud, *Proc. of the Active experiments in Space Symposium.* – Alpbach, Austria, 299, 1983.
14. Scholer, M., On the motion of artificial ion clouds in the magnetosphere, *Planet. Space Sci.*, **18**, 977, 1970.

15

MHD After-Effects of a Sound Impact

Ionospheric HF-heating and releases of chemically active products discussed in the preceding chapters are only of interest to us because of the option to use them to study the magnetosphere–ionosphere plasma and some features of the MHD-wave generation and propagation. As a matter of fact, these mechanisms are purely anthropogenic. It is difficult to imagine situations in extra-terrestrial space in which naturally occurring strong natural radio-emissions that provide noticeable radio-heating, or natural plasma injections much denser than the plasma densities exist.

The third mechanism of MHD-wave generation, which we will discuss next, is an important part of the lithosphere–magnetosphere coupling. Here, we are dealing with an atmospheric wave due to large-scale high-energy releases, either in the low atmosphere or on Earth. Some of the processes contributing to the dynamics of the atmosphere, ionosphere and magnetosphere are caused by weather processes, earthquakes, volcanic eruptions, winds in mountain regions, etc. All these processes create acoustic and acoustic-gravity waves that can reach ionospheric heights. For a deeper understanding of neutral wave propagation in the atmosphere, however, we would refer the reader to [21] where a review and an analysis of the current status of both theoretical and experimental studies of long-range sound propagation in the atmosphere are given. Here we restrict our consideration to a basic sketch of a neutral wave propagation, describing the key facts of the process.

Bombarding the ionosphere with acoustic waves causes macroscopic motion of ionospheric layers and perturbation of the chemical kinetics of ionized and neutral ionospheric species. The acoustic wave reaching the ionosphere also gives rise to temperature and pressure perturbations. In principle, it can perturb the chemical equilibrium between ion-neutral ionospheric species. And finally, the wave entrance can be accompanied by displacements of ionospheric layers.

Let us estimate velocities and displacements within the ionosphere at the acoustic wave beam axis without considering a dissipation. In the case of the isothermal atmosphere, the energy density $(\rho_0 + \delta\rho)v^2$ must be conserved. Here ρ_0 is the atmospheric density at height z, $\delta\rho$ is the density perturbation in the acoustic wave and v is the gas velocity in the wave. With an accuracy up to terms of second order, $\rho_0(z)v(z)^2 = \text{const}$ or

$$\frac{v_{ionosphere}}{v_{ground}} \sim \left(\frac{\rho_{ground}}{\rho_{ionosphere}}\right)^{1/2} .$$

Let us suppose that a 100 s weak ground wave with 1 mm/s initial gas velocity reached the 100 km altitude. Then the ratio of the gas velocity in the wave to the initial wave is $\delta v_{ionosphere}/\delta v_{ground} \sim 10^4$ and the displacement of the ionospheric gas is ~1 km. Due to such unique feature, even rather small large-scale ground perturbations can become visible at the ionospheric heights. Moreover, the wave amplitudes may be so strong that nonlinear effects will become remarkable.

The entrance of the acoustic wave into the ionosphere causes a change of the ionospheric parameters, disturbs chemical equilibrium between the ion-neutral ionospheric species, and results in a change of the ionospheric turbulence scale. An important point is that eventually it can change the Pedersen conductivity and as a sequence of this, a change of the reflected ionospheric properties and Q-factor of the magnetospheric Alfvén resonator.

The motion of the neutral gas in the acoustic wave results in an appearance of the ionospheric electric fields and currents producing an Alfvén pulse going up into the magnetosphere to the conjugate ionosphere. Besides the dynamo field, appearing small-scale irregularities increase the Pedersen effective conductivity (see Chapter 10). Permanently existing large-scale ionospheric current systems are redistributed in the vicinity of the region with anomaly Pedersen conductivity that in turn results in the longitudinal currents of the Alfvén wave.

Thus, as a third possibility, in addition to the sources discussed in previous sections, there may be mechanical motions of the ionosphere convertible to the MHD-waves. We will consider features of the acoustic-gravity waves and ULF-magnetic variations caused by an oscillating and impulsive on-ground and under-ground sources like earthquakes, large-scale atmospheric phenomena, air and underground explosions, etc.

First, an elementary model of a plain low-frequency pulse is considered propagating vertically upwards over a non-uniform atmosphere. This pulse in turn excites an electrical current and magnetic pulse. Estimations of a magnetic pulse initiated by a local source are also given. The consideration of the 1D case of the wave propagating upward is especially useful in estimates of the maximal magnetic signal which can be get from various acoustic sources. The results of an experiment that dealt with acoustic action on the ionosphere and magnetosphere are shown in the last section.

15.1 Foundation of the Theory

Basic Equations

In order to describe the propagation of a neutral wave in an inhomogeneous partially ionized atmosphere and interaction of this wave with the ionized part is to use the MHD-equations with the gravity force.

Let us introduce a mean velocity, $\mathbf{v}$, the total pressure, p, and density, ρ of the ionospheric plasma

$$\mathbf{v} = \frac{m_e N_e \mathbf{v}_e + m_i N_i \mathbf{v}_i + m_n N_n \mathbf{v}_n}{m_e N_e + m_n (N_i + N_n)},$$
$$p = p_e + p_i + p_n, \quad \rho = m_e N_e + m_n (N_i + N_n)$$

Summing equations of momentum with gravity force for the electron (1.44), ion (1.45) and neutral (1.46) components, yields

$$m_e N_e \frac{\mathrm{d}\mathbf{v}_e}{\mathrm{d}t} + m_i N_i \frac{\mathrm{d}\mathbf{v}_i}{\mathrm{d}t} + m_n N_n \frac{\mathrm{d}\mathbf{v}_n}{\mathrm{d}t} = -\boldsymbol{\nabla} p + \rho \mathbf{g} + \frac{1}{c}[\mathbf{j} \times \mathbf{B}] + e(N_i - N_e).$$

Suppose first that $v_e \sim v_i \sim v_n$ and condition of quasi-neutrality $N_i = N_e$ is fulfilled, then the equation of the magnetohydrodynamics become

$$\rho \frac{\mathrm{d}\mathbf{v}}{\mathrm{d}t} = -\boldsymbol{\nabla} p + \rho \mathbf{g} + \frac{1}{c}[\mathbf{j} \times \mathbf{B}]. \tag{15.1}$$

Multiplying (1.44) by m_e, (1.45) by m_i and (1.46) by m_n and supposing that $m_i \approx m_n$ and summing them, we obtain the equation of the mass conservation

$$\frac{\partial \rho}{\partial t} + \boldsymbol{\nabla}(\varrho \cdot \mathbf{v}) = 0. \tag{15.2}$$

Combining (1.44) and (1.45) supplemented by the gravity force and omitting non-linear terms, we obtain ([14], [15])

$$m_i N_i \nu_{in} (\mathbf{v}_n - \mathbf{v}_i) - \frac{m_e \nu_{en}}{e} \mathbf{j} + \frac{1}{c}[\mathbf{j} \times \mathbf{B_0}] - \boldsymbol{\nabla}(p_e + p_i) + m_i N_i \mathbf{g} = 0, \tag{15.3}$$

$$\nu_e \mathbf{j} - \frac{\omega_{ce}}{B_0}[\mathbf{j} \times \mathbf{B}_0] = \frac{N_e e^2}{m_e}\{\mathbf{E} + \frac{1}{c}[\mathbf{v}_i \times \mathbf{B}_0]\} + e N_e \nu_{en} (\mathbf{v}_n - \mathbf{v}_i) - \frac{e}{m_e} \boldsymbol{\nabla} p_e. \tag{15.4}$$

In deriving these equations we have used the inequalities $m_e \ll m_n$, $|\partial \mathbf{j}/\partial t| \ll \nu_e |\mathbf{j}|$ and $\nu_{en} \sim (m_n/m_e)^{1/2} \nu_{in}$. The gravitation force can not produce essential currents [14]. Besides, $\boldsymbol{\nabla} p_e$ and $\boldsymbol{\nabla} p_e$ can be neglected for the large-scale perturbations.

It follows from the definition of the mean velocity that it is equal approximately to the velocity of the neutral molecules: $v \approx v_n + (N/N_n)\, v_i \approx v_n$.

Equations (15.3) and (15.4) can be used to eliminate v_i and to get relationship between the current density $\mathbf{j}$ and $\mathbf{E}$. Quasi-neutrality demands that $N_e = N_i = N$. The final result is

$$\mathbf{j} = \hat{\sigma}\mathbf{E}' \quad \text{where} \quad \mathbf{E}' = \frac{1}{c}\left[\mathbf{v} \times \mathbf{B_0}\right] \quad \hat{\sigma} = \begin{pmatrix} \sigma_P & \sigma_H & 0 \\ -\sigma_H & \sigma_P & 0 \\ 0 & 0 & \sigma_0 \end{pmatrix},$$

with

$$\sigma_0 = \frac{N_0 e^2}{m_e \nu_e}, \quad \sigma_P = \sigma_0 \frac{1 + \beta_e \beta_i}{1 + {\beta_e}^2 {\beta_i}^2 + {\beta_e}^2}, \quad \sigma_H = \sigma_0 \frac{\beta_e}{1 + {\beta_e}^2 {\beta_i}^2 + {\beta_e}^2}$$

For perturbed values of the velocity $(\mathbf{v})$, density (ρ), pressure (p) and electrical current $(\mathbf{j})$ taking into account the gravity force and pressure, the equation system of one-fluid magnetohydrodynamics (4.17a)–(4.17d) becomes

$$\begin{aligned} \rho_0 \frac{\partial \mathbf{v}}{\partial t} &= -\boldsymbol{\nabla} p + \rho \mathbf{g} + \frac{1}{c}\left[\mathbf{j} \times \mathbf{B}_0\right], \\ \frac{\partial \rho}{\partial t} &= -\rho_0 \boldsymbol{\nabla} \cdot \mathbf{v} - \mathbf{v} \boldsymbol{\nabla} \rho_0, \\ \frac{\partial p}{\partial t} &= -\mathbf{v} \boldsymbol{\nabla} p - c_s^2 \rho_0 \boldsymbol{\nabla} \cdot \mathbf{v}, \quad c_s^2 = \gamma \frac{p_0}{\rho_0}. \end{aligned} \tag{15.5}$$

Let p_0 and ρ_0 depend on the height z as $\exp(-z/H)$, where $H = kT/(m_n g)$ is the atmospheric scale height. Then from (15.5) we get

$$\frac{\partial^2 \mathbf{v}}{\partial t^2} = \boldsymbol{\nabla}\left(\mathbf{v}\mathbf{g} + c_s^2 \boldsymbol{\nabla} \cdot \mathbf{v}\right) + \mathbf{g}(\gamma - 1)\boldsymbol{\nabla} \cdot \mathbf{v} + \frac{1}{\rho_0 c}\left[\frac{\partial \mathbf{j}}{\partial t} \times \mathbf{B}_0\right] \tag{15.6}$$

with $c_s^2 = \gamma g H$, where c_s is the sound velocity. Maxwell's equations together with (15.6) describe the interaction of neutral gas with plasma.

Plane Acoustic Waves

Let us consider low-frequency waves emitted by a large-scale ground-based source and propagating over ionized atmosphere with equilibrium pressure p_0 and density ρ_0. For the sake of simplicity, let both parameters are functions of the vertical coordinate z.

Let the horizontal magnetic field $\mathbf{B}_0$ be along the y-axis and the wave propagates along the z-axis of the right-handed Cartesian coordinate system $\{x, y, z\}$ with the z-axis pointed upwards from the Earth. All variables depend only on z. It can be written as $\partial_x = 0$, $\partial_y = 0$. Then the projections of Ampere's and Faraday's laws on z-axis give

$$j_z = 0, \quad b_z = 0. \tag{15.7}$$

Ohm's law and the first equation of (15.7) lead to

$$\sigma_P E'_x + \sigma_H E_z = j_x, \quad -\sigma_H E'_x + \sigma_P E_z = 0, \tag{15.8}$$

where

$$E'_x = E_x - \frac{v_z}{c} B_0$$

and E_x is the electric field in the stationary coordinate system. Let us eliminate E_z from (15.8). The result is

$$j_x = \sigma_C E'_x, \qquad E_z = \frac{\sigma_H}{\sigma_P} E'_x \tag{15.9}$$

where the Cowling conductivity σ_C is

$$\sigma_C = \sigma_P + \frac{\sigma_H^2}{\sigma_P}.$$

Substituting (1.93) for Pedersen σ_P and (1.94) for Hall σ_H conductivities, we obtain

$$\sigma_C = \sigma_0 \frac{1 + \beta_i/\beta_e}{1 + \beta_e \beta_i}, \qquad \sigma_0 = \frac{\omega_{pe}^2}{4\pi \nu_e}.$$

Omitting the term $\beta_i/\beta_e \sim (m_e/m_i)^{1/2}$, we rewrite the last equation in the form

$$\sigma_C = \frac{\sigma_0}{1 + \beta_e \beta_i}. \tag{15.10}$$

It follows from projections of the equations onto the coordinate axis, that the system supports two uncoupled propagating modes. For the first mode variables $v_z, j_x, b_y, E_x, E_z \neq 0$ and

$$\begin{aligned}
&\frac{\partial^2 v_z}{\partial z^2} - \frac{1}{c_s^2}\frac{\partial^2 v_z}{\partial t^2} - \frac{1}{H}\frac{\partial v_z}{\partial z} = -\frac{B_0}{c c_s^2 N_n m_n}\frac{\partial j_x}{\partial t}, \\
&\frac{\partial b_y}{\partial z} = -\frac{4\pi}{c} j_x, \qquad \frac{\partial E_x}{\partial z} = -\frac{1}{c}\frac{\partial b_y}{\partial t}, \\
&j_x = \sigma_C \left(E_x - \frac{B_0}{c} v_z \right).
\end{aligned} \tag{15.11}$$

E_z-component of this mode is

$$E_z = \frac{\sigma_H}{\sigma_P \sigma_C} j_x = \frac{\beta_e}{\sigma_0}\frac{1 - \beta_i/\beta_e}{1 + \beta_i/\beta_e} j_x.$$

And again, if to omit terms $\sim (m_e/m_i)^{1/2}$, we obtain

$$E_z = \frac{\beta_e}{\sigma_0} j_x.$$

For the second mode b_x , E_y , $j_x \neq 0$ and

$$\frac{\partial b_x}{\partial z} = \frac{4\pi}{c} j_y \, , \qquad \frac{\partial E_y}{\partial z} = \frac{1}{c}\frac{\partial b_x}{\partial t} \, , \qquad j_y = \sigma_0 E_y \, . \tag{15.12}$$

A wave propagating over the neutral atmosphere can only couple to the first mode because the second mode is not connected with hydrodynamic motions. For the first mode, we use the vector-potential $\mathbf{A} = A_x \hat{\mathbf{x}}$ of the electromagnetic field:

$$b_y = \frac{\partial A_x}{\partial z} \, , \qquad E_x = -\frac{1}{c}\frac{\partial A_x}{\partial t} \, , \qquad j_x = -\frac{c}{4\pi}\frac{\partial^2 A_x}{\partial z^2} \, . \tag{15.13}$$

Substituting (15.13) into the first and last equation of (15.11) we find the wave equations for coupling of the neutral wave motions to electromagnetic waves. For periodical small perturbations A_x , v_z , etc. $\propto \exp(-i\omega t)$, we obtain [1]

$$\begin{aligned} &\frac{\partial^2 v_z}{\partial z^2} - \frac{1}{H}\frac{\partial v_z}{\partial z} + \frac{\omega^2}{c_s^2}\left(1 + i\frac{\omega_1}{\omega}\right) v_z + \frac{\omega^2}{c_s^2}\frac{\omega_1}{B_0} A_x = 0, \\ &\frac{\partial^2 A_x}{\partial z^2} + i\omega\frac{4\pi\sigma_C}{c^2} A_x - \frac{4\pi\sigma_C}{c^2} B_0 v_z = 0 \end{aligned} \tag{15.14}$$

with

$$\omega_1 = \frac{\nu_{ni}}{1 + (\beta_e \beta_i)^{-1}}, \qquad \nu_{ni} = \nu_{in}\frac{N_i}{N_n}\frac{m_i}{m_n} .$$

Equation (15.14) describes the process of the MHD-wave propagation in an inhomogeneous medium.

Let us introduce dimensionless variables

$$\bar{z} = \frac{z}{2H}, \qquad \Omega = \frac{\omega}{\omega_1}, \qquad \omega_* = \frac{c_s}{2H} .$$

Then (15.14) becomes

$$\begin{aligned} &\frac{\partial^2 V}{\partial \bar{z}^2} - 2\frac{\partial V}{\partial \bar{z}} + \Omega^2 \eta(\bar{z}) A_x + \Omega\left[\Omega + i\eta(\bar{z})\right] V = 0, \\ &\frac{\partial^2 A_x}{\partial \bar{z}^2} + i\Omega\zeta(\bar{z}) A_{\bar{z}} - \zeta(\bar{z}) V = 0, \end{aligned} \tag{15.15}$$

where the function $V = B_0 v_z / \omega_*$. The coefficients in (15.15) are expressed as dimensionless functions:

$$\zeta(\bar{z}) = \frac{4\pi\sigma_C(\bar{z})\, c_s^2}{c^2 \omega_*}, \qquad \eta(\bar{z}) = \frac{\omega_1(\bar{z})}{\omega_*} . \tag{15.16}$$

Figure 15.1 shows height dependencies of functions ζ and η with $H =$ 10 km, $c_s = 0.4\,\mathrm{km/s}$, $\omega_* = 2 \times 10^{-2}\,\mathrm{s}^{-1}$, $\omega_e = 9 \times 10^6\,\mathrm{s}^{-1}$. Here $\eta(\bar{z})$ is the height distribution of the normalized ν_{ni}. The dimensionless conductivity $\zeta(\bar{z})$

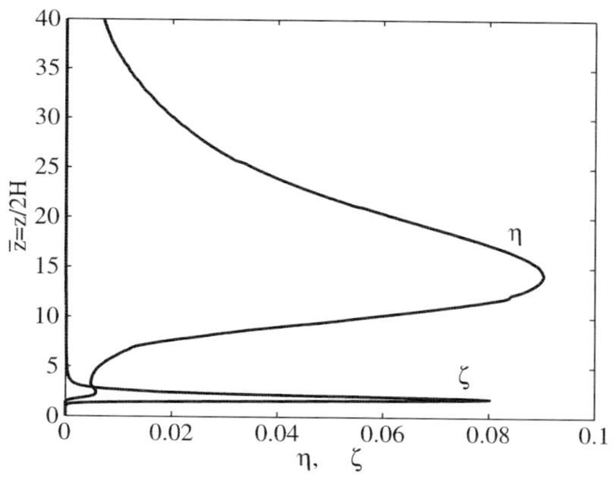

Fig. 15.1. Dependencies of $\eta(\bar{z})$ and $\xi(\bar{z})$ on the dimensionless height $\bar{z} = z/2H$ for $H = 10\,\text{km}$

is proportional to the height distribution of currents which produce magnetic perturbations (see the second equation in (15.15)).

As it is seen from Fig. 15.1, the current is localized within a layer of thickness $\bar{l} = l/(2H) \sim 1$ at height $\bar{z}_1$. If the dimensionless scale of the wave $\bar{l}_s = l_s/(2H) = c_s/(2H\omega) = \Omega^{-1}$ is large in comparison with the current layer thickness $\bar{l} \sim 1$, then it can be used instead of the real thick current layer. As $\bar{l} \sim 1$, the condition $\bar{l} \ll \Omega^{-1}$ may be rewritten $\Omega \ll 1$.

Let the plain $z = z_1$ be a horizontal plane boundary between the atmosphere and ionosphere. The atmosphere is nonconductive (i.e. $\eta(\bar{z}) = 0$ for $\bar{z} < \bar{z}_1$). Above the plane $\bar{z} = \bar{z}_1$ the conductivity changes with a normalized scale size of the order of 10. Then (15.15) becomes

$$\begin{aligned}
&\frac{\partial^2 V}{\partial \bar{z}^2} - 2\frac{\partial V}{\partial \bar{z}} + \Omega^2 \eta(\bar{z}) A_x + \Omega\left[\Omega + i\eta(\bar{z})\right] V = 0, \\
&\frac{\partial^2 A_x}{\partial \bar{z}^2} = 0, \qquad \text{for } \bar{z} > \bar{z}_1, \\
&\frac{\partial^2 V}{\partial \bar{z}^2} - 2\frac{\partial V}{\partial \bar{z}} + \Omega^2 V = 0, \\
&\frac{\partial^2 A_x}{\partial \bar{z}^2} = 0 \qquad \text{for } \bar{z} < \bar{z}_1.
\end{aligned} \tag{15.17}$$

As to the equation for A_x for the region $\bar{z} > \bar{z}_1$, it should be noted that neglecting the terms $i\Omega\zeta(\bar{z}) A_x$ and $\zeta(\bar{z}) V$, we lose some effects such as the excitation of upper ionospheric and magnetospheric resonances because their influence on the thin layer currents are small and can be ignored in the estimations of the induced ground magnetic fields.

Integrating (15.17) over the thin current layer of the thickness $\bar{l}$, we find

$$\left\{\frac{dV}{d\bar{z}}\right\}_{\bar{z}_1} = 0,$$
$$\left\{\frac{dA_x}{d\bar{z}}\right\}_{\bar{z}_1} + il\Omega\zeta_1 A_{x1} - l\zeta_1 V_1 = 0. \tag{15.18}$$

where $\{\}$ denotes the discontinuity of functions at the boundary $\bar{z} = \bar{z}_1$ and $\zeta_1, A_{x1}, V_1 = \zeta, A_x, V\,(x = x_1)$. The velocity V on both sides is approximately equal, the difference being of the order of $\Omega\bar{l} \sim 0.1$.

Let the boundary coincide with the plane $\bar{z} = \bar{z}_1$ and let the speed and potential in the rest of space obey the equations

$$\frac{\partial^2 V}{\partial \bar{z}^2} - 2\frac{\partial V}{\partial \bar{z}} + \Omega^2 V = 0,$$
$$\frac{\partial^2 A_x}{\partial \bar{z}^2} = 0.$$

Let us consider magnetic perturbations caused by an upward propagating acoustic wave. This is

$$V = V_0 \exp\left[(1+k)\,\bar{z}\right],$$
$$k = \sqrt{1-\Omega^2}\,, \quad \operatorname{Im} k > 0. \tag{15.19}$$

Assume that the ground is a perfect conductor. Then

$$A_x = \begin{cases} (2\bar{z}_1 - \bar{z})\,d, & \text{for } \bar{z}_1 < \bar{z}, \\ d\bar{z}, & \text{for } 0 < \bar{z} < \bar{z}_1. \end{cases} \tag{15.20}$$

Substituting (15.20) into the second relationship of (15.18), we can express the integration constant d in terms of the gas velocity on the boundary $\bar{z} = \bar{z}_1$:

$$d = \frac{\zeta_1 \bar{l}}{(i\Omega\zeta_1\bar{l}\bar{z}_1 - 2)} V\,(\bar{z}_1)\,. \tag{15.21}$$

A uniform gas flux, i.e. $\Omega = 0$ produces a magnetic perturbation $b_y = 4\pi\sigma_1 v H B_0/c^2$.

Plane Acoustic Pulse

Let the initial speed of gas displacement be

$$V\,(t) = V_0 t_0^2 \frac{d\delta\,(t)}{dt}\,,$$

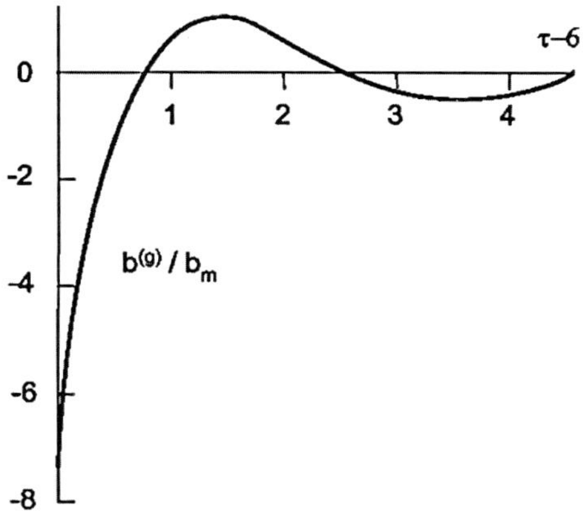

Fig. 15.2. The magnetic variation $b^{(g)}/b_m$ versus dimensionless time τ

where $\delta(t)$ is the delta-function, V_0 and t_0 are compression-depression amplitude and time-scale, respectively.

Referring the reader to [1] for the actual calculations we present here the final expression for the magnetic field induced by the plane acoustic pulse:

$$b = B_0 q \left\{\alpha \exp\left[-\alpha\left(\tau - \bar{z}_1\right)\right] - S \exp\left[-\Omega_0\left(\tau - \bar{z}_1\right]\right) - I\left(\tau\right)\right\}, \qquad (15.22)$$

where

$$q = \frac{\zeta\left(\bar{z}_1\right) \bar{l} V_0 t_0^2 \omega_*^2}{\bar{z}_1 c_s} \left(\frac{\rho(0)}{\rho(\bar{z}_1)}\right)^{1/2} = \frac{4\pi\sigma_C(\bar{z}_1)\, l\, c_s^2 V_0 t_0^2}{z_1 c^2} \left(\frac{\rho(0)}{\rho\left(\bar{z}_1\right)}\right)^{1/2}, \qquad (15.23)$$

$$I(\tau) = \frac{k^3}{2\pi} \int_{-1}^{1} \Omega \left\{\Psi_+\left(\Omega, \tau\right) \exp\left(k\bar{z}_1\right) + \Psi_-\left(\Omega, \tau\right) \exp\left(-k\bar{z}_1\right)\right\} d\Omega. \qquad (15.24)$$

Here

$$\Psi_\pm\left(\Omega, \tau\right) = \frac{n\Omega \cos\left(\Omega\tau\right) - k^2\left[\Omega^2 \pm 2k\right] \sin\left(\Omega\tau\right)}{n^2\Omega^2 + k^2\left[\Omega^2 \pm 2k\right]^2},$$

$$\alpha = 2/\zeta\left(\bar{z}_1\right), \quad S = \frac{\Omega_0^4}{4\left(\Omega_0^2 - 2\right)} \exp\left(\bar{z}_1 \frac{\Omega_0^2 - 2\Omega_0}{2}\right),$$

$$n = \frac{1}{4}\frac{d\eta(\bar{z}_1)}{d\bar{z}_1}, \quad \tau = t\omega_*, \quad \Omega_0 = \sqrt{2\sqrt{2} + 2}.$$

Figure 15.2 shows $b^{(g)}/b_m$ as a function of τ for $\alpha = 20$, $\Omega = 2.2$ and $S = 7.5$. The origin corresponds the time $t \approx 6\,\mathrm{min}$ when the sound wave with $c_s = 3.3 \times 10^4\,\mathrm{cm/s}$ coming to the conductive layer. One can see that the magnetic field reaches the maximal value $b_m = B_0 q S$.

In order to estimate the intensity of the magnetic pulse, we need to define q from (15.23). Let $\zeta = 0.1$, $\bar{l} = 1$, $\omega_* = 2 \times 10^{-2}\,\mathrm{s}^{-1}$, $\bar{z}_1 = 6$ (see Fig. 15.1), corresponding to $z = 120\,\mathrm{km}$ (height of the maximal ionospheric conductivity for the dayside ionosphere). Let also $c_s = 3.3 \times 10^4\,\mathrm{cm/s}$, $(\rho/\rho_0)^{1/2} \approx 10^4$, then

$$q \approx 2 \times 10^{-6} V_0 t_0^2.$$

Let us illustrate this through a simple example. Here, ground motions during strong earthquakes produce an acoustic pulse. The vertical ground displacement during a strong quake is approximately 1 cm for a typical time of 1 s. Then the intensity of the magnetic variations caused by such vertical motion is $b_m = 10^{-6}\,\mathrm{G} = 0.1\,\mathrm{nT}$. The intensity of the induced field is proportional to the square of the signal duration. This means that long-period oscillations can produce visible magnetic variations with intensity comparable with background magnetospheric and ionospheric variations.

For our estimations, we assume that the incident wave is a plane wave. The considered model and therefore all conclusions found here are, strictly speaking, only valid for the equatorial regions and for spatially extended sources. The ionospheric parameters used in the calculations are typical for the dayside ionosphere in the period of maximal solar activity.

Localized Impulsive Source

The paragraphs above have given some indications of the small influence, at least at ionosphere levels, of charged particles and Ampere's force on the wave propagation over neutral particle gas. Therefore, as is usually done in such problems, the neutral component perturbations are calculated first. Then, the induced electrical dynamo-field, current and, finally, the magnetic field are estimated.

Let us consider only qualitatively a situation when the local source is an explosion carried out in the atmosphere. The linear scale l_0 of the explosion of the energy Q_0 is $l_0 = (Q_0/p_0)^{1/3}$, where p_0 is the undisturbed atmospheric pressure at the explosion height. The process goes to the linear stage when the radius of the perturbed region is more than $R_0 \approx 3l_0$. Therefore, we assume that the effective source of disturbances in the atmosphere is a sphere of radius R_0. For example, if the energy Q_0 of a ground blast is $Q_0 = 100$ t TNT$= 4.2. \times 10^{18}$ erg then $l_0 \approx 2 \times 10^4$ cm and $R_0 \approx 6 \times 10^4$ cm for $p_0 = N_m k T_0 = 4.2 \times 10^5\,\mathrm{dyn} \cdot \mathrm{cm}^{-2}$.

The pulse produced by the blast of radius R_0 and energy Q_0 propagates in an inhomogeneous atmosphere. The problem of wave propagation in a neutral atmosphere from a local impulsive source reduces to consideration of two wave modes – acoustic and acoustic-gravity waves. The acoustic gravity branch exists at frequencies of less than the Brūnt – Vāisālā frequency $\omega_B = [(\gamma - 1)\, g/\gamma H]^{1/2}$. Here γ is the polytropic exponent. For $H = 8\,\mathrm{km}$, the Brūnt – Vāisālā period is 300 s. The frequencies of the acoustic branch are more than $\omega_a = (\gamma g/4H)^{1/2}$ corresponding to periods of less than 300 s.

The maximal signal amplitude is to be on the front of the pulse formed by high-frequency components. However, atmospheric thermal conductivity and viscosity cause the initial pulse to broaden. The damping factor α of sound is defined as [22]

$$\alpha = \frac{\omega^2}{2\rho_0 c_s^3}\left[\left(\frac{4}{3}\eta + \zeta\right) + \frac{\kappa(\gamma-1)}{c_p}\right],$$

where η and ζ are coefficients of the dynamic and the second viscosity, respectively, κ is the thermal conductivity coefficient. Hence, one can find the pulse duration:

$$\tau_p^2 = \int_0^{\bar{R}} \frac{dr}{2\rho_0 c_s^3}\left[\left(\frac{4}{3}\eta + \zeta\right) + \frac{\kappa(\gamma-1)}{c_p}\right]. \tag{15.25}$$

If the viscosity is taken into account, then the expressions for the spectral harmonics of velocity and pressure should be multiplied by $\exp\left(-\omega^2\tau_p^2\right)$. Then, for instance, the vertical velocity of the neutral gas in the wave can be presented as follows ([9], [11])

$$w = \frac{H^3 z}{8\pi^{3/2}\gamma c_s \tau_p^3 R^2}\left(\frac{R}{c_s} + \frac{2c_s\tau_p^2}{R} - t\right)\exp\left[\frac{z}{2H} - \frac{(t - R/c_s)^2}{4\tau_p^2}\right]. \tag{15.26}$$

For distances larger than the length of the pulse $R \gg c_s\tau_p$, we get from (15.26) that, at $t = R/c_s + 2^{1/2}\tau_p$, the velocity reaches the maximal value

$$w_m = \frac{1}{\gamma\sqrt{32\pi^3 e}}\exp\left(\frac{z}{2H}\right)\frac{z}{R^2}\frac{H^3}{\tau_p^2 c_s}. \tag{15.27}$$

From (15.25) the duration of the initial 1 s-pulse at the heights of 120, 150 and 200 km is 1, 1.5, 3 and 6 s respectively. Below 100 km, 1 s-pulse is virtually constant. Relation (15.27) allows us to estimate the value of the maximal velocity in the sound pulse. For blast energy of 100 t TNT and $z = 120$ km the maximal velocity $w_m \approx 4 \times 10^2$ cm/s. Then the maximal magnetic field induced produced by such local source is of ~ 0.1 nT.

Internal gravity waves (IGW) are other possible transmission channels of the explosion effects over long distances. Behind the front, the acoustic mode is damped much more sharply than the IGW-mode which can excite low-frequency geomagnetic disturbances through dynamo action. The numerical computations show that 10 min after the explosion, the IGW-front reaches a height of ≈170 km. About 1 h after the explosion, the wave field fills the space with a radius of about 1000 km.

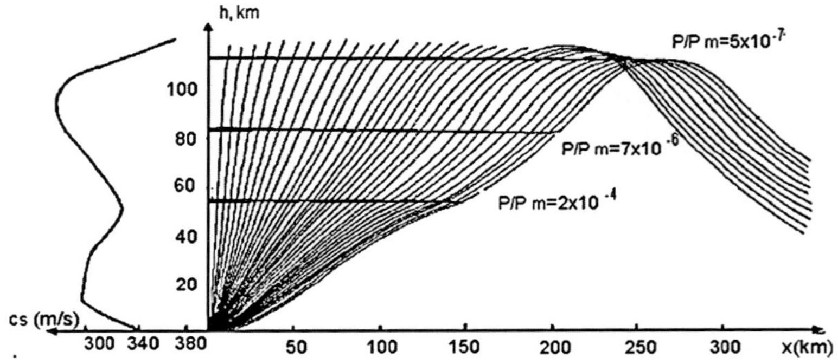

Fig. 15.3. A set of acoustic rays from a point source located on the ground surface. After [23]

Nonlinear Acoustic Wave

The nonlinearity can have a substantial effect on pulse damping. When the wave vertically ascends in the lossless inhomogeneous atmosphere, the wave's energy $\rho v^2 = \text{const}$ is conserved in the first approximation. There is a height where the gas speed in the wave becomes comparable to the sound speed and at this level, a shock wave occurs [31].

The interesting nonlinear effect of an acoustic wave with finite amplitude was considered in ([23], [31]). The numerical calculations in the ray approximation showed that nonlinear distortions of a spherical wave do not depend on the ray's initial angle (i.e. that angle at which the ray originates from a point source) [23]. The shock front is formed at the same height for all the rays [24]. The field amplitude at a fixed height weakly depends on the ray's initial angle, whereas the formation of shock waves only depends on the height. The right side of Fig. 15.3 demonstrates a set of acoustic rays produced by a point source located on the ground surface. The height distribution of sound velocity used in the calculations is shown in the left side of the figure. For the calculated ray structure, pressure peak values were defined as a function of the initial angle for various heights. The location of the isobars is shown in Fig. 15.3 by horizontal lines. One can see that the isobars become horizontal. It decreases the spherical acoustic beam divergence and can increase a value of the magnetic pulse at the distances of ≈100–200 km from the acoustic beam axis.

When propagating, the pulse undergoes changes in its shape and duration (see e.g. [7]). The pulse duration is about three times longer than the duration found above with the linear theory formulae. It lasts for about 3, 10 and 20 s, respectively, for 100, 150 and 200 km.

At the front of such a nonlinear wave there can be an appreciable change of chemical kinetics of both neutral and ionized components. Moreover, electromagnetic radiation from the shock front can penetrate deeply into the magnetosphere.

Induced Magnetospheric MHD-Wave

Let us suppose that the perturbation of the neutral gas propagates in the meridional plane in the middle latitude E-layer. Denote the neutral velocity transversal to $\mathbf{B}_0$ as $v_{n\perp}$. Then the dynamo-field $E_{dy} = B_0 v_{n\perp}/c$ produces the Pedersen $\mathbf{j}_{Pd} = j_{yd}\hat{\mathbf{y}}$ and Hall $\mathbf{j}_{Hd} = j_{xd}\hat{\mathbf{x}}$ currents (see (2.8) for $\sigma_{\parallel} \to \infty$)

$$j_{xd} = \frac{\sigma_H}{\sin I}\frac{v_{n\perp}}{c}B_0, \quad j_{yd} = \sigma_P\frac{v_{n\perp}}{c}B_0.$$

Since $v_{n\perp}$ is independent on y, then $\nabla \cdot \mathbf{j}_{Pd} = 0$ and the current does not generate the longitudinal currents and does not excite an Alfvén wave. However, for the Hall current $\nabla \cdot \mathbf{j}_{Hd} = \partial_x j_{xd}(x,z) \neq 0$. From the continuity equation for the total current it follows that the Hall current, induced by the neutral motion, is closed by the longitudinal current.

The MHD-equation for the field-aligned current $\mathbf{j}_{\parallel}$ of an Alfvén wave in the magnetosphere can be obtained from (7.136)

$$\frac{\partial}{\partial s}c_A^2\frac{\partial}{\partial s}j_{\parallel} - \frac{\partial^2 j_{\parallel}}{\partial t^2} = 0, \tag{15.28}$$

where s is the coordinate along the field-line.

For approximate estimates we take the geomagnetic field-lines to be equipotential in the E-layer. This assumption holds for disturbances with large enough horizontal scales. A typical spatial scale, $l_{\parallel}$ of the electric potential along the field line is related to the horizontal scale, $l_{\perp}$, across the magnetic field as

$$l_{\parallel} \approx l_{\perp}\left(\frac{\sigma_{\parallel}}{\sigma_{\perp}}\right)^{1/2}. \tag{15.29}$$

In the E-layer, $(\sigma_{\parallel}/\sigma_{\perp} \approx 10^4)$ then the perturbations with scales of $l_{\perp} \approx 1\,\text{km}$ are equipotential in the ionosphere of thickness of $l_{\parallel} \approx 100\,\text{km}$. Since $\sigma_{\parallel}/\sigma_{\perp}$ grows rapidly with height, condition (15.29) does not impose any significant constraints on $l_{\perp}$ at E-layer heights, or higher.

Let us now estimate $j_{\parallel}$ the emitted Alfvén pulse. We replace the conductive layer with a thin film of integral conductivity $\Sigma_{P,H}$, and external surface current

$$I_{xd} = B_0\frac{\Sigma_H}{\sin I}\frac{v_{n\perp}}{c}. \tag{15.30}$$

The acoustic pulse, while propagating in the ionosphere, can generate surface current (15.30). This surface current is not zero as long as the vertical scale of the change in Hall conductivity is less than, or comparable with, the vertical scale of the acoustic pulse. Then the current (15.30) will be generated by the compression and rarefaction part of the acoustic pulse. At shorter scales, the magnetic fields are produced by the thin currents , interfere with each other and the resulting field vanishes and (15.30) is inapplicable.

The boundary condition on the top of the ionosphere follows from the continuity equation for the total current:

$$\left(\frac{\partial}{\partial t} - c_A \frac{\bar{\Sigma}_P}{\sin I}\frac{\partial}{\partial s}\right) j_{\parallel}(x,s,t) = -\frac{\partial^2}{\partial x \partial t} I_{xd}(x,t)\,|_{x=0}, \tag{15.31}$$

where $\bar{\Sigma}_P = \Sigma_P/\Sigma_A$ and $\Sigma_A = c^2/4\pi c_A$.

Let the Alfvén velocity, c_A, be constant along the field-lines. Then, by substituting the solution of (15.28) in the form of a traveling wave $j_{\parallel}(t - s/c_A)$ into (15.31), we obtain

$$j_{\parallel} = -\frac{\partial}{\partial x}\frac{I_{xd}(x, t - s/c_A)}{1 + \bar{\Sigma}_P/\sin I}. \tag{15.32}$$

There is no Σ_H in this relation because, as it was shown in Chapter 7, Σ_H has no influence on the Alfvén wave for not so large horizontal scales of $\lesssim 100\,\mathrm{km}$. However, the external ionospheric current I_{xd} is defined by Σ_H.

The current induced by the pulse in the conducting layer is then channeled into the flux tube about that of the fore front of the pulse. This field-aligned current produces a magnetic pulse. Let us substitute (15.30) in (15.32). Integration of the obtained relation gives magnetic field of the current pulse

$$\frac{b_y}{B_0} = -\frac{\Sigma_H}{\Sigma_P}\frac{1}{1 + \sin I/\bar{\Sigma}_P}\frac{v_n(x, t - s/c_A)}{c_A}. \tag{15.33}$$

The transversal distribution of the magnetic field b_y in the Alfvén pulse is proportional to the horizontal distribution of macroscopic velocity of the ionospheric neutral gas in the pulse. The current is defined by its spatial derivative.

Since $\bar{\Sigma}_P \gg 1$ in day then (15.33) reduces to

$$\frac{b_y}{B_0} \approx -\frac{\Sigma_H}{\Sigma_P}\frac{v_n}{c_A}.$$

and in night $\bar{\Sigma}_P \ll 1$

$$\frac{b_y}{B_0} \approx -\frac{\Sigma_H}{\Sigma_A}\frac{v_n}{c_A}.$$

Let us estimate the magnetic field above the ionosphere. Let $v_n = 5 \times 10^2\,\mathrm{cm/s}$, $c_A = 1 \times 10^8\mathrm{cm/s}$, $\Sigma_P \approx \Sigma_H$, then in day

$$b = 0.25\ nT \approx 5 \times 10^{-6}\ \mathrm{G} = 0.5\,\mathrm{nT}$$

and in night

$$b = 0.025\ nT \approx 0.5 \times 10^{-6}\ \mathrm{G} = 0.05\,\mathrm{nT}.$$

Divergence of the field-lines in the magnetosphere decreases the magnetic disturbance by 5–10 times.

The theory just described deals with the propagation of low-frequency waves in the atmosphere and the ionosphere and also with ground and magnetospheric generation of magnetic variations by such waves. This theory only reflects the basic features of this process, which under real conditions is much more complicated. Actually, the wave propagates not in an isothermal atmosphere but in a medium with an intricate height temperature profile. This profile can lead to the appearance of waveguides. As a result, a portion of the wave energy can be trapped and carried away.

We now turn to results of magnetospheric observations of the strong acoustic impact caused by an explosion on the ground.

15.2 Acoustic Shock Experiment

In 1985, acoustic action (Project MASSA[1]) has been proved to be a very powerful tool for artificial generation ionospheric and magnetospheric perturbations. Acoustic action offers the possibility of simultaneous detection of all elements of interest. In this project, in 1981–1982, the responses of the atmosphere, the ionosphere, and the magnetosphere to surface explosions were studied (e.g. [2], [3], [4], [5], [6]).

The 251 tTNT blast was detonated on November 28, 1981, at 02h31m UT (08h31m LT) at a point with coordinates 43°48′N, 76°51′E. The measurement and monitoring equipment were located both in the immediate vicinity of the blast (for the registration of shock wave, seismographs, and optical observations) and at distances of tens of kilometers from the blast: magnetometers, acoustic measurements, Doppler units, ionosphere sounders. At large distances from the blast (hundreds and thousands of kilometers), oblique and return-oblique radio sounders were used on trajectories passing through the region of the blast. Ionic probes in a high-frequency regime, magnetic measurements, and measurements of incoherent scattering of radio waves, as well as other observations were all made.

Virtually all of the instruments used in the experiment registered perturbations of geophysical parameters associated with the blast [2]. Based on these findings, the following picture of the effects can be drawn. The shock wave that was formed at a distance of 10 km from the blast converted into a sound wave which was traceable along the Earth's surface up to a distance of 2000 km. A portion of the energy of the acoustic wave traveling upward was captured in the waveguide formed by the near-Earth temperature inversion; then a partial reflection of the wave took place at altitudes of 40 km and 100 km. However, the acoustic wave, carrying substantial energy, reached

[1] Russian acronym for Magnetosphere-Atmosphere Relationships with Seismo-Acoustics.

the ionosphere in about 6 min, causing tangible perturbations of the electron concentrations in the E- and F-layers. On its path, the pulse became much broader: from 1 s at the distance where the shock wave was transformed into an acoustic wave to 1 min at the altitudes of the F-layer.

The entry of the acoustic wave into the ionosphere was accompanied by a noisy electromagnetic signal in the frequency band of tens of kHz, which traveled along the geomagnetic field-lines into the magnetosphere. At the same time, relatively thin current jets arose in the magnetosphere, inside which a magnetic disturbance of tens of nT was formed. At the time of entry of the acoustic wave, a perturbation arose in the ionosphere which traveled with a speed of 1–10 km/s (depending on the direction relative to the geomagnetic field); this produced crescent-shaped features on the ionograms of stations situated hundreds of kilometers from the explosion.

At this time, the magnetic field variations in the middle latitudes attained magnitudes of ~10 nT. Against this background, no clear variations of the geomagnetic field that could be reliably linked with the blast have been observed.

It is assumed that the generation of electromagnetic effects in the magnetosphere is determined by electrodynamic processes in the E-layer. Here the motion of neutral particles, which is dependent on acoustical waves, should cause a local generation of an electrical field and currents. Thus, near the explosion's field tube, longitudinal currents may be created and plasma noise may be generated.

With this goal in mind, the time of the MASSA experiment blast was chosen such that at the moment when the acoustic wave would reach the altitude of the E-layer (100–120 km), the AUREOL-3 satellite would be located at a minimum distance from the magnetic field tube (see Fig.15.4).

As a result, the satellite detected a spot of increased electromagnetic noise intensity over a wide range of frequencies in the explosion's field tube in the ionosphere ([5], [6]). Figure 15.5 shows the results of EZ-component measurements (in a direction making a 30° angle with the main magnetic field) and the transverse EH-component of the electrical field in a range of 10–20, 100–200, 200–450, and 450–1000 Hz in relative percent (an approximate logarithmic scale). There is a pronounced rise in the noise level above the background level (the long and short dotted lines) in the vicinity of the explosion's field tube (at distances ±200 km) in the frequency range 0.1–2 kHz, and especially with respect to the EZ-component. The magnetic field of these oscillations is insignificant and may be they are almost electrostatic ones.

There was also a significant rise in the range of 15 kHz in intensity on this same segment of the trajectory (cf. left and right panels). The polarization of the electrical oscillations was transverse, and the magnetic component was also small. The ratio of the field components enable us to assume that these were swift waves, propagated near the resonant cone. An increase in the noise intensity can be also noticed in the range of 4.5–16 kHz (Fig. 15.5).

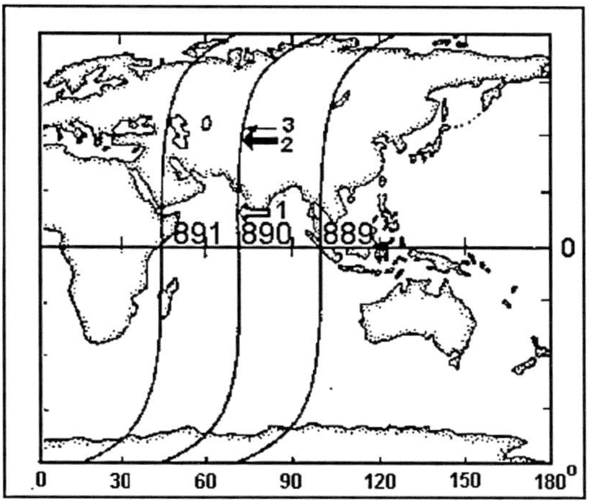

Fig. 15.4. Trajectory of satellite AUREOL-3 during the MASSA experiment: 1) location of the satellite at time of the explosion, 2) point where the satellite trajectory intersects the explosion's field tube, and 3) location of the satellite above the shot size

The Alfvén Pulse

Measurements by an airborne magnetometer showed a significant magnetic variation at $t = 298\,\mathrm{s}$ at a distance of about 700 km from the explosion's field tube, and $L = 1.31$ (see Fig. 15.6) ([5], [6]). Characteristic features of this

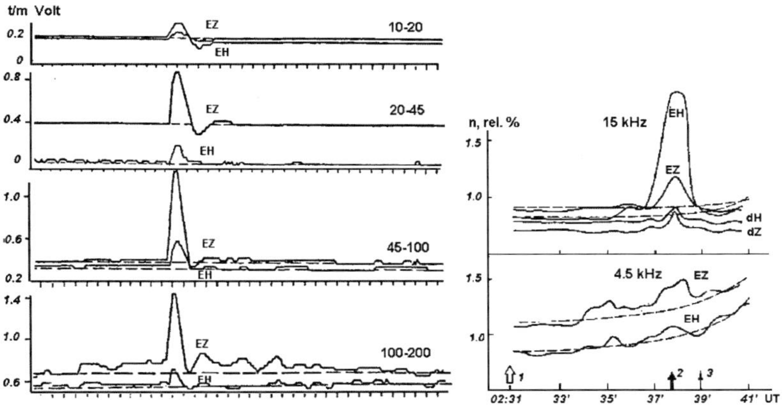

Fig. 15.5. The electrical field for a series of band filters in the range 10 to 200 Hz at the intersection of the explosion's magnetic field tube (left panel. Arbitrary units). The dashes are the background of the EZ-component (approximately along the magnetic field) and the EH-component (across the field). EH and EZ at frequencies 15 and 4.5 kHz along the orbit is shown in the right panel. The background noise level is shown by the dashed lines

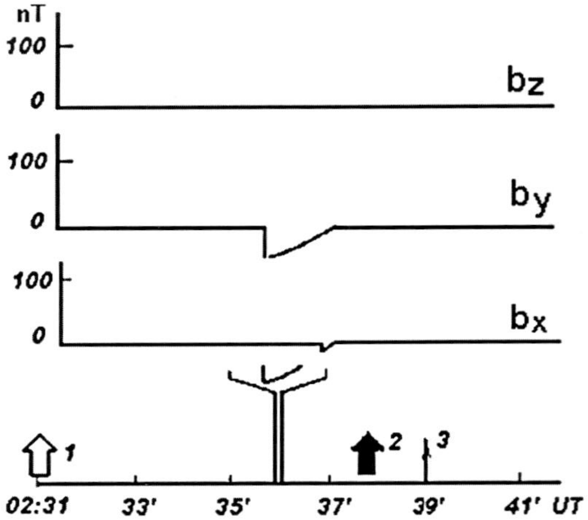

Fig. 15.6. Results of magnetic pulse measurements

variation are a sharp surge or pulse in the magnetic field lasting $t < 0.08$ s, having components in the satellite's coordinate system: $b_{xs} = -25$, $b_{ys} = -114$, $b_{zs} = 0 \pm 5$, and $|\mathbf{b}| = 117$ nT, with vector **b** directed almost exactly west. This pulse is accompanied by a pulse of the electric field having component: $\Delta E_{xs} = 97.7$, $\Delta E_{ys} = 26.4$, $\Delta E_{zs} = 0 \pm 5$ and $|\Delta\mathbf{E}| = 101.2$ mV/m, with vector $\Delta\mathbf{E}$ directed almost exactly north. It is evident that within the limits of measurements accuracy, $\Delta\mathbf{B} \perp \Delta\mathbf{E}$. Hence, the phase velocity of the given electromagnetic wave is 860 ± 140 km/s. The Poynting vector is directed upward along the magnetic field.

According to measurements, at this point the local frequency of the upper hybrid resonance $f_{BHy} = 2.49$ MHz and the magnetic field is $|\mathbf{B}_0| = 0.342$ G, from which $N_e = 6.56 \times 10^4$ cm^{-3}. From data of the mass spectrometer thermal ions, one can assume that most of the ions at this point (at an altitude of about 800 km) are 0^+ ions. Hence, the magnitude of the Alfvén velocity $c_A = 1000$ km/s (for 50% 0^+ and 50% H^+) or 745 km/s (for 100% 0^+. Within the limits of accuracy, this concurs with the measured value v of an electromagnetic wave. Therefore, these data are direct evidence in favor of the movement of the electromagnetic pulse in the magnetosphere in the form of an Alfvén wave.

An important point is that the Alfvén pulse has been recorded within 300 s after the blast and at a distance of $\approx$700 km from the explosion's field tube. An acoustic wave could not reach the E-layer for this time and traveled only to 100 km altitude. The early arrival the Alfvén pulse can be indirect indication that the strong acoustic wave changes the low ionosphere. The results of the observations of electromagnetic noise in the wide frequency range 10–1000 Hz made by the same satellite during a series of subsequent active experiments employing powerful industrial explosions, confirmed the emergency of a 'noisy' spot in the upper atmosphere over the blast point.

The observations of the ionospheric effects provoked by a blast of 4.8 kT [8] were carried out using a network with the slightly oblique Doppler sounding and pulse vertical radio sounding units. The experiments demonstrated that the acoustic front generates a small-scale turbulent structure. The excitation of the ionospheric turbulence was observed both over the blast point and at distances of the order of several hundreds of kilometers from it. At the sporadic E_s-layer a fragmentation of the inhomogeneities into a spatial-temporal scale of 2–10 km and 1–2 s was observed.

Small-scale random inhomogeneities in the lower E-layer significantly increase the effective Pedersen conductivity, Σ_P. Theoretical consideration and laboratory experiments of this problem have been discussed already in Chapter 10, especially the relation of the increase with β_e, possible reasons for the heights of this effect and relative contributions to the integral Pedersen conductivity. It follows from (10.37) that the ratio of the effective Pedersen conductivity σ_P^{eff} to the mean Pedersen conductivity $\langle\sigma_P\rangle$ is given by

$$\frac{\sigma_P^{eff}}{\langle\sigma_P\rangle} = \beta_e \frac{\delta N_e}{\langle N_e\rangle}.$$

$\beta_e \approx 10^2$ at heights 90–100 km (see Figure 2.5). Then the ratio $\sigma_P^{eff}/\langle\sigma_P\rangle \sim 10$. Σ_P of the perturbed layer increases in the order of magnitude. Contribution of the undisturbed lower E-layer into the total integral conductivity of the whole E-layer is $\approx$5–10%. Under changes of the lower ionosphere conductivity 10 times, its contribution into the integral Pedersen conductivity consists of $\approx$50%, i.e. $\delta\Sigma_P/\langle\Sigma_P\rangle \approx 50\%$.

The flickering high-conductive ionospheric inhomogeneity results in an appearance of the longitudinal Alfvén current (e.g. [19]) excited by the background electric field. Here we examine the field-aligned current arising at the perturbed ionosphere region. Let the acoustic wave causes $\sim$10% perturbations δN_e of the electron background ionospheric concentration N_{e0}.

The transversal inhomogeneity results in the ionospheric current redistribution, part of them flows out the ionosphere along the geomagnetic field. The ratio of the total field-aligned current $I_\parallel$ outflowing from the perturbed region to the background horizontal ionospheric current I_0 is

$$\left|\frac{I_\parallel}{I_0}\right| \sim \frac{\Sigma_A}{\Sigma_A + \Sigma_{P0}} \frac{\delta\Sigma_P}{\Sigma_P},$$

where $\Sigma_A = c^2/(4\pi c_A)$ is the Alfvén wave conductivity; Σ_P is the effective integral conductivity of the ionospheric layer perturbed by the acoustic wave; Σ_{P0} is the undisturbed ionospheric integral Pedersen conductivity; $\delta\Sigma_P = \Sigma_P - \Sigma_{P0}$. The background integral current I_0 in the middle latitudes is ~ 0.05 A/m. Let $\Sigma_A \sim \Sigma_{P0}$ since the experiment has been performed during the early morning. Then $I_\parallel \sim 0.25 I_0$. The background integral current I_0

in the middle latitudes is ~50 mA/m. Thus, the field-aligned current $I_{\parallel} \sim$ 10 mA/m. Let the transversal scale be $l_{\perp} = 10^4$ m. Then the field-aligned current density is $j_{\parallel} \sim 2.5\,\mu\text{A/m}^2$. The measured magnetic pulse ≈100 nT and its transversal scale of 10 km give the longitudinal current $j_{\parallel} \approx 10\,\mu\text{A/m}^2$.

It follows from the estimations that the directed electron velocity in the observed thin Alfvén current jet can exceed the electron thermal velocity. This, in turn, can cause an anomalous longitudinal resistivity that can also create a magnetospheric Alfvén wave.

Independent of the question of whether or not the explanation given here for the appearance of the strong MHD-wave in the magnetosphere is true, we can say with confidence that the acoustic action on the ionosphere and the magnetosphere is extremely effective in the sense of the artificial generation of an MHD-wave. Moreover, the managed acoustic impacts are a significant tool to study the isolated components and the entire chain of the interconnected effects responsible for a transmission of the large-scale perturbations in the medium surrounding the Earth: the atmosphere, the ionosphere and the magnetosphere.

References

1. Alperovich, L., M. Gohberg, V. Sorokin, and G. Fedorovich, Generation geomagnetic variations generation by acoustic oscillations caused by earthquakes, *Physics of the Solid Earth*, **58**, No.3, 1979.
2. Alperovich, L., B. Vugmeister, M. Gohberg, V. Drobjev, A. Eruchenkov, E. Ivanov, V. Kudriavtsev, S. Kulichkov, V. Krasnov, A.Matveev, M. Morduchovich, P. Nagorskii, E. Ponomarev, O. Pochotelov, Yu. Tarachuk, V. Troitskaya and G. Fedorovich. Experience in modelling of magnetosphere-ionosphere effects connected with seismic phenomena, *Dokl. Akad. Nauk USSR*, **17**, No.3, 269, 1983.
3. Alperovich, L., M. Gohberg, V. Drobjev, V. Troitskaya, and G. Fedorovich, Project MASSA: an investigation of magnetosphere atmosphere relationships during seismoacoustic events, *Physics of the Solid Earth*, **21**, No.11, 813, 1985.
4. Alperovich, L., E. Ponomarev, and G. Fedorovich, Geophysical phenomena modelled by an explosion, *Physics of the Solid Earth*, **21**, No.11, 816, 1985.
5. Alperovich, L., E. Afraymovich, B. Vugmeister, M. Gohberg, V. Drobjev, A.I. Yeruchenkov, E. Ivanov, A. Kalikhman, M. Morduchovich, A. Matveyev, E. Ponomarev, V. Troitskaya, and G. Fedorovich, The acoustic wave of an explosion, *Physics of the Solid Earth*, **21**, No.11, 835, 1985.
6. Alperovich, L., B. Vugmeister, M. Gohberg, V. Drobjev, V. Kazakov, P. Nagorsky, E. Ponomarev, V. Troitskaya, and G. Fedorovich, Ionospheric effects correlated to geomagnetic field variations as observed during a MASSA experiment, *Physics of the Solid Earth*, **21**, No.11, 873, 1985.
7. Besset, G., and E. Blanc, Propagation of vertical shock waves in the atmosphere, *J. Acoust. Soc. Am.*, **95**, 1830–1839, 1994.

8. Blanc, E. and D. Rickel, Nonlinear wave fronts and ionospheric irregularities observed by HF sounding aver a powerful acoustic source, *Radio Science*, **24**, 279, 1989.
9. Cole, J. D. and C. Greifinger, Acoustic-gravity waves from an energy source at the ground in an isothermal atmosphere, *J. Geophys. Res.*, **74**, 3693, 1969.
10. Danilov, A. V. and S. F. Teselkin, The structures of shock wave's precursors in the weakly ionized nonisothermic plasma, *Sov. J. Plasma Phys.*, **10**, 426, 1984.
11. Dikii, L. A. *The Theory of Oscillations of the Terrestrial Atmosphere*, Gidrometeoizdat, Leningrad, 1969.
12. Galperin, Yu., V. Gladyshev, N. Dzhordjio, R. Kovrajkin, Yu. Lisakov, L. Maslov, R. Nikolaenko, R. Sagdeev, O. Molchanov, M. Mogilevsky, L. Alperovich, M. Gohberg, O. Pokhotelov, K. Begin, Zh.. Bertel'ye, Zh.M. Boske, and A. Ram, VLF and ELF effects in the upper ionosphere caused by large scale acoustic waves in the lower ionosphere observed from Aureol-3 satellite. Chapter in the book: '*Results of the ARCAD-3 Project and of the other Programs in Magnetospheric and Ionospheric Physics*' CNES, CEPADYES, Toulouse, 976, 1985.
13. Galperin, Yu., V. Gladyshev, N. Dzhordjio, R. Kovrajkin, Yu. Lisakov, L. Maslov, R. Nikolaenko, R. Sagdeev, O. Molchanov, M. Mogilevsky, L. Alperovich, M. Gohberg, O. Pokhotelov, K. Begin, Zh. Bertel'ye, Zh. M. Boske, and A. Ram, The Alfvén wave excited in the middle-latitude magnetosphere by a large scale acoustical wave which is propagated in the lower ionosphere, *Physics of the Solid Earth,* **21**, No.11, 877, 1985
14. Gershman, B. N., *Dynamics of the ionospheric plasma*, (in Russian), Nauka, Moscow, 1974.
15. Hines, C. O., Hydromagnetic resonance in ionospheric waves, *J. Atmos. Terr. Phys.*, **7**, 14, 1955.
16. Jeffreys, H., and B. Swirles, *Methods of mathematical physics*, Cambridge University Press, Cambridge, 1966.
17. Jacobson, A. R., A model for conjugate coupling from ionospheric dynamos in the acoustic frequency range, *J. Geophys. Res.*, **91**, A4, 4404, 1986.
18. Kamide Y. and W. Baumjohann, Magnetosphere-Ionosphere Coupling. - Springer- Verlag, Berlin, Heidelberg, 1993. 178.
19. Kozlovsky, A. E. and W. B. Lyatsky, Alfvén wave generation by disturbance of ionospheric conductivity in the field-aligned current region, *J. Geophys. Res.*, **102**, 17297, 1997.
20. Korsunsky, S. V., The propagation of sound beams of finite amplitude in electroconductive media, *J. Acoust.*, **36**, 48, 1990.
21. Kulichkov, S. N., Long-range propagation and scattering of low-frequencysound pulses in the middle atmosphere, it Meteorol. Atmos. Phys., **85**, 47, 2004.
22. Landau, L. D. and E. M. Lifshitz, *Course of Theoretical Physics, **6**: Hydrodynamics*, Pergamon, N.Y., 1985.
23. Novikov, Y. V., A. V. Razin, and V. E. Fridman, The level of nonlinear distortion of acoustic wave from the point source in standard atmosphere, *Preprint No 243, NIRFI.* Gorkiy, 1987 (in Russian).
24. Ostrovsky, L. A., and V. E. Fridman, Dissipation of intensive sound in isothermal atmosphere, *Acoust. J.*, **5**, 625, 1985.

25. Orlov, V. V., and A. N. Uralov, Reaction of the atmosphere to ground weak explosion, *Izv. Akad. Nauk SSR. Fiz. Atmos. Okeana*, **20**, 476, 1984.
26. Pitteway, M. L. V., D. G. Rickel, J. W. Wright, and M. M. Al-Jarrah, Modeling the ionospheric disturbance caused by an explosion on the ground, *Ann. Geophys.*, **3**(6), 695, 1985.
27. Pogorel'tsev, A. I., Propagation of electromagnetic field disturbances caused by acoustic-gravity waves into the magnetically conjugate ionosphere, *Geomagnetism and Aeronomy*, **35**, No. 2, 178, 1995.
28. Pokhotelov, O. A., V. A. Pilipenko, E. N. Fedorov, L. Stenflo, and P. K. Shukla, Induced Electromagnetic Turbulence in the Ionosphere and the Magnetosphere, *Physica Scripta*, **50**, 600, 1994.
29. Pokhotelov, O. A., M. Parrot, E. N. Fedorov, V. A. Pilipenko, V. V. Surkov, V. A. Gladychev, Response of the ionosphere to natural and man-made acoustic sources, *Ann. Geophys.* **13**, 1197, 1995.
30. Romanova, N. N., Nonlinear propagation of acoustic and gravity waves in an isothermal atmosphere, *Izv. Akad. Nauk USSR, Atmos. Oceanic Phys.*, **7**, 1251, 1971.
31. Romanova, N. N., Vertical propagation of acoustic waves of arbitrary shape in the isothermal atmosphere, *Izv. Akad. Nauk USSR, Atmos. Oceanic Phys.* **11**, 233, 1975.
32. Ryutov, D. D., and M. P. Ryutova, Generation of magnetic field by sound oscillations in solar atmosphere, *Sov. Phys. JETP*, **96**, 1708, 1989.

Index

Symbol	Meaning	Page
Φ_0	potential of the polarization electric field	285
$\Phi_0(x,y)$	electric potential in the ionosphere	279
Ψ	current function	286
Ψ	scalar potential of a displacement	88
Σ_H	integral Hall conductivity	42
Σ_P	integral Pedersen conductivity	42
Σ_A	Alfvén wave conductivity, $\Sigma_A = c^2/(4\pi c_A)$	108
Σ_{ij}	i,j-components of integral layered conductivity	42
$\alpha_\mp$	$c/\omega l_z X_\mp$	112
$\bar{\Sigma}_H$	$\bar{\Sigma}_H = \Sigma_H/\Sigma_A$	250
$\bar{\Sigma}_P$	$\bar{\Sigma}_P = \Sigma_P/\Sigma_A$	250
$\bar{\rho}$	normalized field-aligned plasma distribution	164
β	c_s^2/c_A^2	86
β	$k_0 h \varepsilon_a \sin^2 I/\tilde{X}$	255
β_e	electron magnetization	38
β_{in}	coefficients of the ion induced polarization	37
β_{in}^0	reaction rate for charge interchange	37
β_i	ion magnetization	38
$\mathbf{u}$	hydrodynamic velocity	81
$\mathbf{S}$	Pointing vector	103
$\boldsymbol{\Omega}$	velocity vortex	87
$\boldsymbol{\nabla}_\perp^2$	transverse Laplacian	279
$\boldsymbol{\nabla}_\perp$	transversal to $\mathbf{B_0}$ projection of a gradient	279
$\boldsymbol{\xi}$	displacement vector	85
$\boldsymbol{\xi}_\perp$	transverse displacement	87
$\delta(x,y)$	Dirac delta function	89
δ_n	half-width of n-th FLR-harmonics	119
δ_i	half-width of the FLR above the ionosphere	243
δ_{jk}	Kronecker symbol	12
γ_n	decrement of n-th FLR- mode	114
$\hat{\mathbf{x}}, \hat{\mathbf{y}}, \hat{\mathbf{z}}$	unit vectors of the coordinate system	83
$\kappa_{1,2}$	principal curvatures of equipotential surface	158
$\mathbf{0}$	zero matrix	100
$\mathbf{1}$	unit matrix	100
$\mathbf{B}$	macroscopic magnetic field	10
$\mathbf{D}$	electric induction	17, 18
$\mathbf{E}_\perp^*$	complex conjugate to $\mathbf{E}_\perp$	103
$\mathbf{E}_\perp, \mathbf{E}_\parallel$	transversal and longitudinal electric components	87
$\mathbf{E}$	macroscopic electric field	10
$\mathbf{I}_\tau$	horizontal ionospheric current	285
$\mathbf{R}', \mathbf{T}'$	matrices of reflection and transmission	240
$\mathbf{R}_{(\alpha)}$	friction force	12
$\sigma(\mathbf{r},\omega)$	tensor of complex conductivity	18
$\mathbf{b}_\perp, \mathbf{b}_\parallel$	transversal and longitudinal magnetic components	87

List of Notations

Symbol	Meaning	Page
$H_1^{(1)}$	Hankel function of the first kind	264
I	magnetic inclination	40
Kp	geomagnetic activity index	57
L	McIllwain parameter	154
N_e	electron concentration	11
N_i	ion concentration	13
P	gas kinetic pressure	81
P	total pressure	15
P_α	pressure	12
P_m	magnetic pressure	83
Q	Q-factor	55
$Q_n(z)$	eigenfunction of n-th FLR-harmonics	111
R	reflection coefficient	96
R_E	Earth's radius	32
T	plasma temperature (in Kelvins, K)	8
T	transmission coefficient	96
T_A	Period of fundamental harmonics	138
T_e	electron temperature	14
T_i	ion temperature	14
T_n	neutral temperature	14
V_{ph}	phase velocity	91
W_e	energy of chaotic electron motion	10
W_i	energy of chaotic ion motion	10
X	$X = 4\pi\Sigma_P/c$	108
X_K	$X_K = X + Y^2/\widetilde{X}$	219
Y	$Y = 4\pi\Sigma_H/c$	203
$Z_\alpha e$	charge of the multiple charged ions	9
$P_{jk(\alpha)}$	pressure tensor	12
Φ	wave phase	95

Symbol	Meaning	Page
$\mathbf{b}^{(g)}$	fields transmitted to the Earth	240
$\mathbf{b}^{(r)}(k)$	fields reflected from the ionosphere	240
$\mathbf{b}$	small perturbation of magnetic field	16
$\mathbf{b}_{\tau}^{(i)}(k)$	spatial harmonic of an incident $\mathbf{b}_{\tau}$-component	240
$\mathbf{g}$	acceleration of gravity	12
$\mathbf{j}$	total electric current density	11
$\mathbf{j}_{\mathrm{eq}}$	equivalent current density	286
$\mathbf{j}_{ext}$	external current density	10
$\mathbf{k}$	wave vector	91
$\mathbf{v}_{(\alpha)}$	particle velocity	16
$\mathbf{E}$	electric field	16
μ	reduced mass	13
ν_{ei}	electron-ion collision frequency	35
ν_{en}	electron-neutral collision frequency	34
ν_{in}	ion-neutral collision frequency	36
ν, φ, μ	dipole coordinates	153
ω_{An}	frequency of n-th FLR- mode	114
ω_A	$\pi c_A/l_z$	114
ω_{ci}	ion cyclotron frequency	14
ω_n	FLR-frequencies	114
Ei	exponential integral	245
erfc	complementary error function	250
ρ	mass density	15, 81
ρ_α	charge density of particles of the type α	8
ρ_{ext}	external charge density	10
ρ_e	plasma distribution in equatorial plane	164
$\rho_{m\alpha}$	mass density of of the α species	8
σ^{eff}	effective conductivity	303
σ_C	specific Cowling conductivity	43
σ_a	specific atmospheric conductivity	45
σ_g	specific ground conductivity	45
σ_H	specific Hall conductivity	38
σ_P	specific Pedersen conductivity	38
$\sigma_{\mathbf{k}}$	Fourier harmonic of the conductivity	305
σ_{ij}	i, j-components of specific layered conductivity	41
σ_{ij}^{eff}	effective conductivity tensor	305
τ_{jk}	tensor of viscous stresses	12
θ	angle θ between $\mathbf{k}$ and $\mathbf{B}_0$	91
$\theta(t)$	step-function of time t	270
ε_m	transversal dielectric permeability	33
ϱ,φ,z	cylindrical coordinate system	88
$\widetilde{X}$	$\widetilde{X} = X + \sqrt{\varepsilon_m}\,\lvert\sin I\rvert$	242
$\xi_{\perp\omega}(x, z)$	Laplace transform from $\xi_{\perp}(x, z, t)$	126

Symbol	Meaning	Page
$\xi_\perp, \xi_\parallel$	plasma displacements across to and along $\mathbf{B}_0$	87
$b_{\parallel\omega}(x,z)$	Laplace transform from $b_\parallel(x,z,t)$	126
$b_x^{(i)}$	incident magnetic wave field	240
c_A	Alfvén velocity	32, 90
c_A^T, c_A^P	modified Alfvén velocities	160
c_s	adiabatic sound velocity	82
f_α	distribution function	7
$j_\parallel$	field line current	279
k	Boltzmann constant	8
k_A	Alfvén wavenumber	95
$k_\parallel$	longitudinal component of $\mathbf{k}$	91
$k_\perp$	transversal component of $\mathbf{k}$	91
l	particle free path	14
l_x, l_y, l_z	ribs of the 'hydromagnetic box'	107
m_i	ion mass	37
m_n	molecular mass	37
n_α	small perturbation of particle number density	16
p_α	small perturbation of particle pressure	16
p_c	percolation threshold	304
q	$q = kl_z/\pi$	112
r_D	Debye radius	8
r_{ci}	Larmour radius	15
v_T	thermal velocity	14
w	Kramp function	259
x_r	resonance point	97
x_t	turning point	98